IET SECURITY SERIES 08

Hand-Based Biometrics

IET Book Series in Advances in Biometrics – Call for authors

Book Series Editor: Michael Fairhurst, University of Kent, UK

This Book Series provides the foundation on which a valuable library of reference volumes on the topic of Biometrics is build. *Iris and Periocular Biometric Recognition, Mobile Biometrics* and *User-centric Privacy and Security in Biometrics* are the first published volumes in the Series, with further titles currently being commissioned. Proposals for coherently integrated, multi-author edited contributions are welcome for consideration. Please email your proposal to the Book Series Editor, Professor Michael Fairhurst, at: m.c.fairhurst@kent.ac.uk, or to the IET at: author_support@theiet.org.

Published Titles in This Series:

- ***Iris and Periocular Biometric Recognition*** (Christian Rathgeb and Christoph Busch, Eds.): Iris recognition is already widely deployed in large-scale applications, achieving impressive performance. More recently, periocular recognition has been used to augment biometric performance of iris in unconstrained environments where only the ocular region is present in the image. This book addresses the state of the art in this important emerging area.
- ***Mobile Biometrics*** (Guodong Guo and Harry Wechsler, Eds.): Mobile biometrics aim to achieve conventional functionality and robustness while also supporting portability and mobility, bringing greater convenience and opportunity for deployment in a wide range of operational environments. However, achieving these aims brings new challenges, stimulating a new body of research in recent years, and this is the focus of this timely book.
- ***User-centric Privacy and Security in Biometrics*** (Claus Vielhauer, Ed.): The rapid emergence of reliable biometric technologies has brought a new dimension to this area of research, allowing the development of new approaches to embedding security into systems and processes, and providing opportunities for integrating new elements into an overall typical security chain. This book provides a comprehensive overview of leading edge research in the area.

Hand-Based Biometrics

Methods and technology

Edited by
Martin Drahanský

The Institution of Engineering and Technology

Published by The Institution of Engineering and Technology, London, United Kingdom

The Institution of Engineering and Technology is registered as a Charity in England & Wales (no. 211014) and Scotland (no. SC038698).

First published 2018

The Institution of Engineering and Technology
Michael Faraday House
Six Hills Way, Stevenage
Herts, SG1 2AY, United Kingdom

www.theiet.org

British Library Cataloguing in Publication Data
A catalogue record for this product is available from the British Library

ISBN 978-1-78561-224-4 (hardback)
ISBN 978-1-78561-225-1 (PDF)

Typeset in India by MPS Limited

Contents

Preface

This book is the latest in the recently established '*IET Book Series on Advances in Biometrics*'. The history of this Series dates back to the publication by the IET a few years ago of the book '*Age Factors in Biometric Processing*', which provided the impetus and set the pattern for an on-going Series of books, each of which focuses on a key topic in biometrics. Each individual volume will bring together different perspectives and state-of-the-art thinking in its topic area, shedding light on academic research, industrial practice, societal concerns and so on and providing new insights to illuminate and integrate both specific and broader issues of relevance and importance.

The availability of different biometric modalities on which to base the identification of individuals is the key to successfully matching performance to the requirements of any particular application domain. The history of biometrics shows a continuing search for new modalities and a corresponding effort to optimise the extraction of features which can uniquely and robustly characterise different individuals. But there has also been a constant renewing of the development of established modalities to improve performance and to gain the twin advantages of public acceptance and greater effectiveness in deployment.

The fingerprint is one of the oldest and most widely adopted biometric modalities, yet remains the focus of much current research. In recent years, the palmprint has emerged as a related human characteristic well suited to biometric identification, while vein patterning in the hand has also emerged as a viable source of identity information, providing modalities that are in many ways complementary to the fingerprint. Similarly, hand shape has retained a value in practical biometric applications and, as with all modalities, offers particular benefits in appropriate circumstances. The fact that all these modalities are based on observations and measurements of properties of the hand also suggests that a capture infrastructure based on extracting information about hand characteristics could be extremely useful, generating information which can support both choice and preference in relation to an adopted individual modality while also supporting multimodal processing.

This book addresses this important area, but rather than focusing on any specific hand-based modality, it draws together a number of important topics which collectively offer a view of the wide range of information which the hand provides, thereby capturing and surveying the diversity of hand-based biometrics. This breadth of coverage sets this book apart from others which address individual hand-based modalities but also allows the inclusion of informative contributions on the physiology of the hand, providing insights into the very foundations of hand-based biometrics. This ground-breaking approach makes the book a particularly welcome addition to the Series.

The contributors come from a variety of backgrounds, and the volume overall represents an integration of views from across the spectrum of stakeholders, including academia and the medical profession. We hope that the reader will find this a stimulating and informative approach, and that this book will take its place in the emerging Series as a valuable and important resource which will support the development of influential work in this area for some time to come.

Other books in the Series are in production, and we look forward to adding regularly new titles to inform and guide the biometrics community as we continue to grapple with fundamental technical issues and continue to support the transfer of the best ideas from the research laboratory to practical application. It is hoped that this Book Series will prove to be an on-going primary reference source for researchers, for system users, for students and for anyone who has an interest in the fascinating world of biometrics where innovation is able to shine a light on topics where new work can promote better understanding and stimulate practical improvements. To achieve real progress in any field requires that we understand where we have come from, where we are now, and where we are heading. This is exactly what this book and, indeed, all the volumes in this Series aim to provide.

Michael Fairhurst
Series Editor, Advances in Biometrics Book Series

About the editor

Martin Drahanský (born on 4 April 1978 in Brno) started his study at the Brno University of Technology (BUT), Faculty of Electrical Engineering and Computer Science in 1996, after his graduation at the grammar school tř. Kpt. Jaroše 14 with specialization in mathematics and physics. He completed his study at BUT in the field of Computer Technology and Informatics and obtained the MSc degree 'Ing.'. In the same year, he completed a parallel study at the FernUniversität in Hagen, Germany, where he was awarded the degree 'Dipl.-Ing.'. Immediately after finishing his engineering studies, he joined the postgraduate study program of information technology at the newly founded Faculty of Information Technology (FIT) of the BUT and finished his Ph.D. in *Biometric Security Systems – Fingerprint Recognition Technology* in 2005 and obtained the Ph.D. degree. In 2009 he was appointed to associate professor (work title: *Fingerprint Recognition Technology: Image Quality, Skin Diseases and Liveness Detection*) at FIT BUT. In 2017 he has obtained the academic position of a full professor at FIT BUT. From 2005 to 2015, he served as the international mobility coordinator at FIT BUT. From 2016, he is the Dean's advisor for technology transfer at FIT BUT, member of the BUT Commercialization council and also has a Siemens professor position at BUT. From 2010, he has been head of the STRaDe (*Security Technology Research and Development*) research group at FIT BUT.

He began his career as a computer assistant at the grammar school tř. Kpt. Jaroše 14 from 1996 to 2002. From 2001 to 2003, he worked as an external co-worker in the company Oprox a.s. and participated in the MagicGel® project for the genetic laboratory. From 2002 to 2005, he worked as an external research fellow at the Fraunhofer Gesellschaft, Institut für Graphische Datenverarbeitung (IGD) in Darmstadt (Germany) where he participated in BioFinger and BioCompress projects. In the same period (2002–2005), he was a scientist at the University in Siegen, Faculty of Electrical Engineering and Informatics (Germany), where he taught two courses in electrotechnics. Between 2011 and 2015, he worked in parallel with the position of associate professor at FIT BUT and as a senior researcher at the International Clinical Research Center, St. Anne's University Hospital in Brno (biomedical engineering platform). In 2013, he spent two months at the Zürcher Hochschule für Angewandte Wissenschaften, Winterthur (Switzerland) under the Erasmus+ program and in 2014 a half year at the Tokyo Institute of Technology, Tokyo (Japan) with financial support from the Matsumae International Foundation. He visited and is still visiting foreign universities and research institutions many times a year.

He is a senior member of IEEE, member of the International Association for Identification, European Association for Identification, European Association for Biometrics, Gesellschaft für Informatik, BioAPI Consortium, Czech and Slovak Simulation Society, Technical Normalization Committee No. 20 of the Czech Office for Standards, Metrology and Testing. From 2011 to 2013, he was a member of the P103 Panel of the Czech Science Foundation GAČR.

At present, he is the principal researcher of the STRaDe research group at FIT BUT. He is a guarantee and lecturer in two courses in the master's degree program – *Biometric Systems* and *Intelligent Sensors* and one in the doctoral study program – *Advanced Biometric Systems*. He has teleconference lectures of biometric systems at the Strathmore University (Kenya) since 2016. You can find more information on the web page: http://www.fit.vutbr.cz/~drahan.

Chapter 1
General introduction

Martin Drahanský[1] and Svetlana Yanushkevich[2]

1.1 Introduction

It is not necessary to discuss the biometrics, because this is well known not only in research area but nearly the whole population uses it. From the pioneer electro-mechanical systems moved the biometrics to digital world. At the beginning, we still have to present our physical biometric characteristics to the biometric system; however, the acquirement is immediately done into digital representation, i.e., we work with digital data. In most cases, biometric recognition is realized on image or video data.

On our body, we have many possibilities for biometric recognition. Of course, not all of our body parts are suitable for this task. When we neglect behavioral biometric characteristics, we can find two very important areas on our body, which offer a lot of biometric characteristics—these two areas are hand and head. On our hands we can use fingerprints, palmprints, 2D and 3D hand geometry, thermal image of the hand, finger veins, hand veins and nail structure. This topic will be discussed in the scope of this whole book. Considering our head, we can recognize 2D and 3D face, thermal image of the face, eye iris, eye retina, ear shape, lips movement and dental information. All these biometric characteristics predestine these areas to a really very important part of our body, because some of the abovementioned characteristics surely belong to the mostly used, including biometric travel documents (e-passports), where fingerprint, face and eye iris appear. The discussion is opened for other possible technologies, which are suitable for biometric travel documents; however, these are on hand or on head.

1.2 Hand physiology and its suitability for biometrics

As mentioned in the introductory part of this chapter, *hand* is really very interesting, because of having a lot of unique features, which could be extracted and used for comparison of people. Our hands are unique—this has been proven in the whole

[1]Faculty of Information Technology, Centre of Excellence IT4Innovations, Brno University of Technology, Czech Republic
[2]Faculty of Electrical and Computer Engineering, University of Calgary, Canada

year in which the biometric characteristics from hand are in use. It is essential that the human hand features are different for identical twins. The trouble arises especially for faces—identical twins have so similar faces that these cannot be distinguished from each other. Nevertheless, the most of other biometric characteristics on our body are not based only on DNA structure and similarity with our twin but are based on preterm development in the mother's womb. The position of fetus, inner pressure and other factors inside the mother's womb play an important role for building the structure of our cells. These factors influence nearly all of our biometric characteristics on hand. Only the hand geometry (especially in 2D) could be very similar to the twin ones. Fingerprints, veins structure, etc. are very unique for every person on the planet.

The inner hand physiology is discussed in the second chapter, the outer hand physiology in the third chapter and nail structure in the fourth chapter of this book. The *inner hand physiology* covers bones, tendons, muscles, cardiovascular system, etc. The most important diseases appearing in the inner hand structure, having influence on biometric characteristics, are discussed in the second chapter as well. The inner hand structure influences 2D and 3D hand geometry, hand thermal image, finger and hand veins structure. All these biometric characteristics could be influenced by the diseases appearing inside the hand. In many cases, the disease just slightly changes the structure, i.e., we nearly cannot detect any change, however sometimes (e.g., just simple edema) can make the biometric characteristics unusable for biometric recognition purposes.

When we speak about *outer hand physiology*, we mean especially skin. In the third chapter, there is discussed skin structure and dermatologic diseases as well. These diseases could be divided into three main categories [1,2]—(a) histopathologic changes, (b) color change and (c) combination of histopathologic and color changes. Just color change has a slight influence to just optical scanning technology, but no influence to other scanning technologies. On the other hand, the histopathologic changes cause change of the structure of ridges, which are crucial for fingerprint and palmprint recognition. The change of ridge structure could be caused by medicaments [3] as well; the medical drugs based on capecitabine are responsible for finger ridge structure disappearance. Luckily, after the use of this medicament is stopped, the finger ridge structure will be recovered. Unluckily, the patients who use capecitabine-based medicaments cannot use fingerprint recognition technology during the medical treatment.

The *nail* structure belongs to the outer hand physiology; however, this is such specific biometric feature that we devoted a separate chapter space to this topic, concretely the fourth chapter. The nail structure and diseases connected to nail are discussed. Nail structure is not used in biometrics, because there are many factors making this technology not very popular, especially troubles during scanning [the finger nail has to be placed into a chamber where camera and light source are placed in a certain position, the influence of surrounding light has to be blocked, the uniqueness (biometric entropy) of nail is not very high, some diseases influence the nail structure and the nail lacquer or artificial nails are totally making this technology unusable]. On the other hand, during scanning the

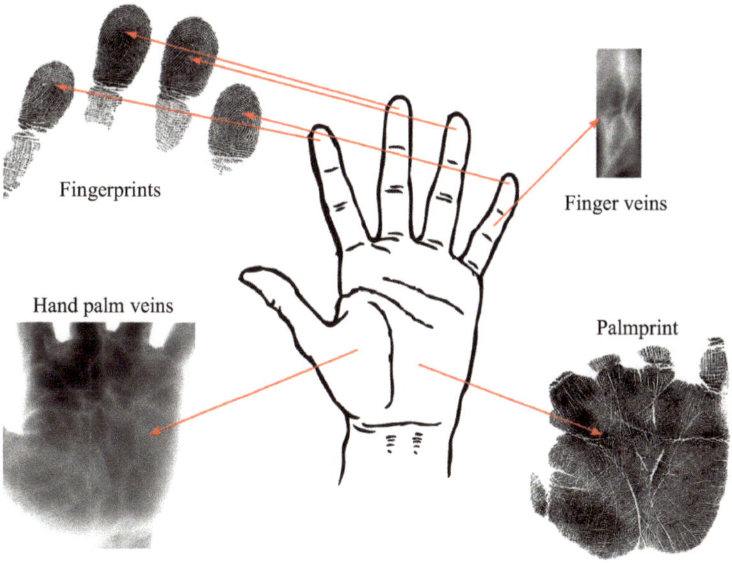

Figure 1.1 Biometric characteristics on hand palm

hand back, we can get a very nice nail structure, which is suitable for biometric recognition.

When we look on our hand palm, we can acquire and use the following biometric characteristics (see Figure 1.1):

- Fingerprint
- Palmprint (sometimes fingerprints are visible in the complete palmprint)
- Hand palm veins (under infrared illumination)
- Finger veins (under infrared illumination)
- Thermal image of the hand palm and fingers (will be discussed in the part of hand back)

Fingerprints are the mostly used biometric characteristic from the hand. In this book, you can find the introduction to the state of the art in Chapter 5. The basic information about fingerprint acquirement, (pre)processing and recognition are summarized here. Sometimes, the users have a connection of fingerprints with criminal police (they search for latent fingerprints on a crime scene); however, most people do not have any problem with the use of their fingerprints for biometric purposes. A good question is what happens with all fingerprints that are stored in large-scale national databases, because they are stored there forever and in some cases of lightly protected databases these fingerprints could be misused. Anyway, we will come to this topic later on. Because these large-scale databases include millions of fingerprints, it is necessary to speed up the process of recognition and search in these databases. This topic is covered in the sixth chapter. Furthermore, the troubles with fingerprints (the same could be applied to palmprint) have to be discussed, because they are very prone to be influenced by dermatologic diseases. This topic is covered in the seventh chapter. Our team invested a lot of time into getting a unique database with annotated diseased fingerprints by a dermatologist. At the moment, we use these diseased fingerprints for analysis. However, this topic includes injuries, dirtiness on finger or scanner surface and other troubles, which can cause an end effect their inapplicability for automatic processing and recognition.

As mentioned in the previous paragraph, if an attacker or generally an impostor gets an unauthorized access to the database with fingerprints, he/she can misuse this stolen data. It is not necessary to copy the images of fingerprints (in most cases, they are not stored in the databases, just only extracted features); just a template with extracted features is suitable for production of a synthetic fingerprint with minutiae points on correct positions, with correct types and angles. This fingerprint has a different pattern and run of ridges, what we can observe by a naked eye immediately; however, for the algorithms based on minutiae recognition, the global pattern (fingerprint class) is unimportant, because they are searching just only for minutiae positions, types and angles. The description of generation of synthetic fingerprints could be found in Chapter 8. This chapter covers not only the generation of a synthetic fingerprint in the nice black and white representation, followed by how could be this image changed into the real world representation. Some of the works related to this chapter, e.g., [4,5], describe the use of simulation techniques for putting the skin diseases or other disturbing artifacts into synthetic fingerprints. Indeed, in the connection with synthetic fingerprints, we should not neglect the use of spoofs. The trouble is if we can generate a synthetic fingerprint that very closely simulates the real finger, or in the worst case, we can get a real fingerprint in a good quality, the creation of finger(print) spoof is not so difficult. This topic is described in Chapter 15. Of course, the spoof created using nearly the same materials could be applied on the whole hand, especially on characteristics based on outer hand physiology. The antispoofing techniques are discussed in Chapters 14 and 16.

Palmprint is, generally speaking, a very big fingerprint. In the case when we use a complete scan of the hand, we get fingerprints together with the palmprint scan (see Figure 1.2). Therefore, all the positive and negative aspects discussed in the previous paragraphs could be applied to palmprint. Palmprint technology is not so often used in the praxis, because the scanner area is too big and the processing time and memory requirements are higher in comparison with fingerprint recognition. Palmprint is of interest for criminal police, because if the offender did not use gloves, he would very probably leave some part or even the whole palmprint, which is very representative for a concrete person. There could be found a big amount of minutiae points and moreover the run of lines of life is very representative as well. The palmprint is covered in Chapter 12. At the moment, the mobile devices (phones and tables) have so high resolution that they could be used for acquirement of palmprint and fingerprints. The quality is really so good that the ridges are visible and usable for the automatic processing. This can cause troubles, because if we post our photos on webpages or social networks, the full structure of our finger and palm ridges could be seen.

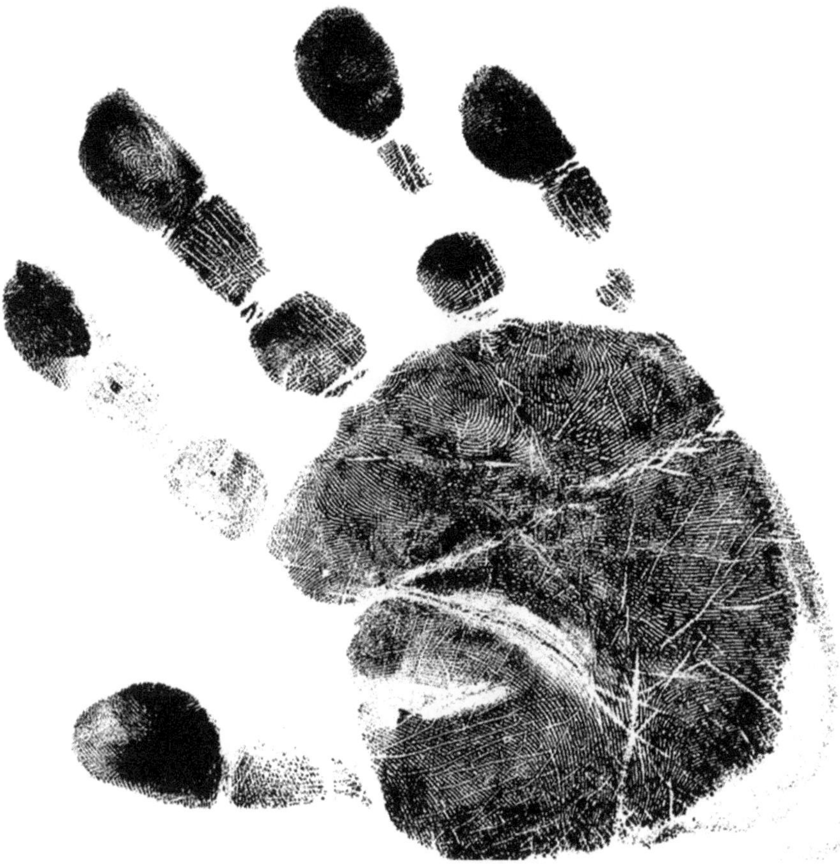

Figure 1.2 Palmprint with fingerprints [6]

If we speak about *hand palm veins*, we can apply the same for *hand back veins*. Just in the praxis, the use of hand back veins is rare. Most technologies are oriented on hand palm veins, because the user just positions his/her hand above the device and the scanner can acquire this biometric characteristic. Hand veins belong to the inner hand physiology and have a big advantage—it is very difficult to change their structure. In general, it is possible, however so difficult, that probably no one will take this surgery into account. Because the veins are inside our hands, we need an infrared illumination to make them visible. The illumination belongs to the near infrared (due to ISO 20473 scheme); however, the concrete wavelength plays an important role. If we are closer to red (above 650 nm), we can see better oxygenated blood, i.e., arteries. If we come closer to 950 nm, we can acquire better deoxygenated blood, i.e., veins. This principle is used in medical oximeters [7]. The arteries and veins on hand palm or back are very nicely visible and thick enough to be acquired without any big trouble. The surrounding light could be filtered out; just only fluorescent lamps can cause troubles, because they produce a strong infrared illumination, which can have an impact to the acquirement process. These devices use infrared filters to filter out the surrounding light, and only a selected wavelength(s) can pass, but if the light source produces radiation on the same wavelength, this can bring a lot of problems. The hand palm vein topic is discussed in Chapter 10.

Finger vein recognition is comparable with hand vein recognition. The illumination principle is the same. The pulse oximeters are used on fingers—they can measure not only the heart pulse but also the oxygenation of blood as well. A small disadvantage arises here—the arteries and vein structure in fingers is not intertwined enough, i.e., the amount of information (biometric entropy) is low. On the other hand, if we acquire a video and the user rotates the finger around (e.g., from one side to the another one), we can get a very precise space structure of the finger veins, i.e., the amount of information (biometric entropy) will be very high, and furthermore, we can avoid the use of finger veins spoofs, because to construct a 3D finger vein structure is not simple, especially if we consider the use of both wavelengths to make visible oxygenated and deoxygenated blood. In that case, just a 3D model printed from an appropriate material (e.g., metal powder) on a 3D printer will be not enough, because this will represent only the deoxygenated blood, anyway the oxygenated scan will be empty. The finger vein recognition is covered in Chapters 9 and 11.

Now we come to the hand back. Here we can acquire and use the following biometric characteristics (see Figure 1.3):

- Thermal image of the hand palm and fingers
- 2D hand geometry
- 3D hand geometry
- Finger nail
- Hand back veins (were discussed in the previous part)
- Finger veins (were discussed in the previous part)

Using a thermal imaging camera, we can get a *thermal image* of our *hand palm* or *hand back*. There are two drawbacks in using this technology—a thermal imaging camera with a good resolution (more than 240×180 pixel) is very expensive. A thermal imaging camera with resolution 640×480 pixel costs above approx. 15,000 EUR. Such devices are not suitable for a common biometric market. The second drawback is the usability of the thermal scan of the hand for biometric recognition. There is very low amount of thermobars [areas with the same temperature building stains with the same color, because each color on the image represents a concrete corresponding temperature (if the emission coefficient was set to the appropriate material)], which represents our hands. Furthermore, these thermobars are strongly influenced by environment (temperature and humidity) and if the person smokes. Smoking causes the narrowing of blood vessels that causes the change of the hand temperature in general. Therefore, this biometric characteristic is not used for biometric recognition of people. Thermal images of fingers are not usable for any form of biometric recognition, because they do not include sufficient amount of information and are totally strongly influenced by surrounding

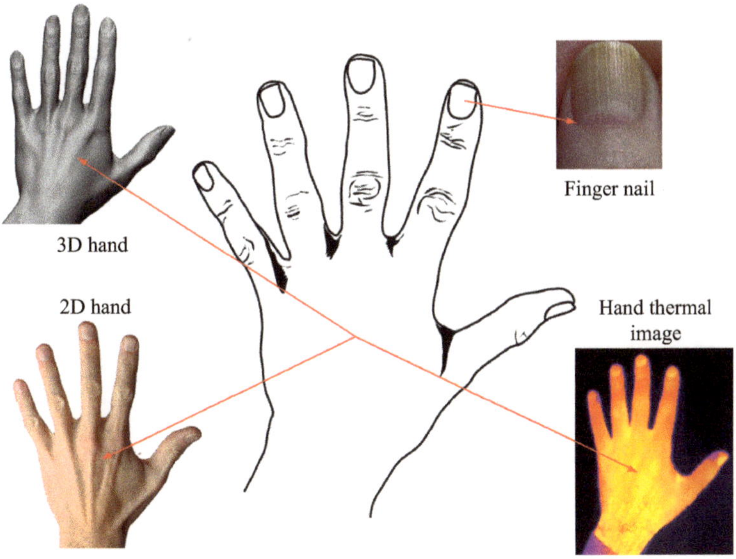

Figure 1.3 Biometric characteristics on hand back

environment and smoking. From this reason, the technology is very shortly mentioned in Chapter 14.

By using a 2D classic camera or line scan camera, we are able to scan and process the *2D hand geometry* (shape). The hand shape could be acquired in two main scenarios—hand is placed on a pad or is scanned moving above the camera. The first scheme brings the hand into a concrete position, and if a fixation pins for fixing the fingers into concrete positions are used as well; we get a very nice images which could be very quickly and automatically processed. In this case, the orthogonal scanning is used, where we acquire the shape of hand from above, but from side as well. The second approach scans the moving hand. In this case, a 2D camera with high framerate or line scan camera could be used. The reason is that the hand is moving and if we use a camera with low framerate, the images will be blurred, i.e., unsuitable for automatic processing and extraction of 2D hand features. However, this second approach enables the use of the technology "on the fly," i.e., where users just only pass the biometric scanning station and do not need to stop and place the hand to a scanner surface. This "on the fly" technology is applicable on fingerprints as well; however, other biometric characteristics are not very suitable for scanning in this style of use. In comparison to 3D hand geometry, 2D hand includes lower amount of information (biometric entropy) that is suitable for smaller groups of people. This 2D hand geometry (shape) technology is discussed in Chapter 12.

If we use any 3D scanning technology (structured light, time of flight or any technology producing depth map or point cloud), we are able to scan and process the *3D hand geometry* (shape). We can select from a big amount of devices on the market, starting by low-cost devices, going through middle class till very expensive scanners. One exception, pretty new on the market, is the use of 3D line scan camera. All these cameras are suitable for "on the fly" scenario. Many of these cameras enable to get not only the point cloud (3D surface) but also the texture information (images of a real hand surface) as well. This texture information could be used as an additional biometric characteristic, especially if we scan hand palm, e.g., the 3D line scan camera is so precise that we can see ridges structure. It is surprising that this technology is still in research form, and no professional devices are on the market. The 2D technology has dominated the market for a very long time. The introduction and explanation of the 3D hand geometry technology could be found in Chapters 12 and 13.

As mentioned in the beginning of this chapter, the *nail structure* is interesting; however, no devices for this technology are available on the market and the research is not very spread in this area. Just fragments in the literature could be found, e.g., [8–10]. Therefore, no space in this book is devoted to this research area, because it will be very difficult to prepare a chapter that covers such topic. Anyway, the abovementioned sources cover this topic and could be used for further study in this biometric recognition area.

At the end of this subsection, we have discussed troubles with *spoofs* and *antispoofing* methods. In some cases, it is not difficult to produce a working spoof. Especially for fingerprints and palmprints, there could be found a lot of articles and webpages, which describe the style of creation of such spoofs. This topic is very interesting for criminal police as well. If an offender leaves spoofed fingerprints in a crime scene, this can lead to arrest of a wrong person. Spoofs of 2D or 3D hand geometry are known as well—the casting of a hand is possible and is not expensive, or the 3D print using a 3D printer is possible as well. The quality of a low-cost 3D printer is sufficient for fooling the biometric system in 2D and 3D version. The vein structure in 2D version could be cheated as well; just only a metallic powder is sufficient to reconstruct the paths of veins. However, if the oxygenated blood (arteries) is taken into account, the production of such a fake mixture of oxygenated and deoxygenated blood is very difficult and nearly impossible. The spoof of finger nail is unknown and could not be found in literature. However, a very precise casting of the nail structure could be useful. The last technology using a thermal imaging camera is very easy to be cheated. Our research group did trials on making the spoof of thermal face image, which is more demanding, and we were successful. Therefore, the production of a hand or finger spoof is very easy, effective and cheap. The spoofs are discussed in Chapter 15. The methods against these spoofs are called antispoofing or liveness detection. These methods are based on various mechanisms, which are connected to a living human body, e.g., pulse, skin properties and perspiration. In the past, there were successful attacks (spoofs) on nearly every antispoofing technology. The most promising technology is the multispectral one, which is based on illumination of living human skin using ultraviolet, visible and infrared light. The skin reflects, absorbs and scatters the incoming illumination. If the skin is dead or any other material is used, the reaction of the skin differs from a live skin. Therefore, this method is very reliable. For fingerprint recognition, an optical tomography or ultrasound technology could be used—these two technologies cannot be overcome easily, i.e., they are promising for the future as well. The antispoofing methods are summarized in Chapters 14 and 16, whereas Chapter 14 is oriented also on optical skin properties.

The last very important property of each of the above-discussed biometric characteristics is its interclass and intraclass variation [11,12]. These measures represent the changes inside my own biometric samples (*intraclass variation*) and among my biometric samples and all other samples not belonging to my person (*interclass variation*). We wish to have very low intraclass variation, i.e., that all acquired biometric samples from our one biometric characteristic are very similar, nearly the same, and otherwise, we want to have the interclass variation so high as possible, i.e., that all biometric samples of the same biometric characteristic, but from various users, differ a lot. Very low intraclass variation has especially biometric characteristics belonging to the inner hand physiology (hand and finger veins); all other hand-based biometric characteristics have a middle one. Regarding the interclass variation, the highest is for fingerprints and palmprints; hand veins, nail and 3D hand geometry have the middle one; and 2D hand shape, thermal hand image and finger veins have the lowest one. The highest biometric entropy (amount of information) could be found in palmprint, fingerprint and 3D hand geometry; middle in 2D hand geometry, hand veins and nail; the lowest in thermal hand image and finger veins. However, it strongly depends on the end application, where the concrete biometric characteristic will be used. If we expect just only a small group of users, all of these technologies could be used. If we expect a high security and a big group of users, just only those biometric characteristics could be used, which have high amount of information (biometric entropy), low intraclass variation and high interclass variation.

1.3 Use of biometrics for ABC systems—watchlists

In this section, we will discuss one very interesting topic, which is connected to every biometric technology, especially those biometric characteristics, which are suitable for electronic travel documents (e-passports)—watchlist for e-boarders.

Biometric-enabled watchlist itself as well as watchlist check procedure are the specific security technologies which include various aspects of profiling of person of interest in both physical and virtual world. It is related to the performance of distributed databases, acquisition of biometric traits, risk assessment of watchlist check errors, manifestation, correction, updating and respective public responses to the watchlist mismatch. Moreover, watchlist technology is a part of national and international security infrastructure that dictates various application constraints. For simplification, we follow the above-specified requirements (time constraints and reliability of decisions) to the biometric-enabled watchlists in mass-transit systems. Based on this understanding, we provide below the analysis of the publications in this area, as the synergy of research efforts in the following directions:

- *Impact of quality of biometric traits on performance:* Effects of real-world biometric databases for e-passport holders are studied in [13,14]. The study

[15] addresses the problem of degraded biometric traits, typical for watchlist, as well as recognition of unconstrained facial images. The watchlist technology includes the age estimation, age progression and template aging tools, as well as detectors of factors that impact the quality and performance including rare features. Intensive integration of forensic experience is urgent trend in biometric-enabled watchlist technology. It is well documented in forensics that errors can occur in profiling, searching, matching and identification using biometric traits [16]. Ideally, the watchlist should contain synthetic facial images of persons of interest constructed by composite machines.

- *Impersonation phenomenon:* Impersonation is the key problem of border passage from ancient time including contemporary e-borders based on biometric-enabled tools. It is well understood that phenomenon of impersonation (or passive, or zero-effort attack) experimentally detected in speech recognition by Doddington *et al.* [17] can be also detected in other biometric modalities. In design of biometric systems, Doddington phenomenon should be taken in account. Unfortunately, not only passive attacks are an inherent property of watchlist technology, impersonation from social networks can affect the watchlist reliability. Impersonation can be mitigated via multibiometrics. For example, this approach has been implemented in [18] (face and fingerprints), as well as in other pioneering project by the Department of Homeland Security (DHS). Particular interest in watchlist technology is mitigating effects of plastic surgery and makeup, as well as spoofing detection [19].
- *Profiling and risk-assessment technologies* [20], including simulation and modeling, aiming to optimize and harmonize logistic of transit hubs, potential threats, throughput and requirements to supporting IT infrastructure (surveillance, authentication, early warning, risk assessment) [21], as well as via situational awareness. Because of high complexity, various proving grounds are created, such as DHS testing areas.
- *Breakthrough approaches:* Recognition at a distance and on the move [22] including detection of the health-relating features and mobile distributed systems. One of the goals is the acquisition of biometric traits through the interview [23]. The state of the art of the contemporary e-border technologies is given in overviews [24,25].

Definition 1: *The e-border infrastructure is a networking tool for traveler authentication and risk assessment.* The automated border control tasks include checking the document for authenticity and verifying the biometric templates stored in the traveler's e-passport/ID against the probe photo or fingerprint taken at the console. The proper technology must account for age progression, in particular, and ensure certainty of traveler biometric appearance (individual can change his/her evidence using plastic surgery, color lenses or/and make-up).

Definition 2: *Given the traveler, his/her risk assessment is defined via evidence accumulation paradigm which results in a certain security indicator for making decision at acceptable level of reliability and credibility under specified time* *constraints*. One of the elements of such evidence accumulation process is the check of the trustworthiness of the Advanced Passenger Information which is provided by the traveler about him/herself. The crucial phase of risk assessment of this data is the watchlist check. Watchlist is defined as a mandatory component of e-borders. It enables identification of individuals of interest using the biometric traits and related contextual information. No biometric-enabled watchlist technology can be 100% effective—there is always a chance that the watchlist may be compromised. Compromising watchlist may, however, require higher amounts of equipment and management resources relative to disabling the watchlist. Thus, the question becomes not "Is the watchlist technology effective?" but "Is watchlist technology A more effective than watchlist technology B?"

Definition 3: *Watchlist check known also as screening, or negative identification, establishes whether a traveler is not on the watchlist. It is characterized by false negative (FN) (miss-match, or miss-identification) errors and false positive (FP), or false alarm (false detection, or impersonation).* The corresponding error rates are called an FN Rate and an FP Rate. In general terms, and in positive identification, those are called false rejection rate (FRR) and false acceptance rate. An FP results in convenience problem, as an innocent traveler is denied access and need to be manually checked or examined to get access.

Definition 4: *Watchlist architecture is defined as a distributed infrastructure for (a) collection (capture biometric and related contextual data from real and virtual world for the purpose of matching), (b) matching (identifying or verifying the identity of an individual), (c) storing (enrolling, maintaining and updating biometric related contextual data) and (d) sharing (exchange biometric and related contextual data, as well as match results among government agencies accordance with national regulations) of personal data of individuals of interest.* Sharing unclassified biometric data with other agencies having a counter terrorism mission is a high priority task. In e-borders, misidentification may not only offend or hurt travelers (wrongly suspected as terrorists) but can also create a bottleneck situation favorable for terrorist attacks. As a result of the watchlist screening process, the travelers may complain that they were adversely affected and seek relief. In such mass-application as e-borders, government agencies involved in watchlist screening have a certain redress process to resolve the complaint and respond to the complainant.

Definition 5: *The redress complaint disposition (RCD) metric is defined as a traveler's complaint that indicates that he/she was adversely affected and seeks a relief. In the RCD-metric, the traveler can be in one of the following states [21]: (a) nonrelated, (b) positive match or (c) misidentified.* Face, fingerprints and iris are the most suitable for watchlist biometric modalities. For fingerprint recognition in the identification mode, the false nonmatch rate is calculated as a proportion of samples from genuine attempts that cannot be matched against enrolled templates of genuine users. For face recognition in verification mode, the FRR is defined in a scenario when a genuine user is incorrectly rejected.

FRR determines the manual workload since such users will most likely complain and will have to be handled manually. For example, one fingerprint is not acceptable for identification mode but can be accepted for verification. However, a single fingerprint together with face leads to a better performance in terms of accuracy. In addition, useful statistics on face recognition in authentication gates is provided in [26].

Definition 6: *Performance of the watchlist is defined as a quantifiable indicator used to assess how well the watchlist is achieving its desired objectives.* The number of transactions per day is a system characteristic for a watchlist. For example, DHS managed over 160,000 search queries a day (for visa, visitors, naturalization, etc.) due to Government Accountability Office of Department of Defence. In addition to the known RCD measures of watchlist efficiency, we defined the following performance metrics: the throughput (the number of served travelers per hour), operational reject rate [expressed as "one in N travelers (1:N) is wrongly directed to special control"] and life-cycle performance assessment which combines theoretical (algorithmic limit), predicted (vendor-reported) and operational (real) false reject and accept rate or accuracy. The main focus is on estimation of risk of a miss when searching for a wanted person, as well as a mismatch of an innocent traveler against someone on a watchlist.

Definition 7: *Social embedding addresses the mapping of the watchlist technology onto social infrastructure, including privacy issues. The depth of social embedding is the key criterion for development the watchlist technology.* Example of a deep embedding watchlist technology is DHS's ADVISE that aims at finding and tracking relationships in data available about the traveler. Searches and evidence accumulation result to as semantic graphs which help to detect activities that threaten the United States by facilitating the analysis of large amounts of structured data (such as information in a database) and unstructured data (such as e-mail texts, reports and news articles). The information is then analyzed and used to monitor social threats such as community-forming, terrorism, political organizing or crime.

Definition 8: *Watchlist inference is defined as a mechanism for traveler risk assessment using a watchlist components such as biometric traits and relevant records or metadata.* Currently, no commonly agreed set of factors exist upon which to base an evaluation, regardless of the watchlist purpose or requirements. We propose the comprehensive criteria and taxonomy for watchlists and focus on the most vulnerability in border crossing biometrics known as impersonation.

Definition 9: *A model of a watchlist technology for traveler risk assessment in e-borders includes technical part (databases, mechanism for searching, collecting, filtering and identification) and management part (logistics, redress code, sharing and privacy policy).*

Definition 10: *The performance of the watchlist is defined as a quantifiable indicator used to assess how well the watchlist is achieving its desired objectives.* The number of transactions per day is a system characteristic for a watchlist.

For example, DHS managed over 160,000 search queries a day (for visa, visitors, naturalization, etc.) due to Government Accountability Office of Department of Defence. In addition to the known RCD measures of watchlist efficiency, we defined the following performance metrics: the throughput (the number of served traveler per/hour), operational reject rate (expressed as "one in *N* travelers (1:*N*) is wrongly directed to special control") and life-cycle performance assessment which combines theoretical (algorithmic limit), predicted (vendor-reported) and operational (real) ones.

Definition 11: *Doddington metric is defined as the four type classification of recognition process*:

- *Category I ("sheep"), recognized normally;*
- *Category II ("goats"), hard to recognize;*
- *Category III ("wolves"), good at impersonating; and*
- *Category IV ("lambs"), easy to impersonate.*

Typical examples of the watchlist check in Doddington metric are given in Figure 1.4: left pair of images results the effect of misidentification and the right pair images addresses the impersonation effect. Our experiments concern the following scenarios of the border crossing passage:

- Scenario 1: The traveler is a person of interest, he/she belongs to the class of "goats" and, thus, poses a risk of not being matched against the watchlist, and, consequently, passing the border control.
- Scenario 2: The traveler is not a person of interest, he/she belongs to the class of "lambs" (and "wolves" in the symmetric matcher) and, thus, might match against someone on the watchlist and generate a false alert and the likely RCD procedure.
- Scenario 3: The traveler is a person of interest, he/she belongs to the class of "wolves" and may create a lot of matches against the watchlist, thus creating logistic problems; therefore, the case, however, may not be what the attacker intends as this event will still alert border control.

Figure 1.4 Images of subjects in Doddington metric: (a) and (b) are "goat" subject; (c) an image of a "wolf/lamb" subject that looks similar to subject (d) (images are from the LFW database [27])

Our previous studies [28,29] and actually running study address the risks of future generation of automated watchlist check in mass-transit systems, such as e-borders. We consider risks of biometric-enabled technology under the following critical constraints: (a) limited operating time (few minutes) and (b) impersonation phenomenon when using face traits of person of interest. These operating conditions are radically different from Entry–Exit systems such as US VISIT in the United States and Smart Borders in EU in which information about the visitor is available in advance.

In contemporary border crossing automation, only alphanumeric data from e-passport/ID are used in traveler risk assessment via watchlist. It is understood that nonbiometric traveler documents can be forged, stolen or even worst; they can be genuine but issued using the false borne certificates. In practice, this drawback means that authentication and risk-assessment machine cannot identify such persons of interest.

It is well documented that vulnerability of recognition process using biometric traits can be measured using Doddington metric. Similar to human, recognition algorithm can mistake one person for another due to resemblance or poor light conditions. In practice of border crossing, this means that machine can identify terrorist as innocent person and vice versa. It is possible to mitigate and even suppress these effects by multibiometrics. For example, in Entry–Exit systems for visitors traveling on visa, high-quality fingerprints are used in addition to high-quality face biometric in database. In mass-transit systems, which operate under time constraints and deal with unknown traveler, the Entry–Exit paradigm cannot be applied directly. In this application, fast and reliable watchlist check is the core of traveler risk assessment. This leads to the need of embedding of this technology into social infrastructure, in particular, via bridging the gap between the watchlist technology and forensics.

However, in practice of e-borders, any extension of biometric modalities in traveler authentication and risk assessment requires creation of a costly supporting infrastructure. Facial biometric is the privilege of most documents and government databases for creation of watchlists. Weakness of facial biometric can be alleviated by fingerprint traits acquired at distance. Motivated by this fact, our study leads to the following key conclusions:

1. Application of Doddington detector in watchlist check provides additional information of critical importance. We showed using Doddington metric that there are always risk of impostors among persons of interest. In terms of security, this means that machine may mistakenly provide border crossing passage to a wanted terrorist.
2. Application of the evidence accumulating paradigm can be considered useful in detection of persons of interest who are hard to recognize, thus increasing the likelihood of detecting a wanted person; this may, however, add few innocent travelers to the pool of suspects.
3. It is in agreement with the reported studies that "Wolves/Lambs" category is the most sensitive Doddington class. Our study confirms that the proposed two-phase watchlist inference is inefficient to identify the subject of interest who

can easily impersonate, compared with the results of using solely the cooperative traveler's biometrics.

The general conclusion is that the proposed watchlist inference is an efficient extension of watchlist technology. The accumulated evidence paradigm allows for bridging the gap between forensics and biometric-enabled watchlists for e-border applications.

The future steps will involve mitigating the effects caused by high variability of Doddington categories. For this, we are investigating Dempster–Shafer measures of uncertainty to be used for the watchlist inference. We also experiment with other biometric modalities, suitable for two-phase evidence accumulation.

1.4 Conclusion

This chapter includes the overview of all described technologies in this whole book. At the beginning, we address medical point of view to the hand, i.e., inner and outer hand physiology, including nail structure. We continue with very well-known fingerprint recognition, continued by palmprint recognition, recognition of hand and finger veins and finished by 2D and 3D hand geometry recognition. Because of lack of interest and availability, we neglect nail structure recognition for biometric purposes and recognition of thermal images of hand and finger. However, we discuss a very important topic, which is liveness detection, i.e., spoofing and antispoofing methods for various hand-based biometric characteristics, especially fingerprints.

ABC systems, watchlists for e-boarders and use of electronic travel documents (e-passports) play an important role for biometric systems based on recognition of hand features, especially for fingerprints because they are used in biometric e-passports. This topic is discussed in the second section of this chapter.

Acknowledgment

This work was supported by The Ministry of Education, Youth and Sports of the Czech Republic from the National Programme of Sustainability (NPU II); project IT4Innovations excellence in science—LQ1602 and by the BUT project "Secure and Reliable Computer Systems" FIT-S-17-4014.

References

[1] Doležel M., Drahanský M., Urbánek J., Březinová E. and Kim T.H. *Influence of Skin Diseases on Fingerprint Quality and Recognition*. New Trends and Developments in Biometrics. Rijeka: InTech – Open Access Publisher, 2012, pp. 275–303. ISBN 9789535108597.

[2] Drahanský M., Březinová E., Lodrová D. and Orság F. *Fingerprint Recognition Influenced by Skin Diseases*. International Journal of Bio-Science and Bio-Technology, 2010, Vol. 3, No. 4, pp. 11–22. ISSN 1976-118X.

[3] Chavarri-Guerra Y. and Soto-Perez-de-Celis E. *Loss of Fingerprints*. New England Journal of Medicine, 2015, Vol. 372, No. 16, p. 22.

[4] Bárta M. and Drahanský M. *Generation of Skin Diseases into Synthetic Fingerprints*. International Journal of Image Processing, 2016, Vol. 10, No. 5, pp. 229–248. ISSN 1985-2304.

[5] Kanich O. and Drahanský M. *Simulation of Synthetic Fingerprint Generation Using Petri Nets*. IET Biometrics, 2017, Vol. 6, No. 6, pp. 402–408. ISSN 2047-4938.

[6] Web page: http://academyofhandanalysis.org/tag/hand-printing/page/2/.

[7] Sinex J.E. *Pulse Oximetry: Principles and Limitations*. The American Journal of Emergency Medicine, 1999, Vol. 17, No. 1, pp. 59–66.

[8] Kumar A., Garg S. and Hanmandlu M. *Biometric Authentication Using Finger Nail Plates*. Expert Systems with Applications, 2014, Vol. 41, No. 2, pp. 373–386.

[9] Premakumari L.T. and Jothi A.S. *Multimodal Biometric Endorsement for Secure Internet Banking using Skin Spectroscopy, Knuckles Texture and Finger Nail Recognition*. International Research Journal of Engineering and Technology, 2016, Vol. 3, No. 2, pp. 1086–1090.

[10] *Automated identification through analysis of optical birefringence within nail beds*, U.S. patent US 6631199 B1, 1998, available on https://www.google.com/patents/US6631199.

[11] Ross A. and Jain A.K. *Multimodal biometrics: An overview*, Signal Processing Conference, 2004 12th European. IEEE, 2004, pp. 1221–1224.

[12] Li S.Z. and Jain A.K. *Encyclopedia of Biometrics*. Springer Publishing Company, Incorporated, 2015, ISBN 1489974873.

[13] Butt M., Marti S., Nouak A., Koplin J., Raghavendra R. and Li G. *Towards e-passport duplicate enrollment check in the European Union*, Proc. European Intelligence and Security Informatics Conf., 2013, pp. 247–251.

[14] DeCann B. and Ross A. *Has This Person Been Encountered Before? Modeling an Anonymous Identification System*, Proc. IEEE Workshop on Biometrics at the Computer Vision and Pattern Recognition Conf., 2012, pp. 89–96.

[15] Bourlai T., Ross A. and Jain A.K. *Restoring Degraded Face Images: A Case Study in Matching Faxed, Printed, and Scanned Photos*. IEEE Transactions on Information Forensics and Security, 2011, Vol. 6, No. 2, pp. 371–384.

[16] Cowell R.G. *FINEX: A Probabilistic Expert System for Forensic Identification*. Forensic Science International, 2003, Vol. 134, No. 1, pp. 196–206.

[17] Doddington G., Liggett W., Martin A., Przybocki M. and Reynolds D. *Sheep, goats, lambs and wolves a statistical analysis of speaker performance in the NIST 1998 speaker recognition evaluation*, Proc. Int. Conf. Spoken Language Processing, 1998.

[18] Cantarero D.C., Herrero D.A.P. and Mendez F.M. *A multi-modal biometric fusion implementation for ABC systems*, Proc. IEEE European Intell. and Security Informatics Conf., 2013, pp. 277–280.

[19] Menotti D., Chiachia G., Pinto A., *et al. Deep Representations for Iris, Face, and Fingerprint Spoofing Detection*. IEEE Transactions on Information Forensics and Security, 2015, Vol. 10, No. 4, pp. 864–879.

[20] Easley D. and Kleinberg J. *Networks, Crowds, and Markets: Reasoning About a Highly Connected World*, Cambridge University Press, New York, NY, 2010.

[21] Lee A.J. and Jacobson S.H. *The Impact of Aviation Checkpoint Queues on Optimizing Security Screening Effectiveness*. Reliability Engineering and System Safety, 2011, Vol. 96, No. 8, pp. 900–911.

[22] Tistarelli M., Li S.Z. and Chellappa R. *Handbook of Remote Biometrics*. Advances in Pattern Recognition, Springer, New York, NY, 2009.

[23] Nunamaker J.F., Derrick D.C., Elkins A.C., Burgoon J.K. and Patton M.W. *Embodied Conversational Agent-based Kiosk for Automated Interviewing*. Journal of Management Information Systems, 2011, Vol. 28, No. 1, pp. 17–48.

[24] Eastwood S.C., Shmerko V.P., Yanushkevich S.N. *et al. Biometric-enabled Authentication Machines: A Survey of Open-set Real-world Applications*. IEEE Transactions on Human-Machine Systems, 2016, Vol. 46, No. 2, pp. 231–242.

[25] Labati R.D., Genovese A., Muñoz E., Piuri V., Scotti F. and Sforza G. *Biometric Recognition in Automated Border Control: A Survey*, ACM Computing Surveys, 2016, Vol. 49, No. 2, pp. A:1–A:39.

[26] Spreeuwers L.J., Hendrikse A.J. and Gerritsen K.J. *Evaluation of automatic face recognition for automatic border control on actual data recorded of travelers at Schiphol Airport*, Proc. of the Int. Conf. Biometrics Special Interest Group (BIOSIG), 2012, pp. 99–110.

[27] Huang G.B., Ramesh M., Berg T. and Learned-Miller E. *Labeled Faces in the Wild: A Database for Studying Face Recognition in Unconstrained Environments*, University of Massachusetts, Amherst, Technical Report 07-49, October 2007.

[28] Lai K., Kanich O., Dvořák M., Drahanský M., Yanushkevich S. and Shmerko V.P. *Biometric Enabled Watchlists Technology*, IET Biometrics, 2017, Vol. 6, No. 6, pp. 1–10. ISSN 2047-4938.

[29] Eastwood S.C., Shmerko V.P., Yanushkevich S., Drahanský M. and Gorodnichy D. *Biometric-Enabled Authentication Machines: A Survey of Open-Set Real-World Applications*. IEEE Transactions on Human-Machine Systems, 2016, Vol. 46, No. 2, pp. 2168–2291. ISSN 2168-2291.

Chapter 2

Inner hand physiology and diseases

Marie Nováková[1] and Ivana Hradilová Svíženská[2]

For description of anatomical structures in the hand, it is important to know that the anatomical position of the hand is hanging alongside the trunk with fingers pointing downward and the palm pointing forward (Figure 2.1).

Hand represents the utmost peripheral part of the upper extremity. In humans, it accomplishes fine movements incomparable with other species, like monkeys, which use their hands for handling items and performing simple movements. In contrast, human hand is able to perform the most intricate movements—for example see piano players, whose hands smoothly and quickly change among all sorts of the movements in numerous joints forming their hands: flexion, extension, abduction, adduction, circumduction, pronation and supination. It is a concert of movements parallel to the concert of sounds. Bones, bound together by joints and operated by muscles, represent the structural conditions for performing these movements. Proper function is ensured by appropriate blood flow in the hand muscles. Blood flow in the hand skin plays moreover a crucial role in control of body temperature.

2.1 Skeleton of the hand

2.1.1 Bones and joints

The skeleton of the hand (Figure 2.2) consists of 8 carpal (wrist) bones, 5 metacarpal bones and 14 phalanges. **Carpal bones** are arranged into two rows: the proximal row is formed by the **scaphoid**, **lunate**, **triquetrum** and **pisiform**; the distal row is composed of the **trapezium**, **trapezoid**, **capitate** and **hamate**. The scaphoid and trapezium have tubercles that project toward the palm to form the eminence on the side of the thumb (laterally). The hamate has the hook, an unciform palmar process. The hook forms, together with the pisiform, the eminence on the side of the little finger (medially). The carpal bones form an arch convex toward the dorsum of hand. The **flexor retinaculum** is a thick connective tissue ligament that connects both eminences to bridge the space of the carpal arch converting the arch into the carpal tunnel [1]. Nine tendons of flexors as well as one

[1]Department of Physiology, Faculty of Medicine, Masaryk University, Czech Republic
[2]Department of Anatomy, Faculty of Medicine, Masaryk University, Czech Republic

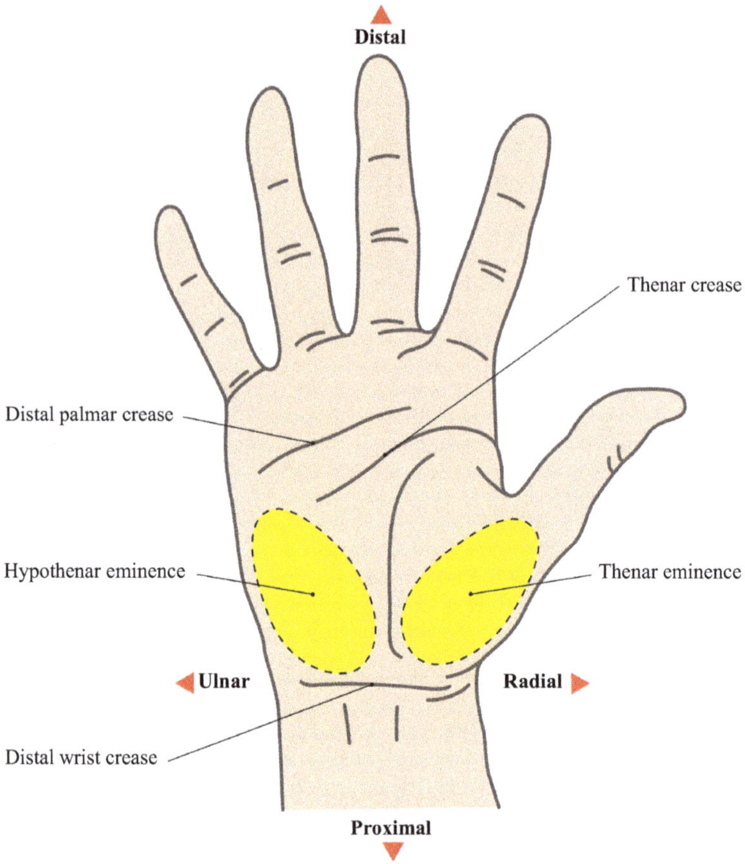

Figure 2.1 Anatomical terms in the palm

nerve pass through the carpal tunnel (see Section 2.2.1). The scaphoid, lunate and triquetrum form the **radiocarpal articulation** (wrist joint) with the distal end of the **radius**, one of two bones of the forearm, and cartilaginous disc overlying the distal end of the **ulna**, the other bone of the forearm. The oval shape of articular facets allows flexion, extension, abduction, adduction and circumduction. The joints between carpal bones share a common articular cavity. The movements at the **intercarpal joints** contribute to the flexion, extension, abduction and adduction of the hand. There is the **distal radio-ulnar joint**, by which the distal ends of both forearm bones articulate with each other, that allows the radius to move over the ulna positioning the hand into the pronation and supination.

Proximal ends of **metacarpal bones**, the bases, articulate with the distal carpal bones at five **carpometacarpal (CM) joints**. The first of them, between the trapezium and the first metacarpal, is much more mobile than other four CM joints that allow only very limited gliding movements. Distal ends of metacarpals, the heads, form five joints with proximal phalanges of digits. These **metacarpophalangeal**

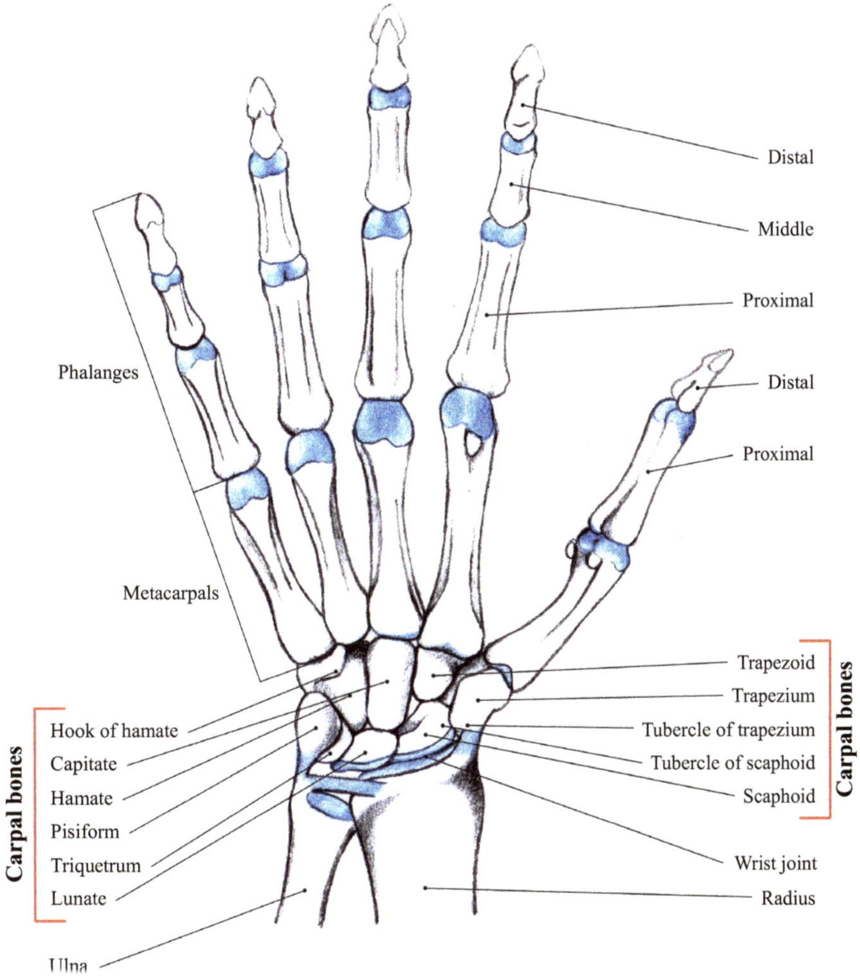

Figure 2.2 Skeleton of the hand

(MP) joints allow flexion, extension, abduction, adduction and circumduction. The second to fifth MP joints are linked together with the deep transverse metacarpal ligaments that restrict movements of metacarpals relative to each other.

Phalanges are bones of the fingers. The thumb consists of two, a proximal and a distal, phalanx, whereas the rest of digits have three phalanges, a proximal, a middle and a distal. There are **interphalangeal (IP) joints** between phalanges of each finger that allow only flexion and extension.

2.1.2 Palpable bony structures of the hand

The scaphoid tubercle and pisiform are palpable at the distal wrist crease from the palmar aspect (Figure 2.1). All MP and IP joints are also palpable from the front.

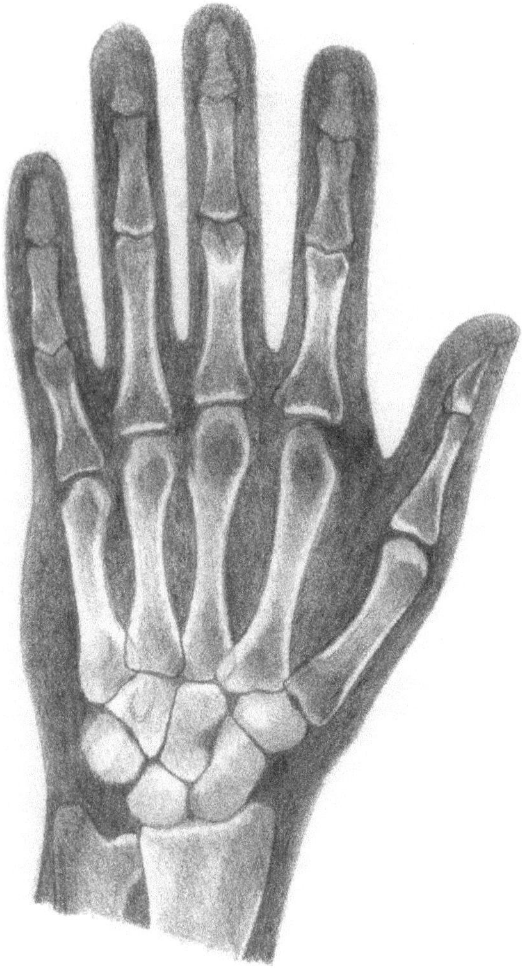

Figure 2.3 X-ray image of the adult hand

The trapezium, scaphoid, capitate and triquetrum are palpable from the dorsal aspect, as well as the distal ends of both forearm bones, especially when the hand is flexed. All the metacarpals, proximal and middle phalanges are palpable from the back, too. We can palpate only proximal ends of distal phalanges from the dorsal side because their distal ends are covered with nails.

2.1.3 X-ray image of the hand (Figure 2.3)

Radiographic assessment of skeletal age in pediatric patients is a common practice among orthopedic surgeons (Figure 2.4). There are two main applications for evaluations of skeletal maturation: the diagnosis of growth disorders and the prediction of final adult height [2]. Wrist and hand contain 45 separate distinguishable

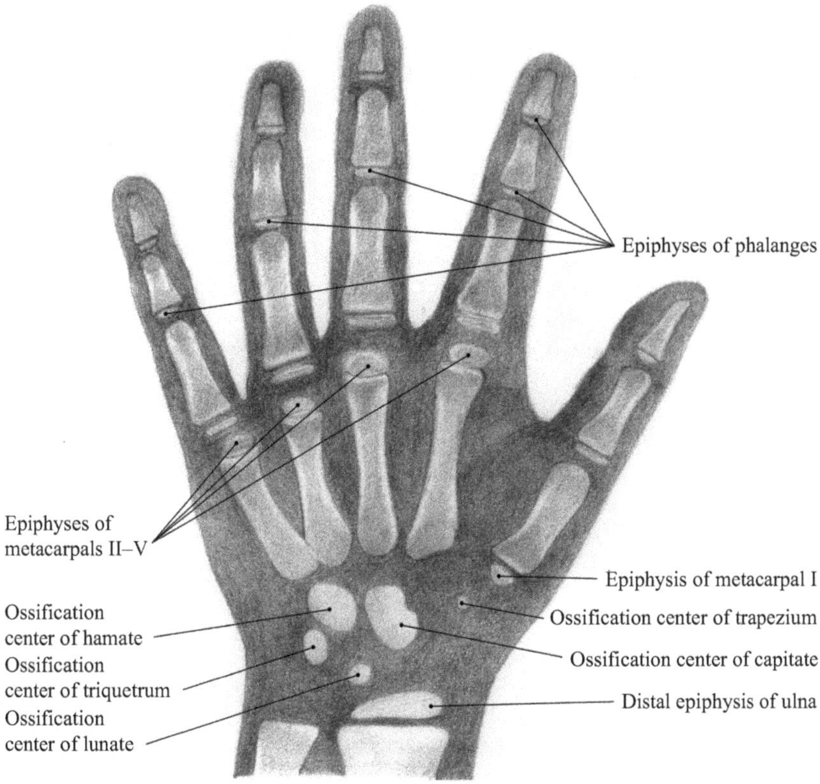

Figure 2.4 X-ray image of the hand of 5-year-old child

centers of ossification. The appearance and fusion of these centers provide an excellent assessment of biological development of a growing individual. Complete union of epiphysis for metacarpals and phalanges in males occurs between 14 and 19 years and in females between 13 and 17 years [3].

2.2 Muscles of the hand

Muscles of the hand can be divided into the extrinsic and intrinsic muscles. Extrinsic muscles have their origins in the arm or forearm, and their tendons are inserted into the bones of the hand. Intrinsic muscles of the hand are short muscles that may be divided into three groups: muscles of the thenar, hypothenar and central group. Because they occur entirely in the hand, they execute precision movements with the fingers and thumb. The surface of all muscles is covered by a thin connective tissue membrane, fascia, that is reinforced in thicker retinacula, septa or aponeuroses in some places.

2.2.1 *Extrinsic muscles of the hand*

From the palmar aspect, the extrinsic muscles flex the hand at the wrist and some of them also flex the fingers at MP and IP joints. The most superficial of them is the **palmaris longus** that has a tendon running in the middle of the wrist on the surface of the flexor retinaculum and widening in the palmar aponeurosis. The **palmar aponeurosis** is a triangular membrane lying deep to the subcutaneous tissue of the central part of the palm. The tendon of the **flexor carpi radialis** runs lateral to the tendon of the palmaris longus and is attached to the bases of the second and third metacarpals. The tendon of the **flexor carpi ulnaris** runs medial to the tendon of the palmaris longus and is attached to the pisiform. From this point, pisohamate ligament continues to the hook of hamate, and pisometacarpal ligament reaches the base of the fifth metacarpal (Figure 2.5). Tendons of flexors of the digits traverse the **carpal tunnel** secured in place with the flexor retinaculum. There are tendons of three muscles in the carpal tunnel (Figure 2.6): (i) the **flexor digitorum superficialis** has four tendons that are running immediately behind the flexor retinaculum and directing toward the second to fifth digits. At the level of the proximal phalanx, each tendon splits into two arms that pass around the tendon of flexor digitorum profundus to attach to both margins of the middle phalanx; (ii) the **flexor digitorum profundus** divides also into four tendons that pass behind the tendons of the flexor digitorum superficialis and insert into distal phalanges of the second to fifth digits; (iii) the **flexor pollicis longus** forms a single tendon, which traverses the carpal tunnel lateral to flexor digitorum tendons, and directs toward the thumb where it attaches to the distal phalanx. Free movement of the tendons in the carpal tunnel and within the fingers is facilitated by **synovial sheaths**, fluid-filled tubular double-layered sacs, which surround the tendons (Figure 2.7). Eight flexor digitorum tendons are surrounded by the **common sheath**, whereas the flexor pollicis longus has a separate sheath in the carpal tunnel. The synovial sheath of the flexor pollicis longus then continues to the thumb. In contrast, the common sheath typically does not communicate with the digital flexor tendon sheaths of the index, middle and ring fingers. However, it is usually continuous with the synovial sheath of the little finger. The digital sheaths of the index, middle and ring fingers usually extend proximally no further than the heads of their metacarpals. There are **fibrous sheaths** on the palmar aspect of digits superficial to the synovial sheath. These fibrous sheaths are formed by fibrous arches and cruciate ligaments, which are attached to the margins of phalanges and joints of fingers. They hold the tendons to the bones.

Extrinsic muscles from the dorsal aspect are referred to as extensors because they extend the hand and digits. Their tendons are segregated into six extensor compartments in the wrist that are between the bones and **extensor retinaculum** (Figure 2.8), the strengthened part of the forearm fascia. The compartments are designated by roman numerals from lateral to medial (Figure 2.11):

- I—abductor pollicis longus, extensor pollicis brevis
- II—extensor carpi radialis longus, extensor carpi radialis brevis
- III—extensor pollicis longus

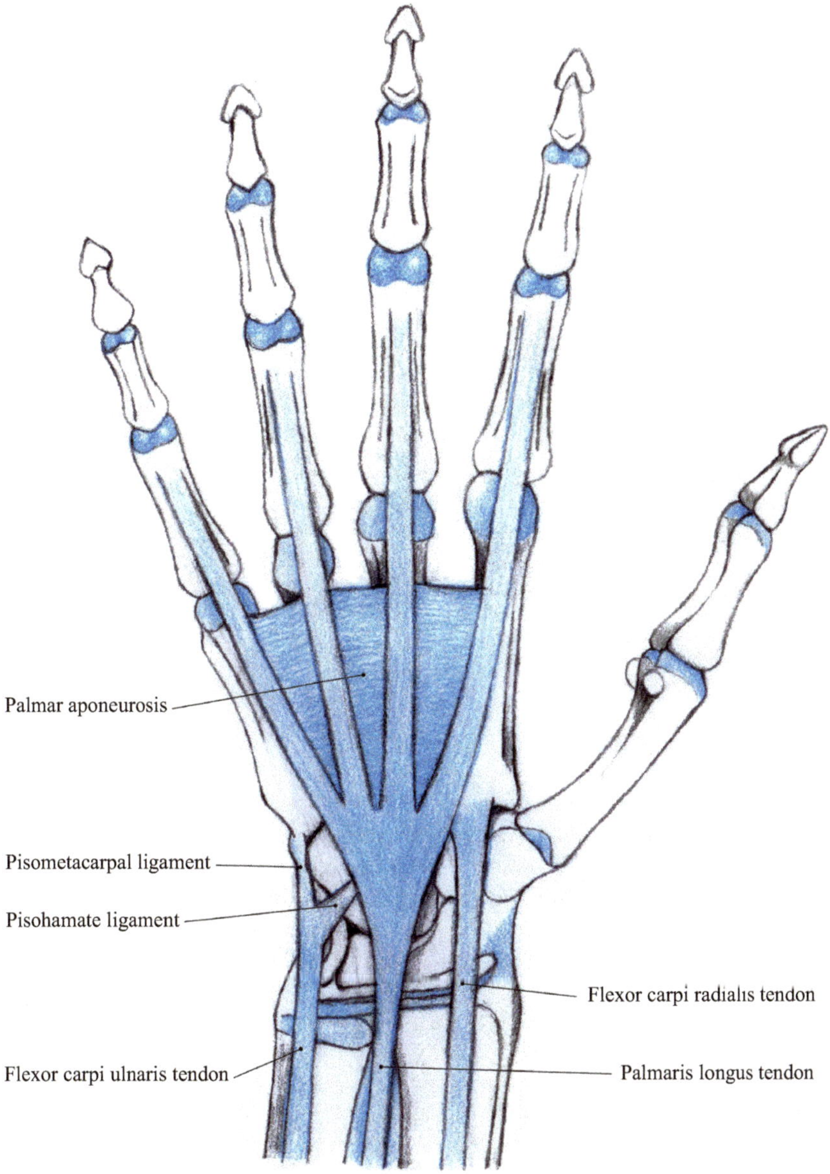

Figure 2.5 Superficial flexors of the hand

- IV—extensor digitorum, extensor indicis
- V—extensor digiti minimi
- VI—extensor carpi ulnaris

The tendons are surrounded with the synovial sheaths when passing through the compartments that usually end proximal to the metacarpals. Four tendons of the

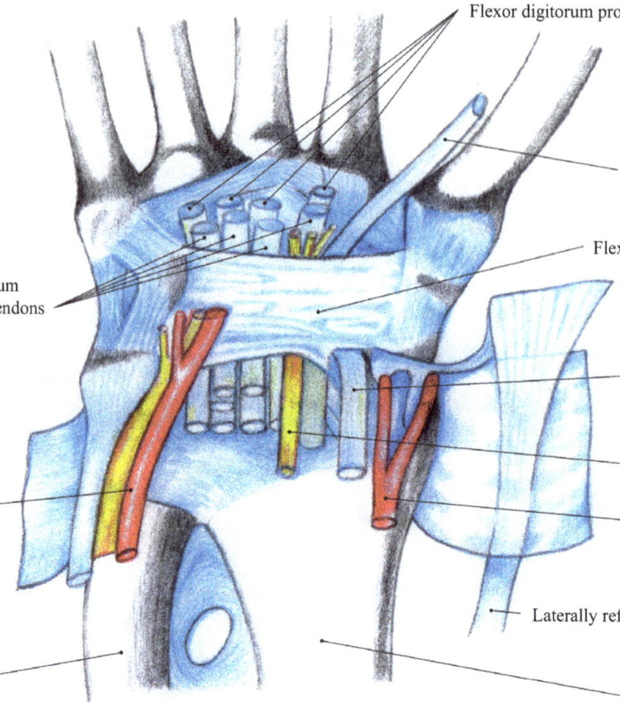

Figure 2.6 Carpal tunnel

extensor digitorum have a common tendon sheath. The **abductor pollicis longus** is inserted into the first metacarpal bone, and the **extensor pollicis brevis** into the proximal phalanx of the thumb. The **extensor carpi radialis longus** inserts into the base of the second metacarpal and the **extensor carpi radialis brevis** to the base of the third metacarpal. The **extensor pollicis longus** is attached to the distal phalanx of the thumb. The **extensor digitorum** forms four tendons for the index, middle, ring and little fingers. On the dorsal surface of the hand, adjacent tendons are interconnected. In the fingers, each tendon inserts, via a triangular-shaped connective tissue aponeurosis, the **extensor hood**, into the base of the dorsal surface of the middle and distal phalanges. The **extensor indicis** muscle is an accessory extensor of the index finger. Its tendon inserts into the extensor hood of the index. The **extensor digiti minimi** is an accessory extensor of the little finger. Its tendon inserts into the extensor hood of the little finger. The **extensor carpi ulnaris** inserts into the base of metacarpal V (Figure 2.8).

2.2.2 Intrinsic muscles of the hand (Figure 2.7)

The intrinsic muscles of the hand, which are associated with opposition of the thumb to the fingers and with delicate movements of the thumb, produce the prominent swelling on the lateral side of the palm, the thenar eminence. The **thenar**

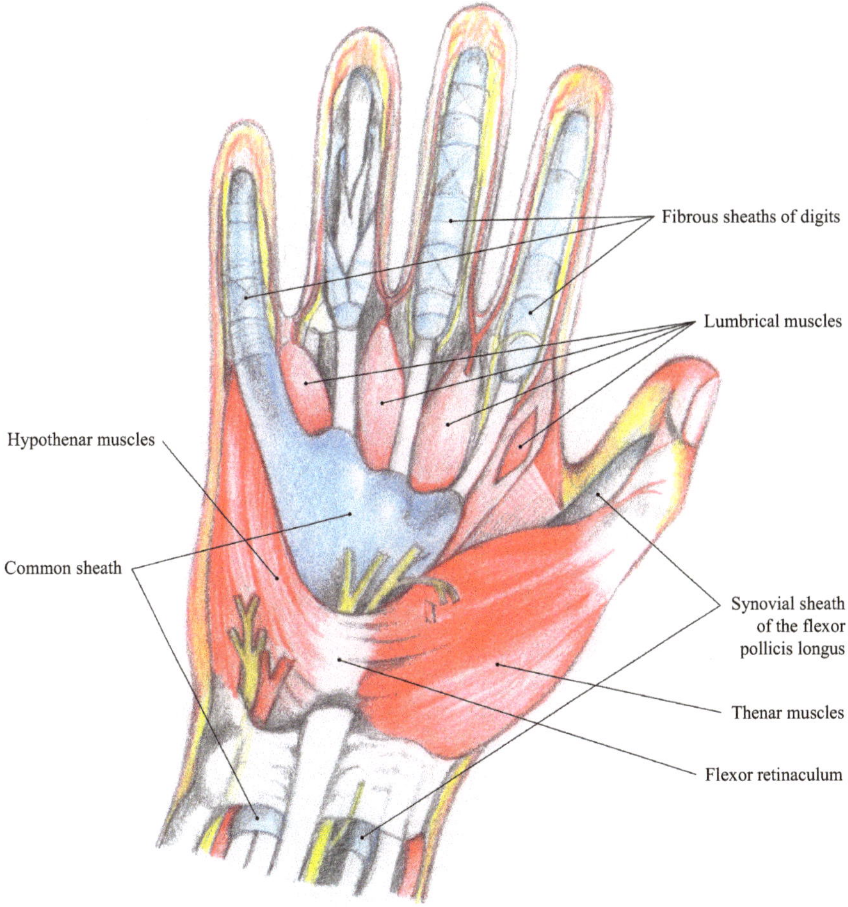

Figure 2.7 Synovial sheaths of flexors

muscles include the abductor pollicis brevis, flexor pollicis brevis, opponens pollicis and adductor pollicis. The **abductor pollicis brevis** is the most superfical one; it originates from the scaphoid and trapezium and inserts into the proximal phalanx of the thumb and into its extensor hood. The **flexor pollicis brevis** originates from the trapezium and inserts into the proximal phalanx of the thumb. The **opponens pollicis** originates from the trapezium and inserts into the lateral margin of the metacarpal I. The **adductor pollicis** originates as two heads: a transverse head from the metacarpal III and an oblique head from the capitate and bases of metacarpals II and III. The two heads converge to form a tendon that inserts into the medial side of the proximal phalanx of the thumb and into its extensor hood. The muscle is a powerful adductor of the thumb and opposes the thumb to the rest of digits. There are the lateral and medial **sesamoid bones** in the tendons of these muscles.

The **hypothenar** muscles (abductor digiti minimi, flexor digiti minimi brevis and opponens digiti minimi) execute precision movements with the little finger and

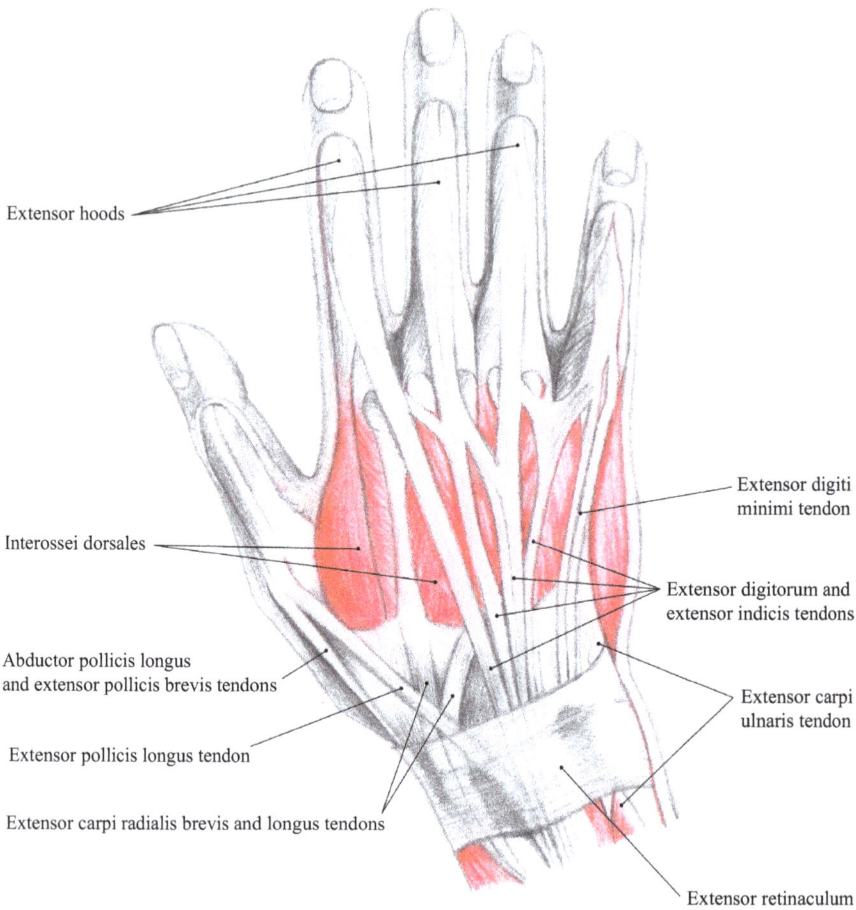

Figure 2.8 Extensors of the hand

contribute to the hypothenar eminence on the medial side of the palm. The **abductor digiti minimi** originates from the pisiform bone and pisohamate ligament and inserts into the proximal phalanx of the little finger and its extensor hood. It is the principal abductor of the little finger. The **flexor digiti minimi brevis** originates from the hook of hamate and inserts into the proximal phalanx of the little finger. The **opponens digiti minimi** is the deepest of the hypothenar muscles. It originates from the hook of hamate and inserts into the medial margin of the metacarpal V. The **palmaris brevis** is a subcutaneous muscle that originates from the palmar aponeurosis, inserts into the dermis of the skin on the medial margin of the hand and wrinkles this skin when contracted.

The **central group** of intrinsic muscles includes lumbricals and interossei muscles. There are four **lumbrical** muscles that originate from tendons of the flexor digitorum profundus. The small intrinsic muscles of the hand are mostly attached to the extensor hoods distal to MP joints; thus, they flex these joints but through the hood extend the IP joints. In addition, the lumbricals tilt the fingers toward the thumb. Four **interossei palmares** are deeper in the palm, originating from metacarpals I, II, IV and V. They are inserted into the extensor hoods of the respective fingers. In addition to flexion and extension movements of the fingers, the palmar interosseous are also major adductors of the thumb, index, ring and little fingers at the MP joints. **Dorsal interosseous** are the most dorsally situated of all of the intrinsic muscles and can be palpated through the skin on the dorsal aspect of the hand. There are four dorsal interosseous muscles arising between adjacent metacarpal bones. Each muscle inserts both into the base of the proximal phalanx and into the extensor hood of the index, middle and ring fingers. In addition to generating flexion and extension movements of the fingers through their attachments to the extensor hoods, the dorsal interossei are the major abductors of the index, middle and ring fingers at the MP joints.

Muscles of the hand are covered by the thin connective tissue membrane, the palmar and dorsal fascia. The **palmar fascia** is thicker in the center of the palm to form the triangular palmar aponeurosis. The proximal apex of the palmar aponeurosis is continuous with the flexor retinaculum and palmaris longus, and distal four slips, one to each finger, are continuous with fibrous flexor sheaths. A septum passes dorsally from each border of the palmar aponeurosis. The **lateral palmar septum** reaches the first metacarpal bone, and the **medial palmar septum** reaches the fifth metacarpal bone. Lateral and medial parts of the palmar fascia cover muscles of the thenar and hypothenar. The **dorsal fascia** of the hand is thin and continuous with the extensor retinaculum proximally.

2.2.3 Palpable and visible muscles and tendons of the hand

Thenar and hypothenar muscles form distinct eminences in the palm. Tendons of extrinsic muscles are readily visible as they pass from the forearm to the hand. From the palmar aspect, the tendon of the flexor carpi radialis passes in the lateral third of the distal forearm and the radial artery, which is immediately lateral to this tendon, is used for taking a radial pulse. The palmaris longus tendon lies medial to the flexor carpi radialis tendon. The tendon of the flexor carpi ulnaris is easily palpated along the medial margin of the distal forearm. Tendons of the extensor digitorum can be seen from the dorsal aspect radiating into the index, middle, ring and little fingers from the wrist. The tendon of the extensor carpi ulnaris can be felt on the medial side of the wrist between the distal end of the ulna and the wrist. Distal ends of tendons of the extensor carpi radialis longus and brevis muscles are on the lateral side of the wrist and can be accentuated by making a tight fist. Extension and abduction of the thumb reveals the **anatomical snuffbox**, a triangular depression formed on the dorsolateral side of the wrist and metacarpal I, bordered by tendons of abductor pollicis longus and extensor pollicis brevis laterally, and extensor pollicis longus medially. Historically, ground tobacco (snuff) was placed in this depression before being inhaled into the nose. The base of the triangle is at the wrist, and the apex is directed into the thumb [1].

A progressive volume loss of intrinsic muscles of the hand can be seen during aging. Moreover, extensor tendons, the extensor retinaculum or intertendinous connections are seen in older patients more frequently [4].

2.3 Nerves of the hand

The hand is supplied by the ulnar, median and radial nerves. All three nerves contribute to cutaneous innervation. The ulnar nerve innervates all intrinsic muscles of the hand except for three thenar muscles and two lateral lumbricals, which are innervated from the median nerve. Digital branches of all nerves also innervate the joints of the digits.

2.3.1 Ulnar nerve (Figure 2.9)

The **ulnar nerve**, which contains fibers of spinal segments C7, C8 and T1, enters the hand lateral to the pisiform. Immediately distal to the pisiform, it divides into the superficial branch, which is mainly cutaneous, and the deep branch, which is mainly muscular. The **superficial branch** innervates the palmaris brevis muscle and then continues as the **palmar digital nerves** to the palmar aspect of the skin of the little finger and medial half of the ring finger. The **deep branch** penetrates and supplies hypothenar muscles to reach the deep layer of the palm, arches laterally deep to tendons of the flexor digitorum superficialis and profundus and supplies the two medial lumbricals, the interossei and adductor pollicis. In addition, the deep branch of the ulnar nerve contributes small articular branches to the wrist joint. The palmar and dorsal branches of the ulnar nerve originate from the trunk of the ulnar nerve in the forearm. The **palmar branch** innervates the skin of the medial side of the palm while the **dorsal branch** innervates the skin of the dorsomedial side of the hand and then divides into three **dorsal digital nerves** for the skin of the little, ring and medial side of the middle fingers.

2.3.2 Median nerve (Figure 2.9)

The **median** nerve contains nerve fibers of spinal segments C5, C6, C7, C8 and T1. It enters the palm through the carpal tunnel where it lies between the flexor retinaculum and tendons of long flexors and may become compressed here (the carpal tunnel syndrome; Figure 2.6). Distal to the retinaculum, the nerve flattens and divides into the **recurrent branch** for the thenar muscles and **palmar digital nerves** for the palmar aspect of the skin of the thumb, index and middle fingers as well as the lateral half of the ring finger together with the nail beds of the same fingers. A small **palmar branch** arises from the median nerve in the distal forearm proximal to the flexor retinaculum, passes superficial to the retinaculum into the hand and innervates the skin over the lateral side of the palm.

2.3.3 Radial nerve (Figure 2.10)

The **superficial branch of the radial nerve**, which enters the dorsal aspect of the hand, arises in the arm. It enters the hand over the tendons of extensors pollicis

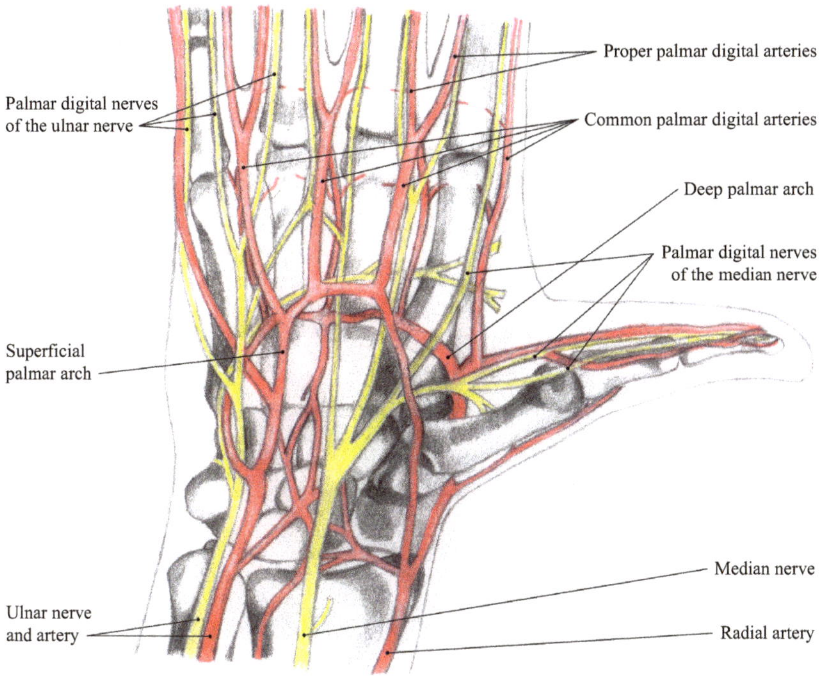

Figure 2.9 Nerves and arteries of the hand

brevis and longus and anatomical snuffbox and gives off **dorsal digital nerves** that innervate the back of the dorsolateral side of the hand and thumb, index and lateral half of the middle finger.

2.4 Arteries of the hand

Arteries, which supply blood to the hand, are the radial and ulnar arteries. The radial artery contributes mainly to blood supply of the thumb and lateral side of the index, while the remaining parts are supplied mainly by the ulnar artery. However, both arteries anastomose to form two carpal networks (palmar and dorsal) and two palmar arches (superficial and deep) that give off most of branches for muscles, joints and digits.

2.4.1 Ulnar artery (Figure 2.9)

The **ulnar artery** enters the hand between the palmaris brevis muscle and flexor retinaculum lateral to the ulnar nerve. Just distal to the pisiform bone, it divides into the superficial palmar arch and deep palmar branch. The **superficial palmar arch** runs laterally across the palm between long flexor tendons of digits and palmar aponeurosis to anastomose with the superficial palmar branch of the radial artery at the halfway of metacarpals. The superficial palmar arch gives off the palmar digital artery for the medial side of the little finger and three **common palmar digital**

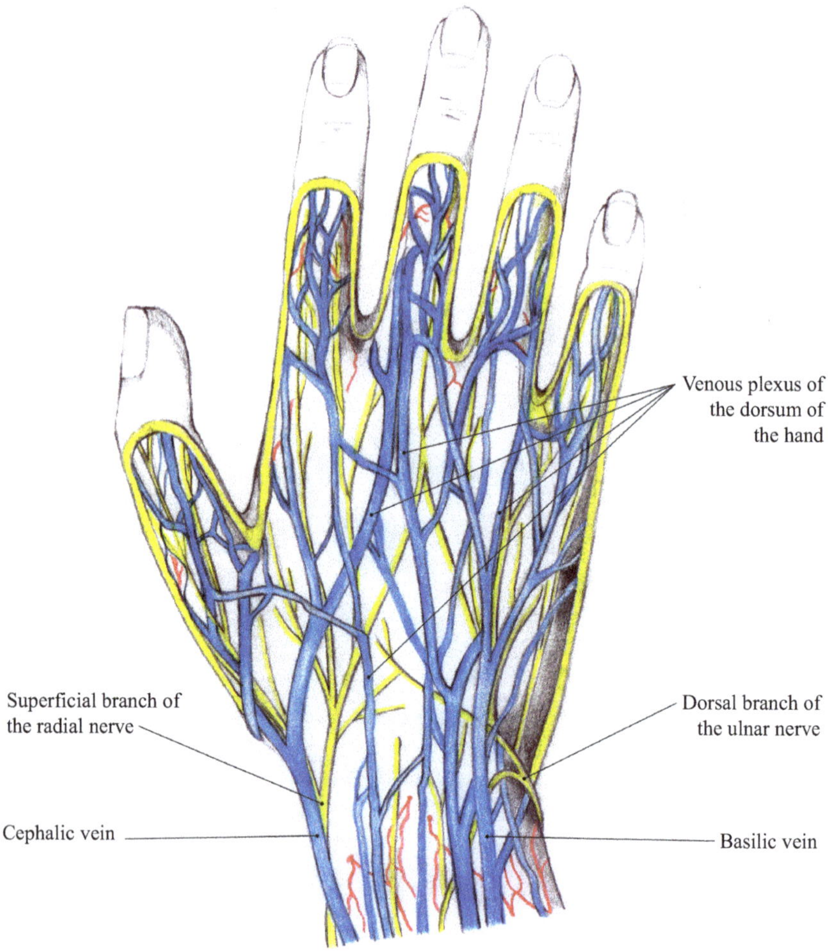

Figure 2.10 Veins and nerves of the hand dorsum

arteries, each of them dividing into two **proper palmar digital arteries** for adjacent sides of the little, ring, middle and index fingers. The **deep palmar branch** passes between hypothenar muscles and anastomoses with the deep palmar arch of the radial artery distal from the bases of metacarpals.

2.4.2 Radial artery (Figure 2.9)

The **radial artery** enters the hand turning around the lateral border of the wrist dorsally, passes in the floor of the anatomical snuffbox and then through the first dorsal interosseous muscle to give off the princeps pollicis artery, radial indicis artery and deep palmar arch. The **deep palmar arch** passes medially between the metacarpal bones and long flexor tendons of digits and anastomoses with the deep palmar branch of the ulnar artery. The deep palmar arch gives off three **palmar metacarpal arteries**, which connect to the common palmar digital arteries, and the

perforating branches, which anastomose with the dorsal metacarpal arteries. The **princeps pollicis artery** and **radial indicis artery** supply the palmar aspect of the thumb and lateral side of the index. The **superficial palmar branch** arises from the radial artery proximal to the wrist, passes through or over the thenar muscles and anastomoses with the superficial palmar arch of the ulnar artery at the halfway of metacarpals. The dorsal aspect of the hand is supplied from the **dorsal metacarpal arteries** that divide into **dorsal digital arteries** for adjacent sides of fingers. The dorsal metacarpal arteries are branches of the **dorsal carpal network** that is supplied from both ulnar and radial arteries. The smaller **palmar carpal network**, which is located behind the long flexor tendons proximal to the carpal tunnel and also receives branches of both arteries, supplies mostly bones and joints of the wrist.

2.5 Veins of the hand

There are deep and superficial veins in the hand interconnected by communicating veins. The deep veins accompany arteries and have the same names as arteries. The superficial veins form the venous plexuses in the subcutaneous tissue (Figure 2.10). The lateral part of the plexuses constitutes the **cephalic vein** that runs over the anatomical snuffbox to the lateral side of the forearm. The medial part of the plexuses constitutes the **basilic vein** that continues to the dorsomedial side of the forearm. The veins of three-phalanx fingers are at both margins of each finger proximally connecting in the arcade on the dorsal aspect of the proximal phalanx. At the interdigital level, these veins are confluent. From the venous arcade, the main venous system is situated centrally from the dorsal aspect of the middle and distal phalanges [5]. The veins of the hand dorsum run relatively straight, whereas the veins of the palm are fine and tortuous. Mean diameter of longitudinal veins of the hand dorsum is 1.27 mm [6]. In the palm, venous blood has a common tendency to flow from the center to the periphery in the subcutaneous venous network, i.e., toward web spaces between the fingers, toward the radial and ulnar margins of the palm and toward the wrist [7]. Aging hands are characterized by ectatic veins that become even more prominent through a loss in dermal thickness and subcutaneous tissue [8].

2.6 Lymph vessels of the hand

Lymph from all structures of the hand is drained to the axillary lymph nodes. Deep lymph vessels of the hand accompany arteries. More numerous superficial vessels form palmar and dorsal cutaneous plexuses. The vessels of both plexuses continue from the vessels arising from fingers. The diameter of dorsal lymph vessels varies from 0.2 to 0.6 mm, and they travel over or under the veins [9]. The palmar lymphatic plexus is denser than the dorsal one.

2.7 Subcutaneous tissue of the hand

The subcutaneous tissue is a fatty connective tissue that provides for the passage of cutaneous nerves, blood and lymph vessels and plays a role in connecting the dermis and fascia of the muscles [10]. In the dorsum of hand, three fascial layers can be identified: (i) the **dorsal superficial fascia** covers the cutaneous nerves and vessels; (ii) the **dorsal intermediate fascia** covers the extensor tendons and is continuous with the antebrachial fascia; and (iii) the **dorsal deep fascia** covers the dorsal aspect of dorsal interosseous muscles and metacarpal bones. The fasciae are separated by the laminae of areolar connective tissue with variable amount of fat [11].

The **palmar fascia** is thicker in the center of the palm to form the triangular palmar aponeurosis (Figure 2.5). Lateral and medial parts of the palmar fascia cover muscles of the thenar and hypothenar.

The hand subcutaneous tissue is traversed by numerous dense bands or skin ligaments, which tie skin to deeper fasciae and hence to bone for obvious functional reasons. These are little developed on the dorsum of the hand but marked in palm and flexor aspects of digits. At well-recognized sites, where cutaneous flexion furrows are usually apparent, skin fixation is firmest. Superficial fibers of the palmar aponeurosis join the dermis of the palm in major palmar furrows and furrows at the roots of digits. Between such sites of maximal anchorage, many smaller fixation bands exist, interspersed by cushioning pads of adipose tissue, which permit a small range of tangential movement. These arrangements have obvious protective value; fat pads in interdigital webs protect digital vessels and nerves from adjacent tendons; thenar and hypothenar pads are clearly useful cushions in applying various tools [1].

2.8 Cross-sections through the hand

2.8.1 Transverse section through the wrist (Figure 2.11)

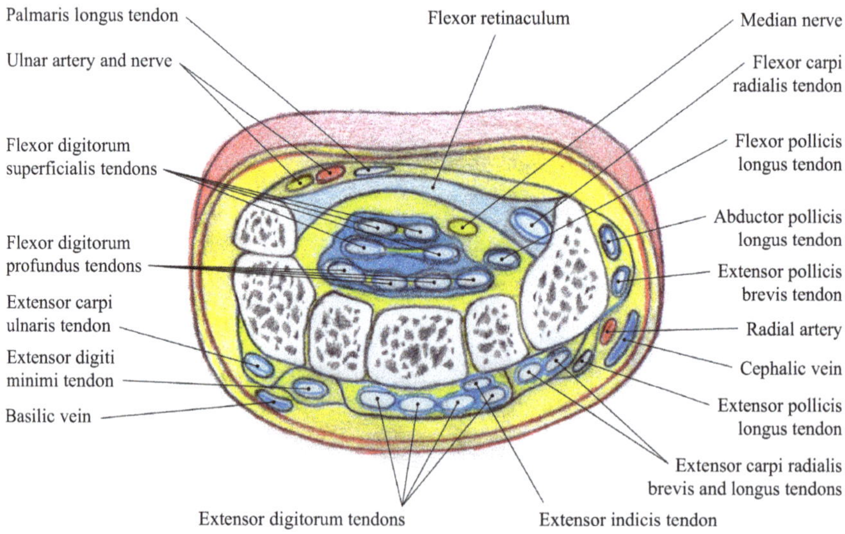

Figure 2.11 Transverse section through the wrist

2.8.2 Transverse section through the metacarpals (Figure 2.12)

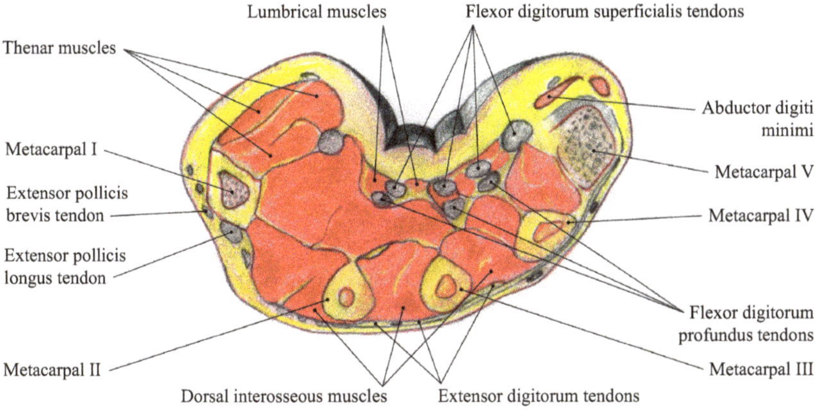

Figure 2.12 Transverse section through the metacarpals

2.8.3 Transverse section through the finger (Figure 2.13)

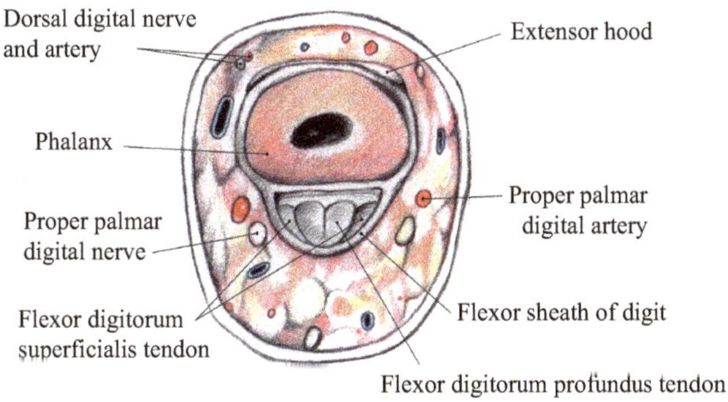

Figure 2.13 Transverse section through the finger

2.9 Gender differences in hand shape

It is possible to distinguish between male and female hands by considering certain geometric features [12]. Morphological sex differences in the absolute length of fingers have been demonstrated in various studies, male fingers being longer as compared to females [13]. In females, index and ring finger tend to be almost equal in length, whereas in males, ring finger tends to be much longer. Thus, the index and ring finger ratio becomes a significant parameter for determining sex [14]. However, Voracek's study did not concur because the ratio varies considerably for different regions and nationalities [15]. The length, breadth, thickness and circumference of various parts of the hand can be used for more accurate sex determination. However, the breadth and circumference of finger joints and wrist only can serve as references in wide age range groups due to the effect of age on hand size [16].

2.10 Blood flow in the hand

The various movements of hand are based on adequate blood flow (**perfusion**) in the hand skeletal muscles. Also, certain aspects of body temperature control (**thermoregulation**) should be considered, since the hand represents the area typical with marked temperature variations. These temperature differences are so prominent that the body periphery resembles more poikilothermic (cold-blooded) organism. Thermoregulation is closely related to changes of blood flow: increased perfusion is responsible for release of heat, decrease of perfusion causes restriction of heat loss.

Blood flow in any vessel of the human body is—in the case of constant blood pressure—determined by diameter of the vessel. **Vasoconstriction**—narrowing of the vessel—decreases blood flow, due to **vasodilation** (**vasodilatation**) blood flow increases. The intensity of perfusion in various regional circulations is regulated at two levels—local and central—and the hand is not an exception. Both ways of regulations—humoral and nervous—are involved (Table 2.1).

As the nervous control is concerned, it is ensured by activation of the sympathetic nervous system, which leads to vasoconstriction; withdrawal of sympathetic tone leads to passive vasodilation [17]. At rest, continuous sympathetic tone is present, which markedly increases during physical exertion. However, during exercise, vasoconstrictive effect of sympathetic system (mediated via alpha-1 adrenergic receptors) serves primarily for diverting the blood from inactive muscles to those involved in exertion. Active vasodilation controlled by nervous system has not been conclusively proven in humans. Vasodilation caused by binding of circulating adrenalin to beta-2 adrenergic receptors is mediated via release of NO from endothelium (which is moreover released by mechanical stimuli due to increased blood pressure during exercise) and does not represent direct neural effect on vessels [18].

Table 2.1 An overview of blood flow regulation in the hand

System	Central	Local
Nervous	Autonomous—sympathetic nervous system	
Humoral	Circulating substances with vasomotor activity	Local humoral substances Metabolic vasodilation

Among circulating vasoactive substances, the prominent place take angiotenzin II (part of renin–angiotensin–aldosterone system, acting as an endocrine, autocrine/paracrine and intracrine hormone) and circulating catecholamines (adrenalin released from adrenal medulla); both act as vasoconstrictors. The primary role of these substances is homeostatic; by adjusting vascular compliance and vascular resistance, they help to keep hemodynamics under control. Abovementioned compounds are mostly part of endocrine regulations in human body. Bradykinin, histamine (released from basophils and mast cells in subendothelial layer of the vascular wall, e.g., during allergic reaction or inflammation) and vasoactive intestinal peptide represent important endogenous vasoactive substances which are locally released, all of them having vasodilatatory effect. Important vasoconstrictive agent is serotonin, released from enterochromaffin cells. In blood, it is collected from plasma by platelets, stored and when platelets bind in damaged tissue, helps to stop bleeding by local vasoconstriction.

Several vasoactive substances are produced by endothelium (Table 2.2, according to Boron and Boulpaep [19]).

Regarding the blood flow in the hand, two circulations have to be taken in consideration: in skeletal muscles of the hand and in its skin. Regardless the fact that the hand represents peripheral part of the body, its perfusion reflects not only local but also whole-body changes, such as physical exercise and consequent need of keeping proper body temperature, which is diverted by heat-producing contractions of skeletal muscles.

2.10.1 Blood flow in hand muscles

At rest, the changes of perfusion in hand muscle tissue contribute to total blood flow changes in the hand less than those in skin circulation. The perfusion of skeletal muscle is closely related to actual, locally specific metabolic needs [17]. Significant changes are observed during the physical exercise, namely, if performed by the hand muscles. In such situation, mainly local vasodilation (the so-called metabolic autoregulation of blood flow) contributes to increased blood flow (*hyperemia*) in the hand. It is caused by local acidosis, hypoxia, hypercapnia, decreased intracellular concentration of ATP and increased concentrations of adenosine, lactic acid and K^+ in extracellular fluid. Eventually, increasing temperature in the muscle tissue directly affects the local blood flow via the peripheral thermoreceptors as well as due to a direct effect of temperature on the vessel tone.

Table 2.2 Main vasoactive agents released from endothelium

Vasodilators	**Vasoconstrictors**
Nitric oxide (NO)	Endothelin
Endothelium-derived hyperpolarizing factor	Endothelium-derived constricting factor 1
Prostacyclin (PGI_2)	Endothelium-derived constricting factor 2

Although adenosine is considered one of the most potent vasodilating agents [20], no single substance or parameter can explain complex change in blood flow in such situation. Moreover, initiation of muscle fiber contraction leads to hyperpolarization (electrical signal being promptly spread via gap junctions) of vascular smooth muscle cells due to a release of K^+ from contracting muscles to extracellular fluid and to vasodilation of nearby arterioles [19].

During whole-body exercise, the changes of autonomous nervous system also affect the perfusion in the hand: at first, vasoconstriction in the skin is observed (due to increased sympathetic tone) followed by vasodilation due to increasing local metabolic and thermoregulatory demands (heat output need).

2.10.2 Blood flow in hand skin

At rest, the hand perfusion reflects mostly the situation in skin circulation. Skin perfusion control is closely related to thermoregulation.

The skin represents the major barrier between the internal and ambient environments. It is the largest organ of the body with various functions—mainly protective, thermoregulatory and immune. In adult human, the skin represents approx. 2–3 kg of body mass. Two types of the skin are distinguished with respect to organization of blood flow: "apical" and "nonapical" one. The apical skin is found on the hand, together with nose, lips, ears and feet. These are typical areas with high surface/volume ratio. Such structural organization favors heat loss [19].

The nutritional demands of the skin are modest, and the blood flow regularly exceeds its metabolic requirements at rest. The resting skin perfusion ranges from 0.2 to 5 L/min in the whole skin (approx. 3 mL/min/100 g of tissue). If necessary, even excessive vasoconstriction appears (at very low temperatures) which decreases the skin perfusion to values little above zero. The evacuation of blood from the skin vessels in such a situation leads to a change of skin color—**pallor**. On the other hand, at high ambient temperatures, the skin perfusion may reach values as high as 7–8 L/min in the whole skin. Blood flow in the skin of finger varies from 0.01 to 1.5 mL/g/min.

In contrast to skeletal muscle perfusion, where the local metabolic control plays a crucial role, the main control of the overall skin perfusion is nervous. At rest or at the thermoneutral ambient temperatures, the local metabolic control of skin perfusion is of little importance, and it is only weakly affected by metabolites and sensory stimuli (such as pain or touch stimuli). The local circulation responds to local temperature stimuli—both cold and hot—with local vasoconstriction or vasodilation, respectively. Local mechanical stimuli can trigger local vascular response, either a so-called white reaction or triple response.

White reaction is a blanched line appearing on the skin after it has been stroked mildly with a sharp instrument. The immediate response is caused by passive removal of the blood by external mechanical pressure. Within next 15–60 s, white (constrictory) dermographism (reaction) develops: its underlying mechanism is contraction of microvascular smooth muscle cells and pericytes due to mechanical incitement. As a result, capillary loops, the collecting venules and the subpapillary venous plexus get emptied.

Triple response (triple response of Lewis) originates from drawing strongly with a sharp instrument across the skin under higher force in comparison with the white response [21]. It may appear at three steps: red reaction, flare reaction and local edema. *Red reaction* (red, dilatatory dermographism) is nerve-independent, and it is brought about by vasodilation due to locally released histamine (from mechanically stimulated mast cells and basophils in the scratched area). A stronger or repeated mechanical stimulus, the *flare reaction*, appears within a few seconds after the red reaction as a local nervous response (so-called axon reflex). Vasodilating neurotransmitters are released from collateral branch of a single nerve fiber stimulated by scratching (the branch containing sensory receptor triggers the action potential traveling then both to spinal cord and to collateral branch after the mechanical stimulation). Even stronger mechanical stimulus triggers a *local edema (wheal)* based on increased capillary permeability due to massive local histamine release, when filtration exceeds absorption. This reaction replaces red reaction and appears within minutes after the mechanical stimulation.

A prominent role of the skin perfusion is observed during thermoregulation. For this purpose, in apical skin, specific structures have developed which enable to change skin perfusion within a wide range—**arteriovenous anastomoses** (a-v anastomoses) called also **glomus bodies** (Figure 2.14). These bodies are most often found on the finger pads and nail beds. Their control is mainly of nervous (sympathetic) origin, and changes in their blood content play a critical role in heat exchange. Maximal increase of sympathetic tone can lead to even complete constriction of a-v anastomosis with consequent enormous reduction of skin perfusion in the area. Withdrawal of sympathetic stimulation leads then to passive vasodilation.

The glomus body consists of an *afferent arteriole* which—after entering to connective tissue capsule of the glomus body—changes into a vessel with a small lumen and a thick, muscular wall—*a-v anastomosis*. The next part of the glomus

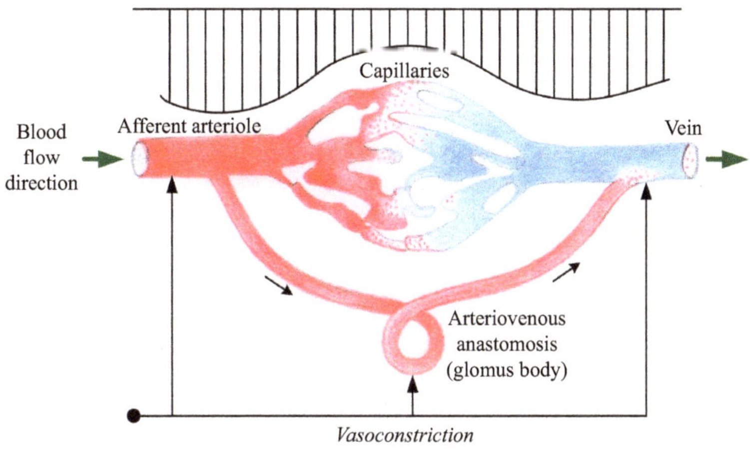

Figure 2.14 Arteriovenous anastomosis

body represents short, thin-walled *vein*, draining eventually the blood into larger skin veins. The whole glomus body is rich in sympathetic innervation. Norepinephrine released constricts the arteriole, anastomotic vessel as well as the venule.

In the case of high core temperature (brain, inner organs of the chest and abdomen), the vasomotor sympathetic tone of the a-v anastomoses decreases and they dilate (passive vasodilatation, see above). Since a-v anastomoses are wider than arterioles or capillaries, they represent low-resistance shunt and thus lead to skin hyperemia, with consequent change of skin color (to reddish, due to the fact the blood in venules is oxygenated). Venular plexuses are filled up with blood and heat loss increases. However, the capillary blood flow is almost not affected, and therefore capillary blood pressure does not increase. Then, there is no tendency to edema formation. Only extremely high blood content in venular plexuses increases the pressure in venules (and as a consequence also the pressure in capillaries) and edemas develop. As heat passes through the walls of skin venous plexuses, local skin temperature—and consequently also heat loss—increases. Low core temperature triggers off the opposite reaction, and the a-v anastomoses are clamped.

2.11 Diseases and disorders of the hand

Disorders and diseases of the hand may happen prenatally or postnatally. Congenital disorders represent the first group of these diagnoses. Postnatally, the hand may be affected by palsy, inflammatory diseases, tumors, infections, vascular disorders and last but not least various traumas.

2.11.1 Congenital disorders

The hand may become a target of prenatally acquired disorders and malformations, caused by either genetically conditioned situation or by malnutrition of the fetus, eventually by trauma during the pregnancy. Most frequently observed congenital disorders of fingers are as follows:

clinodactyly—the curvature of a digit (a finger) in the plane of the palm, most commonly the fifth finger toward the adjacent fourth finger;
syndactyly—two or more digits fused together;
polydactyly (hyperdactyly)—supernumerary finger(s);
brachydactyly—shortness of the finger(s), related to the hand (or other long bones);
macrodactyly—larger than normal size of finger(s), usually part of local gigantism;
absence of fingers
congenital contracture—anything that inhibits normal joint movement before birth can result in joint contractures, intrauterine restriction of fetus movement (environmental factors) and/or genetic disorders can be found behind this situation.

2.11.2 The hand palsy

The hand palsy can originate from local or distant compression of radial, ulnar or median nerves which leads to so-called wrist drop, claw hand and hand of benediction. Another possible reason for paralysis is cerebral palsy due to stroke, tumor or brain injuries.

Ulnar nerve entrapment is a condition where the ulnar nerve becomes physically trapped or pinched, resulting in pain, numbness or weakness. It is caused by the fact that the ulnar nerve passes through a number of small spaces on the medial side of the arm, and it is vulnerable to compression there. Most frequently, the ulnar nerve compression appears at the elbow, followed by the wrist.

2.11.3 The hand inflammatory diseases

Among inflammatory diseases, the impact of which may be observed in the hand are osteoarthritis, rheumatoid arthritis and other connective tissue diseases. These diseases cause mainly deformities of the joints forming the hand.

Dupuytren's contracture (Dupuytren's disease, Viking disease) is a state when one or several fingers are permanently bent in a flexion. Although the risk factors are known (family history, alcoholism, smoking, thyroid gland disorders, liver diseases, diabetes mellitus, previous hand trauma and epilepsy), the cause is unknown. It is characterized by hard nodule(s) just under the skin of the palm, which result from the formation of abnormal connective tissue within the palmar fascia. This disease is typical for men over the age of 50, Caucasians (it is very rare among Asians and Africans). Regardless the treatment, this disorder evinces high recurrence [22].

2.11.4 The hand tumors

Bone tumors in the hand are either primary bone tumors both benign and malignant or metastases of tumors originating from other tissues. Carcinomas of the prostate, breasts, lungs, thyroid and kidneys most frequently metastasize to bones. Metastating malignant bone tumors are 50–100 times more frequent than primary bone tumors.

Benign bone tumors include osteoma, osteoid osteoma, osteochondroma, osteoblastoma, enchondroma, giant cell tumor of bone, aneurysmal bone cyst and fibrous dysplasia of bone.

Malignant primary bone tumors include osteosarcoma, chondrosarcoma, Ewing's sarcoma, fibrosarcoma and other types.

2.11.5 The hand injuries

The hand structures may be damaged to various extent during accident(s) or sport activities. These injuries concern bones (fractures), ligaments, tendons, joints (dislocations and eventually ligament injuries), muscles and various combination(s) of these injuries. Hand and wrist injuries are more common in children (including those caused by sport activities) and may involve the epiphyseal plates with the potential risk of growth impairment.

Out of all sports injuries, wrist and hand injuries represent approximately 3%–9%. Their incidence varies among sports: the wrist injuries are most frequently found in gymnasts (87% of them report it at least once during their career), hand injuries are frequent in rock climbing (44%). Injuries to the hand and wrist are also quite frequent among tennis and golf players (especially the left hand in right-handed golfers). Most of these traumas are soft tissue injuries, but a diagnosis of "wrist sprain" should only be made by exclusion of other more serious injuries [23].

Other injuries of the hand include amputations, contractures, thermal and frostbite injury, chemical burns, etc.

Specific type of injury represents so-called cumulative trauma disorders. They are preceded by occupational risk factors, such as repetitive tasks, force, posture and vibration(s). One of them is *carpal tunnel syndrome* [24]. It is caused by compression of the median nerve as it passes through the wrist at the carpal tunnel. The main symptoms are pain, numbness and tingling. These signs are present in the thumb, index finger, middle finger and the thumb side of the ring fingers, in approx. 50% of patients on both hands. The symptoms start gradually, typically in the night, and may spread into the arm. Weak grip strength may progress gradually into the atrophy of muscles at the base of the thumb. Risk factors include obesity, repetitive wrist work, pregnancy and rheumatoid arthritis. There is tentative evidence that hypothyroidism increases the risk. It is unclear if diabetes plays a role.

Another affection of the wrist, caused by repeated injuries, is ganglion (ganglion cyst, synovial cyst). It is a fluid-filled lump of unknown cause associated with a joint or tendon sheath. The most typical localization is the back of the wrist followed by its front. The underlying mechanism is probably an outpouching of the synovial membrane. Risk factors include gymnastics; however, it is often seen in typists and PC users.

Glossary

abduction	the movement from the axis of the hand passing through the third digit
adduction	the movement toward the axis of the hand
anastomosis	the connection of two vessels
circumduction	the movement which combines abduction, adduction, flexion and extension
cutaneous (sensory) innervation	nerves that supply the skin
distal	the structure points or lies further from the attachment of the limb to the trunk
extension	the movement backward
flexion	the movement forward
hyperemia	increased blood flow

lateral	the structure lies on the side of the thumb
medial	the structure lies on the side of the little finger
muscular (motor) innervation	nerves that supply muscles
perfusion	blood flow (through certain volume of tissue and during certain time period)
pronation	the dorsum of hand points forward
proximal	the structure points or lies closer toward the attachment of the limb to the trunk
sesamoid bone	the bone that develops in the tendon of a muscle
supination	the dorsum of hand points backward
thermoregulation	body responses keeping its temperature within physiological range
vasoconstriction	narrowing of blood vessel (decrease of its diameter)
vasodilation (vasodilatation)	widening of blood vessel (increase of its diameter)

References

[1] Ellis H. Gray's anatomy. 37th ed. P. L. Williams, R. Warwick, M. Dyson, L. H. Bannister. pp. 1598. Illustrated. 1989. Edinburgh: Churchill Livingstone. Br J Surg. 1989 Dec; 76(12): 1359–1359.

[2] Gilsanz V, Ratib O. Hand Bone Age: A Digital Atlas of Skeletal Maturity. Heidelberg: Springer-Verlag; 2005.

[3] Steele DG, Bramblett CA. The Anatomy and Biology of the Human Skeleton. College Station, TX: Texas A&M University Press; 1988. 308 p.

[4] Jakubietz RG, Kloss DF, Gruenert JG, Jakubietz MG. The ageing hand. A study to evaluate the chronological ageing process of the hand. J Plast Reconstr Aesthet Surg. 2008 Jun;61(6):681–6.

[5] Sukop A, Nanka O, Dusková M, *et al.* Clinical anatomy of the dorsal venous network in fingers with regard to replantation. Clin Anat. 2007 Jan;20(1): 77–81.

[6] Kiray A, Ergür I, Tayefi H, Bağriyanik HA, Bacakoğlu AK. Anatomical evaluation of the superficial veins of the upper extremity as graft donor source in microvascular reconstructions: A cadaveric study. Acta Orthop Traumatol Turc. 2013;47(6):405–10.

[7] Kim B-K, Kwon H, Imanishi N, Chang H. Anatomical study of cutaneous venous flow of the palm. Ann Plast Surg. 2017 Apr;78(4):448–454.

[8] Lim A, Mulcahy A. Hand rejuvenation: Combining dorsal veins foam sclerotherapy and calcium hydroxylapatite filler injections. Phlebology. 2017 Jul;32(6):397–402.

[9] Pan W-R, Zeng F-Q, Wang D, Qiu Z-Q. A Method for making a lymphatic specimen of the dorsum of the hand. Lymphat Res Biol. 2015 Dec;13(4): 275–8.

[10] Ishida T, Takeuchi K, Hayashi S, *et al.* Anatomical structure of the subcutaneous tissue on the anterior surface of human thigh. Okajimas Folia Anat Jpn. 2015;92(1):1–6.

[11] Bidic SM, Hatef DA, Rohrich RJ. Dorsal hand anatomy relevant to volumetric rejuvenation. Plast Reconstr Surg. 2010 Jul;126(1):163–8.

[12] Amayeh G, Bebis G, Nicolescu M. Gender classification from hand shape. In: 2008 IEEE Computer Society Conference on Computer Vision and Pattern Recognition Workshops. Anchorage, AK; 2008, pp. 1–7.

[13] Williams TJ, Pepitone ME, Christensen SE, *et al.* Finger-length ratios and sexual orientation. Nature. 2000;404(6777):455–6.

[14] Kanchan T, Krishan K. Anthropometry of hand in sex determination of dismembered remains—A review of literature. J Forensic Leg Med. 2011 Jan;18(1):14–17.

[15] Voracek M. Why digit ratio (2D:4D) is inappropriate for sex determination in medicolegal investigations. Forensic Sci Int. 2009;185(1–3):e29–30.

[16] Jee S-C, Bahn S, Yun MH. Determination of sex from various hand dimensions of Koreans. Forensic Sci Int. 2015 Dec;257:521.e1–10.

[17] Joyner MJ, Casey DP. Regulation of increased blood flow (hyperemia) to muscles during exercise: A hierarchy of competing physiological needs. Physiol Rev. 2015 Apr;95(2):549–601.

[18] Joyner MJ, Dietz NM. Sympathetic vasodilation in human muscle. Acta Physiol Scand. 2003 Mar;177(3):329–36.

[19] Boron WF, Boulpaep EL. Medical Physiology, 3rd Edition. Vol. 2017. Philadelphia, PA: Elsevier; 1312 p.

[20] Hellsten Y, Maclean D, Rådegran G, Saltin B, Bangsbo J. Adenosine concentrations in the interstitium of resting and contracting human skeletal muscle. Circulation. 1998 Jul 7;98(1):6–8.

[21] Bhute D, Doshi B, Pande S, Mahajan S, Kharkar V. Dermatographism. Indian J Dermatol Venereol Leprol. 2008 Apr;74(2):177–9.

[22] Khashan M, Smitham PJ, Khan WS, Goddard NJ. Dupuytren's disease: Review of the current literature. Open Orthop J. 2011;5(Suppl 2):283–8.

[23] Bytomski J, Moorman C, MacAuley D. Oxford American Handbook of Sports Medicine. New York, NY: Oxford Press Inc.; 828 p.

[24] Burton C, Chesterton LS, Davenport G. Diagnosing and managing carpal tunnel syndrome in primary care. Br J Gen Pract. 2014 May;64(622):262–3.

Chapter 3

Outer hand physiology and diseases

Eva Březinová[1,2]

3.1 Introduction

This chapter includes description of outer hand, its physiology, and changes when skin disease is present, everything from medical point of view. The complete skin anatomy of the hand will be described, together with histopathological changes. The skin diseases localized on this special part of the body and their consequences to fingerprint recognition will be mentioned and discussed.

3.2 Anatomic aspects of the skin

The skin is constituted of two distinct compartments: (1) epidermis and epithelial structures of adnexa and (2) dermis and its nonepithelial structures of adnexa. The subcutaneous fat is not truly a part of the skin, as its synonym, hypodermis, denotes, but because it is exceedingly close anatomic relationship to the skin in many pathologic processes, the subcutaneous fat is considered as a part of the skin [1,2] (Figure 3.1).

The epidermis is the thinnest component of skin, varying in thickness from about 0.04 mm on the eyelids to 1.6 mm on the palms of a young adult. The average thickness of epidermis is 0.1 mm. The epidermis is a metabolically active, stratified squamous cornifying epithelium populated at least four different and distinctive types of cells: keratinocytes, melanocytes, Langerhans' cells, and Merkel cells. Within the epidermis, cohesively organized keratinocytes predominate overwhelmingly, in contrast to the dermis, which consists of mostly of relatively noncellular connective tissue composed of collagen bundles, elastic fibers, and ground substance. A variety of cells is scattered in variable numbers throughout the mature dermis: fibrocytes, dermal dendrocytes, histiocytes, Langerhans' cells, mast cells, and, rarely, lymphocytes. Within the dermis are lodged nerves, blood vessels, lymph vessels, smooth muscles, and epithelial structures of adnexa, namely, the folliculosebaceous-apocrine units and the eccrine units. The dermis is 15–40 times

[1]First Department of Dermatovenereology, St. Ann's University Hospital, Czech Republic
[2]Faculty of Medicine, Masaryk University, Czech Republic

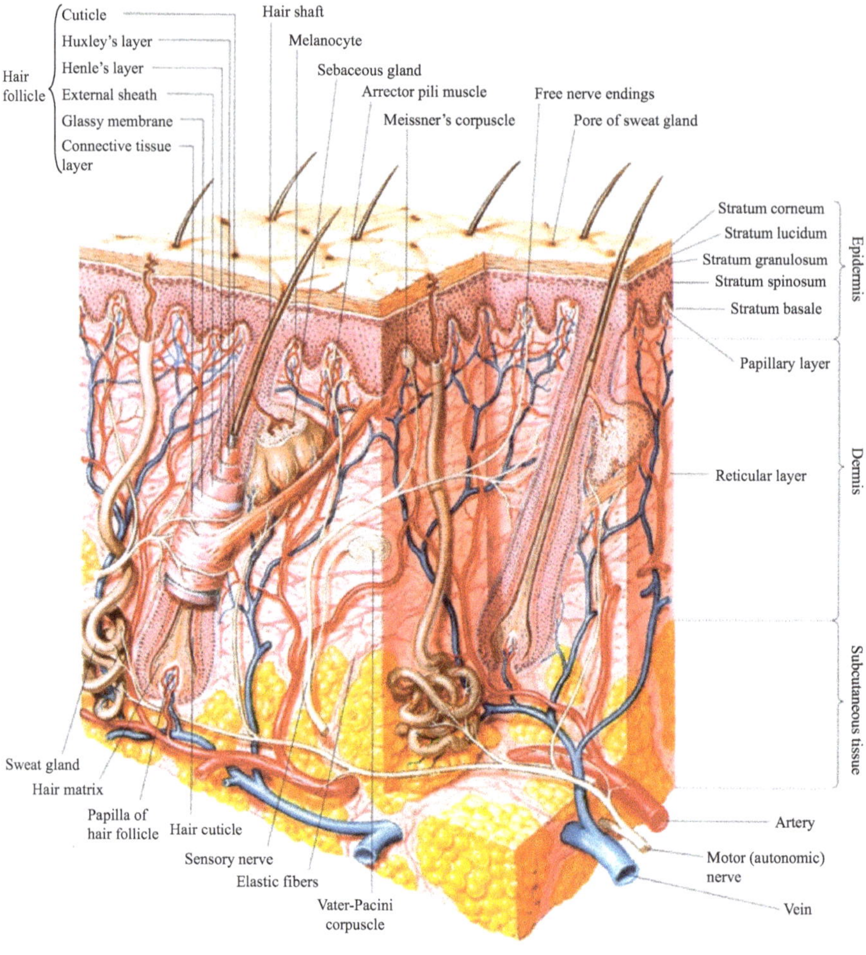

Figure 3.1 Skin structure [3]

thicker than the epidermis, depending on anatomic site, but its energy requirements are far less than those of the epidermis [1].

A fully formed dermis is divisible into two distinct compartments: (1) a thin zone immediately beneath the epidermis (papillary dermis) and around adnexa (periadnexal dermis) and (2) a thick zone that extends from the base of the papillary dermis to the surface of the subcutaneous fat (reticular dermis). The combination of papillary and periadnexal dermis has been called the adventitial dermis. It is characterized by thin, haphazardly arranged collagen bundles, delicate branching elastic fibers, plentiful fibrocytes, abundant ground substance, and highly developed circulation made up mostly of capillaries. The papillary dermis and the epidermis together form a morphologic and functional unit whose interrelatedness is reflected in their alteration jointly in various inflammatory processes, e.g., interface dermatitides such as erythema multiforme and lichen planus, spongiotic dermatitides such as allergic contact dermatitis and pityriasis rosea, and psoriasiform dermatitides such as psoriasis and pityriasis rubra pilaris. A relationship analogous to that of epidermis and papillary dermis exists between adnexal epithelium and periadnexal connective tissue [1].

The reticular dermis is formed predominantly of thick bundles of collagen arranged in an orthogonal pattern. Elastic fibers course among those bundles. Proportionally fewer fibrocytes and blood vessels and less ground substance are present in the thick reticular dermis than in the thin adventitial dermis. Adipocytes of the subcutaneous fat commonly extend in broad columns into the reticular dermis, where they envelop eccrine units and terminate at bases of follicles [1].

3.2.1 Embryologic development of dermoepidermal junction

Early in embryogenesis, the interface between epidermis and dermis is flat. During the first trimester, a basal lamina, synthesized mostly by epidermal basal cells, forms at the junction of dermis and epidermis. At about 12 weeks, the dermoepidermal interface is punctuated at regular intervals by proliferations of germinative cells. These clusters represent anlagen of folliculosebaceous-apocrine units and of eccrine units, with the former being crescentic and the latter being nubbin-like. Starting in the sixth month of fetal life, nipple-shaped insertions of connective tissue, i.e., dermal papillae, fit into hollows of the undersurface of the epidermis [1].

3.2.2 Epidermis

The basic principle of epidermis is process of cornification. This specific form of cellular differentiation causes the formation of the outermost dead layer of skin—stratum corneum or cornified layer. Corneocytes are products of maturation of both epidermal and intraepidermal adnexal keratinocytes. Cornification involves (1) synthesis of lamellar granules and special proteins—differentiation specific keratins, filaggrin, and involucrin and (2) alterations of nuclei, cytoplasmic organelles, plasma membranes, and desmosomes.

Epidermal keratinocytes undergo characteristic changes during about 2 weeks in which they are transformed from undifferentiated basal cells to fully differentiated cornified cells. The epidermis is due to this process differentiated into four cellular layers—basal, spinous, granular, and cornified, which are easily histologically recognized. These layers represent morphologic expressions of successive stages of maturations of germinative keratinocytes to fully cornified ones. The basal row of keratinocytes consists of cuboid or low columnar cells that contain larger oval nuclei and more basophilic cytoplasm than the more mature keratinocytes above them. Suprabasal keratinocytes are polygonal and are called "spinous" cells due to spine-like appearance of processes that, with conventional microscopy, are seen to cross intercellular spaces and form contacts between adjacent keratinocytes and, with electron microscopy, are revealed to be special connecting structures—desmosomes. The spinous zone merges with horizontally oriented diamond shaped cells filled with coarse, basophilic (keratohyaline) granules—cells of granular zone. The outermost part of the epidermis is a layer which consists of flat anuclear, eosinophilic corneocytes.

All at all, as cornification proceeds, vertically oriented, columnar basal keratinocytes become altered gradually into pancake-shaped cornified cells aligned parallel to the skin surface. The corneocytes themselves are stacked in orderly columns that resemble pie plates, and the arrangement varies on different skin sites, but is not discernible at all on palms and soles.

Throughout the process of cornification, keratinocytes are attached to one another by specialized intercellular contact zones just referred to desmosomes, which are intercellular attachments ("intercellular bridges") that dissolve and reform continuously as keratinocytes move outward and mature. Cleavage between desmosomes in the cornified layer results in shedding of corneocytes. Cholesteryl sulfate has been implicated as an intercellular cement substance, hydrolysis of which to resultant free cholesterol coincides with desquamation of corneocytes. Patients with X-linked ichtyosis lack steroid sulfatase, which prevents removal of cholesteryl sulfate and thereby limits shedding of corneocytes and results in hyperkeratosis [1,2].

The unique permeability of cornified layer plays a crucial role in maintenance of the body's fluid and electrolyte balance. The extent to which molecules diffuse through this layer also accounts for the ability of allergic substances to enter the viable epidermis, promote sensitization, and elicit reactions of allergic contact dermatitis on the one hand and the efficacy of topically applied medicaments on the other [1].

The major proliferative population of keratinocytes is localized in the lowest part of the epidermis. This proliferative compartment—the two lower rows of keratinocytes in a normal epidermis—has a cell cycle of 13 days, compared with psoriatic epidermis, which is only 1.5 days. The renewal time of epidermis has been estimated to be about 26 days, divided approximately as 13 days for the time it takes viable keratinocytes to travel from the base of the epidermis to the cornified layer, and another 13 days for the time it takes dead keratinocytes to at last be shed [1].

The epidermis at tips of digits is characterized by a well-developed undulate pattern of epidermal rete ridges and dermal papillae, a prominent granular zone, and a thick, compact, orthokeratotic cornified layer [1].

3.2.3 Dermoepidermal interface

The interface between epidermis and dermis is created by basement membrane. In conventional microscopy, it is seen as a thin band situated immediately beneath basal keratinocytes and stained magenta by periodic acid Schiff. With electron microscopy, it is recognized as a special structure that formed at the junction of epithelial cells and connective tissue adjacent to them. Within the skin, a basement membrane like that at the junction of dermis and epidermis is found around all epithelial structures of adnexa (hair follicles, sebaceous units, apocrine units, eccrine units). Furthermore, the vasculature of the skin and Schwann cells of nerves are surrounded by a basement membrane.

The junction between epidermis and dermis is formed in the 12th week of gestation. The dermoepidermal junction is divided into four parts: (1) basal keratinocytes that constitute the upper boundary of the interface; (2) the lamina lucida, an electron-lucent area that lies beneath basal keratinocytes; (3) the lamina densa, an electron-dense plate situated below the lamina lucida and above the papillary dermis; and (4) the sublamina densa seated immediately below the lamina densa and consisting of the uppermost portion of the papillary dermis. Structural components of these four zones are responsible for binding the epidermis to the dermis. Disruption of any of the components results in a variety of diseases, among them blistering disorders such as porphyria cutanea tarda, dermatitis herpetiformis, and bullous pemhigoid.

The plasma membrane of basal keratinocytes forms the roof of the dermoepidermal junction. Hemidesmosomes, placed at intervals along the plasma membrane, are formed of both intracellular and extracellular components. On the cytoplasmic side of the plasma membrane, a hemidesmosome consists of an electron-dense attachment plaque. Tonofilaments of keratin housed within the cytoplasm of cells converge on the attachment plaque. Another similarly electron-dense plaque, referred to as a subbasal dense plaque, is situated extracellularly. The basal plasma membrane is located between the intracellular attachment plaque and the extracellular subbasal dense plaque. These components form hemidesmosomes. Destruction of hemidesmosomes leads inevitably to development of a subepidermal blister as occurs in bullous pemphigoid, for example [1,2].

Hemidesmosomes are situated at the top of approximately 50-nm-wide lamina lucida that appears to be homogenous, except in loci immediately beneath hemidesmosomes. There, thin anchoring filaments extend from the basal plasma membrane across the lamina lucida through the subbasal dense plaque and end on the lamina densa. Within the homogenous regions of the lamina lucida, glycoproteins, among them laminin and entactin/nidogen, are placed. In healing wounds, fibronectin is laid down and serves as a substrate across which keratinocytes migrate. During embryonic life, fibronectin may provide a scaffold for formation of components that comprise the interface between epidermis and connective tissue [1].

Type IV collagen is the major part of the lamina densa. Anchoring fibrils emanate from the lamina densa and assume a crosshatched configuration. They are present only within specific epithelial basement membranes (skin, oral mucosa, esophagus, and cervix). Anchoring fibrils form an extensive network in the sublamina densa that serves to trap collagen, thereby attaching the lamina densa to the papillary dermis. Anchoring fibrils are made up of several components, including collagen type VII. Genetic defects with abnormality in type VII collagen are seen in patients with recessive and dominant dystrophic forms of epidermolysis bullosa [1].

Dermal microfibril bundles are another important component of the sublamina densa. The role of these fibers is to attach the lamina densa to the network of elastic tissue in the dermis [1].

3.2.4 Melanocytes

Melanocytes are cells that synthesize melanin. They originate in the neural crest and during the fetal life migrate to the epidermis, follicles, dermis, epithelia of various mucous membranes, leptomeninges, inner ear, and a few other tissues.

Melanocytes are located immediately beneath the row of epidermal basal cells and have number of dendrites which branch in all directions among neighboring keratinocytes and even into the uppermost part of the dermis. A single melanocyte is connected approximately to 36 keratinocytes. This relationship is called "the epidermal melanin unit." The unit really consists of a melanocyte and keratinocyte supplied by it. The concentration of melanocytes varies according to the region of the body. Melanin is made in the special particles called melanosomes, located in the cytoplasm of melanocytes. Melanosomes are transported to the tips of dendrites and then transferred to keratinocytes—in a special process in which the tips of dendrites are snipped off and engulfed by keratinocytes [1,2].

Because the absolute number of melanocytes in human skin is approximately the same for both sexes and all races, the amount of melanin in keratinocytes and the distribution of it determine the degree of pigmentation of skin. Differences in color among the races result from differences in the number, size, degree of melanization, distribution, and rate of degradation of melanosomes within keratinocytes. The principal function of melanin is to protect the skin from harmful effects of sunlight, a task accomplished by its scattering and absorbing ultraviolet light [1].

3.2.5 Langerhans' cells

Langerhans' cells are dendritic cells that are placed just above the middle of the spinous zone of the epidermis. The electron microscopy reveals characteristic granules inside Langerhans' cells, with specific shape of a tennis racquet—called Birbeck granules. Origin of Langerhans' cells is in mesenchymal precursors in the bone marrow. Although Langerhans' cells represent about 4% of the entire population of cells in the epidermis, their distribution varies according to anatomic site, with the range being between 460 and 1,000 per mm^2 of epidermis. Fine unmyelinated nerve fibers ascend into epidermis where they touch Langerhans' cells and thereby create a link between the nervous and immune system. They play an integral role in skin immune system—they are important for recognition of antigens and presentations of antigen to T lymphocytes. In sum, the Langerhans' cells have important functions in various immune processes, including allergic contact dermatitis, allograft rejection, immune tolerance, and surveillance for incipient neoplasia [1,2].

3.2.6 Merkel cells

Merkel cells are unique cells at bases of epidermal rete ridges and have contact with nerve fibers. They are considered as touch receptors. The close proximity of the cells to neurites has been recorded in epidermis especially on palms and soles [1,2].

3.2.7 Nerves

Both somatic sensory and autonomic motor nerves are present in skin. The somatic sensory system mediates sensations of pain, itch, temperature, light touch, discriminative touch, pressure, vibration, and proprioception. Nerves of the autonomic motor system control cutaneous vascular tone, pilomotor responses, and sweating. The somatic sensory and autonomic motor parts of the peripheral nervous system are derived mostly from the neural crest, which gives rise to dorsal root and sympathetic ganglia and their nerve processes, Schwann cells, endoneural sheath cells, and laminar cells of specialized sensory end organs. Nerves within the reticular dermis travel along the course of the superficial and deep vascular plexuses, whence derives the term neurovascular plexuses [1].

Cutaneous sensory receptors are divided into two groups: “specialized” receptors (“end organs”) that, viewed by conventional microscopy, are observed to possess terminal nerve endings surrounded by morphologically distinctive lamellar condensations of connective tissue and Schwann cells, e.g., Pacini corpuscles, Meissner’s corpuscles, and mucocutaneous end organs; and the other is of “unspecialized” receptors devoid of distinctive structural features. In the latter group there are most of the sensory nerves that supply the skin, including those linked to Merkel cells [1].

Pacini corpuscles are located especially on weight bearing surfaces in the deep dermis and subcutaneous fat. Pacini corpuscles are mechanoreceptors designed to detect pressure [1].

Meissner’s corpuscles, located at tips of dermal papillae on volar skin and most plentiful on fingertips, are thought to function as mechanoreceptors that adapt rapidly to the sensation of light touch [1].

The most often of all cutaneous symptoms is pruritus, which is defined as a sensation that evokes a desire to scratch. Itching is thought to be either a subthreshold sensation of pain or a particular sensory quality transmitted from a network of nerve endings situated near the dermoepidermal junction [1].

3.2.8 Blood vessels

Vessels are located in dermis in the form of plexuses that are parallel to skin surface. Dermal vasculature consists of a superficial and deep plexus of arterioles and venules, those plexuses being connected to one another by communicating vessels that arise from arteries and lead eventually to veins that reside within septa of subcutaneous fat.

From the superficial plexus, an arcade of capillaries loops into each dermal papilla. A capillary loop within a dermal papilla consists of an ascending arterial component that takes a hairpin turn at its crest to become a descending venous limb. The venous portion of a capillary loop empties into postcapillary venules of the superficial plexus and, successively, into dermal communicating venules, the large venules of the deep plexus, and the small veins of the subcutaneous fat.

Arteriovenous anastomoses known as Sucquet-Hoyer canals are present in skin of the digits especially. Those specialized segments are surrounded by three to six rows of uniform, ovoid, specialized smooth muscle (glomus) cells that serve as sphincters. Each of these shunts, known as a “glomus,” enables blood to bypass the capillaries, thereby accelerating blood flow through skin on acral sites and conserving the core temperature of the body, a boon in frigid weather.

Through the permeable walls of capillaries and venules, oxygen, water, nutrients, and hormone are delivered from the blood to the tissues, and carbon dioxide and other metabolic byproducts are transported from them to excretory organs.

The red color of eruptions in the skin of Caucasians is imparted by blood that flows through dilated vessels in the upper part of the dermis. Many papules of acute inflammatory diseases result in large measure from edema that develops because of an imbalance between leakage of serum into tissues and reabsorption of it by lymphatics. A properly functioning cutaneous vasculature also is crucial to healing of inflammatory reactions and *restitutio ad integrum* [1].

3.2.9 Lymphatics

Lymphatic system is network of lymphatic vessels that is parallel to the skin surface, roughly as the major vascular plexus but remain independent of them. Lymphatic vessels constitute a one-way drainage system that directs its contents into the venous circulation. The structure of cutaneous lymphatics enables clearance of fluids, macromolecules, cells, and foreign material from the interstitium. When the lymphatic system fails, inevitable consequences are impaired immune function, recurrent infection, and fibrosis secondary to chronic lymphedema [1].

3.2.10 Subcutaneous fat

Subcutaneous fat, like the dermis, is derived embryologically from mesenchyme. The basic unit of the subcutaneous fat is a collection of adipocytes that appear to be cohesive, i.e., a primary microlobule. Primary microlobules are aggregated to secondary lobules, which are surrounded by a rim of fibrous tissue known as a septum or trabecula. Arteries and veins in the subcutaneous fat are housed within the major fibrous septa. Vessels branch from large blood vessels within the septa to supply portions of secondary lobule. Every adipocyte in the subcutaneous fat is encircled by a capillary [1,2].

3.2.11 Eccrine units

The eccrine gland is the only true sweat gland in humans. Eccrine sweat is a hypotonic solution that flows from the gland to the skin surface where it cools the body by evaporation. Eccrine units are present nearly everywhere on human skin, they are numerous especially on palms. Approximately 3 million eccrine sweat units are present at birth, and no additional units are formed thereafter. The eccrine unit consists of an irregularly coiled secretory gland proximally, a coiled dermal duct that leads from the secretory gland, a straight duct that passes through the length of the epidermis, and a spiraled intraepidermal duct known also as an acrosyringium [1].

Eccrine glands on the palms, soles, axillae, and forehead, unlike those on the rest of the body, respond predominantly to emotional, rather than thermal, stimuli. A deleterious effect of sweating occurs when allergic contact sensitizers are leached from jewelry or clothing by salts present in sweat. This enhances induction of allergic contact dermatitis in susceptible persons [1].

3.2.12 Regional variation—skin on the hand

The skin of infants is traversed by a subtle maze of ridges that develops during the fourth and fifth months of fetal life. The ridges become increasingly prominent during childhood. Those surface markings remain stable throughout the life. Swirled patterns characterize the palms and soles.

Etchings that cover the entire surface of palms and soles, excluding flexion creases, are termed, collectively, "dermatoglyphic patterns." Parallel ridges and furrows form loops, whirls, and arches on the fingertips in patterns so highly individualistic that fingerprinting is used as a reliable method for identification of persons, even for distinguishing between identical (uniovular) twins. Study of dermatoglyphics has contributed to early detection of genetic abnormalities, e.g., Down's syndrome, and defects caused by infections in utero, e.g., rubella. Palmar skin is characterized histologically by thick cornified and granular layers, a prominent undulate pattern of epidermal rete ridges and dermal papillae, numerous eccrine units and nerve endings, and absence of folliculosebaceous-apocrine units. The corrugated palmar skin is well suited for gripping and grasping objects. The exquisite epicritical sensitivity of the fingertips to tactile stimuli made it possible for Braille to develop his "reading" system for the blind [1].

3.3 Skin diseases with localization on fingertips

There are a lot of skin diseases, which can affect palms and palmar sites of fingers. The pathological processes influence the structure and color of the skin in a different way. The clinical picture of particular skin diseases is the result of many microscopic changes which are precisely described in textbooks of dermatopathology [1,2,4–6]. Dermatopathology is the study and description of structural and compositional changes that occur in skin disease. From a practical point of view, dermatopathology involves the microscopic examination, description, and interpretation of biopsy specimens obtained from the skin. The interpretation of skin specimens can be complicated and difficult, as many diverse inflammatory skin diseases share the same basic inflammatory process or pattern. The final diagnosis requires clinical input and clinicopathological correlation.

The fingerprint recognition process is dependent on these microscopic changes as dermoepidermal junction, thickness of epidermis, or color of the skin has been transformed.

There is evidence that these pathological processes may cause many problems in automatic biometric systems.

The fingerprint recognition systems are usually used only for adults. There is almost no information from appropriate tests with children. Although we know that papillary lines emerge on infant's fingers already in the mother's uterus [7], i.e., we might be able to recognize the fingerprints of infants, the common fingerprint recognition systems are suitable for adults only (due to the area and resolution of fingerprint sensors, etc.). It should not be forgotten that a skin disease in early childhood could have an influence on the skin in adult years (example is incontinentia pigmenti [8] on a small child hand), i.e., there could be some problems with fingerprint acquirement caused by such skin disease in a young age.

The subcategory of skin diseases affecting only the skin color are the least dangerous for the quality of the fingerprint image. In fact, only one fingerprint technology can be considered as sensitive to such diseases—the optical technology [9], but if Fourier transform infrared (FTIR)-based optical sensors are used, the change of skin color may have no influence on the quality of the resulting images. The case of the other two subcategories (influence of skin structure and combination of influence of skin color and structure) is different. If the structure of papillary lines has changed, it is often impossible to recognize the original curvatures of papillary lines, and therefore, it is impossible to decide whether the claimed identity is the user's identity. Unfortunately, there are many such skin diseases which attack papillary line structure. Nearly all sensor technologies, namely, optical, capacitive, e-field, electrooptical, pressure sensitive, and thermal, are exposed to such risk [9]. Only one sensor technology is missing here—the ultrasound technology. This technology has an advantage: the ultrasound waves can penetrate under the upper skin layer to the curvatures in dermoepidermal junction forming the papillary lines structures, and therefore, it might be possible to reconstruct the real fingerprint image, but only if the disease has not attacked this underlying structure. If yes, there is no chance to get an original papillary lines structure.

3.3.1 Dermatopathology of the skin

According to dermatopathology the great majority of skin diseases are diagnosed in a logical approach. The interpretation of skin biopsies requires the identification and integration of two different, morphological features—the "tissue reaction pattern" and the "pattern of inflammation." This is a crude algorithmic approach not only for making the diagnosis from medical point of view but is also substantial for sensitivity of different sensors to different skin diseases.

"Tissue reaction patterns" are different morphological patterns which enable to categorize a group of cutaneous diseases. Within each of these histopathological categories, there are diseases which may have similar or diverse clinical appearances and etiologies. Some diseases can have histopathological features of more than one reaction pattern at a particular time or during the course of their evolution. This may be difficult to categorize [4].

"The pattern of inflammation" refers to the distribution of the inflammatory cell infiltrate within the dermis and/or the subcutaneous tissue. We differentiate several distinctive patterns of inflammation and their recognition helps us to make a specific diagnose [4].

3.3.2 Major tissue reaction patterns

The majority of inflammatory dermatoses can be categorized into six different patterns with following key morphologic features [4]:

1. Lichenoid (epidermal basal cell damage; "interface dermatitis")
    i. Lichen planus, lichenoid drug eruptions, erythema multiforme, superantigen "id" reaction, graft-versus-host disease, subacute radiation dermatitis, lupus erythematosus, dermatomyositis, persistent viral reactions, perniosis, paraneoplastic pemphigus, lichenoid contact dermatitis, late secondary syphilis, drug eruptions, erythroderma, mycosis fungoides, regressing warts and tumors (Figure 3.2) [10].

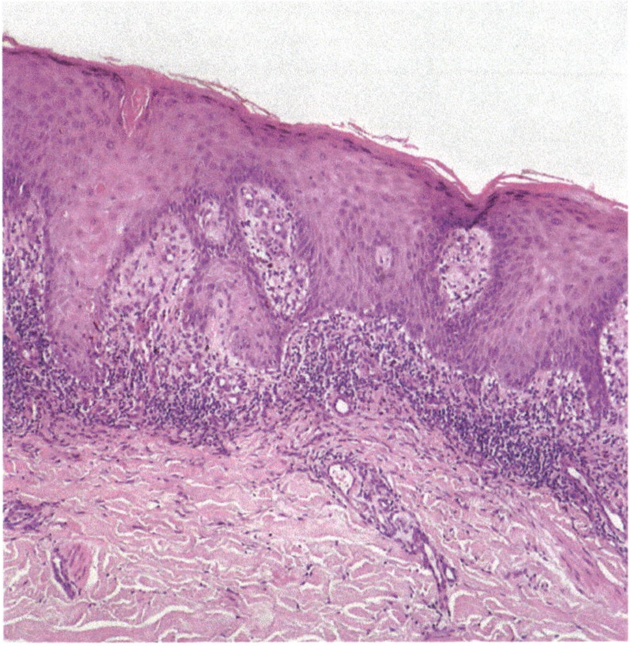

Figure 3.2 Lichen ruber planus (HE stain, 10×) [10]

2. Psoriasiform (regular epidermal hyperplasia = elongation of rete ridges)
   i. Psoriasis, pustular psoriasis, Reiter's syndrome, pityriasis rubra pilaris, parapsoriasis, lichen simplex chronicus subacute and chronic spongiotic dermatitides, erythroderma, mycosis fungoides, chronic candidosis, and dermatophytoses, Norwegian scabies, lamellar ichtyosis, pityriasis rosea, pellagra, acrodermatitis enteropathica, secondary syphilis (Figure 3.3) [10].
3. Spongiotic/eczematous (intraepidermal intercellular edema = spongiosis, presence of widened intercellular spaces between keratinocytes)—acute stages of atopic dermatitis, allergic and irritant contact dermatitis, nummular dermatitis, seborrheic dermatitis, pompholyx.
   i. Neutrophilic: pustular psoriasis/Reiter's syndrome, acute generalized exanthematous pustulosis, palmoplantar pustulosis, Staphylococcal toxic shock syndrome, neisserial infections, dermatophytosis/candidosis, pustular contact dermatitis, amicrobial pustuloses, periodic fever syndromes.
   ii. Eosinophilic: pemphigus, bullous pemphigoid, allergic contact dermatitis, protein contact dermatitis, atopic dermatitis, arthropod bites, eosinophilic folliculitis, incontinentia pigmenti (first stage), drug reactions, "id" reactions (Figure 3.4) [10].

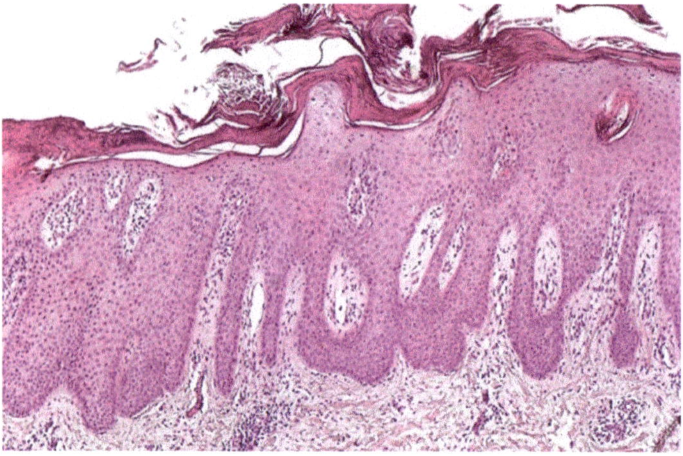

Figure 3.3 Psoriasis vulgaris (HE stain, 10×) [10]

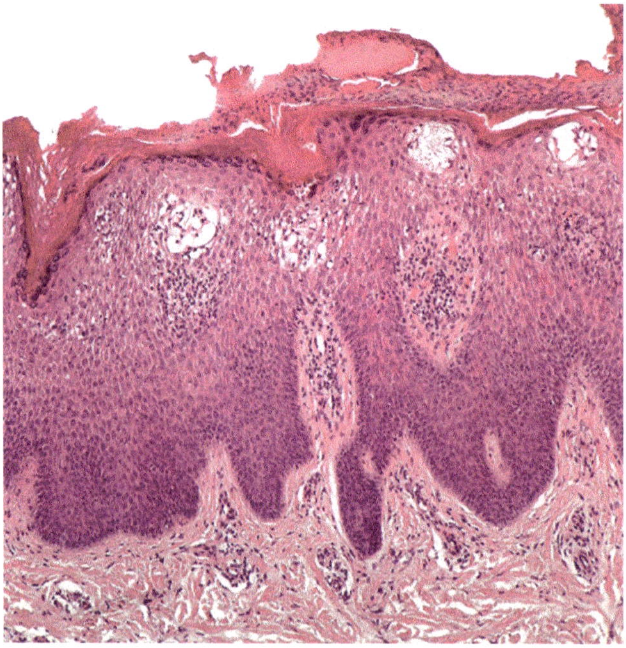

Figure 3.4 Atopic eczema (HE stain, 10×) [10]

4. Vesicobullous (blistering at any level within the epidermis or at the dermoepidermal junction).
   i. Intracorneal and subcorneal blisters: impetigo, staphylococcal "scalded skin" syndrome, dermatophytosis, pemphigus foliaceus and erythematosus, subcorneal pustular dermatosis, acute generalized exanthematous pustulosis, milliaria cristallina.

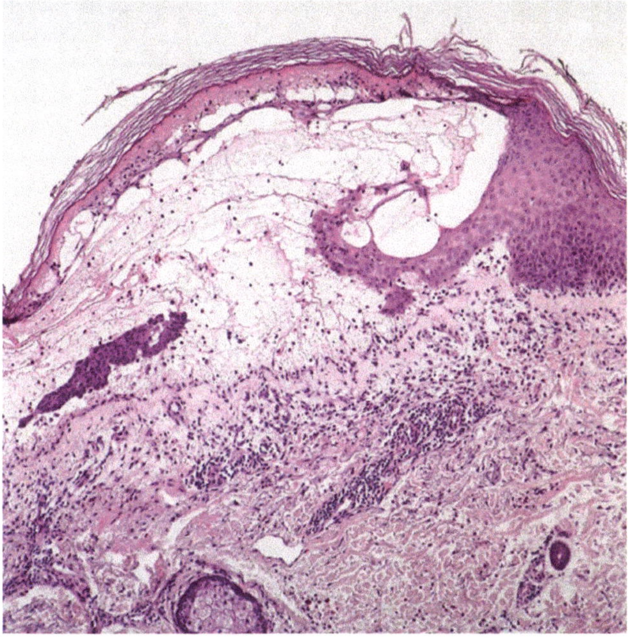

Figure 3.5 Bullous pemphigoid (HE stain, 10×) [10]

ii. Intraepidermal blisters: spongiotic blistering diseases, palmoplantar pustulosis, amicrobial pustulosis of autoimmune diseases, viral blistering diseases, epidermolysis bullosa simplex (localized type), friction blister.
iii. Suprabasilar blisters: pemphigus vulgaris and vegetans, paraneoplastic pemphigus, Hailey-Hailey disease, Darier's disease, Grover's disease.
iv. Subepidermal blisters: bullous pemphigoid, epidermolysis bullosa, porphyria cutanea tarda and pseudoporphyria, burns, toxic epidermal necrolysis, drug reactions, erythema multiforme, dermatitis herpetiformis, linear IgA bullous dermatosis, bullous erysipelas (Figure 3.5) [10].

5. Granulomatous (chronic granulomatous inflammation, that is localized collections of epithelioid cells usually admixed with giant cells, lymphocytes, plasma cells, fibroblasts, and nonepithelioid macrophages). Granulomas can be divided into five histological types according to the constituent cells and other changes within granuloma—sarcoidal, tuberculoid, necrobiotic (collagenolytic), suppurative, and foreign body. Clinically, granulomas presents with a mass that is usually firm and is detectable below the skin surface (epidermis) and usually moveable over the deeper tissues—the clinical differential diagnoses include cutaneous tumors and lymphocytic infiltrates.
   i. Sarcoidal granulomas: sarcoidosis, reactions to foreign materials, secondary syphilis, Sézary syndrome, herpes zoster scars, systemic lymphomas, common variable immunodeficiency

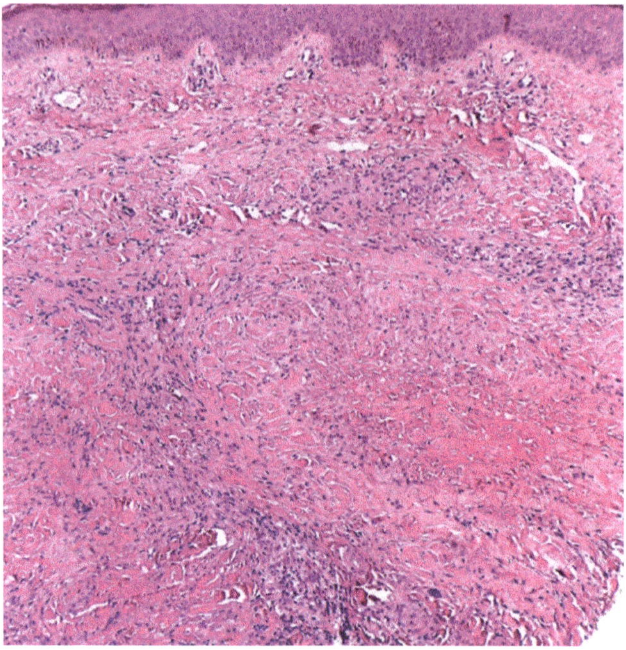

Figure 3.6 Necrobiosis lipoidica (HE stain, 10×) [10]

   ii. Tuberculoid granulomas: tuberculosis and tuberculids, leprosy, fatal bacterial granuloma, late syphilis, leihmaniasis, rosacea, perioral dermatitis, Crohn's disease
   iii. Necrobiotic (collagenolytic) granulomas: granuloma annulare, necrobiosis lipoidica, necrobiotic xanthogranuloma, rheumatoid nodules, rheumatic fever nodules, reactions to foreign materials and vaccines, Crohn's disease (Figure 3.6) [10]
   iv. Suppurative granulomas: deep fungal infections (chromomycosis, sporotrichosis, blastomycosis, coccidioidomycosis, paracoccidioidomycosis, mycetoma), nocardiosis, actinomycosis, nontuberculous mycobacterial infection, cat-scratch disease, lymphogranuloma venereum, pyoderma gangrenosum, ruptured cysts and follicles
   v. Foreign body granulomas (exogenous/endogenous material)
   vi. Xanthogranulomas
6. Vasculopathic (pathological changes in cutaneous blood vessels). The most important category within this tissue reaction pattern is vasculitis, which is defined as an inflammatory process involving the walls of blood vessels of any size. Other categories of vascular disease include noninflammatory purpuras, vascular occlusive diseases, and urticarias.
   i. Noninflammatory purpuras (traumatic purpura, psychogenic purpura, drug purpura, bleeding diatheses, senile purpura).

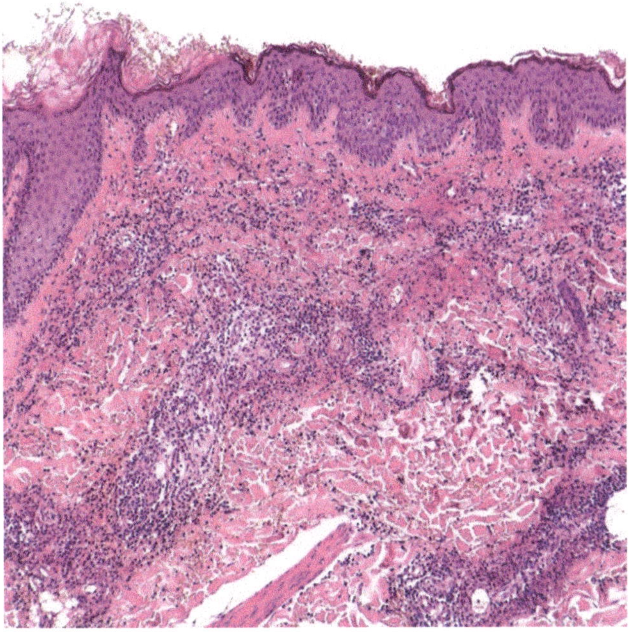

Figure 3.7 Leukocytoclastic vasculitis (HE stain, 10×) [10]

ii. Vascular occlusive diseases: warfarin necrosis, atrophie blanche, disseminated intravascular coagulation, purpura fulminans, thrombotic thrombocytopenic purpura, thrombocythemia, cryoglobulinemia, cholesterol and other types of embolism, antiphospholipid syndrome.

iii. Urticarias.

iv. Acute vasculitis: leukocytoclastic (hypersensitivity) vasculitis, Henöch–Schönlein purpura, eosinophilic vasculitis, rheumatoid vasculitis, urticarial vasculitis, mixed cryoglobulinemia, septic vasculitis, polyarteritis nodosa, Kawasaki disease, superficial thrombophlebitis (Figure 3.7) [10].

v. Neutrophilic dermatoses: periodic fever syndromes, pustular vasculitis of the hands, neutrophilic fixed drug eruptions, Bowel-associated dermatosis-arthritis syndrome, rheumatoid neutrophilic dermatosis, acute generalized pustulosis, Bechcet's disease, abscess-forming neutrophilic dermatosis.

vi. Chronic lymphocytic vasculitis: collagen vascular disease, pityriasis lichenoides, perniosis, rikettsial and viral infections, pyoderma gangraenosum, leukemic vasculitis

vii. Vasculitis with granulomatosis: Crohn's disease, drug reactions, herpes zoster, infectious granulomatous diseases, Wegener's granulomatosis, lymphomatoid granulomatosis, Churg–Strauss syndrome, giant cell (temporal) arteritis, Takayasu's arteritis.

3.3.3 Minor tissue reactions patterns

These "minor tissue reaction patterns" are seen much less frequently than six major patterns already discussed. Like the major reaction patterns, each of the patterns to be considered below is diagnostic of a certain group of diseases of the skin [4].

1. Epidermolytic hyperkeratosis (hyperkeratosis with granular and vacuolar degeneration)
2. Acantholytic dyskeratosis (suprabasilar clefts with acantholytic and dyskeratotic cells)
3. Cornoid lammelation (a column of parakeratotic cells with absence of an underlying granular layer)
4. Papillomatosis (undulations and protrusions of the epidermis)—verruca vulgaris, seborrheic and solar keratosis, stucco keratosis, acanthosis nigricans, acrokeratosis verruciformis
5. Angiofibromas (increased dermal vessels with surrounding fibrosis)
6. Eosinophilic cellulitis with "flame figures" (dermal eosinophils and eosinophilic material adherent to collagen bundles)
7. Transepithelial elimination (elimination of material via the epidermis or hair follicles)

3.3.4 Patterns of inflammation

There are four patterns of cutaneous inflammation characterized on the basis of distribution of inflammatory cells within the skin [4]:

1. Superficial perivascular inflammation (the most often)
2. Superficial and deep dermal inflammation
3. Folliculitis and perifolliculitis
4. Panniculitis

The summary of different histopathologic features mentioned above provide us the information, that skin disease is a result of many complicated changes including all structures of the skin. In the area of fingertips, the changes of papillary lines are critical for fingerprint recognition process.

Sensitivity of different sensors to the persons suffering from skin disease on fingertips is based on all histopathological changes of the disease. Sometimes, it is very difficult to estimate how much and how is the process of fingerprint recognition influenced. One diagnosis can have different clinical types and can differ from person to person. Histopathological changes are also often overlapping.

The situation after successful recovery of a potential user from such skin diseases is, however, very important for the possible further use of fingerprint recognition devices. If the disease has attacked and destroyed the structure of papillary lines in dermoepidermal junction, the papillary lines will not grow in the same form as before (if at all), and therefore, such user could be restricted in his/her future life by being excluded from the use of fingerprint recognition systems, though his fingers don't have any symptoms of a skin disease any more.

3.4 Skin diseases with possible localization on fingertips

Skin diseases localized on fingertips can be present solely on this area or can be part of a systemic involvement. In the following text, the most important skin diseases, which can be localized on fingertips, are described.

Hand eczema [3,11] is an inflammatory itchy noninfectious skin disease with chronic or relapsing course. It is one of the most common dermatological problems. Hand dermatitis causes discomfort and embarrassment and, because of its locations, interferes significantly with normal daily activities, influences the quality of life. Hand dermatitis is common especially in industrial occupations. The prevalence of hand eczema was approximately 5.4% and was twice as common in females as in males. The most common type of hand eczema was irritant contact dermatitis (35%), followed by atopic dermatitis (22%), and allergic contact dermatitis (19%). The most common contact allergies were to nickel, cobalt, fragrance mix, balsam of Peru, and colophony. Hand eczema was more common among people reporting occupational exposure. The most harmful exposure was to chemicals, water and detergents, dust, and dry dirt.

Fingertip eczema [3] is very dry, itchy, or painful chronic form of eczema of the palmar surface of the fingertips, the cause may be an allergic reaction or may occur as an isolated phenomenon of unknown cause. Fingertip eczema is common in adults, also in children. One or more fingers may be involved. In the acute stage, the skin may be moist and then become dry, cracked, and scaly. The skin is peeling from the distal parts of fingertips, exposing a very dry, erythematous, cracked, fissured, tender, or painful surface without skin lines—see Figures 3.8 and 3.10 [3].

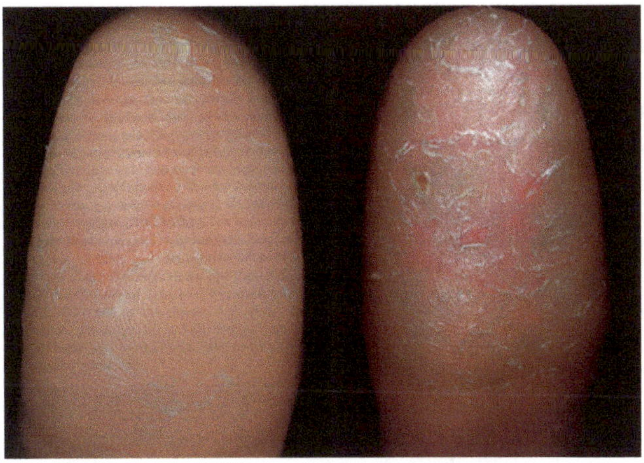

Figure 3.8 Fingertip eczema [3]

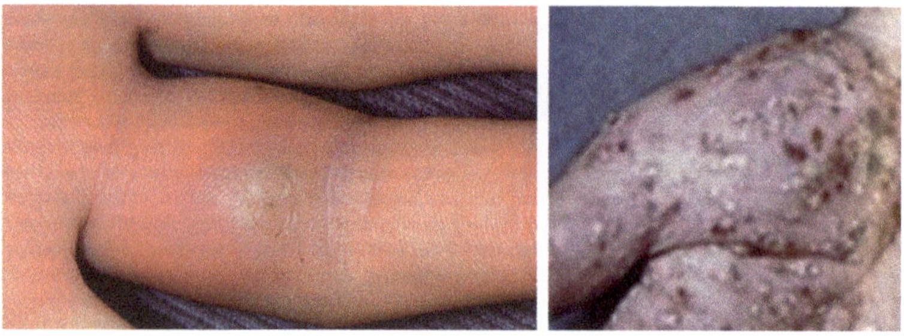

Figure 3.9 Abscess on finger of patient with diabetes [12] and pyoderma [13]

Pomfolyx (*dishydrosis*) [12] is a distinctive reaction pattern of unknown etiology presenting as symmetric vesicular hand and foot dermatitis. Itching precedes the appearance of vesicles on the palms and sides of the fingers. The skin may be red and wet. The vesicles slowly resolve and are replaced by rings of scale. Chronic eczematous changes with erythema, scaling, and lichenification may follow.

Tinea of the palm [3,11] is dry, diffuse, keratotic form of tinea. The dry keratotic form may be asymptomatic, and the patient may be unaware of the infection, attributing the dry, thick, scaly surface to hard physical labor. It is frequently seen in association with tinea pedis which prevalence is 10%–30%.

Pyoderma [11] is a sign of bacterial infection of the skin. It is caused by *Staphylococcus aureus* and *Streptococcus pyogenes*. Some people are more susceptible to these diseases (such as diabetics, alcoholics, etc.)—see Figure 3.9.

Pitted keratolysis [3] is a disease mimicking tinea, especially for people who swelter and wear rubber gloves in the hot, humid environment. Hyperhidrosis is the most frequently observed symptom. The disease is bacterial in origin, characterized by many circular or longitudinal, punched out depressions in the skin surface. The eruption is limited to the stratum corneum.

Keratolysis exfoliativa [3] is a common, chronic, asymptomatic, noninflammatory, bilateral peeling of the palms of the hands. Its cause is unknown. The eruption is most common during the summer months and is often associated with sweaty palms and soles. It is characterized by scaling and peeling, the central area becomes slightly red and tender.

Lichen planus [11] is quite common, unique inflammatory cutaneous and mucous membrane reaction pattern of unknown etiology. Lichen planus of the palms and soles generally occurs as an isolated phenomenon. The lesions are papules aggregated into semitranslucent plaques with globular waxy surface, ulceration may occur.

Acanthosis nigricans [3,12] is nonspecific reaction pattern that may accompany obesity, diabetes, tumors. Acanthosis nigricans is classified into benign and malignant forms. In all cases, the disease presents with symmetric, brown thickening of the skin. During the process, there is papillary hypertrophy, hyperkeratosis, and increased number of melanocytes in the epidermis—see Figure 3.10 [12].

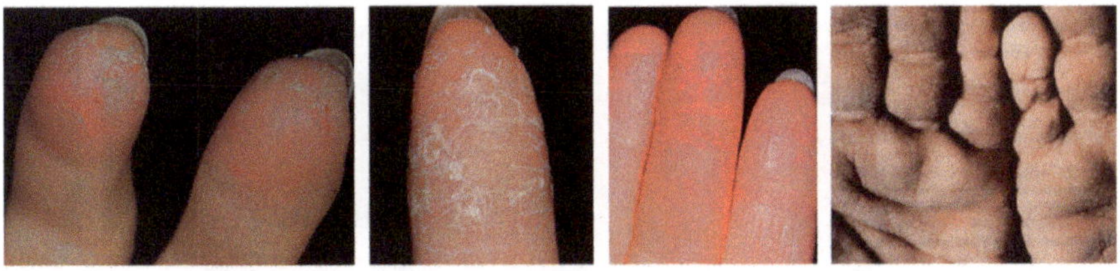

Figure 3.10 Different types of eczema [3] (3× left) and acanthosis nigricans [12] (right)

Pyogenic granuloma [3] is a benign acquired vascular lesion of the skin that is common in children and young adults. It often appears as a response to an injury or hormonal factors. Lesions are small rapidly growing, yellow-to-bright red, dome-shaped.

Systemic sclerosis [3,11] is a chronic autoimmune disease characterized by sclerosis of the skin or other organs. Emergence of acrosclerosis is decisive for fingerprinting. Initially, the skin is infused with edema mainly affecting hands. With the progressive edema stiff skin appears and necrosis of fingers may form. The disease leads to sclerodactyly with contractures of the fingers. For more than 90% of patients is typical Raynaud's phenomenon (see below). The typical patient is a woman over 50 years of age.

Raynaud's phenomenon [3,11] represents an episodic vasoconstriction of the digital arteries and arterioles that is precipitated by cold and stress. It is much more common in women. There are three stages during a single episode: pallor (white), cyanosis (blue), and hyperemia (red). Estimates of the prevalence of Raynaud's phenomenon ranged between 4.7% and 21% for women and 3.2% and 16% for men.

Drug induced skin reactions [3] are among the most common adverse drug reactions. They occur in many forms and can mimic virtually any dermatosis. Occur in 2%–3% of hospitalized patients. Sulfonamides, NSAIDs, and anticonvulsants are most often applied in the etiology.

Herpes simplex virus [12,14] in the host with systemic immune-compromise may cause chronic ulcerations as you can see by the patient with advanced HIV disease in Figure 3.11 (left).

Herpetic infection may uncommonly occur on the fingers or periungually. Lesions begin with tenderness and erythema and deepseated blisters develop 24–48 h after symptoms begin (see Figure 3.11, right).

Scabies [11,15] is highly contagious disease caused by the mite *Sarcoptes scabiei*. It is characterized by red papules, vesicles, and crusts located usually on the areas with tender skin, palms, and soles especially in infants.

Erythema multiforme [11] is quite common skin disorder with multifactorial cause (see Figure 3.12). The most common triggering agents are infections (in the first place herpes virus) and drugs. Minor and major variant of this disease is described. Both forms are characterized by erythematous

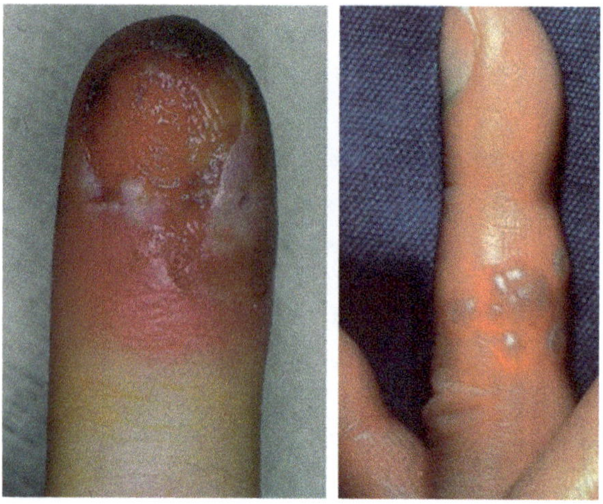

Figure 3.11 *Herpes simplex virus: patient with HIV [14] (left); deepseated blisters [12] (right)*

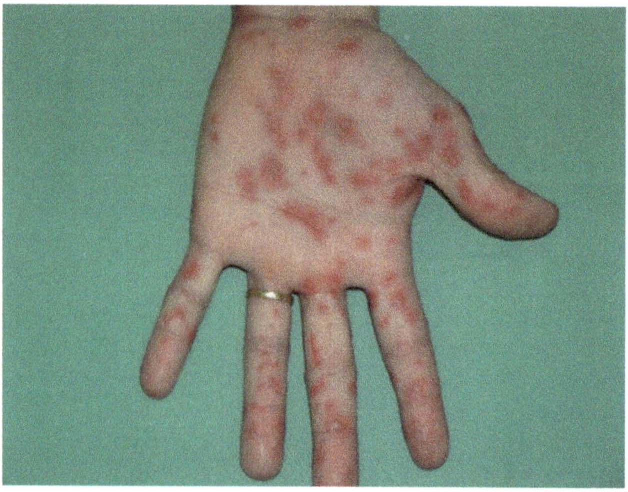

Figure 3.12 *Erythema multiforme [11]*

target-shaped lesions with a center with hemorrhage, blistering, necrosis, or crust. When the trigger is herpetic infection, frequent recurrences come.

Dermatitis artifacta [11] are changes of skin due to the manipulation by patient. Patients often have psychosomatic, psychiatric, or drug abuse problems.

Hand, foot, and mouth disease [3,12] is contagious enteroviral infection occurring primarily in children and characterized by a vesicular palmoplantar eruption. The skin lesions begin as red macules that rapidly become pale, white, oval vesicles with red areola (Figure 3.13) [12].

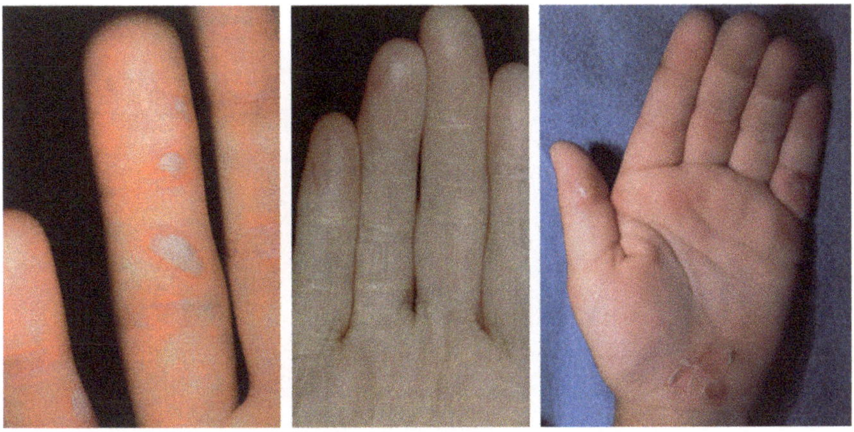

Figure 3.13 Hand, foot, and mouth syndrome [12]; xanthomas [14]; epidermolysis bullosa [15]

Xanthomas [3,11] are lipid deposits in the skin and tendons that occur secondary to a lipid abnormality. These localized deposits are yellow and are frequently very firm (Figure 3.13).

Scarlet fever (*scarlatina*) [3,11] is contagious disease produced by streptococcal, erythrogenic toxin. It is most common in children (ages 1–10 years). In the ending stages of the disease large sheaths of epidermis may be shed from the palms in glovelike cast, exposing new tender and red epidermis beneath (Figure 3.17) [3].

Kawasaki's disease [3] is an acute febrile illness of infants and children, characterized by cutaneous and mucosal erythema and edema with subsequent desquamation, cervical lymphadenitis, and complicated by coronary artery aneurysms (20%). Most cases of Kawasaki's disease in adults represent toxic shock syndrome. Erythematous macules appear 1–3 days after onset of fever, enlarge, and become more numerous, and then desquamation beginning on tips of fingers is highly characteristic (see Figure 3.14).

Secondary syphilis [3,11] is characterized by mucocutaneous lesions, which may assume a variety of shapes, including round, elliptic, or annular. The color is characteristic, resembling a "clean-cut ham" or having a coppery tint (Figure 3.15).

Carotenosis [12] is yellowish discoloration of the skin, especially of the palms and soles that is sometimes seen in diabetic patients.

Hereditary hemorrhagic telangiectasia [3] is an autosomal dominant condition affecting blood vessels, especially in the mucous membranes of the mouth and the gastrointestinal tract. The diagnostic lesions are small, pulsating, macular, and papular, usually punctuate. Telangiectasias are present on different parts of the body, including palms and soles (see Figure 3.16).

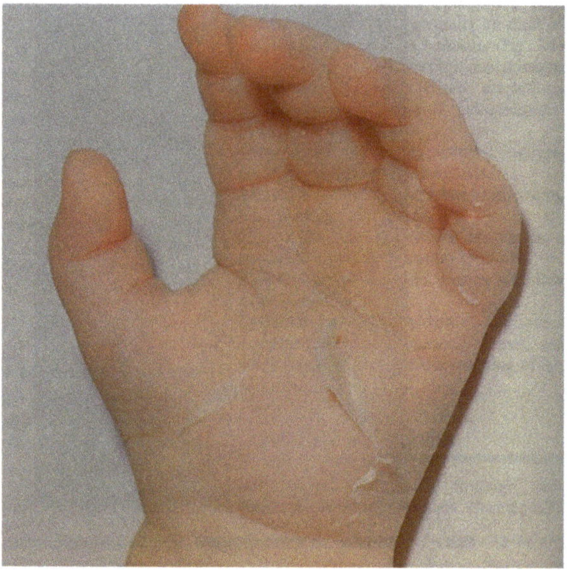

Figure 3.14 *Kawasaki's disease [14]*

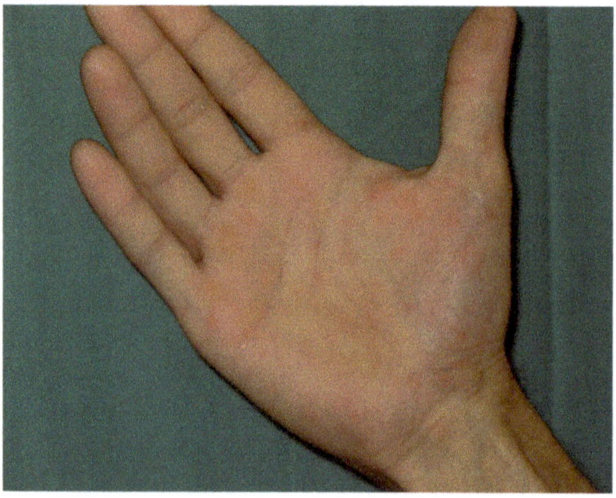

Figure 3.15 *Secondary syphilis [11]*

Warts (*verruca vulgaris*) [11] are benign epidermal neoplasms that are caused by human papilloma viruses (HPVs). Warts commonly appear at sites of trauma, on the hand, in periungual regions. HPVs induce hyperplasia and hyperkeratosis.

Psoriasis [3,11,15] is characterized by scaly papules and plaques. IT occurs in 1%–3% of the population. The disease is transmitted genetically, environmental factors are needed to precipitate the disease. The disease is lifelong

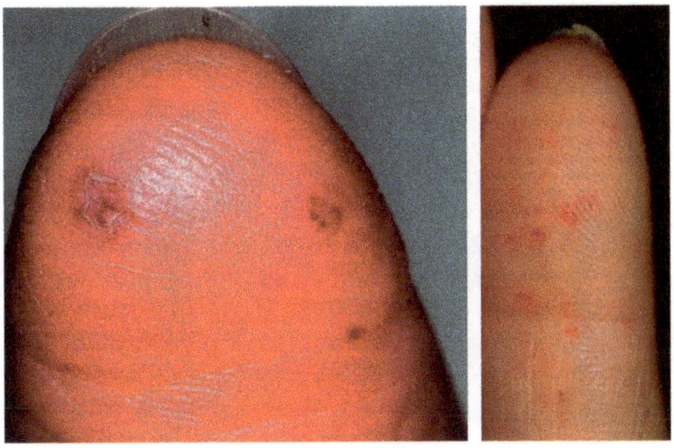

Figure 3.16 Hereditary hemorrhagic telangiectasia (left [12], right [14])

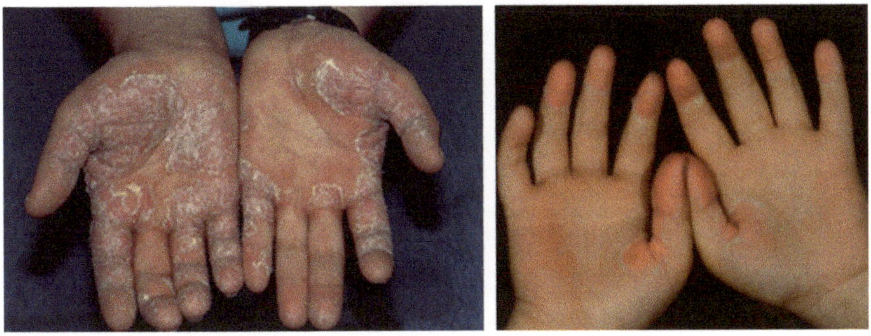

Figure 3.17 Psoriasis [15]; scarlet fever [3]

and characterized by chronic, recurrent exacerbations, and remissions that are emotionally and physically debilitating. Psoriasis of the palms and fingertips is characterized by red plaques with thick brown scale and may be indistinguishable from chronic eczema (Figure 3.18) [15].

Systemic lupus erythematosus [3] is a multisystem disease of unknown origin characterized by production of numerous diverse of antibodies that cause several combinations of clinical signs, symptoms, and laboratory abnormalities. The prevalence of LE in North America and northern Europe is about 40 per 100,000 population. In the case of acute cutaneous LE indurated erythematous lesions may be presented on palms.

Epidermolysis bullosa [3,15] is a term given to groups of genetic diseases in which minor trauma causes noninflammatory blistering (mechanobullous diseases). Repetitive trauma may lead to a mitten-like deformity with digits encased in an epidermal "cocoon." These diseases are classified as scarring and nonscarring and histologically by the level of blister formation. Approximately 50 epidermolysis cases occur per million live births in the United States (Figure 3.13) [15].

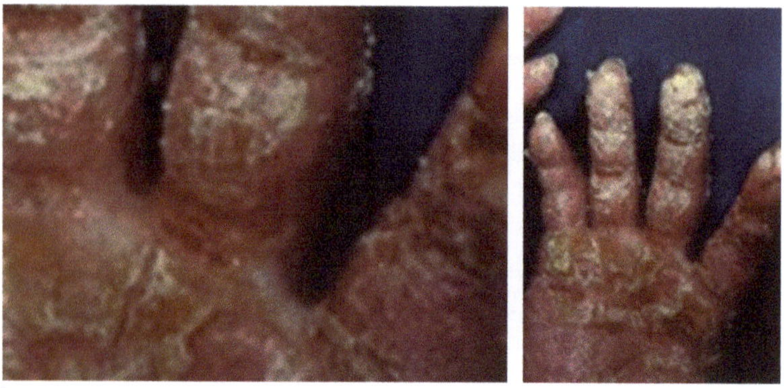

Figure 3.18 Psoriasis vulgaris [13]

Acknowledgments

I would like to thank Assoc. Prof. MUDr. Josef Feit, CSc. and Assoc. Prof. Hana Jedlickova, Ph.D. for kindly providing of histopathological photographs of particular skin diseases from Atlas of Dermatopathology (www.muni.cz/atlases).

References

[1] A. B. Ackerman. Histologic Diagnosis of Inflammatory Skin Diseases: An Algorithmic Method Based on Pattern Analysis. 2 Sub edition. Baltimore: Williams & Wilkins, 1997. 943 p.

[2] E. Zappi and E. A. Zappi. Dermatopathology. London: Springer London, 2013. 204 p. ISBN 978-1-4471-2893-9.

[3] T. P. Habif. Clinical Dermatology. 4th edition. Hong Kong, China: Mosby, 2004.

[4] D. Weedon. An approach to the interpretation of skin biopsies. In: Weedon's Skin Pathology. 3rd edition. Edinburgh: Churchill Livingstone, 2010. pp. 3–18. ISBN: 978-0-7020-3485-5 printed in China.

[5] R. P. Rapini. Practical Dermatopathology. Elsevier Health Sciences, 2012. 474 p.

[6] K. S. Masterpol, A. Primiani, and L. M. Duncan. Atlas of Essential Dermatopathology. London: Springer London, 2013. 126 p. ISBN 978-1-4471-4470-0.

[7] J. Štork, P. Arenberger, K. Pizinger, V. Semrádová, and F. Vosmík. Dermatovenerologie. Prague, Czech Republic: Galén, 2008.

[8] O. Braun-Falco, G. Plewig, H. H. Wolff, and W. H. C. Burgdorf, Eds. Dermatology. 2nd edition. Heidelberg, Germany: Springer, 2000.

[9] W. L. Weston, A. T. Lane, and J. G. Morelli. Color Textbook of Pediatric Dermatology. Hong Kong, China: Mosby Elsevier, 2007.

[10] http://atlases.muni.cz/atlases/kuze/atl_cz/sect_main.html.

[11] E. Brezinova, M. Drahansky, and F. Orsag. Dermatologic diseases and fingerprint recognition. In: Database Theory and Application, Bio-Science and Bio-Technology. Berlin, Germany: Springer, 2010. pp. 251–257.

[12] W. D. James, T. G. Berger, and D. M. Elston. Andrew's Diseases of the Skin—Clinical Dermatology. 10th edition. Ontario, Canada: Elsevier Saunders, 2006.

[13] K. Wolff, R. A. Johnson, and D. Suurmond. Fitzpatrick's Color Atlas and Synopsis of Clinical Dermatology. 5th edition. New York, NY: McGraw-Hill, 2005.

[14] "The Science of the Skin," http://www.naturalrussia.com/natural/skin/structure.html.

[15] M. Drahansky, E. Brezinova, D. Lodrova, and F. Orsag. "Fingerprint recognition influenced by skin diseases," International Journal of Bio-Science and Bio-Technology, vol. 3, no. 4, pp. 11–22, 2010.

Chapter 4

Nail structure and common disorders

Nick Johnson[1], Kanagaratnam Jeyapalan[1], and Joseph Dias[1]

The nail is a complex structure with many important functions; several significant conditions may affect the nail and signs of serious underlying systemic disease may be seen in the nail.

4.1 Gross anatomy and function

The nail is an important part of the fingertip. Its main functions are protection of the underlying nail bed, increasing sensory perception by providing counterforce to the finger pad, thermoregulation of temperature and peripheral circulation along with assisting in picking up objects. Toe nails also contribute to pedal biomechanics.

The nail unit includes the eponychium, paronychium, hyponychium, lunula, nail matrix and proximal nail fold. Together these are known as the perionychium.

The proximal nail fold is formed by a ventral floor and dorsal roof approximately 15 mm distal to the distal interphalangeal joint (DIPJ). The ventral floor contains the germinal matrix and produces 90% of the nail plate thickness [1]. Factors influencing thickness include the size of the matrix, linear rate of nail growth, blood supply, subungual disease and some medications. The dorsal roof is responsible for the shine of the nail. The germinal matrix distal margin is demarcated by the lunula which is a white semicircle at the base of the nail. The sterile matrix is located distal to this and is the secondary site of nail production. The distal aspect of the dorsal fold is adherent to the nail plate and called the eponychium [2].

The nail plate itself is formed of onchyn. This is a keratinised material formed from germinal cell death. The distal two-thirds of the nail plate are closely adherent to the underlying periosteum. A semirigid keratin produced by the sterile matrix acts as an adhesive. The roughened surface of the distal phalanx helps to anchor the overlying nail-bed tissue. At the distal tip of the nail plate, the sterile matrix of the nail bed forms a junction with the skin known as the hyponychium. A keratinised plug acts as a barrier to infection. Either sides of the nail plate are the lateral folds or paronychium.

[1]Academic Team of Musculoskeletal Surgery (AToMS), University Hospitals of Leicester NHS Trust, United Kingdom

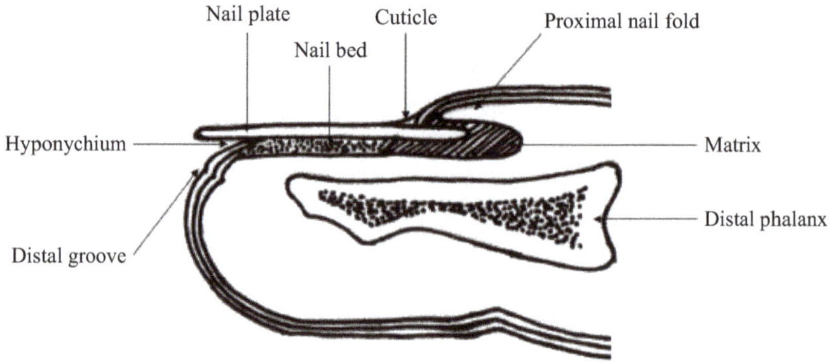

Figure 4.1 Illustration of a sagittal view through a fingertip demonstrating the anatomy of the nail unit

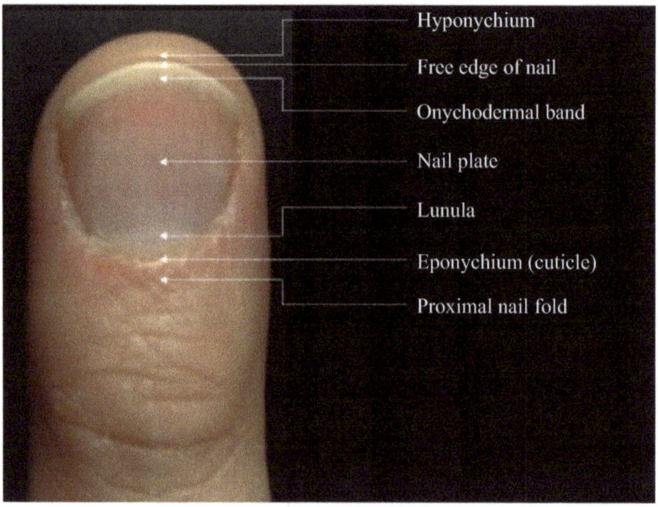

Figure 4.2 Clinical photograph of the normal anatomy of the nail unit in an adult

The skin overlying the pulp has deep papillary ridges and a thick epidermis. Fibrous septa extend from the dermis to the underlying periosteum providing stability and producing separate septal compartments. The pulp itself contains fibrofatty tissue.

The terminal branches of the radial and ulnar proper palmar digital arteries, which originate from the common palmar digital arteries, provide blood supply to the perionychium. A branch travels dorsal to the DIPJ and supplies a superficial arcade that supplies the nail fold and proximal matrix [3].

Nerve supply comes from the digital nerves which divide proximal to the nail fold base and give a branch to the nail bed and another to the pulp. Any contact with the nail plate stimulates the underlying nerves in the nail bed increasing sensitivity (Figures 4.1 and 4.2).

4.2 Imaging

Management of nail bed and nail-related pathology has traditionally relied on clinical examination, scrapings or biopsies of the nail and underlying skin to establish a diagnosis. The presence of the nail plate and the complex anatomy of the nail bed can make it difficult to accurately identify the pathology or exactly define the involvement of the different elements of the nail apparatus. Certain nail conditions, like subungual glomus tumours, may be too small to biopsy and incomplete surgical excision can lead to recurrence [4]. When clinical examination is equivocal, non-invasive imaging techniques may be needed to aid diagnosis and assess the anatomy of the nail apparatus. Options include plain radiography, magnetic resonance imaging (MRI), computerised tomography (CT) or ultrasound scanning.

Plain radiography has traditionally been the complementary imaging method for the nail and distal phalanx but provides little information on soft tissue structures [5]. MRI, and less commonly CT, can be used in the assessment of nail pathology. With advances in transducer technology, ultrasonography has become a useful adjunct in the diagnosis and management of select group of patients with nail disease [6,7]. The nail apparatus is composed of tissues of varying echogenicities well suited to ultrasonographic assessment. Ultrasound is a non-invasive, non-ionizing method that is readily available. Conditions such as glomus tumours, cysts, radiologically occult bone erosions and radio-transparent foreign bodies can be detected by ultrasonography and the site and dimension of the lesions can be defined (Figure 4.3).

4.3 Age-related nail changes

With increasing age, several changes are seen in the nail. The main responsible cause for these changes is impaired circulation. Other factors include trauma, radiation exposure, infection and systemic disease.

Nails in older people are often dull and opaque. Calcium content in the nail plate decreases and iron decreases. Keratinocytes are larger and the nail-bed dermis demonstrates thickened blood vessels with degeneration of elastic tissue. Nail plate thickness may increase, decrease or be unchanged. From the age of 25, the rate of nail growth falls by 0.5% per year [8].

4.4 Common nail conditions

4.4.1 Infection

Paronychia is an inflammatory reaction caused by bacterial invasion of the nail folds of tissue. It may be acute or chronic. Paronychial infections develop due to a disruption between the seal of the proximal nail fold and the nail plate which allows entry of infective organisms.

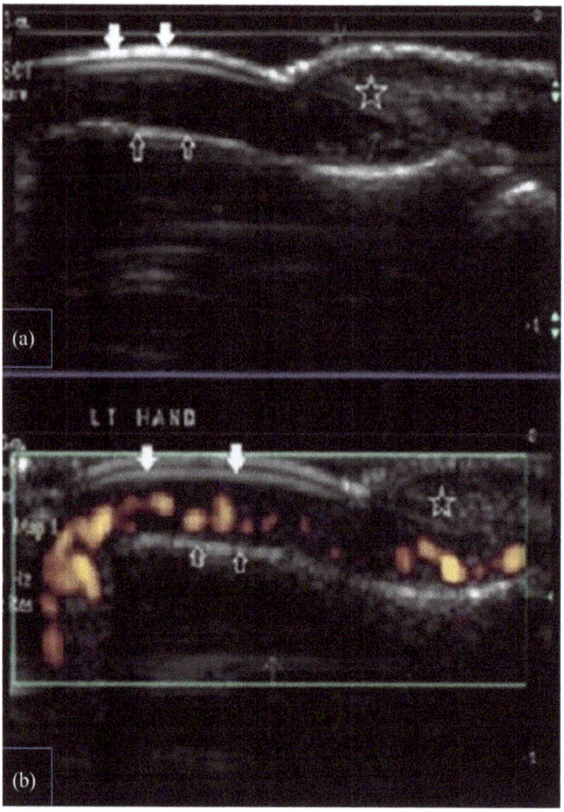

Figure 4.3 *(a) Sagittal ultrasonographic image through the nail complex demonstrating smooth and clearly defined superficial and deep surface of the normal nail plate. Uniform intermediate echo signal from underlying nail bed is seen. Origin of the nail plate is seen within the normal thickness of the germinal matrix. Proximal subcutaneous fat in the nail fold demonstrates higher but uniform echogenicity compared to that of the nail bed. The smooth surface of the dorsum of the distal phalangeal cortex is clearly seen. (b) Sagittal power Doppler mode ultrasound image demonstrates normal vascularity of the nail bed. Solid arrow: nail plate, open arrow: dorsal aspect of distal phalanx and star: eponychium*

4.4.2 Acute paronychia

The most common infecting organism is *staphylococcus aureus* followed by streptococci and pseudomonas. Causes and associated factors include nail biting, trauma, artificial nails and excessive manicuring [9]. Symptoms include pain, erythema and swelling. Nail distortion and discolouration may occur. This may progress to abscess formation with nail bed involvement.

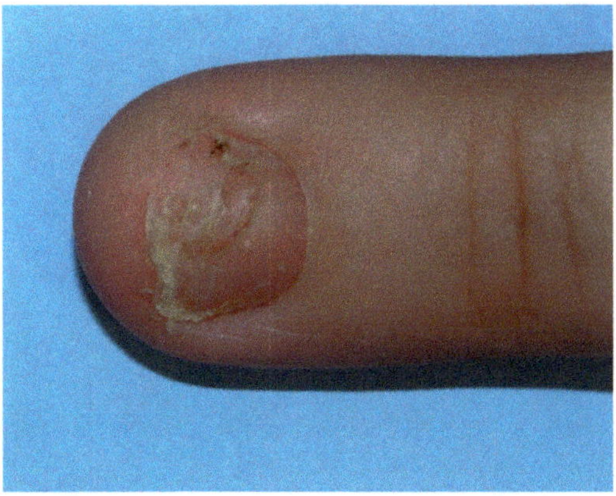

Figure 4.4 Changes of chronic paronychia with thickened nail folds and irregular nail growth

If an abscess has not formed management with regular warm soaking and antibiotics may be effective. If an abscess forms, surgical drainage is required. If pus has collected below the nail plate, then this should be removed to allow drainage.

4.4.3 Chronic paronychia

Chronic paronychia is similar to acute paronychia but has been present for 6 weeks or more. Immunosuppressed patients are at risk, and it is seen more frequently in swimmers, bar staff and dish washers [10]. The persistent infection may cause swollen, boggy nail folds with thickened nail plate ridges. Abscess formation is unlikely. *Candida albicans* is cultured in 95% of cases [11]. Treatment includes avoiding causative factors, topical steroids and an antifungal agent. Surgery may be performed if medical treatment fails. The nail can be removed to allow direct topical treatment and good results have been reported with eponychial marsupialisation [12,13]. In cases refractory to treatment, malignancy must be considered (Figure 4.4).

4.4.4 Onychomycosis

Onychomycosis is a fungal, yeast or non-dermatophytic mould infection of the nail. 10% of the population worldwide are affected and it accounts for 50% of nail problems [14]. Nails become yellow or white with keratinaceous deposits under the nail plate. The infection is difficult to eradicate as it occurs deep within the nail plate. It usually occurs in the distal and lateral portion of the nails. Classification is descriptive according to the anatomic location of the involved nail bed. The most common variety is distal subungual onychomycosis, often accompanied with spread into the palmar skin. Proximal subungual onychomycosis is generally seen in immunocompromised patients, usually with fungal elements in the deep layers of the nail plate (Figure 4.5).

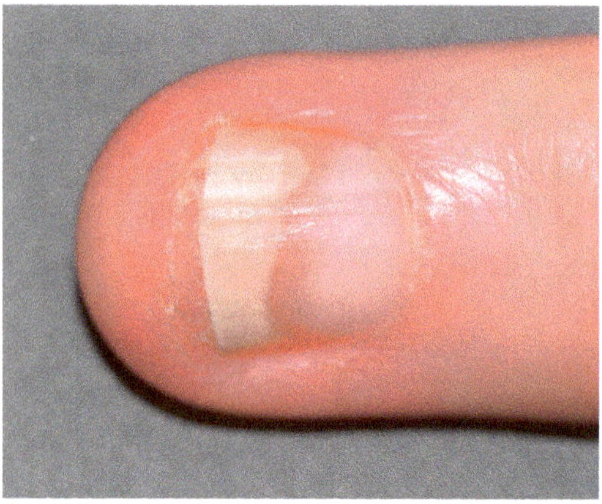

Figure 4.5 Characteristic signs of distal subungual onychomycosis with white keratinaceous deposits under the distal nail plate

4.5 Trauma

Fingertip injuries are the most common injuries in the hand [15]. The fingertip is used for exploration and is the most distal part of the limb so is at high risk of injury. In adults, injuries are most commonly lacerations followed by crush and avulsion injuries. They frequently occur during occupational activities. In children, the mechanism of injury is usually crushing the finger, often in a door. Nail-bed damage is seen in 15%–24% of fingertip injuries [16].

4.5.1 Nail-bed injury

Injury to the fingertip may cause compression of the nail matrix between the nail plate and the underlying distal phalanx, producing a laceration of the nail bed [17].

Normally, the nail plate grows longitudinally and takes between 70 and 160 days to cover the complete nail plate length. Following injury, nail plate growth is reduced or absent for up to 21 days. There is then a period of rapid growth for approximately 50 days. This then slows before a normal, sustained growth rate resumes [18]. These changes in nail plate growth create the abnormalities that are often seen when nails regrow following trauma. Scar development after non-anatomic alignment can result in nail deformity, leading to an unsatisfactory cosmetic outcome [19].

Scar tissue is not capable of synthesising nail material. Injury to certain parts of the perionychium leads to characteristic defects. An injury and subsequent scar of the germinal matrix can create a split or absent nail. A sterile matrix scar may cause a split or non-adherent nail distal to the scar. Nail fold dorsal roof injury that creates a scar leads to a dull marking on the nail plate. To prevent these problems, nail-bed lacerations are usually treated with nail plate removal and repair of the nail bed.

4.5.2 Subungual haematoma

Subungual haematomas are common nail-bed injuries caused by blunt or sharp trauma. Bleeding from the rich vascular nail bed below the nail plate rapidly leads to haematoma and increased pressure under the nail.

Management of this common condition is controversial. Subungual haematoma drainage by trephination can be performed to relieve the pressure and discomfort. Some authors have suggested this is sufficient treatment for a subungual haematoma which involves 25%–50% of the nail bed, whereas others have recommended removal of the nail plate and repair of any underlying laceration [20,21].

Simon and Wolgin noted that nail-bed lacerations greater than 3 mm in size were seen in 60% of patients with a subungual haematoma greater than 50% of the nail plate and in 95% of patients who also had a distal phalanx fracture. They therefore advised that if a subungual haematoma greater than 50% is present, then the nail plate should be removed and the nail bed repaired [21]. Others have disputed these findings. Seaberg *et al.* treated 48 patients with subungual haematoma with trephination only, regardless of the size of the haematoma, and reported no infections or major nail abnormalities. They suggested trephination alone was a satisfactory treatment [22].

4.6 Nail tumours

4.6.1 Subungual melanoma

Subungual melanoma is a rare but very serious nail disorder. Prognosis is poorer than melanoma in other locations. Melanoma is a malignant neoplasm of melanocyte cells, which are responsible for the synthesis of melanin pigment. Only 0.7%–3.5% of melanomas are subungual [23]. It is usually seen in patients aged 60–70 years and is seen equally in men and women [24]. Ninety percent of cases are in the thumb or great toe [25].

Delayed diagnosis is common, often only after secondary problems such as infection or ulceration have developed, leading to poor outcome. Fifty-three percent of patients have locally advanced disease at presentation. Five-year survival rates have been reported as 74% in those with stage 1 disease and 40% with stage 2 disease [26].

The tumour arises mainly from the nail matrix as this is the area containing the most melanocytes but may also be found in the nail folds, nail bed or hyponychium. Hutchinson's sign comprises longitudinal pigmentation of the nail, known as melanonychia, which has progressed into the proximal nail fold [27]. This is a useful sign but not the most common presentation.

Often a pigmented band may be initially seen developing from the nail matrix. A black border of the nail edge is pathognomonic. With progression of the lesion, a fungating ulcer will progressively involve the matrix and lift the nail. Treatment involves wide local excision or amputation with adjuvant chemotherapy.

4.6.2 Squamous cell carcinoma

Squamous cell carcinoma (SCC) is another rare but important condition. It is difficult to differentiate from other nail conditions such as paronychia, onychomycosis, traumatic nail changes, warts and pyogenic granuloma. Time to diagnosis has been reported to range from a few months to 35 years [28]. The spectrum of clinical features that may be seen is wide including leukonychia, hyperkeratosis, melanonychia, chronic pain, swelling and onycholysis.

It usually affects those aged 40–70 and most commonly involves the thumb [29]. SCC appears to be less aggressive than SCC in other areas with metastasis being unusual. It is strongly associated with human papillomavirus (HPV) infection [30]. Other associated factors include sun and radiation exposure, trauma and chronic paronychia.

Treatment depends on the extent of the tumour. Surgical excision is usually satisfactory with low recurrence rates.

4.6.3 Glomus tumours

Glomus tumours are rare and account for 1% of all hand tumours [31]. The glomus body is an arteriovenous anastomosis responsible for thermoregulation of skin temperature. They are found in the fingertips and especially in the subungual region. Glomus body tumours are benign vascular hamartomas of the glomus body [32].

Patients classically describe a triad of pain, sensitivity to temperature and point tenderness [33]. Signs seen include a blue discolouration, small palpable lump and nail deformity. Several clinical tests are suggestive of a glomus tumour. Love noted that pain will be reduced after compressing the site of maximum tenderness with a small object such as a pin head. Placing the hand in cold water will aggravate pain [34]. Treatment is by surgical excision which generally leads to resolution of all symptoms (Figure 4.6).

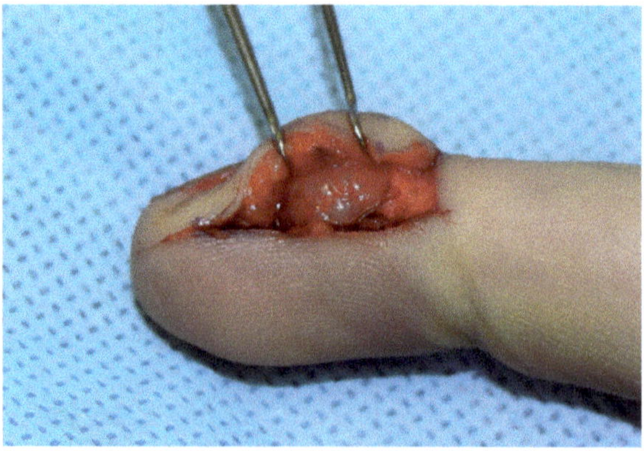

Figure 4.6 Intra-operative photograph demonstrating a glomus tumour prior to surgical excision

4.7 Systemic disease

Evaluation of the structure of a patient's nails may reveal important signs suggestive of an underlying systematic condition.

4.7.1 Beau lines

Beau lines are horizontal grooves seen on the nail plate. They are caused by an interruption in mitosis usually due to significant illness, pemphigus, Raynaud's disease or chemotherapy treatment.

4.7.2 Clubbing

Thickening of the proximal nail bed soft tissue leads to clubbing. Opposition of digits reveals a loss of the normal diamond shape gap. An angle greater than 180 degrees between the nail plate and proximal digit soft tissue is diagnostic. Many diseases can cause clubbing including lung cancer, chronic obstructive pulmonary disease and liver cirrhosis. It is thought the underlying mechanism is changes in the soft tissues related to increased microvascular blood flow.

4.7.3 Koilonychia

Koilonychia is often referred to as spoon nail as the nail becomes increasingly thin and concave. It may be a normal finding in infants but in adults is associated with iron deficiency anaemia (Figure 4.7).

4.7.4 Mees lines

Mees lines are white horizontal lines which pass across the whole width of the nail plate. As they occur in the nail plate itself, they progress distally with nail plate

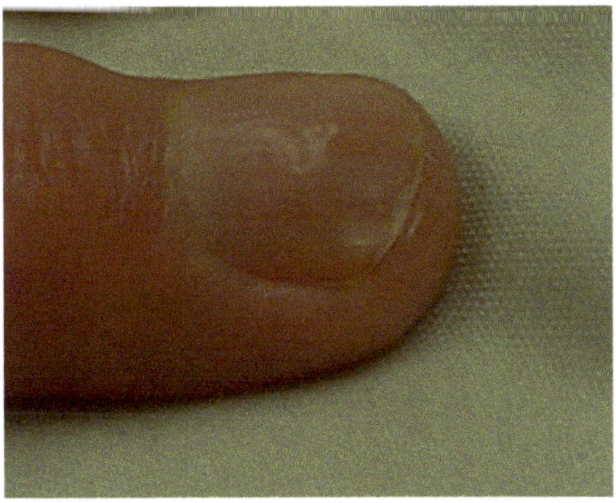

Figure 4.7 Early koilonychia with concavity of the distal nail plate

growth. They are mainly diagnostic of arsenic poisoning but may be seen in renal failure or other heavy metal poisoning.

4.7.5 Muehrcke's lines

Muehrcke's lines are paired white, horizontal lines. They are caused by pathology within the nail bed and therefore do not move distally with nail plate growth. Abnormal vascularity is present which is obliterated with direct pressure causing the lines to disappear. They are non-specific but may be related to hypoalbuminaemia.

4.7.6 Splinter haemorrhages

Splinter haemorrhages are linear browny red markings seen in the nail bed. They are due to leaky capillaries and not obliterated by pressure. Typically they are associated with endocarditis but are more commonly seen with trauma. They may be seen with many systemic conditions such as renal, lung, endocrine and skin diseases.

References

[1] Yeo CJ, Sebastin SJ, Chong AK. Fingertip injuries. Singapore Med J. 2010, 51: 78–86.
[2] Fleckman P, Allan C. Surgical anatomy of the nail unit. Dermatol Surg. 2001, 27: 257–60.
[3] Strauch B, de Moura W. Arterial system of the fingers. J Hand Surg. 1990, 15: 148–54.
[4] Matsunaga A, Ochiai T, Abe I, *et al*. Subungual glomus tumour: evaluation of ultrasound imaging in preoperative assessment. Eur J Dermatol. 2007, 17: 67–9.
[5] Wortsman X, Jemec GB. Ultrasound imaging of nails. Dermatol Clin. 2006, 24: 323–8.
[6] Wortsman X, Gutierrez M, Saavedra T, Honeyman J. The role of ultrasound in rheumatic skin and nail lesions: a multi-specialist approach. Clin Rheumatol. 2011, 30: 739–48.
[7] Drap J. Imaging of the tumors of the perionychium. Hand Clin. 2002, 18: 655–70.
[8] Cohen PR, Scher RK. Geriatric nail disorders: diagnosis and treatment. J Am Acad Dermatol. 1992, 26: 521–31.
[9] Roberge RJ, Weinstein D, Thimons MM. Perionychial infections associated with sculptured nails. Am J Emerg Med. 1999, 17: 581–2.
[10] Rockwell PG. Acute and chronic paronychia. Am Fam Physician. 2001, 63.
[11] Jebson PJ. Infections of the fingertip. Paronychias and felons. Hand Clin. 1998, 14: 55, viii.
[12] Hochman LG. Paronychia: more than just an abscess. Int J Dermatol. 1995, 34: 385–6.

[13] Canales FL, Newmeyer 3rd WL, Kilgore Jr ES. The treatment of felons and paronychias. Hand Clin. 1989, 5: 515–23.
[14] Allevato MAJ. Diseases mimicking onychomycosis. Clin Dermatol. 2010, 28: 164–77.
[15] Fassler PR. Fingertip injuries: evaluation and treatment. J Am Acad Orthop Surg. 1996, 4: 84–92.
[16] Chang J, Vernadakis AJ, McClellan WT. Fingertip injuries. Clin Occup Environ Med. 2005, 5: 22, ix.
[17] Gellman H. Fingertip-nail bed injuries in children: current concepts and controversies of treatment. J Craniofac Surg. 2009, 20: 1033–5.
[18] Johnson M, Shuster S. Continuous formation of nail along the bed. Br J Dermatol. 1993, 128: 277–80.
[19] Kleinert HE, Putcha SM, Ashbell TS, Kutz JE. The deformed finger nail, a frequent result of failure to repair nail bed injuries. J Trauma Acute Care Surg. 1967, 7: 177–90.
[20] Roser SE, Gellman H. Comparison of nail bed repair versus nail trephination for subungual hematomas in children. J Hand Surg. 1999, 24: 1166–70.
[21] Simon RR, Wolgin M. Subungual hematoma: association with occult laceration requiring repair. Am J Emerg Med. 1987, 5: 302–4.
[22] Seaberg DC, Angelos WJ, Paris PM. Treatment of subungual hematomas with nail trephination: a prospective study. Am J Emerg Med. 1991, 9: 209–10.
[23] Levit EK, Kagen MH, Scher RK, Grossman M, Altman E. The ABC rule for clinical detection of subungual melanoma. J Am Acad Dermatol. 2000, 42: 269–74.
[24] Rigby HS, Briggs JC. Subungual melanoma: a clinico-pathological study of 24 cases. Br J Plast Surg. 1992, 45: 275–8.
[25] Haneke E, Baran R. Longitudinal melanonychia. Dermatol Surg. 2001, 27: 580–4.
[26] O'leary JA, Berend KR, Johnson JL, Levin LS, Seigler HF. Subungual melanoma: a review of 93 cases with identification of prognostic variables. Clin Orthop. 2000, 378: 206–12.
[27] Molina D, Sanchez JL. Pigmented longitudinal bands of the nail: a clinicopathologic study. Am J Dermatopathol. 1995, 17: 539–41.
[28] Lumpkin LR, Rosen T, Tschen JA. Subungual squamous cell carcinoma. J Am Acad Dermatol. 1984, 11: 735–8.
[29] Dalle S, Depape L, Phan A, Balme B, Ronger-Savle S, Thomas L. Squamous cell carcinoma of the nail apparatus: clinicopathological study of 35 cases. Br J Dermatol. 2007, 156: 871–4.
[30] Ashinoff R, Li JJ, Jacobson M, Friedman-Kien AE, Geronemus RG. Detection of human papillomavirus DNA in squamous cell carcinoma of the nail bed and finger determined by polymerase chain reaction. Arch Dermatol. 1991, 127: 1813–18.

[31] Achauer BM, Eriksson E, Guyuron B, Coleman III JJ, Russell RC, Vander Kolk CA. Plastic Surgery: Indications, Operations, and Outcomes. 2000. Philadelphia, PA: Mosby Publishing.
[32] Tomak Y, Akcay I, Dabak N, Eroglu L. Subungual glomus tumours of the hand: diagnosis and treatment of 14 cases. Scand J Plast Reconstr Surg Hand Surg. 2003, 37: 121–4.
[33] Bhaskaranand K, Navadgi BC. Glomus tumour of the hand. J Hand Surg. 2002, 27: 229–31.
[34] Hazani R, Houle JM, Kasdan ML, Wilhelmi BJ. Glomus tumors of the hand. Eplasty. 2008, 8:e48.

Chapter 5
State of the art in fingerprint recognition

Ondřej Kanich[1] *and Martin Drahanský*[1]

5.1 Introduction

This chapter deals with fingerprint recognition technology. Each of us has *papillary lines* (elevated skin reliefs, called often *ridges* and the gaps between ridges are called *valleys*) that are uniquely shaped for each person so that they can be distinguished on the basis of their passing through each other on the surface of their fingers (hands and feet). Today, it is the most widespread technology that can be met in almost everyday life.

5.1.1 User acceptance

Due to the relatively massive expansion of fingerprint technology to everyday life, there is a problem with the user's acceptance of this technology. Nowadays almost everyone saw or used fingerprint readers—they are integrated into laptops, mobile phones and other mobile devices, various attendance or access systems are based on fingerprints, etc. However, some users are negatively linking fingerprints with criminal police and investigations. Another part of the people has (legitimate) fear of fingerprint abuse—the fingerprint itself is easily obtainable because everything someone touches contains his/her fingerprint. This is a relatively easy way to get it, to make an artificial fingerprint fake and use it for spoofing the biometric system.

Another danger that exists here is the abuse of a template with stored minutiae, that is, a template that does not contain a fingerprint image but only extracted features. Even from this data, it is possible to generate a synthetic fingerprint and create a fake fingerprint, which is most likely to be used with most fingerprint systems. On the other hand, the user needs to be reassured—the templates are cryptographically protected, i.e., the data cannot usually be read directly, and many existing sensors already have built-in liveness detection (antispoofing methods) that reveals fake or dead fingers, i.e., the security of this technology is really high.

[1]Faculty of Information Technology, Centre of Excellence IT4Innovations, Brno University of Technology, Czech Republic

5.1.2 Reliability

First, let's look at fingerprint technology representation on the biometric market (see Figure 5.1—the trend did not change significantly since 2011, and the prognosis till 2021 is still the same, i.e., people will still trust to the same technologies [1]). The fingerprint technology itself, which includes access and attendance systems in particular, occupies 17% of the total market, excluding automated fingerprint identification systems (AFISs) (fingerprint recognition systems for identification and verification purposes in forensic practice). The main market share belongs to iris (20%), followed by fingerprints (17%), face (17%), hand/palm (16%) and voice (15%). The vein structure has 10% and other biometric technologies reach together 5% of the market share. However, AFISs would change the role, because they are used in nearly each country for forensic purposes (crime investigation), i.e., fingerprints have probably still the majority on the market.

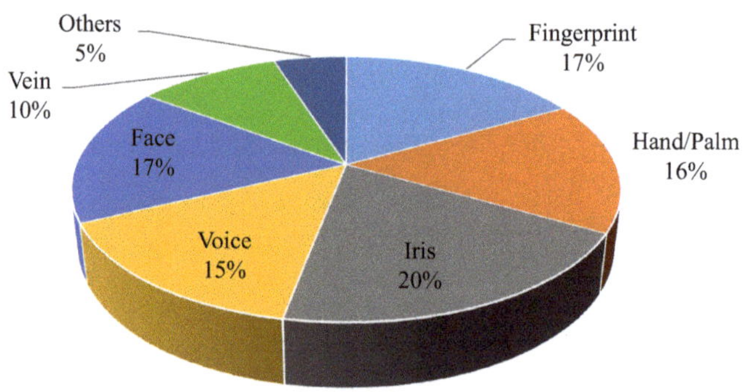

Figure 5.1 Distribution of biometric technologies on the market [1]

One reason for the massive spread of fingerprints is the use for criminal investigation, where fingerprints are taken as the main evidence, which associates a sense of confidence in this technology, when it clearly demonstrates the identity of a person in such a sensitive area as criminal investigation and subsequent law case.

One important factor is the amount of information that is hidden in the fingerprint, so how many people can be theoretically recognized from each other just using their fingerprints. This quantity, even if it is not fully correct from the viewpoint of information theory, is usually called biometric entropy. The creation of the fingerprint model and the subsequent calculation of the amount of information contained therein could be found, e.g., in these works [2,3]. Table 5.1 shows that, according to the set parameters (N, R and M), the probability of matching two fingerprints is in the order of 10^{-7} to 10^{-80}. On the contrary, the model proposed in [3] posed a different question—how large is the maximum amount of information contained in the fingerprint, i.e., how many people can be recognized on the basis of this information. The result is in the order of 2^{122}, i.e., 10^{36}. The range is of course larger—depending on the parameters used in the computational model, but in real conditions, the idealized results cannot be taken seriously.

Table 5.1 Comparison of probability of fingerprint matching for different models [2]

Author	p (fingerprint configuration)	$N=36$, $R=24$, $M=72$	$N=12$, $R=8$, $M=24$
Galton (1892)	$(1/16) \times (1/256) \times (1/2)^R$	$1.45 \cdot 10^{-11}$	$9.54 \cdot 10^{-7}$
Pearson (1930)	$(1/16) \times (1/256) \times (1/36)^R$	$1.09 \cdot 10^{-41}$	$8.65 \cdot 10^{-17}$
Henry (1900)	$(1/4)^{N+2}$	$1.32 \cdot 10^{-23}$	$3.72 \cdot 10^{-9}$
Wentworth and Wilder (1918)	$(1/50)^N$	$6.87 \cdot 10^{-62}$	$4.10 \cdot 10^{-21}$
Cummins and Midlo (1943)	$(1/31) \times (1/50)^N$	$2.22 \cdot 10^{-63}$	$1.32 \cdot 10^{-22}$
Gupta (1968)	$(1/10) \times (1/10) \times (1/10)^N$	$1.00 \cdot 10^{-38}$	$1.00 \cdot 10^{-14}$
Roxburgh (1933)	$(1/1{,}000) \times (15/(10 \times 2.412))^N$	$3.75 \cdot 10^{-47}$	$3.35 \cdot 10^{-18}$
Trauring (1963)	$(0.1944)^N$	$2.47 \cdot 10^{-26}$	$2.91 \cdot 10^{-9}$
Osterburg (1980)	$(0.766)^{M-N} \times (0.234)^N$	$1.33 \cdot 10^{-27}$	$1.10 \cdot 10^{-9}$
Stoney (1985)	$(N/5) \times 0.6 \times (0.5 \cdot 10^{-3})^{N-1}$	$1.20 \cdot 10^{-80}$	$3.50 \cdot 10^{-26}$

As with other technologies, error rates for the fingerprint technology are expressed using false acceptance rate (FAR)/false rejection rate (FRR), false match rate (FMR)/false non-match rate (FNMR) and receiver operating characteristics (ROC)/detection error tradeoff (DET) values. Normally, industrial fingerprint algorithms move at FAR or FRR (calculated according to equal error rate (EER)) around 10^{-1} [4]. The individual values vary depending on the sensor manufacturer and the algorithm, not least the population that uses the technology. According to [5], the FAR was around 0.01%, while the FRR was around 0.6%. An inappropriate example of a population can be, e.g., a company where employees have a damaged papillary lines due to hard work. They may be, e.g., elderly people, who develop creases (wrinkles) on the fingertips—the papillary lines recede into the background. Generally speaking, fingerprint technology has an almost excellent recognition capability that is close to 100% thanks to very advanced algorithms. Some works are devoted to fingerprint image enhancement, including fingerprints with skin diseases.

5.2 History

The use of biometric characteristics is known since ancient times. Basically, people use biometric recognition on a daily basis—they are able to recognize a concrete person by voice, face, walk, etc. All this information is human biometric—it can be recorded and then process it using a machine.

The oldest documented information regarded to the use of fingerprinting comes from China from the fourteenth century. However, these are indirect proofs of biometrics—the surviving drawings on the rock walls were similar to fingerprints, or the fingerprints of the author on surface (perhaps as evidence of authorship) were found on ceramics.

The first evidence of the use of fingerprints comes from the nineteenth century A.D. At this time fingerprints began to be used in crime investigation. The preserved written materials are specifically the following:

- 1858—*William James Herschel* was the English Governor of India (Hooghly) and began to use fingerprints for track-recorders to confirm people's identity. Most of the local workers could not read and write, and it was impossible for them to expect a signature when they received the paycheck. Mr. Herschel let every laborer print the fingerprint original to his payroll tape, confirming the identity of the worker and at the same time the lawful remittance of money. At the same time, he began to collect fingerprints for his own research, which helped to write a work on the origin of fingerprints [6].
- 1865—*Francis Galton* came up with a study of the inheritance of physical properties [7], in which he discussed the fact that children inherit from their parents some of their properties, including physical characteristics and behavioral characteristics.
- 1869—*Francis Galton* becomes cofounder of science called *eugenics*, which is a doctrine of hereditary diseases and defects in the fetus. This field is based on his previous work. Later, in 1875, Francis Galton became the founder of twin research.
- 1880—*Francis Galton* comes with the science of *anthropometry* [8], which is a method of measurement of human body dimensions.
- 1882—*Alphonse Bertillon* introduced and used the *Bertillionage* [9] since 1879. It is essentially anthropometry, which has led to a collision with Francis Galton's anthropometric approach.
- 1892—*Francis Galton* published a complete work of "Fingerprints" [10], which brought dactyloscopy [11] into practice in 1900. In the following year (1893), Francis Galton compares dactyloscopy with anthropometry, and in 1894 concludes that both methods are good and reliable, so they both provide practical use. In 1896, the dactyloscopy was used as an identification system in Argentina.
- 1900—Francis Galton enforces dactyloscopy for identification and verification purposes. He proved the permanence and uniqueness of the skin reliefs (papillary lines) on his fingers. Dactyloscopy was introduced into police practice.
- 1924—in this year, the Federal Bureau of Investigation (FBI) was established.
- 1965—for the first time, an AFIS [12] with 810,000 fingerprints was used.
- 2000—the AFIS at the FBI contains a total of 47 million ten prints of fingerprints (fingerprints from each hand). On average, there will be 50,000 searches per day. The response to remote search in the database is approximately 2 h.
- 2010—the AFIS at the FBI contains a total of 66 million ten prints of fingerprints (fingerprints from each hand). On average, there will be 162,000 searches per day. The response to remote search in the database is approximately 1 h and 10 min (processing an electronic request in an urgent case takes 10 min).
- 2011—the Advanced Fingerprint Identification Technology was deployed as part of NGI (Next-Generation Identification) project of FBI. It improves matching accuracy from 92% to over 99.6% [13].

- 2017 (September)—the NGI at the FBI contains a total of almost 120 million ten prints of fingerprints. On average, there is 210,358 ten-print fingerprint processed. Criminal response time (urgent) on average is 9 min and 58 s. Civil response time on average is 36 min and 11 s [14].

As mentioned above, each of us has papillary lines on the surface of our hand palms and feet soles, whose structure clearly determines the physical identity of a person. An exception is for people with various types of illnesses and skin disorders. Papillary lines have their own graphic representation—a *fingerprint*. Papillary lines are formed during embryonic development. The structure of the papillary lines is unchanged over time, of course, unless they are damaged by, e.g., cuts in dermis. Each finger is a unique pattern, i.e., there are no two identical fingers in the world [11].

At the beginning of the introduction to the theory of fingerprint recognition, history will be described. The first preserved remains are archeological artifacts and cave paintings. Names associated with the history of fingerprints:

- *Nehemiah Grew*—a pioneer of dactyloscopy, devoted his activity to the study of papillary lines and the location of sweat pores on his hands; in 1684, he published a work on the structure of papillary lines on his fingers and hands.
- *Johann Christoph Andreas Mayer*—published in 1788, a work on the uniqueness of fingerprints for each individual.
- *Thomas Bewick*—was a woodcutter and ornithologist, but since 1809, he used his fingerprint along with the written form of his name as a trademark, confirming the theory of the uniqueness of his fingerprint.
- *Jan Evangelista Purkyně*—in 1823, he published a work on nine classes of fingerprints but did not mention the use of fingerprints for identifying people.
- *Henry Faulds*—in 1880, he published a work on the usability of fingerprints for the purpose of identifying people and suggested a method for fingerprinting using ink.
- *Juan Vucetich*—in 1891, he introduced the use of fingerprints for criminalistics purposes, i.e., their categorization and use of the filing cabinet.
- *Francis Galton*—published a book on fingerprints [10] and calculated the probability of matching two fingerprints as 1–64 billion.

In the field of *dactyloscopy* [11], i.e., fingerprint recognition in forensic applications (for forensic purposes), so-called dactyloscopic laws [15] apply:

- There are no two people in the world whose papillary lines would have the same structure.
- A pattern formed by papillary lines remains relatively unchanged throughout the life of an individual.
- Papillary lines are restored by growing the skin at the finger surface. These cannot be altered or removed unless the dermal layer of the skin is damaged. Then there will be no renewal of papillary lines at this point.
- Configuration types vary individually, but the changes are small enough to lie within tolerance limits and thus allow for systematic classification.

5.3 Fingerprint types and acquisition methods

In dactyloscopic practice, three kinds of fingerprint that differ not only in appearance and location but also in the way of scanning can be seen (Figure 5.2):

- rolled (also colored)
- plain (alive)
- latent (also hidden)

However, there are fingerprints that are totally inappropriate for automated recognition or minutiae comparisons—these are primarily fingers suffering from various skin diseases or disorders on the surface of the fingertips.

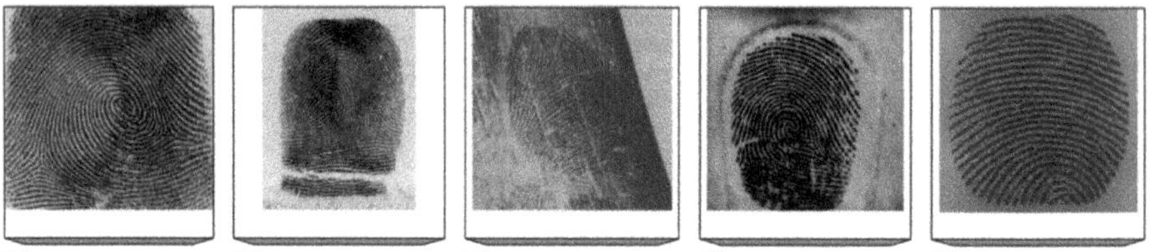

Figure 5.2 Fingerprints: (a) and (b) rolled, (c) latent, (d) and (e) plain

The term fingerprints is related to the basic term—the *papillary line = ridges* (Figure 5.3). A fingerprint is a pattern formed by a structure of papillary lines. The height of the papillary lines lies between 0.1 and 0.4 mm, and the width of the papillary lines is ranging from 0.2 to 0.5 mm.

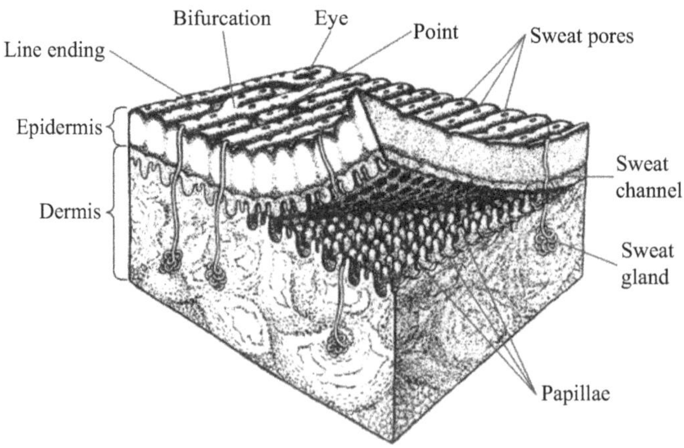

Figure 5.3 Cut of the skin with papillary lines [16]

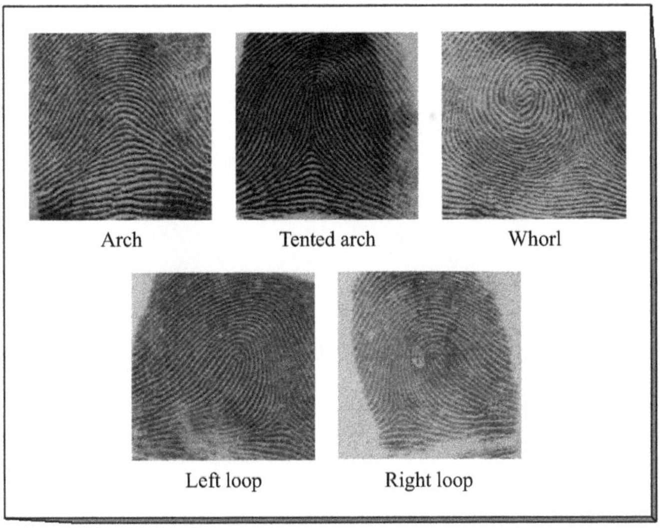

Figure 5.4 Fingerprint classes

From a global perspective, the papillary lines in the fingerprint create a pattern called the *fingerprint class* [17] (see Figure 5.4). The following fingerprint classes are known:

- arch
- tented arch
- whorl (spiral)
- left loop
- right loop

For classification of fingerprints in AFISs [12], *Henry's classification scheme* is used, which is based on the categorization of fingerprints into three basic classes—arch, loop and whorl. Individual derivations, such as tented arches, right or left loops and possibly double loops, divide fingerprints into other classes. In order to create a classification system, it was necessary to classify fingerprints into smaller subclasses so that it would not be necessary to search the whole database to find the identity of the criminal. Henry's classification system is named after Edward Henry, who has been linked to works by well-known researchers—W. Herschel, H. Faulds and F. Galton.

In dactyloscopic systems (for criminal police forensic purposes), all classes for classification are used. The fingerprint classification algorithm is based on the information contained in the fingerprint. Basic terms related to the classification [see Figure 5.5(a)] [18] are as follows:

- *Delta*—a fingerprint location where the papillary line runs in three directions (most of which are on the edge, there may be two deltas).

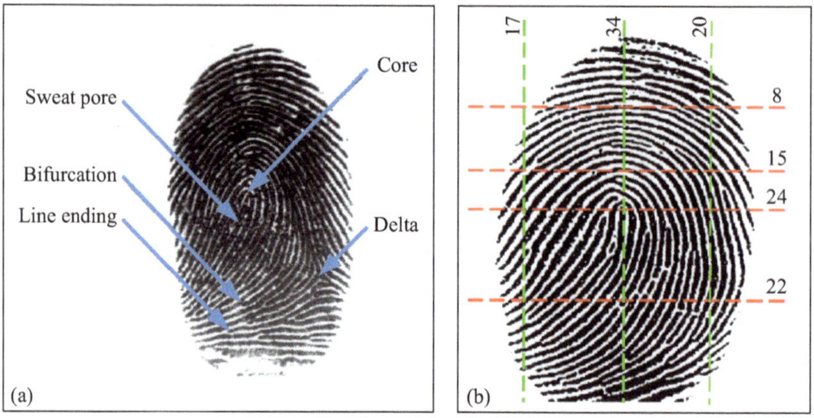

Figure 5.5 (a) Significant features of a fingerprint and (b) numbers of papillary lines

- *Core*—the center of the fingerprint, located at the lowest arch of the run of papillary lines in the fingerprint; the position does not match the actual center of the image.
- *Type lines*—they define the space between the topmost papillary line belonging to the center and the lowest one belonging to the delta.

The amount of papillary lines between two defined fingerprint points (most commonly between core and delta) is used as another metric for fingerprint classification. An example of the number of papillary lines in the vertical and horizontal directions is shown in Figure 5.5(b) (the number of papillary lines increases in the direction to the fingerprint core, similar to the concentric circles).

Fingerprints are distinguished, however, on the basis of special shapes that form papillary lines. These formations are called *minutiae* points. The basic minutiae include (see Figure 5.6, respectively) line ending, single fork, double fork (bifurcation), triple fork, hook, cross and side contact; bottom line of Figure 5.6: point, interval, single loop, double loop, single bridge, double bridge and intersection. In dactyloscopic systems for criminological purposes, much more minutiae are used than those listed here—but these are derived from the basic types of minutiae. Conversely, only two types are used for access systems: line ending and bifurcation.

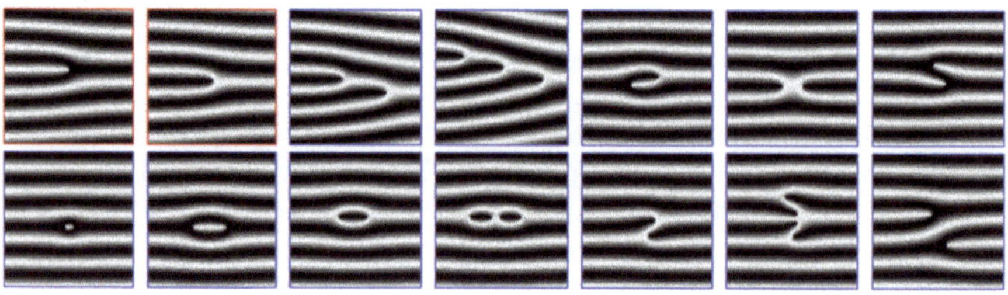

Figure 5.6 Basic types of minutiae

The *gradient* (*orientation*) of the minutia is the direction in which the papillary line would continue at the minutia point. Two notations can be distinguished (Figure 5.7: left image = line ending and right image = bifurcation):

- Mark A: standard notation
- Mark B: FBI/AFIS notation (opposite to standard notation)

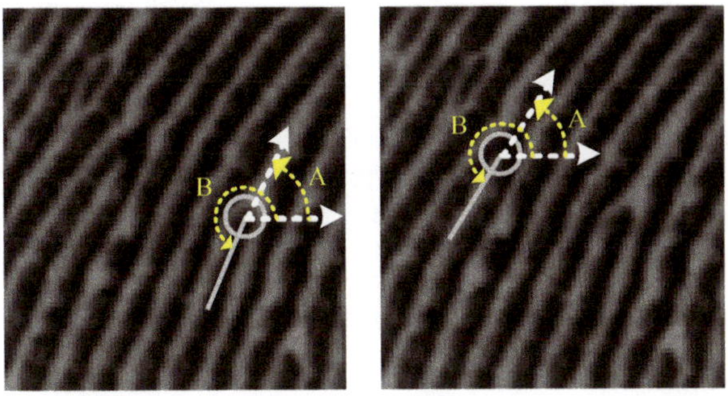

Figure 5.7 Standard (left) and FBI (right) notation of the minutia orientation

Electronic fingerprint readers are used in biometric fingerprint systems to acquire fingerprint images. The only exception being dactyloscopic cards, where the classic fingerprinting still prevails on a paper dactyloscopic card that is later loaded into the computer using a scanner. Some of the important parameters for fingerprint readers are resolution (ranging from 250 up to 1,000 dpi, the most common being around 500 dpi), scanning area (FBI recommends the area of 1 × 1 in., however, typically is around 0.7 × 0.7 cm for access systems and 10 × 6 cm for dactyloscopic fingerprint systems), amount of bits (number of bits for color coding—the standard is 8 bits for gray-scales, but sensors that only use 3 bits can be found), geometric accuracy (the amount of geometric fingerprint distortion versus reality) and image quality (various metrics such as [19] have been proposed for this parameter).

Following sensor technologies are distinguished [19]: *optical*, *capacitive*, *ultrasonic*, *e-field*, *electro-optical*, *pressure*, *thermal* and *microelectromechanical system (MEMS)*, including other less known methods (e.g., optical tomography).

However, exception will be discussed as first—ink fingerprint on a *dactyloscopic card* (see Figure 5.8). The dactyloscopic card consists of the description margins and the fingerprints themselves—the upper part prints all ten fingers in the form of plain fingerprints, and the control fingerprints of the fingertips are printed at the bottom. On the other side of the dactyloscopic card, there are palmprints.

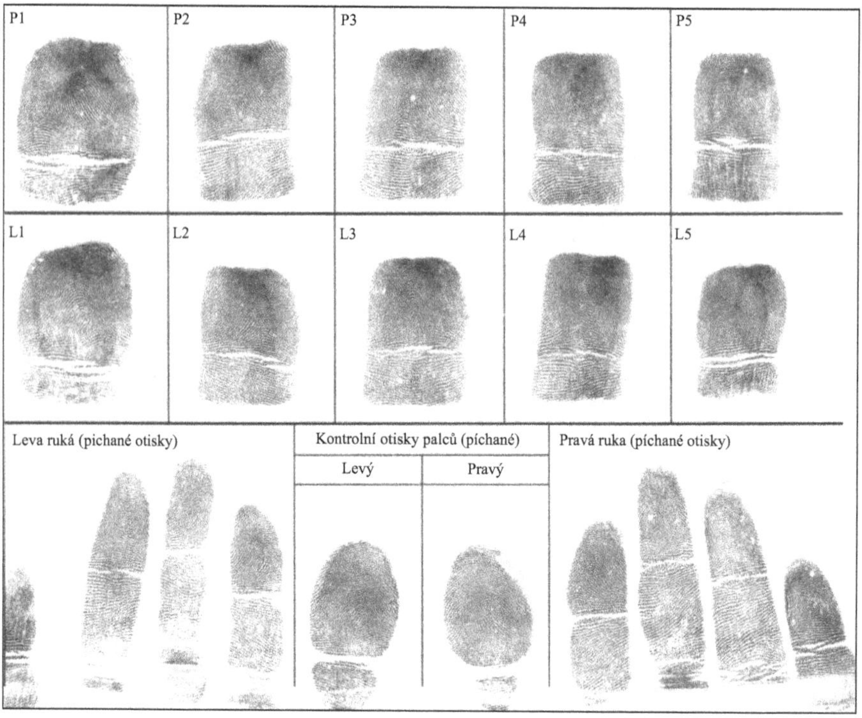

Figure 5.8 Example of a Czech dactyloscopic card (without description margins)

5.3.1 Optical technology

This is a relatively simple optical principle (see Figure 5.9), i.e., the light source (LED) illuminates the finger surface that is attached to the sensor's glass surface (there are also contactless 3D optical sensors, i.e., the finger has to not touch the surface in all cases—one example is the contactless fingerprint reader from the Swiss Company Touchless Biometric Systems) and the camera (CCD/CMOS) scans the image. When protective glass is replaced by a transparent roller tube and optics, the camera and light source are in it, and then the result is simple swipe optical sensor. It is also possible that such a roller functions like an optics and camera with a light source that is beneath it.

Another type of optical sensor uses optical coherence tomography (OCT) [20]. It is very expensive, but it gets the image from a deeper layer of the skin, which is harder to spoof. It can also obtain a view where sweat pores are clearly visible.

5.3.2 Capacitive technology

For capacitive technology (see Figure 5.10), the sensor is composed of a matrix of small conductive areas on which a layer of nonconductive silicon dioxide is put. The fineness of these conductive surfaces is higher than that of the papillary lines.

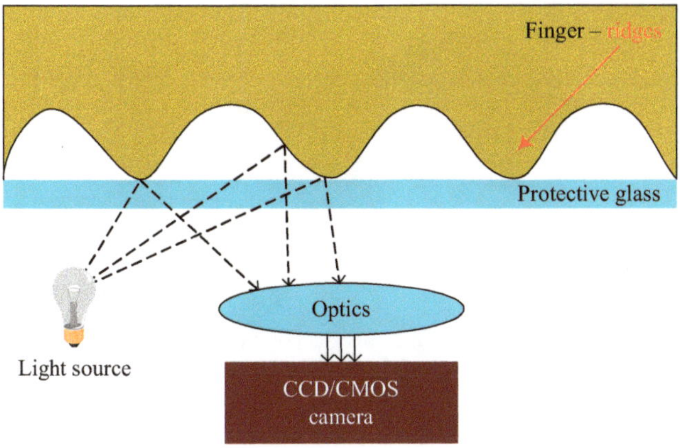

Figure 5.9 Principle of optical technology

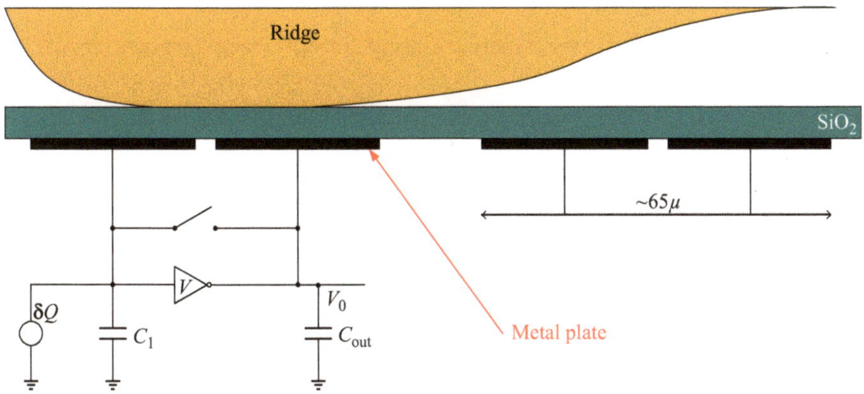

Figure 5.10 Principle of capacitive technology

By attaching a finger, capacitors (one electrode is the papillary line at the point of contact with the face of the sensor matrix and the other is just matrix plane) make the output of which is the value corresponding to the overlap of the surface area (see the capacitor principle). To create swipe sensor, one of the dimension is shortened. A related technology is so-called e-field technology [21].

There is one modification of the capacitive sensor that is worth mentioning. It is a combination of e-field and capacitive technology. It uses a low-radio frequency (RF) signal and because of that it is often known as RF technology. This signal is sent to the skin, and because of that, between the signal references plane and the live (conductive) layer of the skin, an RF electric field is created. Its equipotential contours mimic the shape of the live layer of the skin, so when it is measured by an antennae array, fingerprint image is produced. This principle can be seen in

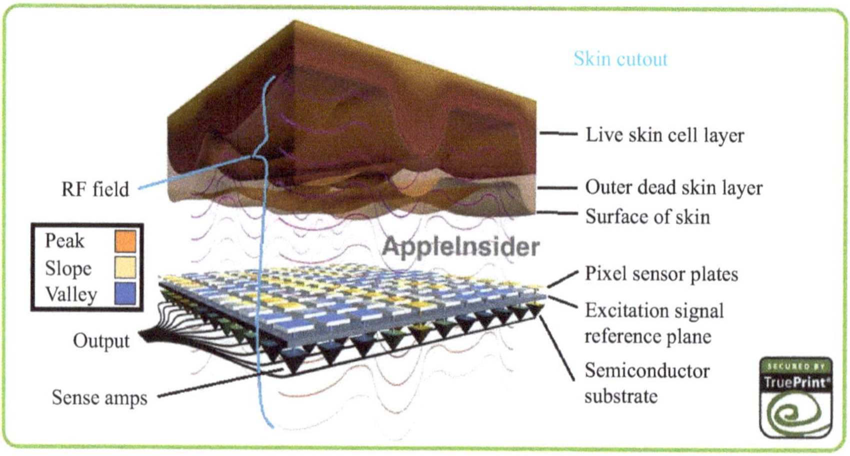

Figure 5.11 Principle of RF technology

Figure 5.11. Another interesting principle is using thin-film transistors on a capacitive touch panel. That way it is possible to have a touch-sensitive area (i.e., a smartphone can be controlled this way) and fingerprint sensing at the same time [22–24].

5.3.3 Ultrasonic technology

The ultrasonic technology is based on a rotating ultrasonic transducer (see Figure 5.12), which incorporates a receiver. This rotates along the circular path and scans the fingerprint. Ultrasonic waves penetrate even under the surface of the skin. This technology can easily detect fake fingers.

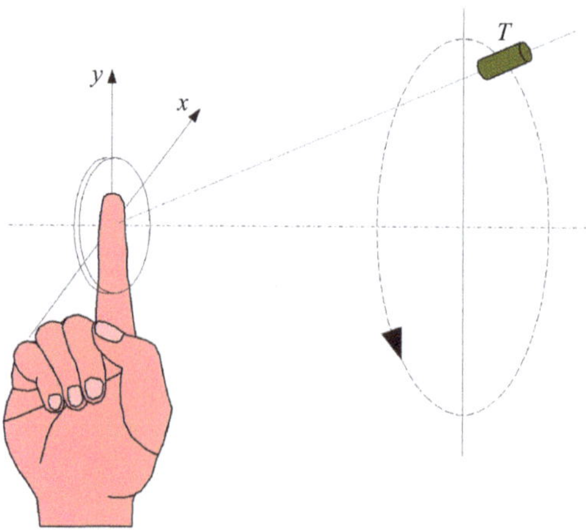

Figure 5.12 Principle of ultrasonic technology

5.3.4 Electro-optical technology

The sensor based on electro-optical technology consists of four layers, with the finger pressure eliciting contact of the black coaxial layer of emitting light in the phosphor layer. This radiation passes through the base layer into the sensor (see Figure 5.13).

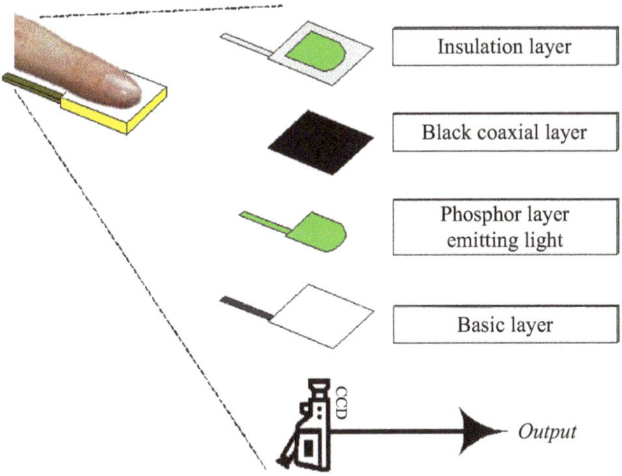

Figure 5.13 Principle of electrophoretic technology

5.3.5 Pressure technology

The pressure technology sensor consists of three layers, with a nonconductive gel inserted between the electroconductive layers (see Figure 5.14). By attaching a finger to the sensor surface, the nonconductive gel is pressed at the point of contact with the papillary lines and brings both electroconductive layers in touch.

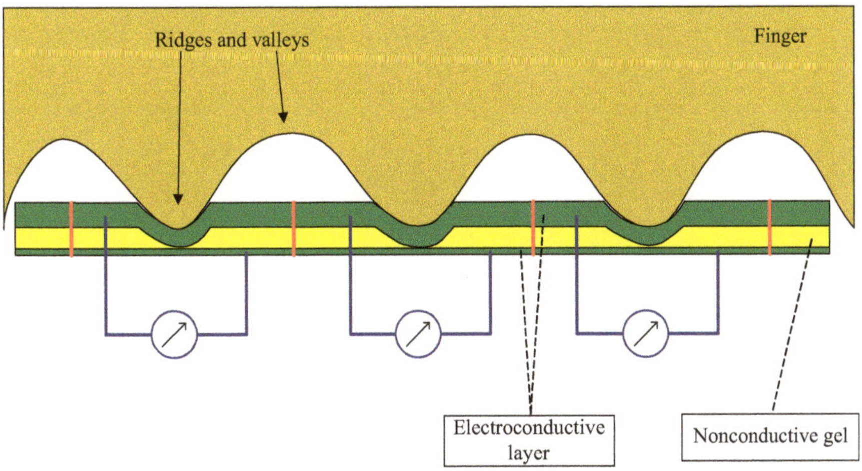

Figure 5.14 Principle of pressure technology

5.3.6 Thermal technology

The thermal technology principle is based on thermal radiation (see Figure 5.15). The papillary lines have a higher heat radiance than the valleys between them. The finger is swept over a pyroelectric cell that senses this thermal radiation.

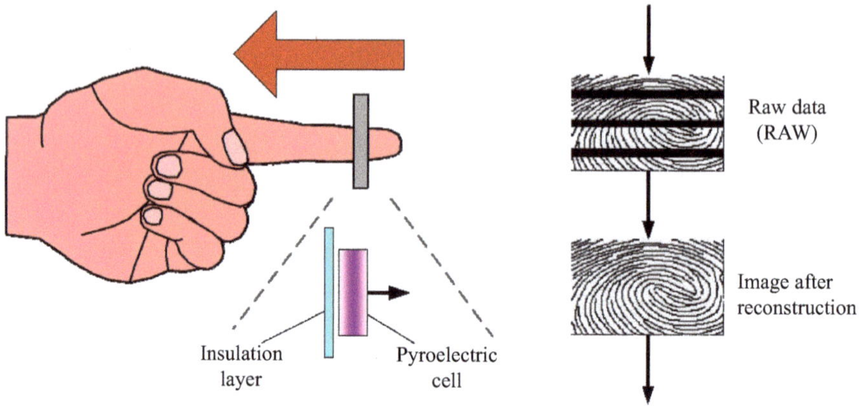

Figure 5.15 Principle of thermal technology

5.3.7 E-field technology

In this technology, the sensor consists of a drive ring and a matrix of antennas. The drive ring generates a sinusoidal radio frequency signal, and the matrix of active antennas receives that signal modulated by the skin structure, or, more precisely, modulated by the dermis structure, because the electric field passes the upper parts of the skin (the epidermis). Similarly to the ultrasonic technology, this technology is also resistant to fake fingers and ignores the dirt and light injuries on the finger. The image quality here is better than the one from capacitive or electro-optical sensors [17,25].

5.3.8 MEMS technology

The MEMS [25] uses micro parts to scan a fingerprint. One of the methods uses piezo-resistive micro beams. The user sweeps his finger along the sensor which consists of three rows of piezo-resistive gauges. Their parallel deflection will create a voltage variation which is measured and transformed into the fingerprint. The resulting image is only binary-colored, which is a big disadvantage of this type of sensor technology. This pressure-based MEMS swipe sensor principle can be seen in Figure 5.16. Another method is using microheaters. This method heats the finger a little bit and measures the change of temperature of the heat element. A ridge works as a heat sink, so the heat element which is connected to the ridge shows less of a rise in temperature [17,23,25].

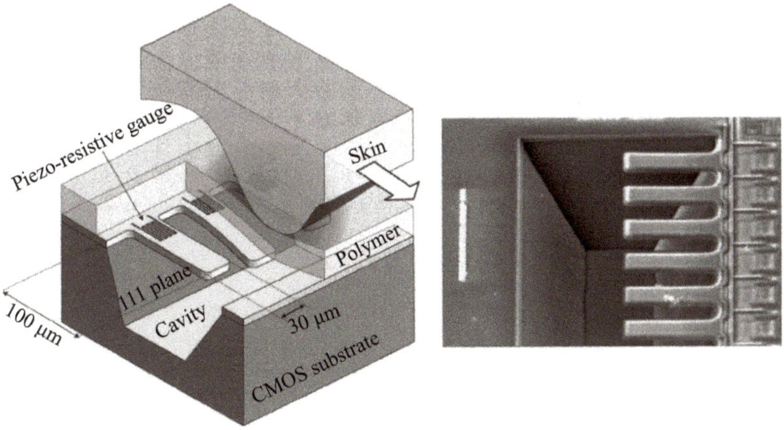

Figure 5.16 Principle of tactile MEMS technology

5.4 Fingerprint recognition

In general, there are two modes of use of the biometric system—registration and verification/identification. Registration usually takes place by the user having to place his or her registered finger three or five times on the sensor surface, and the system calculates the average of these fingerprints to produce the template with a high quality. Comparison (verification/identification) works similarly, but fingerprints can be mistaken either same fingerprints can be compared as different (due to intraclass variability) or similar fingerprints that come from totally different individuals (due to interclass variability). Therefore, the comparison process is not very simple. The fingerprint process is illustrated in Figure 5.17. This process consists of the steps described in the following subsections.

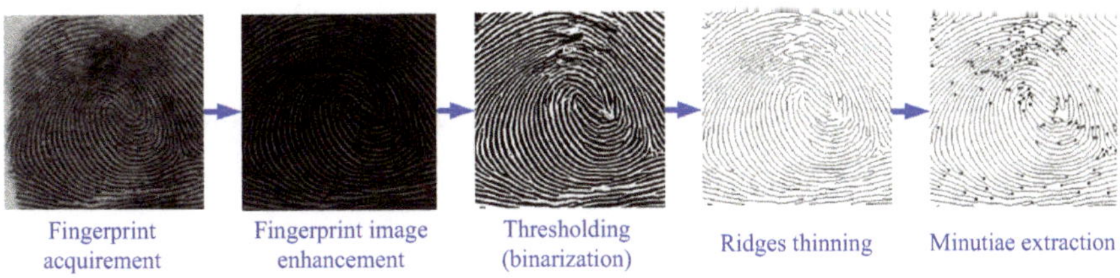

Figure 5.17 Scheme of the fingerprint processing

5.4.1 Fingerprint image and its processing

The first step of fingerprint processing is to capture (acquire) an input image, preprocess and extract papillary lines. Through this process, a proper description of the course of the papillary lines is obtained from the fingerprint image and can be further processed.

5.4.1.1 Image capture

Sensing a fingerprint image, such as getting a fingerprint image from a sensor (optical, capacitive, ...), or another template, gives us a basic, digital form of fingerprint. There is a large amount of noise in the input image, which requires subsequent enhancement. When scanning, it is necessary to distinguish between rolled/plain and possibly latent fingerprints. In addition, it is necessary to take care of the effects of damaged fingerprints, injuries, etc. It is necessary to check the liveness of the finger (antispoofing), whether there is not any spoof used instead of a real finger from a genuine user.

5.4.1.2 Orientation field

For the computation of the orientation field, it is necessary to calculate the direction of the papillary line from the surroundings (according to the gray color tone) at each point of the image. If the point is directly on the papillary line, it will most likely determine its direction. First, the orientation field for each pixel is calculated. In the second step, the transformation to the block orientation field is done. The block orientation field is then mapped on the original fingerprint image. A sample of the (block) orientation field is shown in Figure 5.18. Methods used are usually focusing on local analysis, global analysis or adaptive methods which are trying to combine both approaches [26].

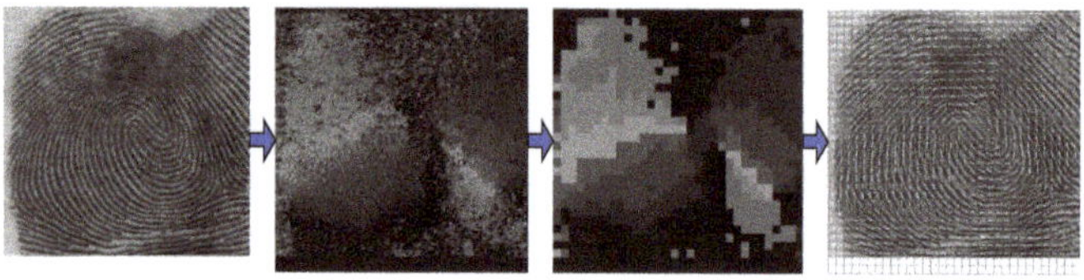

Figure 5.18 Calculation of the orientation field

5.4.1.3 Extraction of papillary lines

Further enhancement of the image and thresholding will provide black papillary lines and white background. This image enhancement includes, e.g., histogram scaling. This is related to the quality control of the input image [19]. The 2D Gabor function is used for filtering (often in the previous step—computation of orientation field). For filtering in the frequency domain (FFT → Filter Application → IFFT), the following filters are used:

Low Pass:

$$H(u,v) = \begin{cases} 1 & D(u,v) \leq D_0 \\ 0 & D(u,v) > D_0 \end{cases} \tag{5.1}$$

where D_0 is the border frequency and $D(u,v) = \sqrt{u^2 + v^2}$.
Filter Butterworth [27]:

$$H(u,v) = \frac{1}{1 + [D(u,v)/D_0]^{2n}} \tag{5.2}$$

Filter Ikonomopoulos [28]:

$$H_i = \begin{cases} 1 & \theta_i < \tan^{-1}(v/u) < \theta_{i+1} \wedge u^2 + v^2 > r_c^2 \\ 0 & \text{otherwise} \end{cases} \tag{5.3}$$

where u and v are frequency coordinates, n is number of directions and $\theta_i = (i-1)\pi/2n$.

For example, the *regional average thresholding* scheme [18], which first divides the image into 8×8 blocks, is used for image thresholding (except for other methods), then calculates the average level of gray in that area, then sets the value of the left 8×4 to this value and moves the operation window 4 points to the right. When the right edge is reached, the window moves 8 points down and starts again from the left.

5.4.1.4 Thinning of papillary lines

From the previous step, image of papillary lines is obtained, each of which could have various widths. The next step is to thin the papillary lines to a 1 point thickness. For thinning, a relatively simple algorithm is used to reduce the number of dots on the image of the papillary line so that its thickness will be only 1 point. The most commonly used method is based on Emyroglu [28], which uses two types of points (*ridge meeting point* and *ridge continuity point*). It must be true that the papillary line must not dwindle in any direction in order to avoid a problem with the position of the minutiae.

5.4.2 Detection and extraction of minutiae

For the detection and extraction of minutiae, the method of detecting papillary lines according to Hong, the so-called Hong method [29], is used (among other possible methods). This method is based on the fact that the papillary lines run parallel to each other and reach the maximum level of gray in the middle of the line (black/dark points). The fingerprint is multiplied by two h_t and h_b masks that have a 180° offset phase [29]:

$$h_t(i,j;u,v) = \begin{cases} -\dfrac{1}{\sqrt{2\pi}\delta} e^{-(u^2/\delta^2)} & \text{for } u = (v \cdot \cot(O(i,j))) - \dfrac{H}{2\cos(O(i,j))} \wedge v \in \Omega \\ \dfrac{1}{\sqrt{2\pi}\delta} e^{-(u^2/\delta^2)} & \text{for } u = (v \cdot \cot(O(i,j))) \wedge v \in \Omega \\ 0 & \text{otherwise} \end{cases} \tag{5.4}$$

$$h_b(i,j;u,v) = \begin{cases} -\dfrac{1}{\sqrt{2\pi}\delta} e^{-(u^2/\delta^2)} & \text{for } u = (v \cdot \cot(O(i,j)) + \dfrac{H}{2\cos(O(i,j))} \wedge v \in \Omega \\ \dfrac{1}{\sqrt{2\pi}\delta} e^{-(u^2/\delta^2)} & \text{for } u = (v \cdot \cot(O(i,j))) \wedge v \in \Omega \\ 0 & \text{otherwise} \end{cases} \tag{5.5}$$

$$\Omega = \left[-\left| \frac{L \sin(\mathrm{O}(i,j))}{2} \right|, \left| \frac{L \sin(\mathrm{O}(i,j))}{2} \right| \right] \tag{5.6}$$

On average, $L \times H$ is 11×7 [29]. Ideally, the mask width should be equal to the width of the local line. Point (i, j) is marked as a point of the papillary line if both values after convolutions (h_t and h_b filters) are larger than the threshold T_R. With respect to the parameter δ, both masks and smoothing are performed. In the resulting image, it is necessary to check and correct the damages of papillary lines and filling breaks in the papillary lines run.

Generally, two basic types of minutiae are detected: papillary *line ending* and *bifurcation*, while other types of minutiae are a combination of these two basic types. Detection decisions are made on the basis of the following conditions:

- If $\sum_{u=-1}^{1} \sum_{v=-1}^{1} T_R(i+u, j+v) = 2$, then it is a *line ending*.
- If $\sum_{u=-1}^{1} \sum_{v=-1}^{1} T_R(i+u, j+v) > 3$, then it is a *bifurcation*.

This situation is schematically illustrated in Figure 5.19. Both conditions mean, in essence, that the sum of points in the neighborhood is made, and if it is equal to 2, it is a line ending, and if it is greater than 3, it is a bifurcation.

The following data is stored for each minutia: the *position* of the minutia (coordinates x and y), the *type* of the minutia (line ending/bifurcation) and the *gradient* (the orientation of the papillary line).

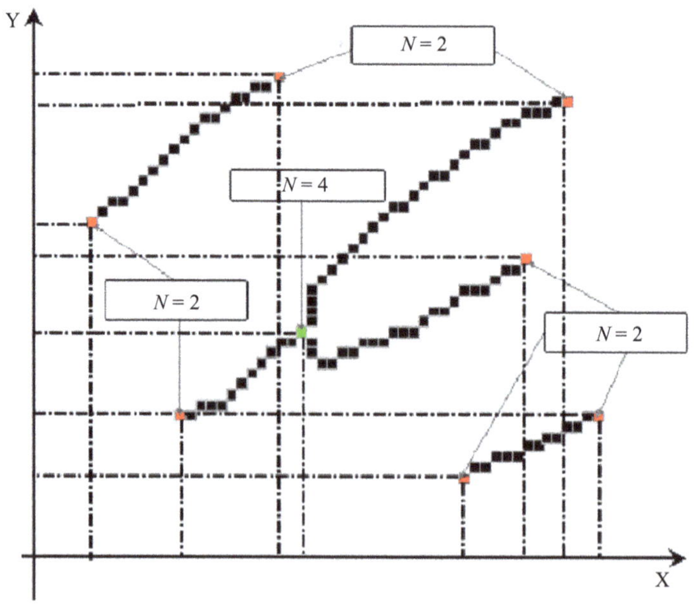

Figure 5.19 Representation of the detection of minutiae on thinned papillary lines

5.4.2.1 Methods of fingerprint recognition

The result of the extraction of minutiae is compared with the stored template from the database, smart cards, etc. The methods for comparing fingerprints are as follows:

- Methods based on minutiae [3,17,30]
  - They use position, type and gradient (direction)
  - Generally, issues of pattern comparison.
- Methods based on correlation [31,32]
  - 2D correlation between input and template
  - Computationally demanding but implementable in hardware.
- Methods based on the properties of papillary lines [33]
  - Orientation and frequency of papillary lines, line shape, texture information, etc.
  - Low resolution, but often used as support information for the minutiae-based method.
- Methods based on high resolution or 3D properties of the finger or papillary lines [34]
  - Usage level of 3D features e.g., sweat pores placement and distribution, ridge edge features, etc.
  - Usage of 3D features such as curvature of the finger, depth values, etc.
  - Can be combined with traditional 2D methods mentioned above.

5.4.2.2 Recognition based on minutiae

The method based on minutiae is the most commonly used method. This is practically a problem of comparing patterns—two sets of minutiae. Two basic methods for comparison using the minutiae are *Hong method* [35] and *Ratha method* [36]. Both methods are based on two main steps: generating a global overlap (alignment) and locating a local overlap (compare)—see Figure 5.20. A schematic representation of the course of the methods is shown in Figure 5.21.

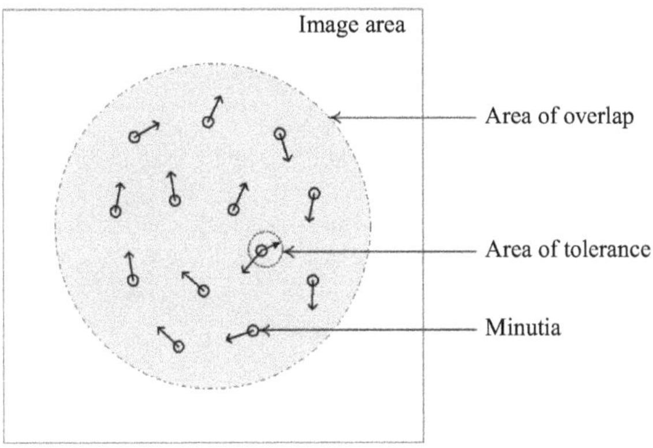

Figure 5.20 Areas of overlap and tolerance

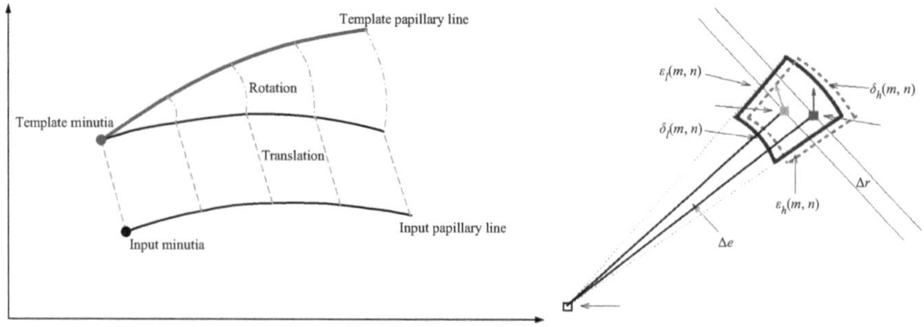

Figure 5.21 A schematic representation of the comparison based on minutiae

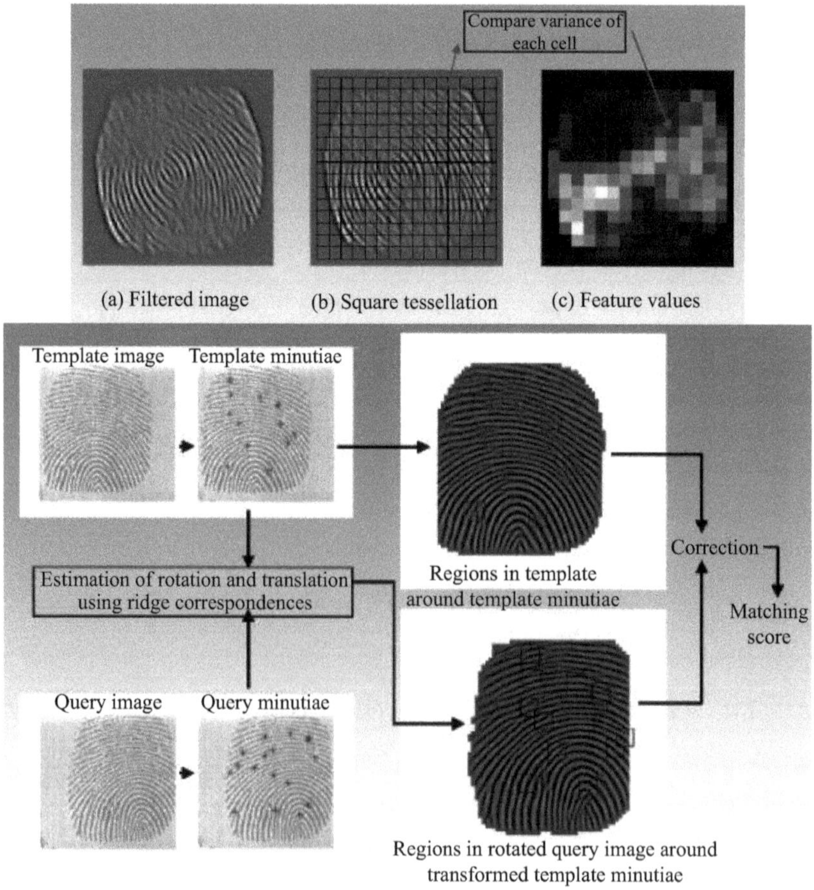

Figure 5.22 Recognition based on the properties of papillary lines (top) and recognition based on correlation (down) [39]

The more recent methods are focused on their portability on light architectures (like smart card or system-on-a-chip) or to suitability for template-protection techniques (necessity of fixed-length, alignment free, noise-tolerant feature coding). Example can be minutia cylinder-code which encodes spatial and directional relationships between minutia and its neighborhood [30].

5.4.2.3 Recognition based on the properties of papillary lines

The procedure for the method based on the properties of the papillary lines is shown in Figure 5.22. This is a description of peculiarities (curvature) of the papillary lines, which are then used for comparison.

5.5 Difficult fingerprints

From the beginning of use of fingerprints, it is fully obvious that sometimes the fingerprint has a very low quality. First of all, phenomena that are connected to the finger itself will be discussed—the *dirt on the finger* which can be caused by a few grains of dust, small particle or a greasy finger from some meal. Liquids or generally conductive materials are one of the most problematic types of dirt. For every sensor technology, the most problematic type of dirt is different. These phenomena can be in conjunction with the *dry or moist finger*. Sweaty fingers for example could be assigned to both groups. The effect of moisture or dryness of the finger is very significant. It is also a very frequent way of damaging the fingerprint. Investigated user can be very nervous which leads to sweaty or extremely dry fingers. On the other hand, in the everyday usage, fingers can be moist after using some lotion or they can be simply recently washed. The skin resistance which is important for some sensors can be up to ten times higher or lower than the average value. The *physically damaged finger* is common in some groups of users, namely, among people who are employed in manual work. In their jobs, some abrasions or cuts are inevitable. If it is only a small injury or everyday wear damage, the papillary lines will regenerate. On the contrary, a deep wound will affect the papillary lines forever. Fingers can be also damaged by *skin diseases*—this will be described in Chapter 7 [37,38].

The second part of this category is the phenomena caused by users when acquiring the image. One of them, usually caused by inexperienced users, is small or insufficient *contact region*. That simply means that the finger was presented to the sensor in a way which allows only a partial acquisition of the fingerprint. The small contact region can also be caused by extremely low *pressure*. The other way around, very high pressure can create a black oval instead of the fingerprint image. When the pressure is somewhere between these extremes, it can cause either very thick or very thin papillary lines. It is very sensor dependent what the "right" pressure is. There is an endless number of movements that users can do when showing the *noncooperative behavior*. It can be dynamic or static behavior which is done on purpose, i.e., it is not an accident. The line between an unintentional wrong usage and a noncooperative behavior is very thin. Nevertheless, there are users who resent biometric systems or just try to push to its limits. They can move the finger or change the pressure while the sensor is acquiring image. The other possibility is the static behavior like a rotation of the finger, a small contact region with just a side of the finger, etc. The details about damaged fingerprints will be discussed in Chapter 7.

5.6 Fingerprint related standards

The most standards generally affect all biometrics, including fingerprints. Still, there are some specific standards or parts of them that are tightly bound only to fingerprints. Specific fingerprint standards are as follows (these are the most important):

- ISO/IEC 19794-2: Information technology—biometric data interchange formats—finger minutiae data, September 2005 and December 2011.
- ISO/IEC 19794-3: Information technology—biometric data interchange formats—finger pattern spectral data, August 2006.
- ISO/IEC 19794-4: Information technology—biometric data interchange formats—finger pattern skeletal data, October 2006.
- ISO/IEC 19794-8: Information technology—biometric data interchange formats—finger pattern skeletal data, December 2011.
- ISO/IEC 20027: Biometrics interoperability profiles—best practices for slap ten-print captures, September 2015.
- ISO/IEC 24779-4: Information technology—cross-jurisdictional and societal aspects of implementation of biometric technologies—pictograms, icons and symbols for use with biometric systems—fingerprint applications, January 2017.
- ISO/IEC 29198: Information technology—biometrics—characterization and measurement of difficulty for fingerprint databases for technology evaluation, December 2013.
- ANSI/INCITS 381-2009: ANSI—finger-image-based interchange format, NIST, 2009 (revision of ANSI/INCITS 381-2004, May 2004).
- ANSI/INCITS 377-2009: ANSI—finger-pattern-based interchange format, NIST, 2009 (revision of ANSI/INCITS 377-2004 from February 2004).
- ANSI/INCITS 378-2009: ANSI—finger minutiae format for data interchange, NIST, 2009 (revision of ANSI/INCITS 378-2004 from February 2004).
- ANSI/INCITS 398/2008: ANSI—common biometric exchange formats framework (CBEFF), NIST, 2008 (revision of ANSI/INCITS 398-2005 from 2005).
- ANSI/NIST CSL 1-1993: Data format for the exchange of fingerprint information, ANSI, November 1993.
- ANSI/NIST ITL 1-2007: Data format for interchange of fingerprint, facial and other biometric information, NIST, May 2007 (revision of ANSI/NIST ITL 1-2000 from September 2000).
- IAFIS-IC-0010: Electronic fingerprint specification, FBI, December 1995.
- IAFIS-IC-0110V2: WSQ—fingerprint script compression specifications, FBI, February 1993.

- CJIS-IC-0020: WAN interface specification (fingerprint transfer), FBI, November 1995.
- Further parts of the ISO/IEC 19795 standards, which deal with biometric systems in general, but also specify some specific requirements for fingerprints.

5.7 Commercial devices and applications

The focus will be on commercial devices that are available on the market. Their distribution will be based on the scanning technologies of the individual devices. Firstly, *optical* technology—examples are shown in Figure 5.23 (left to right: Identix BioTouch® 500; Sagem MorphoSmart Optic 300; BioLink MatchBook 3.5). Other technologies are *capacitive* sensors, as shown in Figure 5.24 (from left to right: Veridicom 5th Sense; Fujitsu MBF200; UPEK TouchStrip™ TCS3-TCD4). Follow the *ultrasound* and *e-field* technology—see Figure 5.25 (left to right: UltraScan; Optel, the right-hand side is Bio-i CYTE). *Pressure* and *temperature* sensitive technologies are also shown—see Figure 5.26 (left to right: BMF BLP-100; Fidelica FIS-3002; Atmel® AT77C104B is thermal (right)). Lastly *RF* sensors and prototype of *OCT technology* can be seen in Figure 5.27 (from left to right: Swipe Sensor Development Kit FPC-SSD with Fingerprints FPC1080A Swipe Sensor, Zvetco Verifi P5100 with AuthenTec TCS1 sensor and lastly OCT prototype from Langevin Institute in Paris [40]).

Fingerprint information can also be used to generate a key that can be used for cryptographic purposes. This is called a biometric security system. Figure 5.28

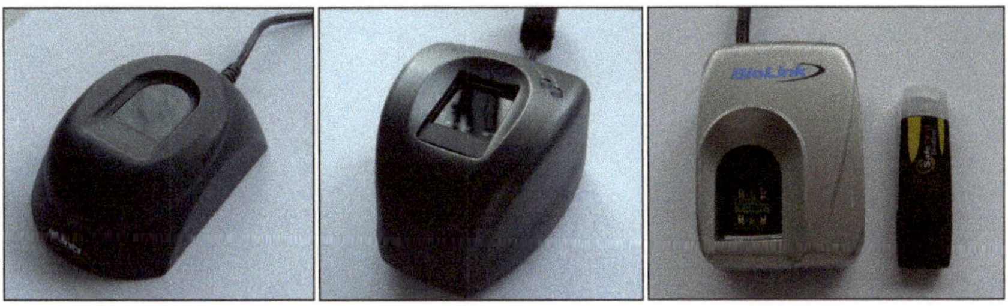

Figure 5.23 Commercial devices—optical technology

Figure 5.24 Commercial devices—capacitive technology

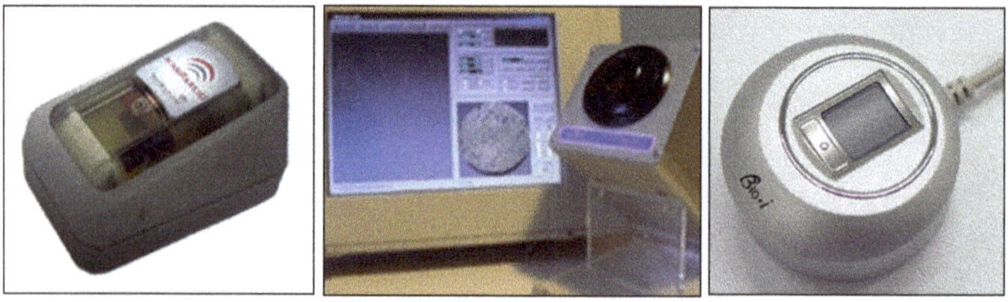

Figure 5.25 Commercial devices—ultrasound and e-field technology

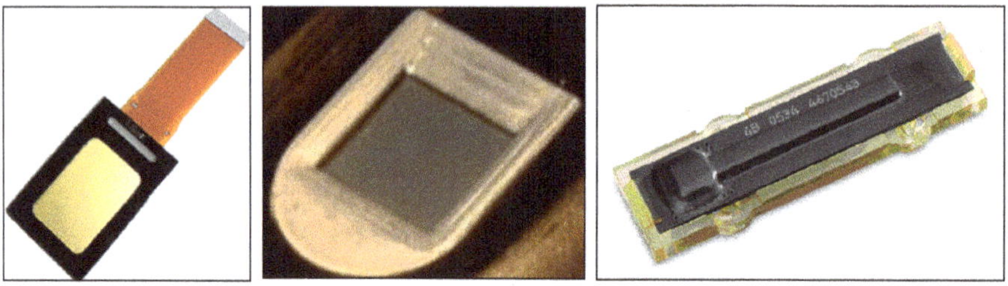

Figure 5.26 Commercial devices—pressure and temperature sensitive technology

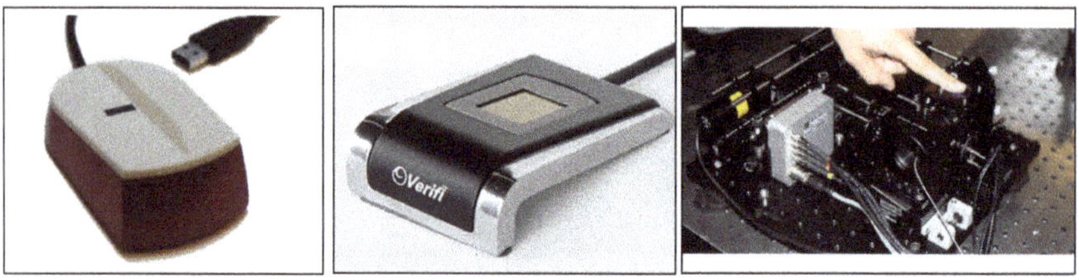

Figure 5.27 Commercial devices—RF and OCT technology

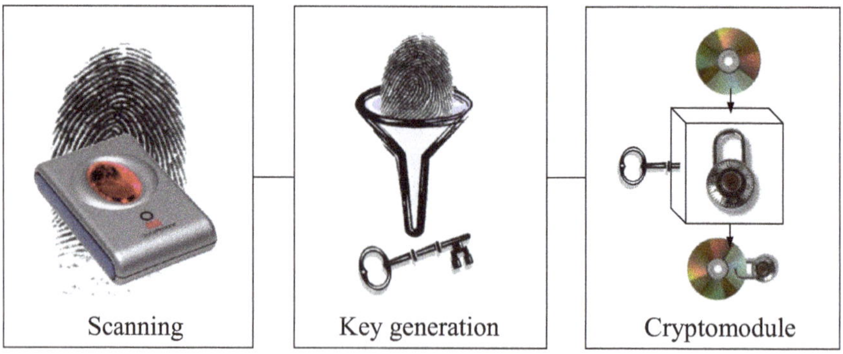

Figure 5.28 Biometric security system

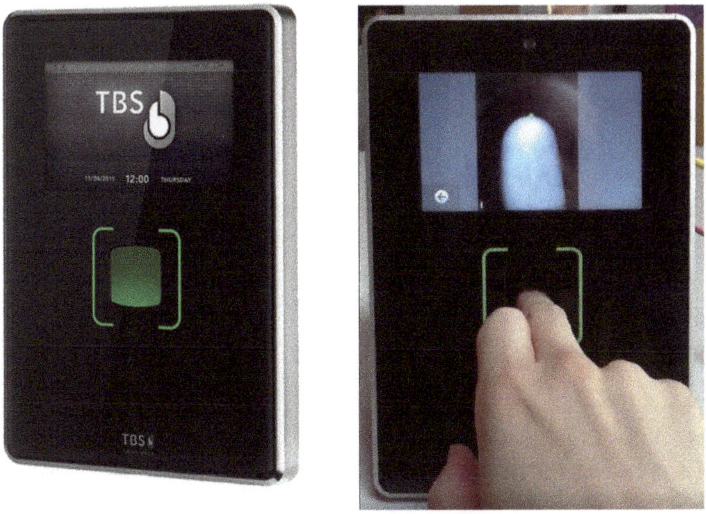

Figure 5.29 TBS 3D terminal

shows the process of generating the key from the fingerprint. First fingerprint is loaded. Then, the key generation process takes place using the above methods, and then the information is transformed into a key that is passed to the last module that uses the key to encrypt. More information can be found in [3].

An interesting application is the 3D fingerprint reader from the company Touchless Biometric Systems AG—see Figure 5.29. This sensor contains three cameras that acquire your finger at different angles, including structured multispectral light. From these three images supported by projected patterns, it is possible to reconstruct a 3D fingerprint image with a 3D profile of the papillary lines visible on the images. It is also a contactless sensor—the finger is only placed on the edge, the fingertip is in the air without touching the sensor. A great advantage is the nondistortion of the course of the papillary lines by pressure, which often causes a distortion to them by a variety of pressures or sweeping of the finger across the surface of the sensor.

Of course, there is another line of exciting industry solutions, such as fingerprint flash drives. However, these specific solutions go beyond the original intention of this book. Some interesting examples can be found, for example, in [3] or [41].

5.8 Conclusion

Fingerprints are currently the most widely used biometric technology. Their enforcement is supported by many factors—they are well accepted by users, they have high biometric entropy (they are suitable for distinguishing large numbers of people), technology is advanced (reliability is high), fingerprints are accepted from legal point of view (fingerprints serve as evidence in court) and, moreover, the sensors are already too small and have little consumption that can be integrated into mobile devices. It is, of course, a question of whether the development of other biometric technologies will cause users to lose interest in this technology or not.

If we neglect forensic (criminal) purposes of use of fingerprints, our fingerprints are often used in special scenarios not belonging to the classical area of identity verification. These exceptions include a variety of fingerprint or fingerprint access for cryptographic purposes for accessing notebooks or external storage media. In this case, fingerprint is used as a key, but it is disadvantageous when fingerprint is revealed—then the key is revealed directly as well. A new trend is to combine fingerprint with cryptographic key so that a mathematical combination is used for encryption, with only the knowledge of the key or knowledge of the fingerprint being insufficient to perform the required cryptographic transaction. Of course, this principle can be also used with other biometric features, but fingerprint was the first use.

A complete conclusion is that the area of fingerprints is one of the most explored and experienced biometrics. When selecting this biometric property for verifying or identifying people, you cannot make a practical mistake—it is of course desirable to choose the right technology vendor who has experience in the field and has undergone independent testing. Integrated liveness detection should be integrated in the sensor, of course. Most manufacturers are already doing this, and liveness detection really becomes an obvious part of fingerprint reader hardware.

Acknowledgment

This work was supported by The Ministry of Education, Youth and Sports of the Czech Republic from the National Programme of Sustainability (NPU II); project IT4Innovations excellence in science—LQ1602.

References

[1] *Global Biometrics Market by Type, by End Use Sector, by Region, Competition Forecast and Opportunities 2011–2021*, TechSci Research, p. 13, 2016. Available online: https://www.slideshare.net/TechSci_Research/biometrics-market-size-share-2021-brochure?from_action=save.

[2] Pankanti S., Prabhakar S., Jain A.K. *On the individuality of fingerprints*. IEEE Transactions on Pattern Analysis and Machine Intelligence, 2002, 24.8: 1010–1025.

[3] Drahanský M. *Biometric security systems fingerprint recognition technology*. Brno: VUTIUM, 2005.

[4] Bundesamt für Sicherheit in der Informationstechnik (Federal Office for Information Security of Germany). *Evaluation of fingerprint recognition technologies—BioFinger*. Darmstadt: Fraunhofer IGD, 2004.

[5] Wilson C.L., Grother P.J., Micheals R.J., *et al. Fingerprint vendor technology evaluation 2003: Summary of results and analysis report*. NIST Technical Report NISTIR, 2004, 7123.

[6] Herschel W.J. *The origin of finger-printing*. London: H. Milford, Oxford University Press, 1916.

[7] Galton F. *Hereditary talent and character*. Macmillan's Magazine, 1865, 12.157–166: 318–327.

[8] Krishan K. *Anthropometry in forensic medicine and forensic science—'Forensic Anthropometry'*. The Internet Journal of Forensic Science, 2007, 2.1: 95–97.

[9] Hauptvogel K.H., Ritzschke M. *Biometrie um die Jahrhundertwende*. Berlin: Humboldt University of Berlin, 2004. Available online: http://www2.informatik.hu-berlin.de/~ritzschk/paper/bertillon_1.pdf.

[10] Galton F. *Fingerprints*. London: Macmillan and Co., 1892.

[11] Straus J., Porada, V., *et al. Kriminalistikcá daktyloskopie (Criminalistic dactyloscopy)*. Prague: Police Academy of the Czech Republic, 2005.

[12] Komarinski P. *Automated fingerprint identification systems (AFIS)*. Burlington, MA: Academic Press, 2005.

[13] U.S. Department of Justice, Federal Bureau of Investigation, Criminal Justice Information Services Division. *Next Generation Identification Flyer*, [cit. 2017-08-11]. Available online: https://www.fbi.gov/file-repository/about-us-cjis-fingerprints_biometrics-biometric-center-of-excellences-ngi-one-pager-final.pdf.

[14] U.S. Department of Justice, Federal Bureau of Investigation, Criminal Justice Information Services Division. *Next Generation Identification (NGI) Monthly Fact Sheet*, [cit. 2017-08-11]. Available online: https://www.fbi.gov/file-repository/ngi-monthly-fact-sheet.

[15] Jozefek A. *Principy některých daktyloskopických klasifikačních systémů* (*Principles of some dactyloscopic classification systems*). Prague: Ústav kriminalistiky Právnické fakulty University Karlovy (Department of Criminology, Faculty of Law, Charles University), 1972.

[16] Collins C.G. *Fingerprint science*. USA: Custom Publishing Company, 1985. Available online: http://www.handanalysis.co.uk/EdCampbell-PalmD-History.htm.

[17] Maltoni D., Maio D., Jain A.K., Prabhakar S. *Handbook of fingerprint recognition*. London: Springer Science & Business Media, 2009.

[18] Drahanský M. *Fingerabdruckerkennung mittels neuronaler Netze*. MSc. thesis, FernUniversität in Hagen, 2001.

[19] Drahanský M., Dřezinová E., Orság F., Lodrová D. *Classification of skin diseases and their impact on fingerprint recognition*. In: BIOSIG, 2009. p. 173–176.

[20] Auksorius E., Boccara A.C. *Fingerprint imaging from the inside of a finger with full-field optical coherence tomography*. Biomedical Optics Express, 2015, 4465–4471. DOI 10.1364/BOE.6.004465.

[21] Setlak D. *Electric field fingerprint sensor apparatus and related methods*. International patent US 5963679, May 10, 1999.

[22] Koundinya P., Theril S., Feng T., Prakash V., Bao J., Shi W. *Multi-resolution touch panel with built-in fingerprint sensing support*. In: Design, Automation and Test in Europe Conference and Exhibition (DATE), 2014. DOI 10.7873/DATE.2014.258.

[23] Mainguet J.F. *Personal Website—Fingerprint*, [cit. 2015-26-02]. Available online: http://fingerchip.pagesperso-orange.fr/biometrics/types/fingerprint.htm.

[24] Apple Insider. *Apple Preparing Software Update to Enhance Functionality of iPhone 5s Touch ID*, [cit. 2016-13-12]. Available online:

http://appleinsider.com/articles/14/02/28/apple-preparing-software-update-to-enhance-functionality-of-iphone-5s-touch-id.

[25] Drahanský M. *Fingerprint recognition technology—related topics*. LAP LAMBERT Academic Publishing GmbH & Co. KG, 2011, p. 172.

[26] Turroni F., Maltoni D., Cappelli R., Maio D. *Improving fingerprint orientation extraction*. IEEE Transactions on Information Forensics and Security, 2011, 6.3. DOI 10.1109/TIFS.2011.2150216.

[27] *Butterworth filter*. Available online: https://en.wikipedia.org/wiki/Butterworth_filter.

[28] Emyroglu Y. *Fingerprint image enhancement & recognition*. Turkey: Yldyz Technical University, 1997.

[29] Hong L., Wan Y., Jain A. *Fingerprint image enhancement: Algorithm and performance evaluation*. IEEE Transactions on Pattern Analysis and Machine Intelligence, 1998, 20.8: 777–789.

[30] Cappelli R., Ferrara M., Maltoni D. *Minutia cylinder-code: A new representation and matching technique for fingerprint recognition*. IEEE Transactions on Pattern Analysis and Machine Intelligence, 2010, 32.12. DOI 10.1109/TPAMI.2010.52.

[31] Bazen A.M., Verwaaijen G.T.B., Gerez S.H., Veelenturf L.P.J., Zwaag van der B.J. *A correlation-based fingerprint verification system*. In: ProRISC 2000 Workshop on Circuits, Systems and Signal Processing, Veldhoven (NL), 2000.

[32] Koichi I., Hiroshi N., Koji K., Takafumi A., Tatsuo H. *A fingerprint matching algorithm using phase-only correlation*. IEICE Transactions on Fundamentals, 2004, E-87A.3: 682–691.

[33] Krivec V., Birchbauer J., Marius W., Bischof H. *A hybrid fingerprint matcher on card*. In: Proceedings of the 1st Conference on Biometrics and Electronic Signatures of the GI Working Group BIOSIG, 2003.

[34] Liu F., Zhang D., Shen L. *Study on novel curvature features for 3D fingerprint recognition*. Neurocomputing, 2015, 168: 599–608. DOI https://doi.org/10.1016/j.neucom.2015.05.065.

[35] Hong C., Jie T., Xin Y. *Fingerprint matching with registration pattern inspection*. In: Proceedings of the 4th International Conference on Audio- and Video-based Biometric Person Authentication, Guildford (UK), 2003.

[36] Ratha N., Bolle R. *Automatic fingerprint recognition systems*. New York: Springer-Verlag, 2004.

[37] Kanich O. *Fingerprint damage simulation*. Saarbrücken: Lambert Academic Publishing, 2014.

[38] Kanich O., Drahanský M. *Simulation of synthetic fingerprint generation using Petri nets*. IET Biometrics, 2017, 6.6: 402–408. DOI 10.1049/iet-bmt.2016.0041.

[39] Jain A.K. *CSE 891—Selected Topics: Biometrics*. 2004. Available online: https://is.muni.cz/el/1433/jaro2013/PV204/um/03bio/NonminutiaeRepresentations.pdf.

[40] Photonics. *OCT System Captures "Internal" Fingerprint*. [cit. 2017-08-11]. Available online: https://www.photonics.com/Article.aspx?AID=58100.

[41] Drahanský M. *Nutzung biometrischer Daten zur Gewinnung personenbezogener krypto-graphischer Schlüssel*. Darmstadt (DE), 2004.

Chapter 6
Large-scale fingerprint recognition accelerated in hardware*

Raffaele Cappelli[1], *Matteo Ferrara*[1], *and Davide Maltoni*[1]

To make Automatic Fingerprint Identification Systems (AFIS) capable of searching across several millions of fingerprints in a few seconds, very powerful (and expensive) distributed computing architectures are typically used. The recent improvement of algorithms and the availability of powerful CPUs and GPUs now make it possible to deploy large-scale fingerprint recognition on low-cost hardware, thus approaching a larger number of applications (e.g., welfare benefits in poor countries). This chapter discusses architectural design, algorithms, and hardware optimization to speed up fingerprint recognition on large databases.

6.1 Introduction

Due to the computational-demanding nature of fingerprint comparison, designing large-scale fingerprint-based identification systems is very challenging in terms of efficiency. The time required to search a fingerprint on a database grows with the size of the database itself: when the database contains millions of fingerprints, very expensive hardware platforms are required to meet the required throughput. This is the case of the AFIS owned by police agencies such as Federal Bureau of Investigation (FBI), or the huge civil identification systems being deployed in emerging countries (e.g., UIDAI project [1]). To increase the speed of fingerprint identification, there are basically two options:

- reducing the processing time (e.g., by designing algorithms with a lower computational complexity [2], or using parallel architectures [3–8]);
- reducing the total number of fingerprint comparisons (through fingerprint classification [9,10], prefiltering or multistage matching [11–13]).

*Portions reprinted from Information Sciences, vol. 306, R. Cappelli, M. Ferrara and D. Maltoni, "Large-scale fingerprint identification on GPU", pp. 1–20, June 2015, Copyright (2017), with permission from Elsevier.

[1]Department of Computer Science and Engineering, University of Bologna, Italy

This chapter discusses both the above alternatives and describes the design, implementation, and experimental evaluation of fingerprint comparison and indexing algorithms that have been optimized for specific hardware architectures. In order to perform experiments in a realistic scenario, the prototype of a complete identification system has been implemented (Figure 6.1). It consists of two main modules:

- *Identification query manager*—maintains a queue of the queries received from the clients; submits each request to the identification engine; provides the results to the clients.
- *Fingerprint identification engine*—performs 1:*N* identification of the query template against the templates in the database and provides the final result (a list of candidates).

The fingerprint identification engine is based on two steps (Levels 1 and 2): the former operates a preselection of the 1% database templates most similar to the query one using a fast technique; the latter applies an accurate fingerprint matcher to produce the final list of candidates. The first step is critical: it has to be very fast but accurate enough to avoid missing candidates. Two different algorithms have been designed and evaluated for this purpose: a fast matcher based on the Local Similarity Sort (LSS) measure (see Section 6.3.1) and an indexer based on the Hamming Distance Similarity (HDS) measure (see Section 6.3.2). For both algorithms, a baseline algorithm and an optimized version targeting specific hardware architectures are described.

The rest of this chapter is organized as follows: Section 6.2 introduces the main features of fingerprint patterns and the Minutia Cylinder-Code (MCC) representation [14], which is common to all the algorithms experimented. Section 6.3 describes the similarity measures adopted by the various algorithms, while the

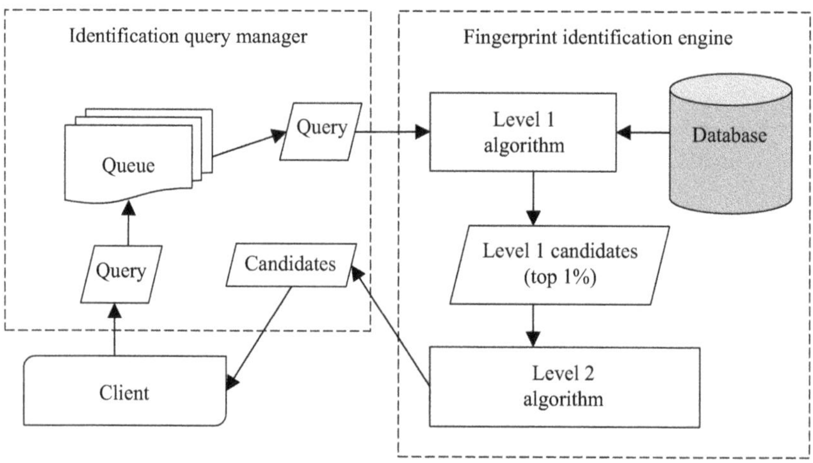

Figure 6.1 A simple schema of the main modules of the prototype system

matching algorithms and the indexer are discussed in Sections 6.4 and 6.5, respectively. Section 6.6 reports experimental results and includes a comparison with existing published approaches. Finally, Section 6.7 draws some conclusions.

6.2 Fingerprint patterns and MCC representation

A *fingerprint* is the representation of the epidermis of a finger: It consists of a pattern of interleaved ridges and valleys (Figure 6.2). Discontinuities in the ridges (e.g., terminations or bifurcations) are called *minutiae* (Figure 6.2). Each minutia can be described by some attributes, including its location in the fingerprint, its direction, its type (termination or bifurcation), and a value representing the quality of the fingerprint pattern in its neighborhood. Most minutia-based comparison algorithms simply consider each minutia as a triplet $m = \{x, y, \theta\}$ that indicates its coordinates and direction.

All the fingerprint comparison algorithms experimented in this chapter are based on minutiae and precisely on a particular representation of minutiae named MCC [14]. Let $MT = \{m_i\}$ be a set of minutiae extracted from a fingerprint: each minutia is a triplet $m_i = (x_i, y_i, \theta_i)$ where (x_i, y_i) is the minutia location and θ_i is the minutia direction (in the range $[0, 2\pi[$). The MCC representation associates a local descriptor (the cylinder) to each minutia m_i: this descriptor encodes spatial and directional relationships between the minutia and its neighborhood of radius R and can be conveniently represented as a cylinder, whose base and height are related to the spatial and directional information, respectively [see Figure 6.3(a)]. The cylinder is divided into N_D sections: each section corresponds to a directional difference in the range $[-\pi, \pi[$; sections are discretized into $N_S \times N_S$ cells (N_S is the

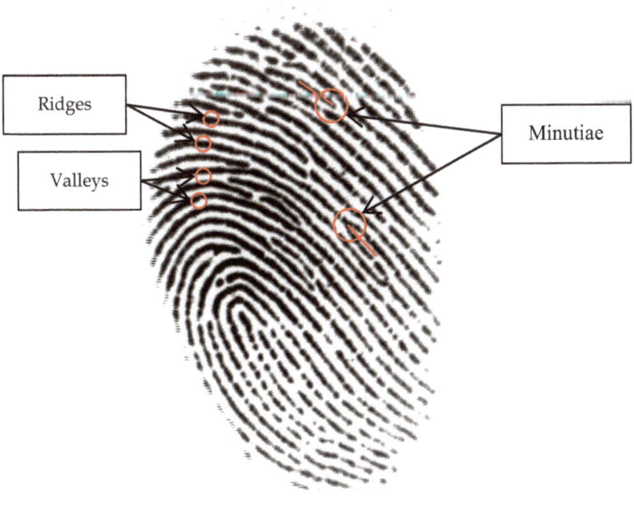

Figure 6.2 A fingerprint with ridges, valleys, and minutiae highlighted

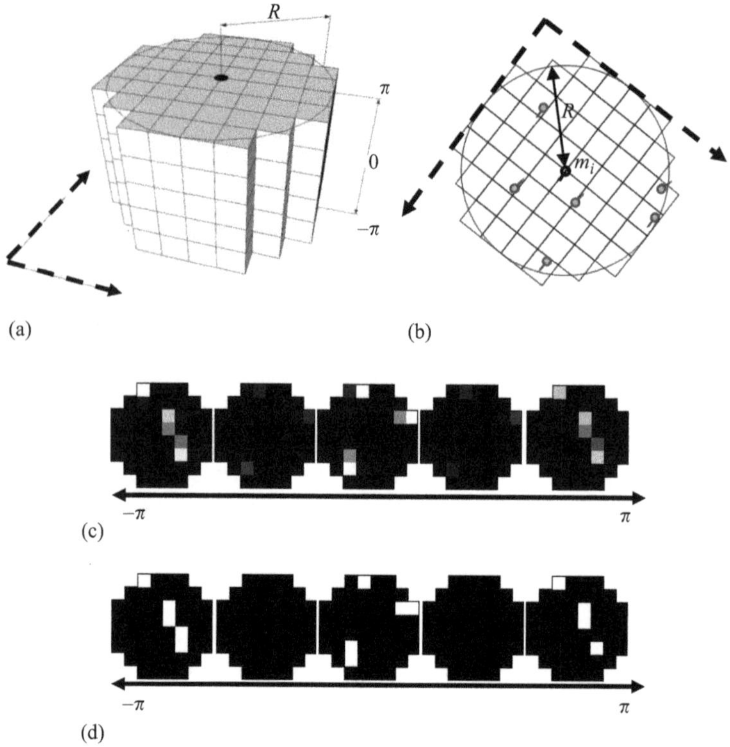

Figure 6.3 *(a) A graphical representation of the local descriptor associated to a minutia in the MCC representation, with $N_S = 8$ and $N_D = 5$; (b) minutiae involved in a cylinder; (c) cell values in the N_D sections (lighter areas represent higher values) of the cylinder built over the minutiae in (b); (d) binarized cell values stored as bits. Note that cylinder sections in (c) and (d) are rotated according to the direction of minutia m_i*

number of cells along the section diameter). During the cylinder creation, a numerical value is calculated for each cell, by accumulating contributions from minutiae in a neighborhood of the projection of the cell center onto the cylinder base. The contribution of each minutia m_t to a cell (of the cylinder corresponding to a given minutia m_i) depends both on

- spatial information (how much m_t is close to the center of the cell) and
- directional information (how much the directional difference between m_t and m_i is similar to the directional difference associated to the section where the cell lies).

In other words, the value of a cell represents the likelihood of finding minutiae that are close to the cell and whose directional difference with respect to m_i is similar to a given value. Figure 6.3(b) and (c) shows the cylinder associated to a minutia with six minutiae in its neighborhood.

Once a cylinder is built from a minutia m_i, it can be simply treated as a single feature vector, obtained by linearizing the cell values. With a negligible loss of accuracy (see [13–15]), each element of the feature vector can be stored as a bit [Figure 6.3(d)]: in the following, $\mathbf{v}_i \in \{0,1\}^n$ denotes an MCC bit-vector obtained from minutia m_i, and $T = \{c_i\}$, denotes an *MCC template* obtained from a fingerprint, where each $c_i = (\mathbf{v}_i, x_i, y_i, \theta_i)$ is a tuple containing a bit-vector and its associated minutia information. Note that although strictly speaking, the term "cylinder" corresponds to $\mathbf{v}_i$ [14], in the following, for simplicity, it denotes the whole tuple c_i. Each bit-vector $\mathbf{v}_i$ is a fixed-length local descriptor that is well suited for implementing hardware-accelerated algorithms; in fact it is:

- invariant for translation and rotation, since (i) it only encodes distances and directional differences between minutiae and (ii) its base is rotated according to the corresponding minutia angle;
- robust against skin distortion (which is small at a local level) and against small feature extraction errors, thanks to the smoothed nature of the functions defining the contribution of each minutia.

6.3 Similarity measures

Among the various similarity measures that can be adopted for comparing cylinders [14], the following three have been selected for this work: LSS (used for fast matching at Level 1), LSS–Distortion-Tolerant-Relaxation (DTR) (used for accurate matching at Level 2), and HDS (used for indexing at Level 1).

6.3.1 LSS and LSS–DTR measures

A simple but effective similarity measure between two cylinders $c_i = (\mathbf{v}_i, x_i, y_i, \theta_i)$ and $c_j = (\mathbf{v}_j, x_j, y_j, \theta_j)$ is

$$s_L(c_i, c_j) = \begin{cases} 1 - \dfrac{\|\mathbf{v}_i \oplus \mathbf{v}_j\|}{\|\mathbf{v}_i\| + \|\mathbf{v}_j\|} & \text{if } d_\phi(\theta_i, \theta_j) \le \delta_\theta \\ 0 & \text{otherwise} \end{cases} \tag{6.1}$$

where $\oplus$ denotes the bitwise XOR operator; $\|\cdot\|$ denotes the Euclidean norm; $d_\phi(\theta_i, \theta_j)$ is the difference between the two angles; and δ_θ is a parameter controlling the maximum rotation allowed between two fingerprints.

Note that (6.1) is a *local* similarity measure between two cylinders: in order to compare two fingerprint templates T_A and T_B, a single value (*global score*), denoting the overall fingerprint similarity, has to be obtained from the pairwise (local) cylinder similarities. Various global similarity measures have been proposed for MCC [14–16]; the simplest and most efficient one is the LSS, which is calculated as the average of the top n_p local similarity scores between cylinders of the two templates. The value of n_P is not an overall constant, but it partially depends on the number of minutiae in the two templates:

$$n_P = \min_{n_P} + \left\lfloor \frac{(\max_{n_p} - \min_{n_p})}{(1 + e^{-\tau_P \cdot (\min\{|T_A|,|T_B|\} - \mu_P)})} \right\rceil \tag{6.2}$$

where $\mu_P, \tau_P, \min_{\min_P}$, and $\max_{n_p}$ are parameters introduced in [14] and $\lfloor \cdot \rceil$ denotes the rounding operator.

A more accurate, but less efficient, similarity measure is the *LSS–DTR* [16]. LSS–DTR adds a consolidation step to LSS, in order to obtain a score that reflects to what extent the local similarities hold at global level. LSS–DTR is based on compatibility measures between candidate pairs of corresponding minutiae and uses spatial and directional features that are invariant for rotation/translation and tolerate skin distortion (Figure 6.4).

6.3.2 HDS measure

Given two templates T_A and T_B, containing n-dimensional MCC binary vectors, due to the invariant nature of the MCC features, a simple but effective similarity measure between T_A and T_B can be defined as follows:

$$hds(T_A, T_B) = \frac{\sum_{\mathbf{v} \in T_A} \max_{\mathbf{v}_j \in T_B} \{ nhs(\mathbf{v}, \mathbf{v}_j) \}}{|T_A|} \tag{6.3}$$

where $|T_A|$ denotes the cardinality of set T_A,

$$nhs(\mathbf{a}, \mathbf{b}) = \left(1 - \frac{d_H(\mathbf{a}, \mathbf{b})}{n}\right)^p \tag{6.4}$$

and $d_H(\mathbf{a}, \mathbf{b})$ is the Hamming distance. *nhs* is a similarity measure between two binary vectors normalized between zero and one, based on the Hamming distance;

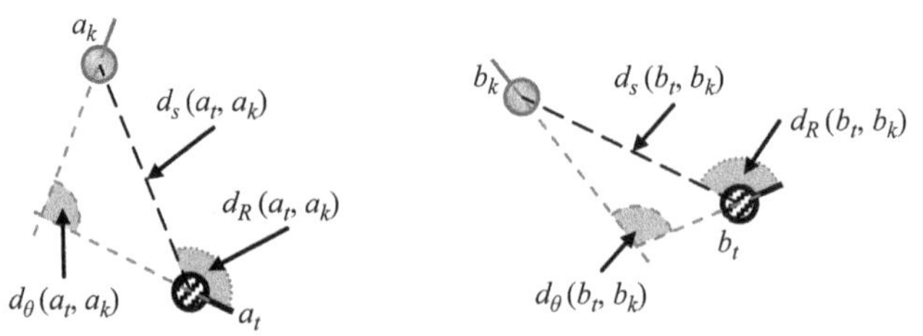

Figure 6.4 During the LSS–DTR consolidation phase, the compatibility between a given candidate pair (a_t, b_t) and another candidate pair (a_k, b_k) depends on: the spatial distances (dashed black lines), the directional differences (gray angles with dashed border), and the radial angles (gray angles with dotted border)

p is a parameter controlling the shape of the similarity function ($p > 0$); in particular, the higher the p, the quicker the *nhs* drops to zero as d_H increases.

Equation (6.3) defines a similarity score in the range $[0, 1]$ (0 means no similarity, 1 means maximum similarity), which is obtained by selecting, for each vector in T_A, the maximum normalized Hamming similarity (6.4) with vectors in T_B, and calculating the average of such values.

According to the studies in [17–19], an effective method for indexing binary vectors using Hamming-based metrics is the Locality-Sensitive Hashing (LSH). LSH is based on the idea that, if two vectors are similar, then after a "projection" into a lower dimensional subspace, they will remain similar. The projection of a binary vector $\mathbf{v}$ into a subspace with h dimensions ($h < n$) simply consists in selecting a subset of h bits from the n bits in $\mathbf{v}$. More formally, let $H = \{i_1, i_2, \ldots, i_h\} \subseteq \{1, \ldots, n\}$: the projection of a given binary vector $\mathbf{v}$ on H is defined as $\mathbf{v}[H] = [v_{i_1}, v_{i_2}, \ldots, v_{i_h}]$. The set of indices H defines a hash function $f_H : \{0, 1\}^n \rightarrow \mathbb{N}$ that maps a binary vector to the natural number whose binary representation is $\mathbf{v}[H]$. In the LSH approach, ℓ hash functions are defined by randomly choosing ℓ subsets $H_1, H_2, \ldots, H_\ell$, and the index consists in ℓ hash tables $\mathbb{H}_1, \mathbb{H}_2, \ldots, \mathbb{H}_\ell$. Given a set of binary vectors to be indexed, each vector $\mathbf{v}$ is placed into bucket $f_{H_k}(\mathbf{v})$ of each hash table $\mathbb{H}_k$, for $k = 1, \ldots, \ell$. To perform a similarity search, the hash functions are applied to the query vector, and all the vectors in the corresponding buckets are retrieved as candidates. The candidates are finally ranked according to their Hamming distance.

Let

$$c_{\mathbb{F}}(\mathbf{v}, \mathbf{v}_j) = \sum_{k=1}^{\ell} \delta\left[f_{H_k}(\mathbf{v}) - f_{H_k}(\mathbf{v}_j)\right] \tag{6.5}$$

be the number of hash functions under which the two vectors collide and

$$\delta[a] = \begin{cases} 1 & \text{if } a = 0 \\ 0 & \text{otherwise} \end{cases} \tag{6.6}$$

is a unit impulse function.

Under some hypothesis (see [13] for details), the similarity measure can be estimated as follows:

$$hds(T_A, T_B) \cong \frac{\sum_{\mathbf{v} \in T_A} \left(\max_{\mathbf{v}_j \in T_B} \left\{c_{\mathbb{F}}(\mathbf{v}, \mathbf{v}_j)\right\}\right)^{p/h}}{|T_A| \cdot \ell^{p/h}} \tag{6.7}$$

According to (6.7), it is possible to estimate the similarity between two templates by simply counting the number of collisions of each pair of binary vectors. Intuitively, if two vectors collide under many hash functions, then their normalized Hamming similarity is likely to be high, while if the number of collisions is small, then probably the two vectors are not very similar.

6.4 Fast matcher for Level 1

6.4.1 Baseline algorithm

Given a database of N MCC templates $DB = \{T_1, T_2, \ldots, T_N\}$ and a query template T_Q, the result of the $1:N$ comparison is a set of N matching scores $S = \{s_1, s_2, \ldots, s_N\}$. Figure 6.5 shows a baseline algorithm for fingerprint identification using MCC with LSS, where:

- each cylinder is simply a pair $c = (\mathbf{v}, \theta)$, since minutiae coordinates x, y are not required by LSS;
- *ComputeNumPairs*$(\cdot)$ computes the value n_P (6.2);
- *TopValues*(L, n_p) returns the top n_p values in L (or L itself if the number of values in L is less than n_p);
- *Sum*(L_p) computes the sum of the values in L_p.

6.4.2 Optimized algorithm for CPU

Starting from the baseline algorithm, an optimized algorithm has been designed to exploit the features of $\times 64$ CPUs with Streaming SIMD Extensions (SSE) [20]. In particular, the following changes and optimizations have been carried out:

- Computational complexity: selecting the top n_p local scores is a time-consuming operation, with a complexity of $O(nc \cdot \log(nc))$, where $nc \approx O(|T_k| \cdot |T_Q|)$ is the number of local similarity scores in L. By quantizing local similarity scores

Input:
- MCC template T_Q (query)
- N MCC templates $DB = \{T_1, T_2, \ldots, T_N\}$

Output:
- N match scores $S = \{s_1, s_2, \ldots, s_N\}$

Algorithm:

ForEach template T_k in DB
  $L = new\ List()$; // List of local scores (initially empty)
  ForEach cylinder $c_i = (\mathbf{v}_i, \theta_i)$ in T_Q
    ForEach cylinder $c_j = (\mathbf{v}_j, \theta_j)$ in T_k
      If $d_\phi(\theta_i, \theta_j) \le \delta_\theta$ Then
        $s_L = 1 - \frac{\|\mathbf{v}_i \oplus \mathbf{v}_j\|}{\|\mathbf{v}_i\| + \|\mathbf{v}_j\|}$ // Computes the local similarity score, see Eq. (6.1)
        $L.Add(s_L)$ // Adds the local score to L
      End If
    End ForEach
  End ForEach
  $n_p = ComputeNumPairs(|T_k|, |T_Q|)$ // pairs to consider, see Eq. (6.2)
  $L_p = TopValues(L, n_p)$
  $s_k = Sum(L_p)/n_p$ // Match score between T_Q and T_k
End ForEach
Return $S = \{s_1, s_2, \ldots, s_N\}$

Figure 6.5 A simple algorithm for MCC 1:N matching

into w values and adopting a counting-sort strategy [21], the complexity can be reduced to $O(nc + w)$. This approach has the further advantage of reducing the space requirements to a vector $\mathbf{b} \in \mathbb{N}^w$, much smaller than the list L of nc scores.

- Optimization of bit vector length: the values chosen for parameters N_s and N_D (see Table 6.1) offer a good trade-off between accuracy and efficiency and result in a bit-vector length $n = 260$ (five sections of 52 bit-valued cells). Preliminary experiments showed that removing one bit from each of the five cylinder sections does not reduce the accuracy and allows $n = 255$; hence, each bit-vector $\mathbf{v}$ can be stored in just two 128-bit SSE registers [20].
- Precomputation of cylinder norms: for each cylinder $c = (\mathbf{v}, \theta)$ of each template, the norm $\eta = \|\mathbf{v}\|$ is calculated once and stored for future use. Hence, each cylinder is a triplet $c = (\mathbf{v}, \theta, \eta)$. The norm of a bit-vector corresponds to the square root of the population count (number of bits with value one), for which an ad-hoc instruction (*popcnt*) is available on recent CPUs [20].
- Bit-wise operations: the bit-wise XOR is computed using SSE instructions [20] operating on 128-bit registers (hence only two SSE instructions are needed to compute $\mathbf{v}_i \oplus \mathbf{v}_j$).
- Look-up table for square roots: since the result of the population-count operation is always an integer number in $[0, n]$, with $n = 255$, it is convenient to use a look-up table ($\mathbf{LUT_{Sqrt}}$) instead of actually computing the square roots.
- Angle quantization: cylinder angles θ are quantized into $z = 256$ values: this allows to implement $d_\theta(\cdot)$ quite efficiently using modulo-256 arithmetic. This does not cause any accuracy degradation since current standards for storing minutiae templates require the same quantization [22].
- A maximum number of cylinders per template are defined ($Max_C = 256$). This allows to precompute any possible output of $ComputeNumPairs(\cdot)$ in look-up table $\mathbf{LUT_p}$. Again, this does not constitute a limitation since current fingerprint minutiae standard templates have the same restriction [22].
- Cylinder norms, $\mathbf{LUT_{Sqrt}}$ values, and similarity scores are stored as integers using fixed-point arithmetic: this allows to remove all floating-point operations.

Figure 6.6 shows the resulting optimized algorithm, where $PopC(\cdot)$ computes the population count of a binary vector using the *popcnt* instruction [20]. The average of the top n_p local similarity scores is computed from $\mathbf{b}$ (lines 11–18), by accumulating bucket values to reach n_p (the *min* operation in line 14 avoids exceeding n_p with the last bucket value). Note that, to save arithmetic operations, s_k is computed (line 18) as one minus the average of the lowest n_p distances (d_L).

6.4.3 Optimized algorithm for GPU

The use of GPUs for general-purpose parallel computing is increasingly attracting researchers' interests, especially in applications where a large amount of data need to be processed, and/or a large number of operations have to be carried out (e.g., [23–25]). The parallel processing capability of a GPU allows it to divide complex computing tasks into thousands of smaller tasks that can run concurrently. A typical hardware configuration consists of a CPU (called host) connected to one or more

Input:
- MCC template T_Q (query)
- N MCC templates $DB = \{T_1, T_2, \ldots, T_N\}$

Output:
- N match scores $S = \{s_1, s_2, \ldots, s_N\}$

Algorithm:

ForEach template T_k in DB
    Reset buckets $\mathbf{b}[i], 0 \leq i < w$
    ForEach cylinder $c_i = (\mathbf{v}_i, \theta_i, \eta_i)$ in T_Q
        ForEach cylinder $c_j = (\mathbf{v}_j, \theta_j, \eta_j)$ in T_k
            If $d_\phi(\theta_i, \theta_j) \leq \delta_\theta$ Then
                $d_L = \left\lfloor w \cdot \frac{\mathbf{LUT_{Sqrt}}[PopC(\mathbf{v}_i \oplus \mathbf{v}_j)]}{\eta_i + \eta_j} \right\rfloor$ // SSE for bitwise XOR
                $\mathbf{b}[d_L] = \mathbf{b}[d_L] + 1$ // Increments the corresp. bucket
            End If
        End ForEach
    End ForEach
    $n_p = \mathbf{LUT_p}[\min\{|T_k|, |T_Q|\}]$ // number of distances to consider
    $sum = 0,\ t = n_p,\ i = 0$
    While $i < w$ and $t > 0$ // the sum of the lowest n_p
        $sum = sum + \min(\mathbf{b}[i], t) \cdot i$ // d_L is computed from $\mathbf{b}$
        $t = t - \min(\mathbf{b}[i], t),\ i = i + 1$
    End While
    $sum = sum + t \cdot w$ // in case there are less than n_p votes in $\mathbf{b}$
    $s_k = 1 - \frac{sum}{n_p \cdot w}$ // match score between T_Q and T_k
End ForEach
Return $S = \{s_1, s_2, \ldots, s_N\}$

Figure 6.6 Optimized algorithm for CPU

GPUs (called devices); given a computation to be done, it is split into several parts, some of which can be executed in parallel on the GPUs (kernel programs); a CPU program (called host program) sends input data and GPU instructions to the GPUs, which execute the given computations in parallel and return the results to the CPU. To this purpose, specific tools are needed to schedule execution of kernels and communicate with the GPUs. The Compute Unified Device Architecture (CUDA) [26] is one of the most widely adopted frameworks for this purpose; CUDA is a hardware and software architecture that enables NVIDIA GPUs to execute parallel kernels written in C/C++. The physical architecture of CUDA-enabled GPUs consists of a set of Streaming Multiprocessors (SM), each containing 32 cores for Single Instruction, Multiple Data (SIMD) execution.

In CUDA programming model, a CUDA kernel is executed in parallel across a set of *threads*, which are organized into *blocks*. All threads of the same block are executed on the same SM and share the limited memory resources of that multiprocessor: for this reason, the maximum number of threads in a block cannot be too large. However a kernel can be executed by multiple, equally sized blocks, forming a *grid*: the total number of threads is then equal to the number of blocks times the number of threads per block [26]. Each SM schedules and executes threads in groups of 32 parallel threads (being 32 the number of cores in a SM) called *warps*.

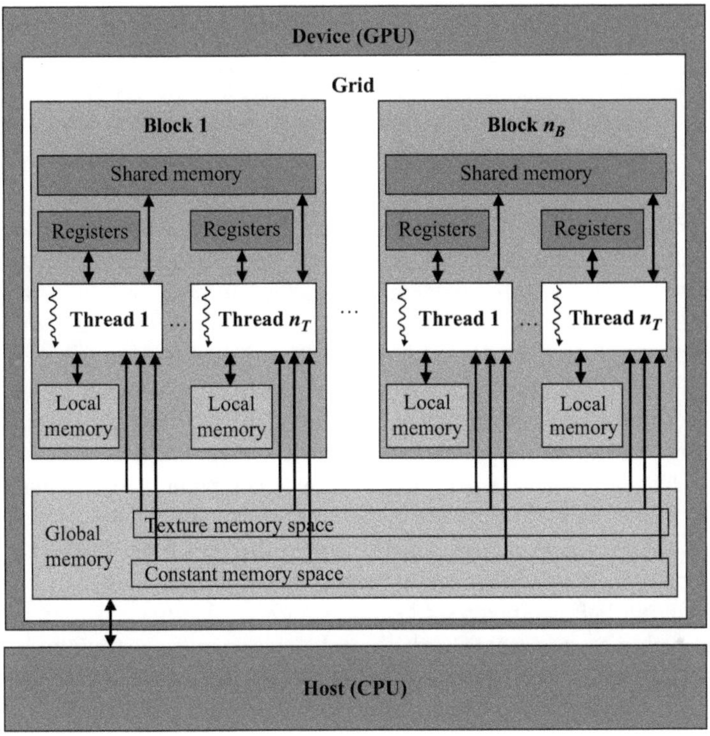

Figure 6.7 CUDA: grid, blocks, threads, and the various memory spaces

A warp executes one common instruction at a time, so full efficiency is realized when all 32 threads of a warp synchronize their execution path. If threads of the same warp take different paths (due to flow control instructions), they have to wait for each other: this situation is called *divergence*. CUDA threads have access to various memory types (Figure 6.7): each thread has its *registers*, which are the fastest memory, and its private *local memory* (which is much slower); each block has a small *shared memory*, accessible to all threads of the block and with the same lifetime of the block; all threads have access to the *global memory*: the largest memory, which is used for communication between different blocks and with the host. A GPU is able to access global memory via 32, 64, or 128-byte memory transactions. When a warp executes an instruction that accesses global memory, it coalesces the memory accesses of the threads within the warp into one or more of these memory transactions, depending on the size of the word accessed by each thread and the distribution of the memory addresses across the threads [27]. Therefore, a very important optimization in CUDA is ensuring that global memory accesses are as much *coalesced* as possible. Finally, all threads have read-only access to the *constant memory space*, a region in global memory that is accessed via *constant cache*, a cache optimized for broadcast, that is accessed to the same address by all the threads in a warp.

The design of the GPU algorithm has been guided by the following optimization goals (see [28] for more details):

- *OG1*—Minimize data transfer between the host and the device.
- *OG2*—Choose the most appropriate types of memory and access patterns.
- *OG3*—Maintain a sufficient number of active threads per multiprocessor.
- *OG4*—Minimize differences in the execution paths within the same warp.

The design of the new GPU algorithm started from the analysis of the optimized algorithm for CPU described in the previous section. Figure 6.8 shows a graphical representation of the main computations involved in such sequential algorithm. The number of local similarity computations (gray circles in the figure) depends on the number of cylinders in the query template and database templates, while the number of LSS computations (gray rectangles in the figure) depends on the number of database templates N. According to our studies, a good strategy to parallelize the computation consists in designing two different kernels that are sequentially executed: the former (named Step-1) is in charge of computing local similarities and the latter (named Step-2) to calculate similarity scores using LSS. Since two kernels can only communicate via global memory, vectors **b** needs to be stored in global memory by Step-1 and read by Step-2. This is less efficient than storing **b** in shared memory, but has important benefits, since (i) it better balances the thread workload within each kernel (OG4); (ii) guarantees a sufficient number of active threads (OG3), and (iii) optimizes coalesced memory accesses (OG2). After several iterations of analysis, implementation, optimization, and performance evaluation,

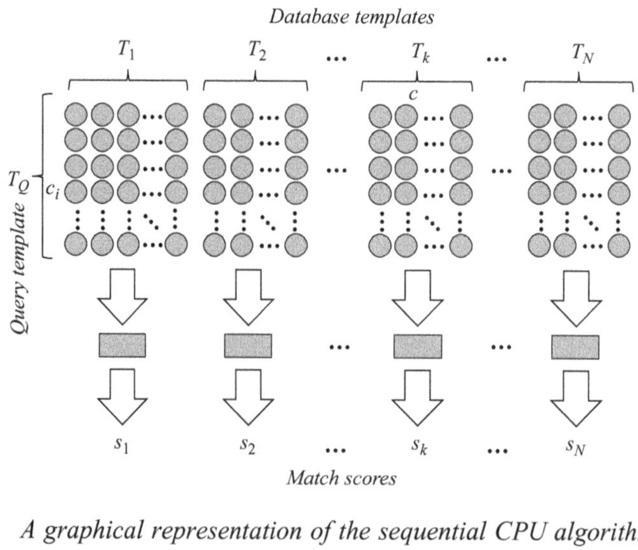

Figure 6.8 A graphical representation of the sequential CPU algorithm in Figure 6.6: Each gray circle represents the computation of a local similarity (between a cylinder c_j of a database template T_k and a cylinder c_i of the query template T_Q) and the increment of the corresponding bucket (lines 5–7 of the algorithm); each gray rectangle represents the computation of a match score s_k between T_Q and T_k using LSS (lines 11–18 of the algorithm)

we find out that an effective allocation of the workload is to assign (i) all computations related to a given database cylinder to a single thread in kernel Step-1 (one "column" of circles in Figure 6.8) and (ii) the computation of each similarity score to a single thread in kernel Step-2 (one rectangle in Figure 6.8). The implementation of this strategy leads to a very efficient GPU algorithm, which is described in detail in [28].

6.5 Indexer for Level 1

6.5.1 Creation of the index

During the creation of the index (Figure 6.9), any binary vector $\mathbf{v}_j$ of each template T_i is given as input to all the hash functions (see Section 6.3.2), and the pair (i,j), which identifies template T_i but also vector $\mathbf{v}_j$ (corresponding to minutia m_j of T_i), is stored in the corresponding buckets.

Since binary vectors obtained with MCC representation tend to be quite sparse (i.e., they have more zeros than ones), buckets whose binary representations have only few one bits are more likely to contain a large number of pairs and then are less selective. For this reason, a parameter (min_{PC}) is used to discard buckets with a low number of one bits (population count).

6.5.2 Searching

At retrieval time, the hash functions are applied to each binary vector of the searched template, and the number of collisions with database vectors are counted using an accumulator matrix (Figure 6.10). In this way, the most similar database templates according to (6.7) are efficiently determined, and the candidate list is easily produced.

Input:
- Database of MCC templates $DB = \{T_1, T_2, ...\}$
- Set of hash functions $\mathbb{F} = \{f_{H_1}, f_{H_2}, ... f_{H_\ell}\}$

Output:
- Hash tables $\mathbb{H}_1, \mathbb{H}_2, ... \mathbb{H}_\ell$

Algorithm:
```
ForEach template T_i in DB
  ForEach vector v_j in T_i
    ForEach hash function f_Hk in F
      b = f_Hk(v_j)
      If PopCount(b) ≥ min_PC then
        Store (i, j) in bucket b of table H_k
      End If
    End ForEach
  End ForEach
End ForEach
```

Figure 6.9 Algorithm for index creation (offline stage). Function PopCount(b) calculates the population count of number b. © 2017 IEEE. Reprinted, with permission, from Reference [13]

Input:
- MCC template T_Q (query)
- Database of MCC templates $DB = \{T_1, T_2, \dots\}$
- Set of hash functions $\mathbb{F} = \{f_{H_1}, f_{H_2}, \dots f_{H_\ell}\}$
- Hash tables $\mathbb{H}_1, \mathbb{H}_2, \dots \mathbb{H}_\ell$
- Maximum number of candidates max_C

Output:
- A set of candidates $CL = \{(i_k, s_k)\}$, where each candidate (i_k, s_k) is a database template index i and a score s

Algorithm:

Initialize the score accumulator $\mathbb{S}$
ForEach vector $\mathbf{v}$ in T_Q
  Let m be the minutia of T_Q corresponding to $\mathbf{v}$
  Reset collision accumulator $\mathbb{A}$
  ForEach hash function f_{H_k} in $\mathbb{F}$
    $b = f_{H_k}(\mathbf{v})$
    If $PopCount(b) \geq min_{PC}$ then
      ForEach pair (i, j) in bucket b of table $\mathbb{H}_k$
        Let m_j be the jth minutia of template T_i
        If $Compatible(m, m_j) = true$ then
          $\mathbb{A}[i, j] = \mathbb{A}[i, j] + 1$ *// Adds a "collision" to minutia j of T_i*
        End If
      End ForEach
    End If
  End ForEach
  ForEach T_i with at least one collision in $\mathbb{A}$
    Let $CF_{max} = \max_j\{\mathbb{A}[i, j]\}$
    $\mathbb{S}[i] = \mathbb{S}[i] + (CF_{max})^{\frac{p}{h}}$
  End ForEach
End ForEach
ForEach template index i in $\mathbb{S}$ *// Normalize scores*
  $\mathbb{S}[i] = \mathbb{S}[i] \Big/ \left(|T_Q| \cdot \ell^{\frac{p}{h}}\right)$ *// Now $\mathbb{S}[i] \cong hds(T_Q, T_i)$, see Eq. (6.7)*
End ForEach
Create CL by selecting at most max_C pairs $(i, \mathbb{S}[i])$ with the highest scores in $\mathbb{S}$

Figure 6.10 Algorithm for searching a fingerprint in the index. © 2017 IEEE. Reprinted, with permission, from Reference [13]

Only minutiae satisfying basic geometric constraints are considered; function $Compatible(m, m_j)$ defines those constraints:

$$Compatible(m, m_j) = \begin{cases} true & \text{if } d_\theta(m, m_j) \leq \delta_\theta \text{ and } d_{xy}(m, m_j) \leq \delta_{xy} \\ false & \text{otherwise} \end{cases} \tag{6.8}$$

where d_θ and d_{xy} are the angular difference and the Euclidean distance between the two minutiae, respectively. Enforcing such constraints results in tolerating a maximum rotation δ_θ and maximum displacement δ_{xy} between the searched template and the database templates. Note that parameters δ_θ and δ_{xy} can be adjusted for each specific search, since they do not affect the indexing phase.

6.5.3 Optimized indexing algorithm for CPU

Starting from the baseline indexing algorithm, an optimized version has been designed to exploit the features of x64 CPUs with SSE extensions [20]. The optimization involved precomputation of any possible value, quantization of angles in (6.8), etc. Unfortunately, this algorithm is less suited to benefit from SSE instructions, with respect to the matching algorithm described in Section 6.4.1, as the results reported in the following section show.

6.6 Experimental results

Although several fingerprint databases are available for evaluating fingerprint verification algorithms (1:1 comparison), e.g., [29–33], they are too small for assessing the performance of large-scale identification systems (1:*N* comparison). For this reason, a specific dataset has been prepared (called FIDELITY IDX) in the context of the FIDELITY European project [34], containing 300,000 different subjects, by combining real and synthetic fingerprint databases.

To simulate an identification scenario (1:*N* comparison), the FIDELITY IDX database contains two subsets: the reference and the probe database.

- The reference database (IDX-R) consists of 250,000 subjects (400 with real biometric samples, the others with synthetic ones). Each subject is identified by an ID (from 1 to 250,000). For each subject, a sample of two fingers is available.
- The probe database (IDX-P) consists of 100,000 subjects: 50,000 subjects with mate (SM) in IDX-R (400 with real biometric samples, the others with synthetic ones) and 50,000 subjects without mate (SNM) in IDX-R (400 with real biometric samples, the others with synthetic ones). Obviously, for SM subjects, the samples are different from the corresponding samples in IDX-R.

For each subject, each database contains the following data:

- Two fingerprint images [see Figure 6.11(a)].
- Two corresponding ISO 19794-2 format minutia templates [22] [see Figure 6.11(b)].
- Two corresponding MCC templates [14] (obtained from the minutiae templates using MCC parameters reported in [16]) [see Figure 6.11(c)].

The databases have been built starting from a synthetic database generated using the synthetic generator software SFinGe [35] and from the following real fingerprint sets: FVC2004 DB1 A [2], BioSec FO [36], and FP_CM_V300 [37].

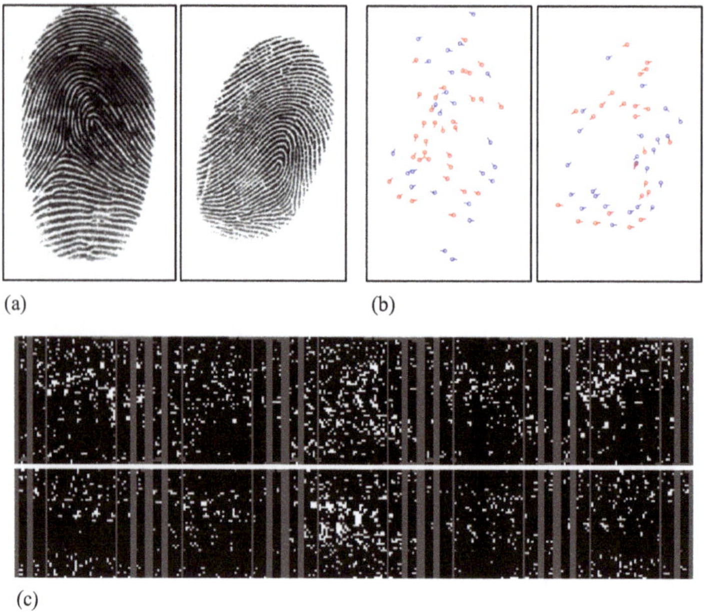

Figure 6.11 (a) An example of two fingerprints from FIDELITY IDX and the corresponding (b) minutia and (c) MCC templates (here linearized for simplicity of graphical representation)

The identification system prototype described in Section 6.1 has been used for all the experiments, with IDX-R as fingerprint database and an identification query executed for each subject in IDX-P. The same identification experiments has been repeated for each Level 1 algorithm described in Sections 6.4 and 6.5 (see Figure 6.1), while the rest of the prototype has not been changed. The following performance indicators have been calculated for each experiment:

- False negative identification rate (FNIR) [1,38]: percentage of queries with a mate in DB that are not correctly identified.
- False positive identification rate (FPIR) [1,38]: percentage of queries without a mate in DB that are mistakenly identified.

All the algorithms have been evaluated on a PC with two Intel Xeon E5-2650 CPUs running at 2.0 GHz (8 cores) and four NVIDIA Tesla C2075 GPUs running at 1.15 GHz (448 cores). The baseline CPU identification and indexing algorithms were implemented in C#, the optimized CPU algorithm in C++ with compiler intrinsics [20] to generate SSE and *popcnt* instructions, and the optimized GPU algorithm in C# and C++ for the host code and CUDA C [26] for the device code. All algorithms were implemented with large use of multithreading to exploit the multiple cores in the CPU.

Table 6.1 MCC parameter values used in the experiments

Parameter[a]	**Description**	**Value**
N_D	Sections in each cylinder	5
N_S	Cells along the diameter of each section	8
n	Bits in each cylinder (bit-vector length)	255
$\min_{n_p}$	Parameter for computing LSS (6.2)	11
$\max_{n_p}$	Parameter for computing LSS (6.2)	13
μ_P	Parameter for computing LSS (6.2)	30
τ_P	Parameter for computing LSS (6.2)	2/5
δ_θ	Max allowed rotation between two fingerprints	$\pi/6$
w	Possible values for quantized local similarities	64
Max_C	Maximum number of cylinders per template	256

[a]For any MCC parameter not listed in this table, the same value of [16] was used.

The following sections report and discuss experiments carried out to measure the performance of the proposed algorithms and compare their results to previous works published in the literature.

6.6.1 Parameters

The MCC cylinders have been derived from the minutia templates contained in the database as described in [14] and summarized in Section 6.2; then they have been stored on disk and used for the experiments reported in the rest of this section. The parameter values reported in Table 6.1 have been used for all experiments.

6.6.2 Results and comparison with the state of the art

The graph in Figure 6.12 shows the FNIR/FPIR trade-off varying the security threshold of the system. All the algorithms proposed in the previous sections have achieved the same accuracy: as to the identification algorithms (baseline and optimized CPU/GPU), this is because the optimizations do not include any relevant approximation or simplification. As to the indexing algorithm, this can be explained by considering that the indexing is used only at the Level 1, while the second level is always the same. At FPIR = 0.1%, FNIR is about 1.0%: this level of accuracy is in line with large-scale identification results using a single finger [1] and with the accuracy of state-of-the-art fingerprint verification algorithms [33].

Table 6.2 reports the total time required by the system to complete all the queries and the corresponding throughput (thousands of comparisons per second). With the proposed GPU algorithm at Level 1, the system completes all the queries in about 10 min using four GPUs and in 39 min with one GPU; about one day is necessary with the baseline parallel CPU algorithm, and about 1 h with the optimized parallel one. The optimized CPU identification algorithm is about 23 times faster than the baseline, proving that the proposed optimizations are very effective.

The baseline indexing algorithm is about six times faster than the baseline matcher doing exhaustive search. The indexer optimization allows to more than

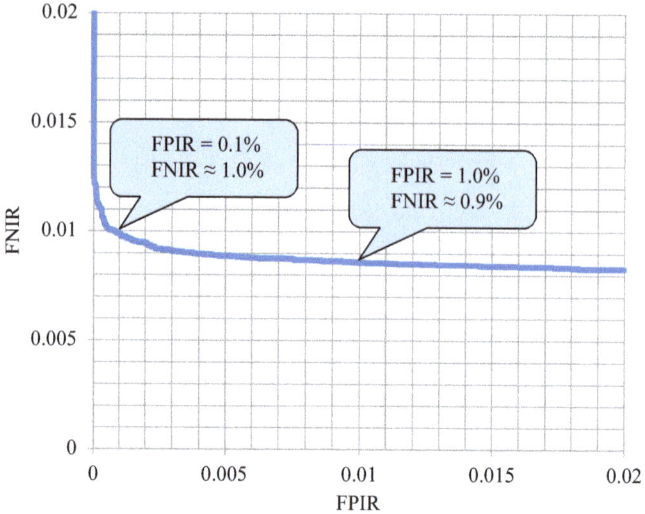

Figure 6.12 FNIR vs FPIR at rank-1 of the proposed algorithms

Table 6.2 Results: execution time, average query time, and throughput

Level 1		Implementation	Speed-up	Execution time (dd:hh:mm)	Average query time (ms)	Throughput (MMPS)	Speed-up
LSS	(Fast matcher)	Baseline [CPU]		01:00:13	871.6	0.3	
		Optimized [CPU]	23x	00:01:00	35.8	7.0	
		Optimized [1 GPU]	35x	00:00:39	23.5	10.6	
		Optimized [4 GPUs]	141x	00:00:10	5.9	42.4	
		Optimized [CPU+4 GPUs]	149x	00:00:09	5.6	44.6	
LSH	(Indexer)	Baseline [CPU]		00:03:33	128.0	1.9	6x
		Optimized [CPU]		00:01:38	58.9	4.2	14x

double efficiency, but the resulting approach is not competitive with the CPU optimized exhaustive matcher, mainly due to the nature of the indexing strategy that is not well suited to SIMD optimization and to maximize the benefits of memory caching.

The efficiency of the proposed algorithm was compared to results published in the literature for other parallel fingerprint-identification systems based on the main hardware architectures (i.e., FPGA, CPU, and GPU). In particular, the following studies were considered:

- Lindoso *et al.* [5], a FPGA-based fingerprint identification system, whose speed-up is measured with respect to a sequential algorithm running on a 3-GHz Intel Pentium IV processor.

Table 6.3 Comparison with previously published parallel algorithms on FPGAs, CPUs, and GPUs

Method	DB size	Throughput of the baseline sequential algorithm on CPU (KMPS)	Hardware	Throughput of the parallel algorithm (KMPS)	Speed-up
Lindoso *et al.* [5]	56	0.3	FPGA Xilinx Virtex-4 LX	7.1	23.7
Jiang and Crookes [4]	10,000	26.2	FPGA Xilinx Virtex-E	1,219.5	46.5
Peralta *et al.* [8]	400,000	4.5	Cluster of 12 nodes with Intel Xeon CPU E5-2620, 12 cores (24 threads) per node	812.7	180.6
Gutierrez *et al.* [6]	100,000	1.6	One GeForce GTX 680 GPU	55.7	34.8
			1 T M2090 GPU	50.0	31.3
			2 T M2090 GPUs	97.7	61.1
Proposed algorithm	250,000	18.1	1 T C2075 GPU	10,638.3	587.8
			4 T C2075 GPUs	42,372.9	2 ,341.0

- Jiang and Crookes [4], a FPGA-based fingerprint identification system; to the best of our knowledge, this is the fastest FPGA-based algorithm reported in the scientific literature; its speed-up is measured with respect to a sequential algorithm running on a 2.8-GHz Intel Celeron processor.
- Peralta *et al.* [8], a CPU-based fingerprint identification system running on a cluster of 12 nodes, each equipped with two Intel Xeon E5-2620 processors (each node can run up to 24 parallel threads); three different algorithms were experimented in this study, but for simplicity in the following comparison, only the fastest one is reported; the speed-up is measured with respect to a sequential execution of the same algorithm on a single CPU core.
- Gutierrez *et al.* [6], a GPU-based fingerprint identification system; to the best of our knowledge; this is the only study published on scientific journals that reports fingerprint identification results on GPUs; results from all the three different GPU hardware configurations experimented in this study are reported in the following comparison; the speed-up is measured with respect to a sequential algorithm running on an Intel Xeon E5-2630 processor.

Table 6.3 compares the performance of the above systems to the proposed algorithm; each row of the table reports:

- The size of the fingerprint database on which identification experiments were carried out.

- The throughput (in thousands of fingerprints comparisons per second) of the sequential algorithm on CPU used to calculate the speed-up of the respective parallel algorithm, as reported in the corresponding paper.
- A short description of the specific hardware used for the parallel algorithm.
- The maximum number of threads that the hardware can execute in parallel (where applicable).
- The throughput (in thousands of fingerprints comparisons per second) of the parallel algorithm, as reported in the corresponding paper.
- The speed-up, simply measured as the ratio between the two throughputs.

It is well evident that the proposed GPU algorithm overcomes all previous approaches, both in terms of absolute performance and in terms of relative speed-up. In particular, the following observations can be drawn:

- Using just one GPU, the proposed algorithm is able to compare more than 10 million of fingerprints per second: it is two orders of magnitude faster than the GPU algorithm described in [6] and one order of magnitude faster than a cluster of 12 PCs running the algorithm described in [8]. On a PC with four GPUs, the throughput is more than 42 million of fingerprints per second: with such a throughput, less than one second would be required to perform 10 queries on a database with 4 million fingerprints.
- The speed-up of the proposed algorithm is remarkable, especially if compared to the results in [6], which were obtained on GPUs with more computational power than those used in this work; for instance, on a GPU with 512 cores, [6] obtained a 50× speed-up, while the proposed algorithm shows a 588× speed-up on a GPU with 448 cores.

6.7 Conclusions

Through the design, implementation, optimization, and test of a large number of fingerprint recognition approaches, we developed some insights on large-scale fingerprint recognition that we try to briefly summarize here:

- Implementing two- (or more-) level systems is advantageous for efficiency. To avoid losing accuracy is crucial that the first level does not miss candidates that cannot be recovered later.
- Vector operation on CPU (e.g., SSE, AVX, etc.) can be very useful to speed up execution on CPU, but do not think the compiler makes the dirty work for you. A proper encoding, alignment, and choice of instructions is necessary to achieve good speed-up.
- Analogously, porting a fingerprint recognition algorithm on GPU is not simple, and a trivial implementation often leads to a worse efficiency with respect to the CPU. Unfortunately, well-known libraries such as cuBLAS for linear algebra (heavily used by deep learning frameworks) does not help much here, and the threading granularity, memory access, and parallelization strategy need

to be carefully tuned by the developer, often based on experience and trial and error strategies.

- Fingerprint indexing is an interesting approach for Level 1 candidate selection, but in our experience, indexing was not competitive with optimized fast exhaustive search. Of course, this could depend on the specific indexing approach (based on MCC and hashing). However, in our experience, a number of nonexpected overheads (difficult to overcome) come out when the initial elegant approach has to be coded in efficient operations. So when estimating the computational complexity of a novel approach from equations or pseudocode, consider that practical implementation issues can significantly alter your best guess.

References

[1] Unique Identification Authority of India, *Role of Biometric Technology in Aadhaar Enrollment*, Planning Commission, Government of India, New Delhi, 2012.

[2] D. Maltoni, D. Maio, A.K. Jain, and S. Prabhakar, *Handbook of Fingerprint Recognition*, 2nd ed. Springer-Verlag: New York, NJ, USA, 2009.

[3] N.K. Ratha, K. Karu, S. Chen, and A.K. Jain, "A real-time matching system for large fingerprint databases," *IEEE Transactions on Pattern Analysis and Machine Intelligence*, vol. 18, no. 8, pp. 799–813, August 1996.

[4] R.M. Jiang and D. Crookes, "FPGA-based minutia matching for biometric fingerprint image database retrieval," *Journal of Real-Time Image Processing*, vol. 3, no. 3, pp. 177–182, September 2008.

[5] Lindoso, L. Entrena, and J. Izquierdo, "FPGA-Based Acceleration of Fingerprint Minutiae Matching," in *Programmable Logic, 2007. SPL '07. 2007 3rd Southern Conference on*, Mar del Plata, Argentina, 2007, pp. 81–86.

[6] P.D. Gutierrez, M. Lastra, F. Herrera, and J.M. Benitez., "A high performance fingerprint matching system for large databases based on GPU," *IEEE Transactions on Information Forensics and Security*, vol. 9, no. 1, pp. 62 71, January 2014.

[7] A.I. Awad, "Fingerprint local invariant feature extraction on GPU with CUDA," *Informatica (Slovenia)*, vol. 37, no. 3, pp. 279–284, 2013.

[8] D. Peralta, I. Triguero, R. Sanchez-Reillo, F. Herrera, and J.M. Benitez, "Fast fingerprint identification for large databases," *Pattern Recognition*, vol. 47, no. 2, pp. 588–602, February 2014.

[9] R. Cappelli and D. Maio, "The State of the Art in Fingerprint Classification," in *Automatic Fingerprint Recognition Systems*. Springer: New York, NY, 2004, ch. 9, pp. 183–205.

[10] R. Cappelli, D. Maio, and D. Maltoni, "Combining fingerprint classifiers," in *First International Workshop on Multiple Classifier Systems (MCS2000)*, Cagliari, 2000, pp. 351–361.

[11] R. Cappelli, "Fast and accurate fingerprint indexing based on ridge orientation and frequency," *IEEE Transactions on Systems, Man, and Cybernetics, Part B: Cybernetics*, vol. 41, no. 6, pp. 1511–1521, December 2011.

[12] R. Cappelli and M. Ferrara, "A fingerprint retrieval system based on level-1 and level-2 features," *Expert Systems With Applications*, vol. 39, no. 12, pp. 10465–10478, September 2012.

[13] R. Cappelli, M. Ferrara, and D. Maltoni, "Fingerprint indexing based on minutia cylinder code," *IEEE Transactions on Pattern Analysis and Machine Intelligence*, vol. 33, no. 5, pp. 1051–1057, May 2011.

[14] R. Cappelli, M. Ferrara, and D. Maltoni, "Minutia cylinder-code: a new representation and matching technique for fingerprint recognition," *IEEE Transactions on Pattern Analysis and Machine Intelligence*, vol. 32, no. 12, pp. 2128–2141, December 2010.

[15] R. Cappelli, M. Ferrara, and D. Maio, "A fast and accurate palmprint recognition system based on minutiae," *IEEE Transactions on Systems, Man and Cybernetics – Part B*, vol. 42, no. 3, pp. 956–962, June 2012.

[16] R. Cappelli, M. Ferrara, D. Maltoni, and M. Tistarelli, "MCC: A Baseline Algorithm for Fingerprint Verification in FVC-onGoing," in *Proceedings 11th International Conference on Control, Automation, Robotics and Vision (ICARCV)*, Singapore, 2010.

[17] X. Shuai, C. Zhang, and P. Hao, "Fingerprint Indexing Based on Composite Set of Reduced SIFT Features," in *Pattern Recognition, 2008. ICPR 2008. 19th International Conference on*, 2008, pp. 1–4.

[18] P. Indyk and R. Motwani, "Approximate Nearest Neighbors: Towards Removing the Curse of Dimensionality," in *13th ACM Symposium on Theory of Computing*, 1998, pp. 604–613.

[19] Gionis, P. Indyk, and R. Motwani, "Similarity Search in High Dimensions via Hashing," in *25th International Conference on Very Large Data Bases*, 1999, pp. 518–529.

[20] Intel. (2017, May) Intel Instruction Set Architecture Extensions. [Online]. http://software.intel.com/en-us/intel-isa-extensions.

[21] T.H. Cormen, C.E. Leiserson, R.L. Rivest, and C. Stein, "Counting Sort," in *Introduction to Algorithms*. MIT Press and McGraw-Hill, 2001, ch. 8.2, pp. 168–170.

[22] ISO/IEC 19794-2, *Information Technology – Biometric Data Interchange Formats – Part 2: Finger Minutiae Data*. 2011.

[23] M. Fort and J.A. Sellarès, "Solving the k-influence region problem with the GPU," *Information Sciences*, vol. 269, pp. 255–269, June 2014.

[24] L. Mussi, F. Daolio, and S. Cagnoni, "Evaluation of parallel particle swarm optimization algorithms within the CUDA^{TM} architecture," *Information Sciences*, vol. 181, no. 20, pp. 4642–4657, October 2011.

[25] Y. Zhang, D. Yi, B. Wei, and Y. Zhuang, "A GPU-accelerated non-negative sparse latent semantic analysis algorithm for social tagging data," *Information Sciences*, vol. 281, pp. 687–702, October 2014.

[26] NVIDIA. (2017, May) CUDA C Programming Guide. [Online]. http://docs.nvidia.com/cuda/cuda-c-programming-guide/index.html.

[27] NVIDIA. (2014, October) CUDA C Programming Guide. [Online]. http://docs.nvidia.com/cuda/cuda-c-programming-guide/index.html.

[28] R. Cappelli, M. Ferrara, and D. Maltoni, "Large-scale fingerprint identification on GPU," *Information Sciences*, vol. 306, pp. 1–20, June 2015.

[29] D. Maio, D. Maltoni, R. Cappelli, J.L. Wayman, and A.K. Jain, "FVC2000: fingerprint verification competition," *IEEE Transactions on Pattern Analysis and Machine Intelligence*, vol. 24, no. 3, pp. 402–412, August 2002.

[30] D. Maio, D. Maltoni, R. Cappelli, J.L. Wayman, and A.K. Jain, "FVC2002: second fingerprint verification competition," *International Conference on Pattern Recognition*, vol. 16, pp. 811–814, 2002.

[31] D. Maio, D. Maltoni, R. Cappelli, J.L. Wayman, and A.K. Jain, "FVC2004: Third Fingerprint Verification Competition," in *Proceedings International Conference on Biometric Authentication (ICBA04)*, Hong Kong, 2004, pp. 1–7.

[32] R. Cappelli, M. Ferrara, A. Franco, and D. Maltoni, "Fingerprint verification competition 2006," *Biometric Technology Today*, vol. 15, no. 7–8, pp. 7–9, August 2007.

[33] B. Dorizzi, R. Cappelli, M. Ferrara, *et al.*, "Fingerprint and On-line signature Verification Competitions at ICB 2009," in *Proceedings 3rd IAPR/IEEE International Conference on Biometrics (ICB09)*, Alghero, 2009.

[34] (2017, May) FIDELITY European Project Web Site. [Online]. http://www.fidelity-project.eu/.

[35] BioLab. (2017, May) SFinGe (Synthetic Fingerprint Generator) Web Site. [Online]. http://biolab.csr.unibo.it/sfinge.html.

[36] Universidad Autonoma de Madrid. (2017, May) Biometric Recognition Group – ATVS Web Site. [Online]. https://atvs.ii.uam.es/atvs/.

[37] X. Jia, X. Yang, Y. Zang, N. Zhang, and J. Tian, "A Cross-device Matching Fingerprint Database from Multi-type Sensors," in *Proceedings of 21st International Conference on Pattern Recognition (ICPR)*, 2012, pp. 3001–3004.

[38] P.J. Grother, G.W. Quinn, and P. Jonathon, *Report on the Evaluation of 2D Still-image Face Recognition Algorithms*. NIST, 2010.

Chapter 7

Processing of fingerprints influenced by skin diseases

Mona Heidari[1], Ondřej Kanich[1], and Martin Drahanský[1]

7.1 Introduction

Human beings have always had a need for a secure world, and with the speed of technology development in the recent past years, this topic is becoming more and more important. Technology affects almost every corner of our lives: work, home, family and leisure. Our society has become more mobile, more electronically connected and less place-dependent. Humans want to have their technological devices, data, bank and other accounts, companies, cars or other possessions secured. However, traditional representations of identity such as passwords or cards no longer offer such security. Passwords are easily breakable and can be forgotten; cards can be stolen or lost. Biometric technologies are based on recognition of biometric traits of individuals, such as face, speech or fingerprint recognition, and they represent the most promising way how to provide security and represent identity in our growing modern world [1].

Fingerprint-based systems are the most widely used biometric technology. Although the individuality of fingerprints was well known already in the ancient times, it was not until 1880 when Henry Faulds published a work that introduced the possibility of using fingerprints for the purpose of human identification [2]. Since that time, fingerprint technology has been evolving and nowadays fingerprint recognition systems have been applied in a variety of areas [1]. They are used not only in forensics for criminalistic purposes but also as an access method to facilities, computers, mobile phones or electronic banking; as a data protection method and for civil identification (passports, driver licenses, national IDs), not to mention applications in government, commercial financial sector, education or health care [1]. This technology has been well accepted by people and we use it on a daily basis. However, there is a significant number of people who cannot use fingerprint systems as easily because their fingertip skin is affected by some kind of skin disease. As these systems count heavily on the structure of an individual's fingertip papillary line pattern that positively determines their identity, people suffering from skin diseases

[1]Faculty of Information Technology, Centre of Excellence IT4Innovations, Brno University of Technology, Czech Republic

might be discriminated against as their papillary patterns may be impaired. It is very likely that fingerprint devices have not been designed to deal with damaged fingerprints, and therefore after scanning the fingerprint, they usually reject it.

In some cases, the condition of an image obtained from the damaged fingerprint is not even good enough for further processing, but in others, the damage is minor and the condition of the fingerprint image should not be an obstacle for papillary lines and minutiae extraction, and further matching. The challenge now is to recognize the presence of skin diseases in fingerprint images, provide sufficient algorithms that will detect them and, if possible, eliminate their influence on the fingerprint recognition process.

7.2 Damaged fingerprints

Undamaged or perfect fingerprint is only a theoretical term. In real life scenario, flawless fingerprint cannot be achieved. The only possible way is to synthetically generate a fingerprint. That is because the original finger is very often somehow damaged, the acquirement method/technology is not perfect, the environmental conditions are not ideal, therefore the resulting fingerprint has some damages. These damages can be basically divided into three groups: *finger and user condition*, *sensor effects* and *influence of the environment*. Information in this subsection is based on [3].

An ideal fingerprint has nicely visible ridges (papillary lines), which are clearly distinguishable from a background—see Figure 7.1 (fingerprint acquired by Sagem MSO300 optical reader). The best choice could be such fingerprint reader technology, which delivers black and white images—e.g., pressure sensitive sensor BLP-100 from the BMF company (seated in Kawasaki, Japan). The majority of sensors delivers fingerprint images in gray-scales, i.e., sometimes it is difficult to distinguish ridges from the background, especially if there are some distortions on finger or coming from the environment (e.g., dirty area of the fingerprint reader). If a histogram of an ideal fingerprint is computed, there should be only two peaks visible—one representing black (ridges) and one representing white (background and valleys). This distribution is not present very often—see Figure 7.2 (left fingerprint acquired by Biolink optical reader, right one by Veridicom capacitive reader).

Figure 7.1 Nearly ideal fingerprint with nicely visible ridges

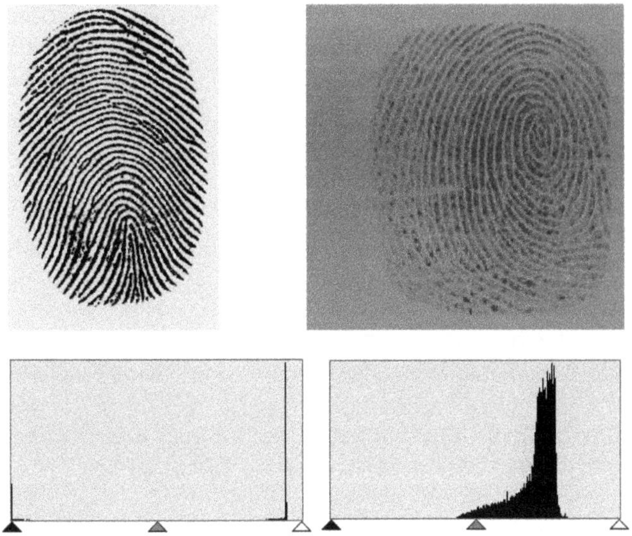

Figure 7.2 Images and histograms of high (left) and low (right) quality fingerprints

All deviations from an ideal fingerprint show low-quality index and the algorithm has to consider whether this fingerprint could be used for further processing or not. This will be described in Section 7.3. Now the focus will be put on description of various distortions, which could be detected in fingerprints and their cause.

7.2.1 Finger and user condition

First phenomena that are connected to the finger itself will be discussed. The **dirt on the finger** can be caused by a few grains of dust, small particle or a greasy finger from some meal. Liquids or generally conductive materials are one of the most problematic types of dirt. For every sensor technology, the most problematic type of dirt is different. These phenomena can be in conjunction with the **dry or moist finger**. Sweaty fingers for example could be assigned to both groups. The effect of moisture or dryness of the finger is very significant. It is also a very frequent way of damaging the fingerprint. Investigated user can be very nervous which leads to sweaty or extremely dry fingers. On the other hand, in the everyday usage, fingers can be moist after using some lotion or they can be simply recently washed. The skin resistance which is important for some sensors can be up to ten times higher or lower than the average value. These two phenomena are discussed more thoroughly in previous part. The **physically damaged finger** is common in some groups of users, namely, among people who are working manually. In their jobs, some abrasions or cuts are inevitable. If it is only a small injury or every day wear damage the papillary lines will regenerate. On the contrary, a deep wound will affect the papillary lines forever. Fingers can be also damaged by **skin diseases**. That will be introduced in Section 7.3.

The second part of this category is phenomena caused by users when acquiring the image. One of them, usually caused by inexperienced users, is small or insufficient **contact region**. That simply means that the finger was presented to the sensor in a way which allows only a partial acquisition of the fingerprint. The small contact region can also be caused by extremely low **pressure**. The other way around, very high pressure can create a black oval instead of the fingerprint image. When the pressure is somewhere between these extremes, it can cause either very thick or very thin papillary lines. It is very sensor dependent what the "right" pressure is. There is an endless number of movements that users can do when showing the **noncooperative behavior**. It can be dynamic or static behavior which is done on purpose, i.e., it is not an accident. The line between an unintentional wrong usage and a noncooperative behavior is very thin. Nevertheless, there are users who resent biometric systems or just try to push to its limits. They can move the finger or change the pressure while the sensor is acquiring image. The other possibility is the static behavior like a rotation of the finger, a small contact region with just a side of the finger, etc.

7.2.1.1 Dirt on fingers

The first category of distortions in fingerprints is caused by dirt on fingers. This could be caused by miscellaneous materials, e.g., oil or fat, detergents or soaps, crumbs, glue and hair. For the acquirement, it is not so important whether the finger was made dirty before touching the scanner surface or whether the finger was made dirty because of touching the scanner surface, which was dirty from the previous user. Important is that the finger surface is dirty during the scanning process.

We acquired a small database (approx. 200 images) with fingerprints, which are influenced by some dirt. The analysis of this database has shown that there are the following groups of changes in acquired fingerprints:

- *Thicker ridges*—if the user has normal skin conditions, i.e., not too dry and not too wet skin, and the pressure to the sensor is average, it is possible to acquire very nice fingerprint images. However, if any of these conditions is not fulfilled then the quality of acquired fingerprint is getting lower. Thicker ridges could be caused, e.g., by increased pressure to the sensor surface [see Figure 7.3(a)—fingerprint acquired by Bergdata thermal reader]. Other cause for thicker ridges is very often wet and oily (sweat coming out from sweat glands is spreading over the ridges) finger surface [see Figure 7.3(b)—fingerprint acquired by SecuGen optical reader]. Nevertheless, other materials in our environment can cause thicker ridges as well—for example, hand cream Nivea [see Figure 7.3(c)—fingerprint acquired by SecuGen reader], detergent Cif for cleaning surfaces in kitchen [see Figure 7.3(d)—fingerprint acquired by SecuGen reader], liquid industrial soap [see Figure 7.4(a)—fingerprint acquired by Sagem MSO300 optical reader], cleaning paste for very dirty hands Solvina [see Figure 7.4(b)—fingerprint acquired by Sagem MSO300 optical reader], shower gel and shampoo Umbro [see Figure 7.4(c)—fingerprint acquired by UPEK EikonTouch 300 capacitive reader] and dish washing detergent Lena [see Figure 7.4(d)—fingerprint acquired by UPEK EikonTouch 300 capacitive reader].

(a) (b) (c) (d)

Figure 7.3 Fingerprints with thicker ridges caused by (from left to right) (a) increased pressure to the sensor surface, (b) wet finger surface, (c) hand cream Nivea and (d) detergent Cif for cleaning surfaces in kitchen

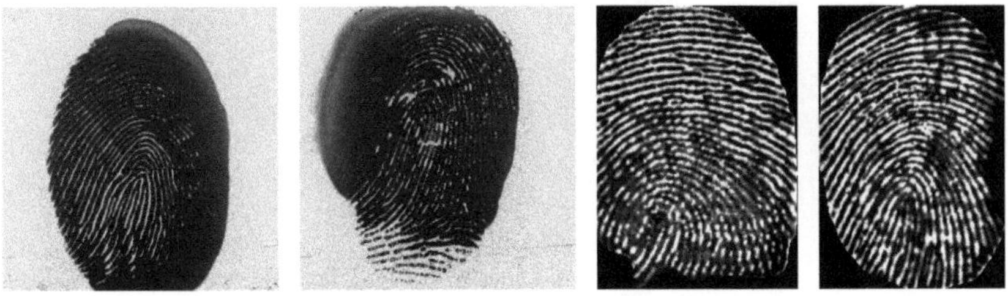

(a) (b) (c) (d)

Figure 7.4 Fingerprints with thicker ridges caused by (from left to right) (a) liquid industrial soap, (b) cleaning paste for very dirty hands Solvina, (c) shower gel and shampoo Umbro and (d) dish washing detergent Lena

- *Broken ridges*—one part, discussed by previous thicker ridges could be continued here—if the finger is very dry, many technologies have troubles to acquire such ridges, especially capacitive and optical technologies (see Figure 7.5(a) fingerprint acquired by Biolink optical reader and Figure 7.5(b) by Veridicom capacitive reader). Very dry fingers could cause troubles for criminal praxis as well, because the latent fingerprint traces left on the crime scene have really low quality. However, there could be other influences which have an impact to the low quality of broken ridges, for example, cleaning paste for very dirty hands Solvina (see Figure 7.5(c)—fingerprint acquired by Sagem MSO300 optical reader; compare this image with Figure 7.4(b)), earth dust (see Figure 7.5(d)—fingerprint acquired by Veridicom capacitive reader), fine sand (see Figure 7.6(a)—fingerprint acquired by Veridicom capacitive reader), metallic dust (see Figure 7.6(b)—fingerprint acquired by Veridicom capacitive reader) or generally any kind of dust or crumbs.

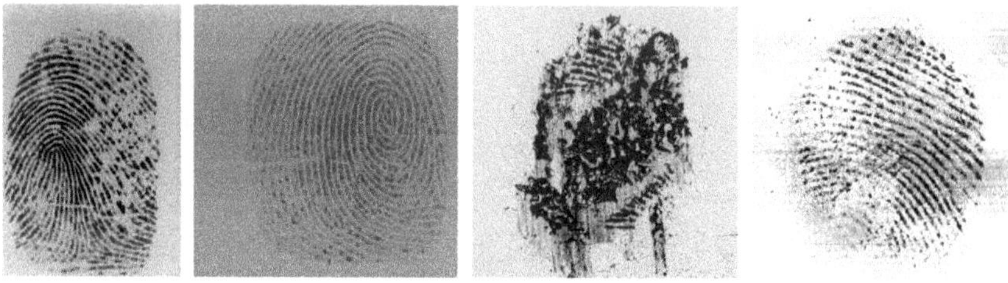

(a) (b) (c) (d)

Figure 7.5 Fingerprints with broken ridges caused by (from left to right) (a) and (b) very dry finger skin, (c) cleaning paste for very dirty hands Solvina and (d) earth dust

(a) (b)

Figure 7.6 Fingerprints with broken ridges caused by: (a) fine sand and (b) metallic dust

- *Additional false minutiae points*—this is a very special case, but not so rare. In this case, the thickness of the dirt has to be approximately same as the ridge. This dirtiness could be hair, fiber or any similar thin and long object. If this object is crossing ridges in these crossings, there could be falsely detected minutiae points, especially false multiple bifurcations (see Figure 7.7—left fingerprint acquired by TBS touchless reader and the right fingerprint by e-field reader). Very small dust pieces or crumbs can connect ridges and on that place there could be detected bifurcation as well. Optical technology is a little bit untypical for thin and long objects on the finger or scanner surface because these objects can have dark or bright color; therefore, they will represent two different minutiae points—dark objects will represent false additional bifurcations (or multiple bifurcations) and bright objects will represent false line endings. Other bigger objects are very often detected as "something wrong" in the image and this area is denoted by low confidence, i.e., these areas are not used for minutiae detection any more. However, smaller objects are taken as ridges and therefore the algorithms for quality estimation do not detect them.

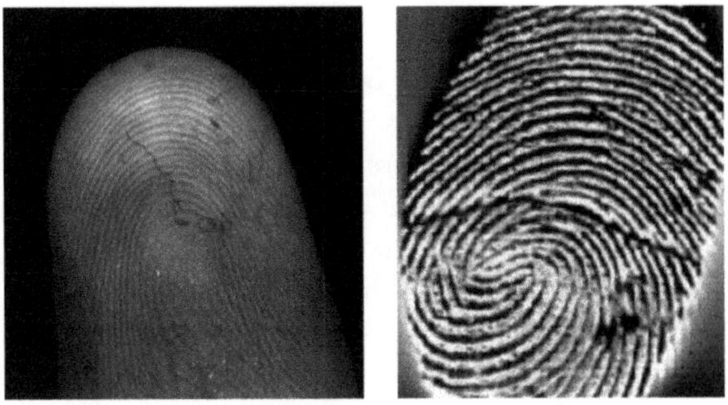

Figure 7.7 Fingerprints with additional minutiae points caused by fiber (left) or hair (right)

One very special case which does not belong to any of the previous categories is the use of medicaments. One very famous example is the use of capecitabine (cancer treatment medicament), which is called chemotherapy-induced acral erythema. This is not the only case—application of acids or alkalis or other corrosive liquids, or use of abrasive materials can destroy the skin structure and therefore ridges can fully or partially disappear.

7.2.2 Sensor effect

Phenomena connected to the sensor are dangerous because they influence all users using the device. In the registration phase, a common error among all users can be made, which is a potential security hazard. In the verification phase, this new common error can make the verification or the identification of the user harder.

The first phenomenon is the **dirt on the surface** of the sensor. While the dirt on the finger is associated with grease or small dust, the surface of the sensor can be polluted much more. There can be grease from several fingers, it can be wooden, metallic or earth dust or fine sand because the sensor is used in a factory or outside. When the amount of dirt is higher, the sensor cannot even recognize that there is a finger on its sensing area. When the dirt on the surface comes from the papillary lines, it is called the **latent fingerprint**. This is fingerprint residues from previous users and not only that it can create fake papillary lines for the current user but sometimes it can be reactivated creating a great security hazard. There is also a possibility of the **physical damage** of the sensor. Whether it is intentional or caused by accident, the sensor can be damaged, but usually not destroyed. The resulting damage is different for every sensor technology. For example, when using the optical technology, the usual damage is a visible crack on the protective glass. The **sensor technology** itself has a large impact on how the fingerprint will look like. For example, some technologies like the ultrasonic one can get the image from a deeper level of skin and the resulting image is then without shallow scars. Some technologies get ridges as a white color, some get ridges as a black color and some technologies get colorized image (basically photo) of the fingerprint.

7.2.3 Influence of the environment

The last group has an influence usually both sensor and user/finger.

The **vibration** can create a blurred image or damage the device internally. It is common in mobile devices when they are used in the means of transport or similar situations. The **temperature** usually influences the fingerprint indirectly by high moisture of the finger or on the other hand, extreme dryness of the finger when it is very cold. But the temperature can also have an impact on the sensor technology. There are technologies that are more sensitive to the surrounding temperature than others. The **surrounding light** is another sensor technology specific phenomenon. It influences only sensors which have a light sensing unit, i.e., the optical or electro-optical technology. A problematic situation occurs when the sensor is large and the finger is small. The uncovered edges of the sensor create the fingerprint image from the surrounding light and not from the finger itself. That can lead to different results when using the sensor in different light conditions. The **electromagnetic radiation** is the last phenomena. In some cases, it can influence the device as a whole. It can even change the information that is transported from one part of the biometric system to the other. This can then in more sensitive sensor technologies lead to a blurred image.

7.2.4 Examples of damaged fingerprints

Figure 7.8 shows a fingerprint image damaged by various factors. We can see wrinkles in the middle of the fingerprint, not ideal contact region (image is not centered) caused by the user, latent fingerprint (right and up side) and the dark parts in the corners are caused by the surrounding light.

Figure 7.8 Fingerprint damaged by various factors

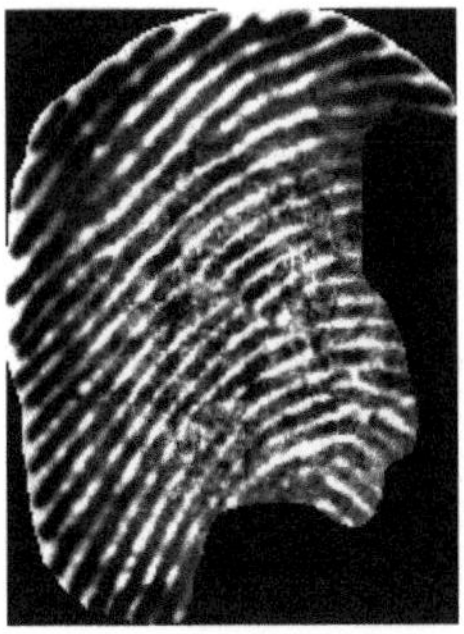

Figure 7.9 Fingerprint damaged by a detergent

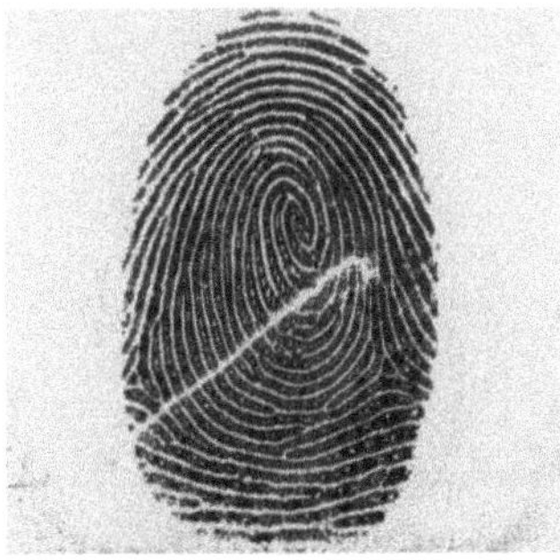

Figure 7.10 Fingerprint with physical damage

Fingerprint in Figure 7.9 is damaged by shower gel which was used before the fingerprint acquirement. Various sensor technologies also create completely different images in comparison with Figure 7.8.

In Figure 7.10, another sensing technology can be seen. On the left bottom part of the image, a small crack in the protective glass of sensor can be seen. In the middle of the fingerprint, there is a long scar. That means sensor and finger are physically damaged in this image.

7.3 Diseased fingers

Skin diseases represent a very important, but often neglected factor of fingerprint recognition. It is not possible to say in general how many people suffer from skin diseases, because there are so many various types of skin diseases [4,5]. The situation after the successful recovery of a potential user from such skin diseases is, however, very important for the possible further use of fingerprint recognition devices. If the disease has attacked and destroyed the structure of ridges in the epidermis and the underlying dermis (the so-called dermoepidermal junction—the connection of the top two layers of the skin), the ridges will not grow in the same form as before (if at all) and therefore this user could be restricted in his future life by being excluded from the use of fingerprint recognition systems, though his fingers no longer have any symptoms of the skin disease. It is almost impossible to get to a fingerprint database with skin diseased users and with information about their disease. Creation of this type of database is even harder because of the cooperation of technicians with medical doctors and patients. On the other hand, there is not a reasonable way of testing how recognition algorithms can cope with skin diseases [8].

Skin is constantly being regenerated. A keratinocyte ("skin cell") starts its life at the lower layer of the epidermis (the basal layer), which is nourished by blood vessels and is supplied with nerve endings from dermis. The cell migrates upward from the basal layer to the stratum corneum (the outermost skin layer). During 4 weeks, the cell undergoes a series of changes, gradually flattening out and moving toward the surface. Then it dies and is shed. This physiological process can be negatively affected in many diseases of the skin. The epidermis is not supplied with blood vessels, but has nerve endings. The shape of the dermoepidermal junction basically forms the structure of ridges [8].

In most cases of dermatological disorders, we find a lot of changes in the ultrastructure of the skin, including the epidermis and dermis. There is often inflammation (inflammatory cells), atrophy or hypertrophy, fibrotisation and many other changes visible under the microscope. These differences result in changes of color (optical characteristics), changes of dermal vessels and capillaries (blood perfusion) and changes of elasticity and thickness of the skin (optical characteristics after pressure change) [8].

The first group represents diseases causing **histopathological changes of the epidermis and dermis**—these diseases usually cause problems for all kinds of fingerprint scanners, because they can influence either the color or the internal structure of the skin. The most common representatives of this group are [4,5]: hand and fingertip eczema, dyshidrosis, tinea, pyoderma, pitted keratolysis, pyogenic granuloma, systemic sclerosis or Raynaud's phenomenon [8].

The second group represents diseases causing **skin discoloration**—these diseases may cause problems for optical fingerprint sensors and also for sensors which use a fingerprint antispoof detection check based on the color or spectral analysis of the human skin. Typical representatives are [4,5]: macular drug eruptions and rashes in infectious diseases (hand, foot and mouth disease, scarlet fever, secondary syphilis, Kawasaki's disease), pitted keratolysis, Raynaud's phenomenon, xanthomas, carotenosis or hereditary hemorrhagic teleangiectasia [8].

The third group represents diseases causing **histopathological changes at the junction of the epidermis and dermis**—these diseases could cause structure changes underneath the skin at the junction between dermis and epidermis—i.e., in the area from which ultrasonic fingerprint scanners acquire fingerprint pattern images. Typical representatives are [4,5]: hand eczema, verruca vulgaris (warts), psoriasis or epidermolysis bullosa [8].

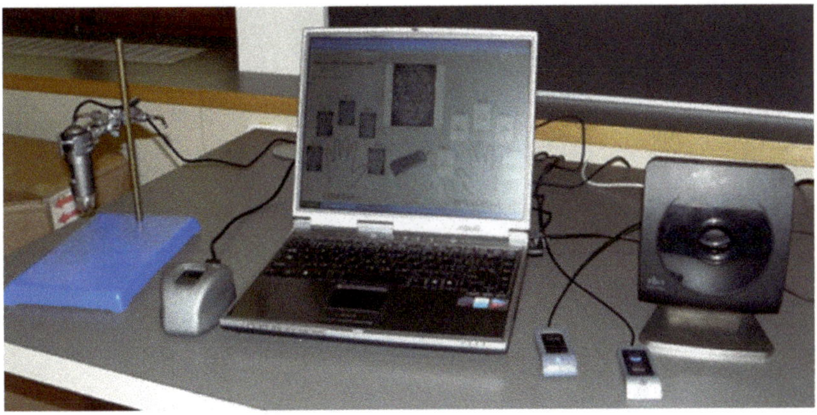

Figure 7.11 Workplace for acquirement of diseased fingerprints

As has been shown before, working with fingerprints that come from users with skin diseases can be very difficult. Our database with diseased fingerprints were created in cooperation with the University Hospital Olomouc (Czech Republic), using the supportive consultations from the St. Anne's University Hospital in Brno (Czech Republic) and the private dermatologic clinic in Darmstadt (Germany)—see the next section. In Figure 7.11, the workspace which was sent to the dermatologists to acquire fingerprints can be seen. It contains a 3D touchless and touch optical sensor, a sweep and touch capacitive sensor and a digital microscope. Some institutions also acquired fingerprints using a dactyloscopic card. Each image in the database has anonymized information about the patient, severity and type of disease. There are thousands of fingerprints in the database. The size of the database and the described unique information about them is the reason why the database is one of the few, if not the only one in the world. Therefore, it is perfect for studying what damage each disease does. In Figures 7.12–7.15, the results of disease-affected fingerprint data collection are shown. The name of the disease and description are given in each figure heading. The codename of applied capturing principle is stated under each subfigure. We use the following codenames: CRD (dactyloscopic card), MSO (Sagem MSO 300), E500 (UPEK Eikon Touch 500), EII (UPEK Eikon II), TBS (TBS 3D Enroll 2011), DIN (Dinolite) [6–8].

7.4 Description of the unique database with diseased fingerprints

The acquired database contains over 2,000 fingerprint images from patients suffering from various kinds of skin diseases. In total, 12 particular skin diseases were obtained [9–11]. The database was thoroughly analyzed in order to find any common features in the damage caused by the diseases. Features that were found

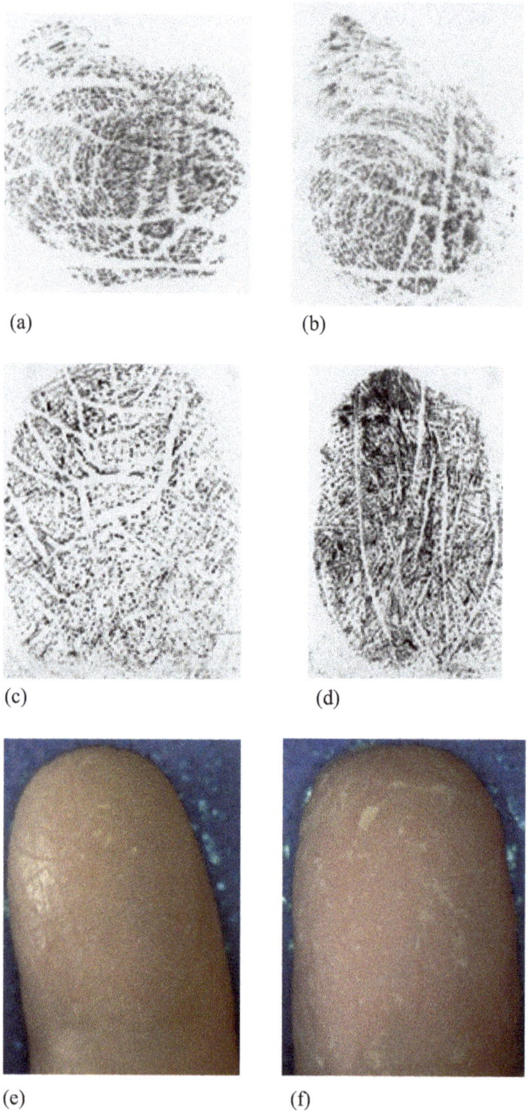

Figure 7.12 Fingertip eczema—a severe form: (a and b) CRD, (c and d) MSO and (e and f) DIN

were classified into five categories that are later used for the disease detection itself [11].

In this section, database analysis is given, as well as characteristics of each skin disease from the database and characteristics of the specific influence they have on the resulting fingerprint images.

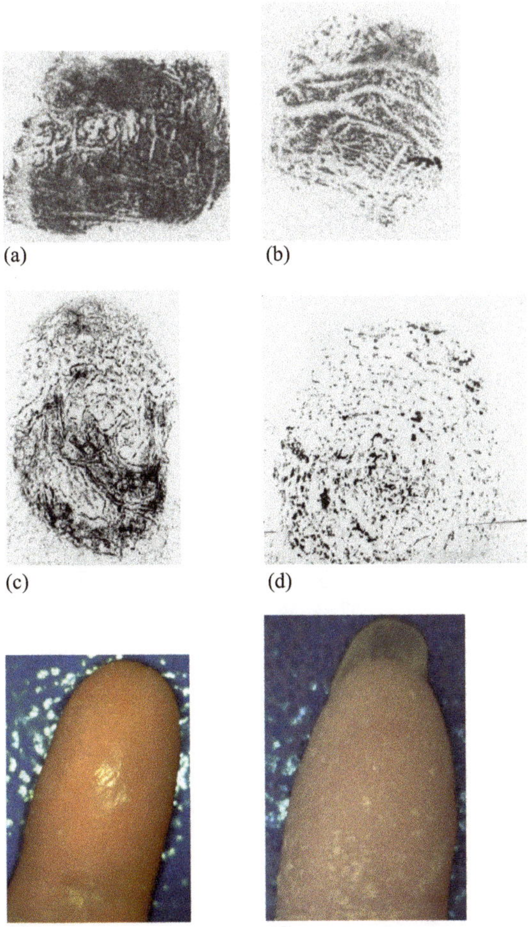

(a) (b) (c) (d) (e) (f)

Figure 7.13 Psoriasis—a full seizure: (a and b) CRD, (c and d) MSO and (e and f) DIN

7.4.1 Database analysis

The raw diseased fingerprint database was first analyzed in order to provide a solid foundation for future research. For every particular disease, common signs among all fingerprint images affected by this disease were found and a general description of each disease and its influences was defined. Based on these descriptions and sets of common signs and their frequencies, the diseased fingerprint images were classified into five categories. These categories are later used in the actual detection of the damaged areas in a fingerprint image and they help to divide the large detection task into smaller bearable parts [11].

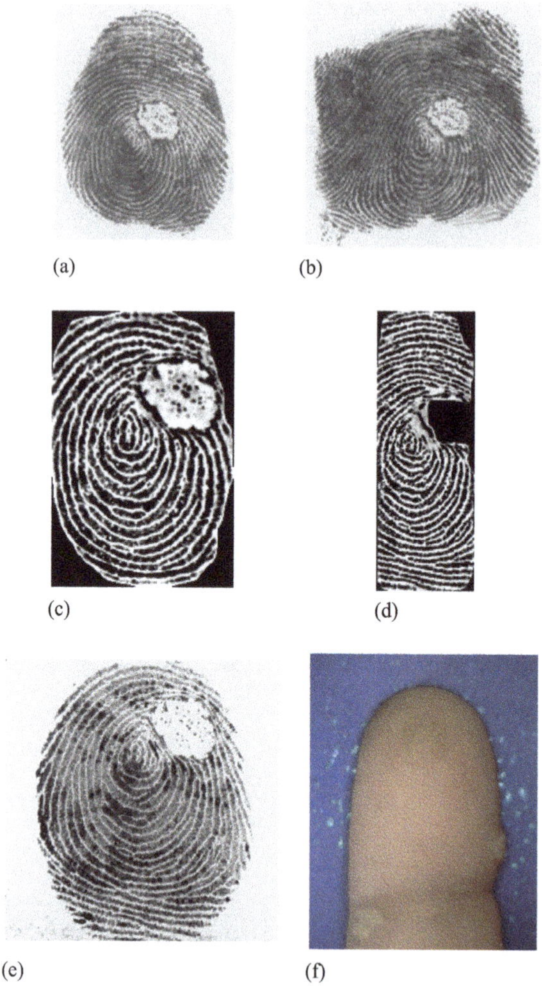

Figure 7.14 Verruca vulgaris (wart): (a and b) CRD, (c) E500, (d) EII, (e) MSO and (f) DIN

Most of the fingerprint images come from a dactyloscopic card. The number of fingerprints of each disease is displayed in Table 7.1 [12].

By observing and comparing the fingerprint images, 12 common features were defined, seven of them are local features [12]:

- straight lines (SL),
- a grid (G),
- small papillary lines disruptions (PLD),
- small "cheetah" spots (CS),
- larger round/oblong spots (ROS),
- large irregular spots (IS) and
- dark places (DP),

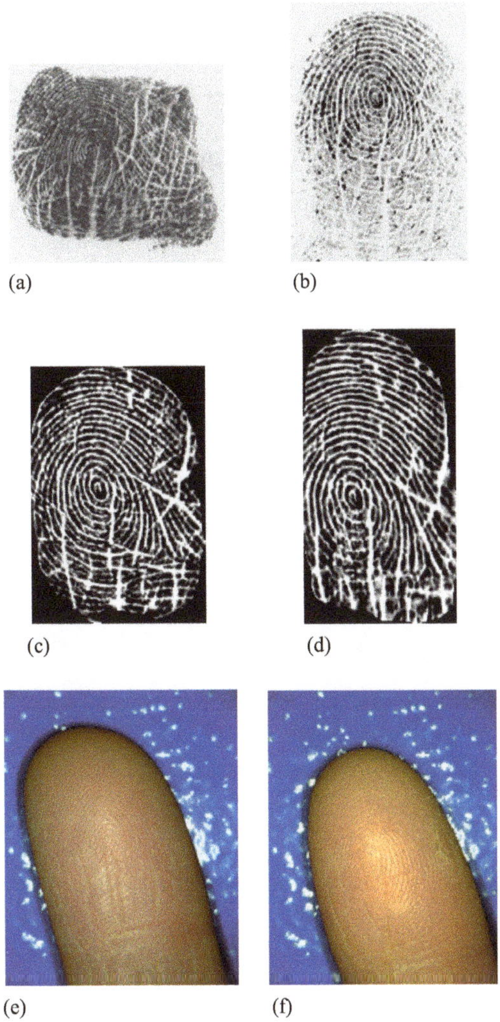

Figure 7.15 Colagenosis: (a) CRD, (b) MSO, (c) E500, (d) EII and (e and f) DIN

the other five were global image patterns [12]:

- blurriness of (parts of) the image (B),
- a significantly high contrast of the image (HC),
- the entire fingerprint area affected (EA),
- total deformation of the fingerprint image (TD) and
- a significantly high quality and healthy fingerprint (HQ).

For every disease, its image features were counted (see Tables 7.2 and 7.3). Fingerprint images obtained from optical scanners were excluded as their character is significantly dissimilar to the others. The actual number of images taken into account is stated in the column "sum."

Table 7.1 Database content [12]

Disease	No. of fingerprints in the DB	Percentages	No. of patients
Fingertip eczema	1,107	51.132	17
Psoriasis vulgaris	326	15.058	9
Dyshidrotic eczema	247	11.409	4
Hyperkeratotic eczema	118	5.450	2
Verruca vulgaris	96	4.434	4
Scleroderma	50	2.310	1
Acrodermatitis continua	40	1.848	1
Colagenosis	36	1.663	1
Raynaud's phenomenon	9	0.416	1
Effusion of fingers	35	1.617	1
Cut wound	18	0.831	2
"Unknown" disease	83	3.834	1
Total	**2,165**		**44**

Table 7.2 Local features of damaged fingerprint images

Disease	Percentages of particular features							Sum
	SL	G	PLD	CS	ROS	IS	DP	
Fingertip eczema	72.03	24.65	15.91	12.24	32.34	16.61	15.73	572
Psoriasis vulgaris	40.37	6.42	2.75	12.84	48.17	32.57	62.84	218
Dyshidrotic eczema	63.11	7.38	14.75	18.03	78.69	29.51	32.79	122
Hyperkeratotic eczema	3.92	0	66.67	15.69	74.51	3.92	5.88	51
Verruca vulgaris	3.17	0	14.29	12.70	74.60	0	25.40	63
Scleroderma	0	0	0	0	0	0	30.43	23
Acrodermatitis continua	14.29	0	0	85.71	60	14.29	65.71	35
Colagenosis	100	78.13	0	0	15.63	0	25	32
Raynaud's phenomenon	0	0	100	0	0	0	0	8
Effusion of fingers	10	0	73.33	43.33	63.33	6.67	13.33	30
Cut wound	93.75	0	0	0	18.75	0	12.50	16
"Unknown" disease	100	86.67	0	0	76.67	30	73.33	30

7.4.2 Characteristics of present diseases

This section gives an overview of all the diseases present in the database, their characteristics and description of their influence on resulting fingerprint images [11,12]. For detailed description of skin diseases mentioned in this section, please refer to [4,5,13]. This section deals with the detection of four of them: *atopic eczema*, *acrodermatitis*, *psoriasis vulgaris* and *verruca vulgaris*.

7.4.3 Fingertip eczema

Fingertip eczema is a very dry, inflammatory, noninfectious disease which occurs on the palmar surface or the fingertips. The skin becomes cracked and scaly, and

Table 7.3 Global features of damaged fingerprint images

Disease	**Percentages of particular features**					**Sum**
	B	**HC**	**EA**	**TD**	**HQ**	
Fingertip eczema	18.01	21.50	40.38	36.36	29.02	572
Psoriasis vulgaris	34.86	27.06	61.93	58.72	18.35	218
Dyshidrotic eczema	30.33	30.33	31.97	29.51	9.84	122
Hyperkeratotic eczema	31.37	29.41	9.80	0	37.25	51
Verruca vulgaris	19.05	80.95	7.94	7.94	76.19	63
Scleroderma	0	0	0	0	100	23
Acrodermatitis continua	48.57	25.71	100	100	0	35
Colagenosis	9.38	40.63	0	0	25	32
Raynaud's phenomenon	0	0	0	0	100	8
Effusion of fingers	23.33	16.67	40	16.67	3.33	30
Cut wound	37.50	68.75	0	0	50	16
"Unknown" disease	30	20	90	83.33	0	30

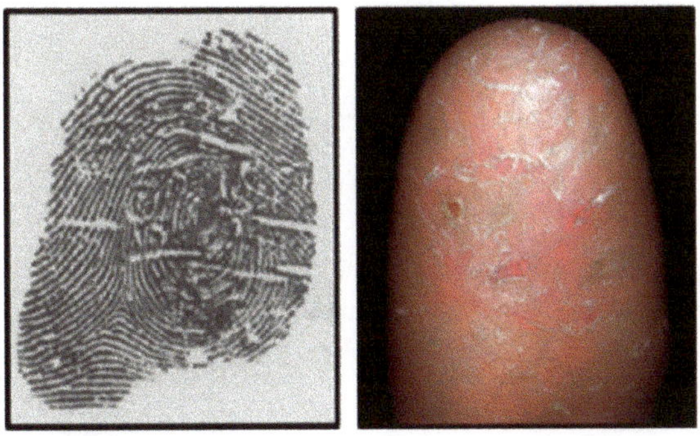

Figure 7.16 Fingertip eczema. Source: Database and [12]

usually starts peeling off which results in exposition of red and tender skin surfaces [4,5,13] (Figure 7.16).

As the number of fingerprints with fingertip eczema in the database is large, a wide range of typical features was observed. There are two groups of these fingerprints: (i) less and (ii) more severely damaged. In the first group of fingerprints, occurrence of thin lines of different directions was typical. These lines often connect or cross each other. In some cases, small round white spots were present, and in others, occasional dark areas make the papillary lines partially unreadable. However, overall, papillary lines of fingerprints of the first group are generally very well readable and it is possible to remove the influence of the disease from the fingerprint.

In the second group, the damage is more severe. Fingerprints are usually almost completely damaged, SL cover the entire fingerprint area and create grids by crossing each other. The background is darker and large IS can be seen. As the papillary lines cannot be seen at all, this type of damage is by no means recoverable.

7.4.4 Psoriasis vulgaris

Psoriasis is a common, chronic and inflammatory disease of the skin which is often indistinguishable from a serious form of hand eczema. It is characterized by dry and scaling plaques covered with dry scales that peel in layers [4,13].

The vast majority of fingerprints affected by psoriasis are completely damaged. Papillary lines are mostly unreadable. The most frequent feature is a large irregular dark spot bounded by a white border. Apart from this feature, the presence of larger dark areas or thick lines is also common, as well as ROS (Figure 7.17).

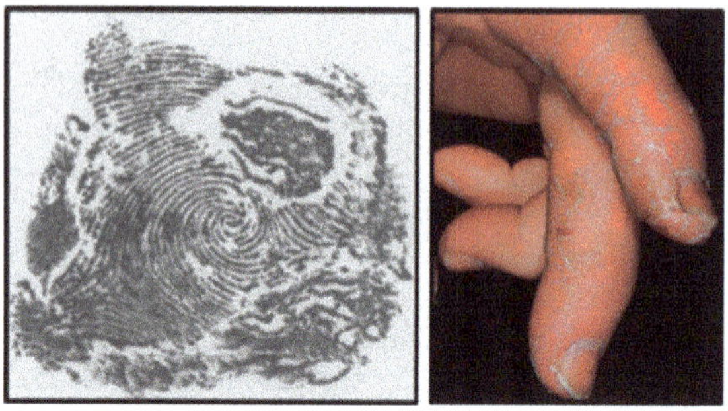

Figure 7.17 Psoriasis vulgaris. Source: Database and [4]

7.4.5 Dyshidrotic eczema

Also known as pompholyx, this disease is a variant of hand and foot dermatitis that makes skin extremely dry. Its typical features are itching vesicles and scales located on the palms and sides of fingers [4] (Figure 7.18).

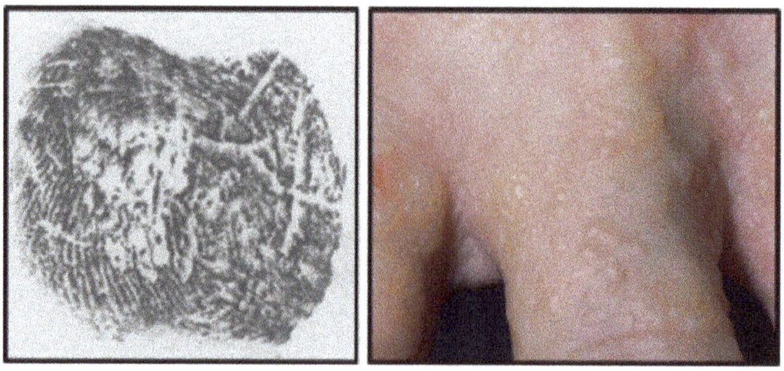

Figure 7.18 Dyshidrotic eczema. Source: Database and [5]

Fingerprint images damaged by dyshidrotic eczema are generally covered with irregular blurred shapes with no specific form. Another typical feature is a thick line. These fingerprints were divided into two groups, according to how severe the damage is. In the first group of less severely affected fingerprints, the entire area of a fingerprint is often covered, but papillary lines remain visible. Papillary lines are usually disrupted at multiple places and irregular blurred white spots may appear. Fingerprints in the second group are seriously damaged and cannot be repaired. The image area is typically covered by thicker lines in combination with large blurred white spots. Papillary lines are not sufficiently visible.

7.4.6 Hyperkeratotic eczema

A chronic form of hand eczema characterized by the occurrence of orange and brown scales with cracks between them [4,13]. Only one-third to one-half of the fingerprint area is usually affected. Sometimes, only the papillary lines are multiply disrupted. In other cases however, papillary lines are distorted and their direction is difficult to determine. Small-to-medium round spots are likely to be present (Figure 7.19).

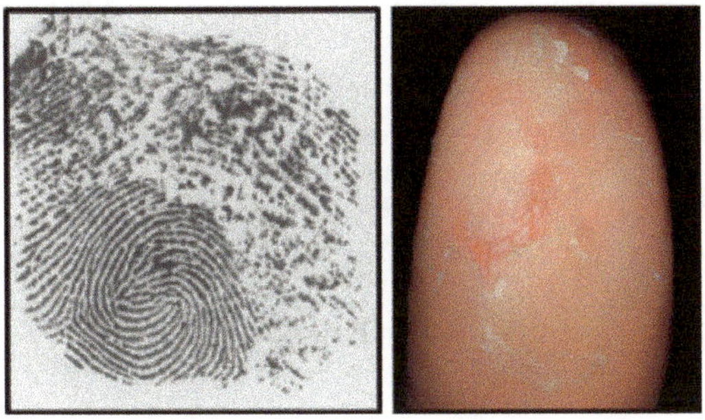

Figure 7.19 Hyperkeratotic eczema. Source: Database and [4]

7.4.7 Verruca vulgaris (warts)

This is a very common skin disease, characterized by the presence of stiff elevated bumps on the skin surface. They grow in size which is in average about 5 mm but can reach up to more than 1 cm. On their surface, tiny black dots may appear [12,13]. The influence of this disease on the fingerprint images is minor and easily removable. Typically, 1–4 round white spots occur, sometimes with black dots in their center (Figure 7.20).

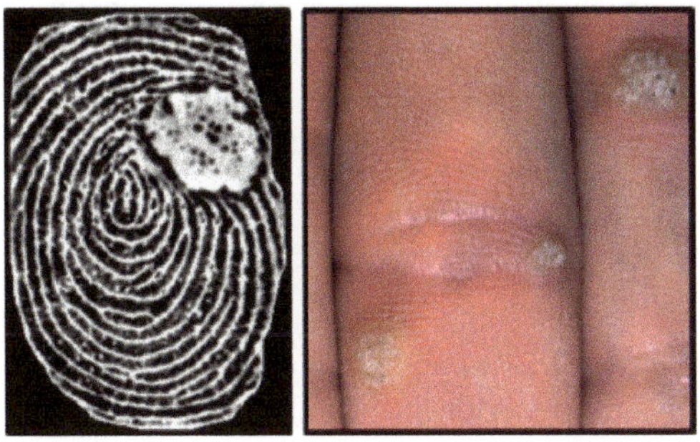

Figure 7.20 Verruca vulgaris. Source: Database and [4]

7.4.8 Systemic scleroderma

Scleroderma is characterized by the appearance of hard, smooth and ivory-colored areas. In the early stage, affected areas are red and swollen; later, they become completely immobile and lose their natural peaked contour [4,13]. The fingerprints in the database did not show any signs of damage. It can be therefore concluded that the number of acquired fingerprints was not sufficient to describe the disease's influence on fingerprint images (Figure 7.21).

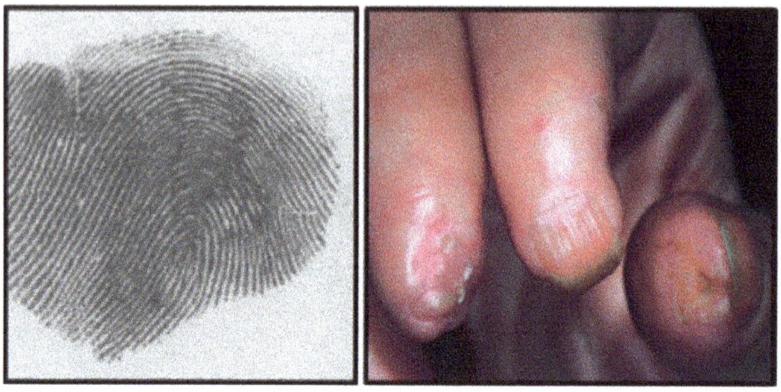

Figure 7.21 Systemic scleroderma. Source: Database and [4]

7.4.9 Acrodermatitis continua

Also known as acrodermatitis continua of Hallopeau or dermatitis repens, this disease is a chronic inflammatory disease of the hands and feet, and one of the less frequent types of psoriasis vulgaris. The outbreak of the disease is accompanied by asymmetric formation of pustules of the fingertips and continues with eruption of fresh pustules with hyperkeratosis and crusting. As the disease progresses, nails can even float away [13].

Fingerprint images are typical for the occurrence of small round spots that look like a cheetah skin and cover usually the whole fingerprint area. Larger oblong or round spots occur as well and SL or cracks are also not uncommon. Papillary lines cannot be recognized at all, and the original structure of the fingerprint is completely covered. Larger dark areas are often present and the spots can be blurred together. Almost in all cases, the fingerprint image is completely damaged and cannot be repaired (Figure 7.22).

7.4.10 Colagenosis

Colagenosis is a connective tissue and inflammatory autoimmune disease. The only typical feature of fingerprints with this disease is thin lines crossing each other. Under these lines, papillary lines are well visible (Figure 7.23).

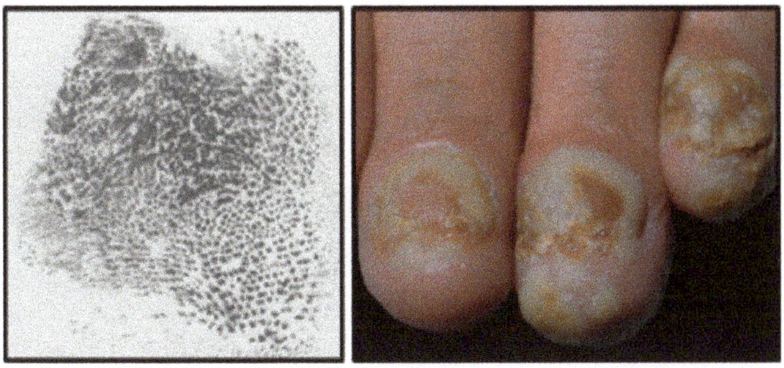

Figure 7.22 Acrodermatitis continua. Source: Database and [5]

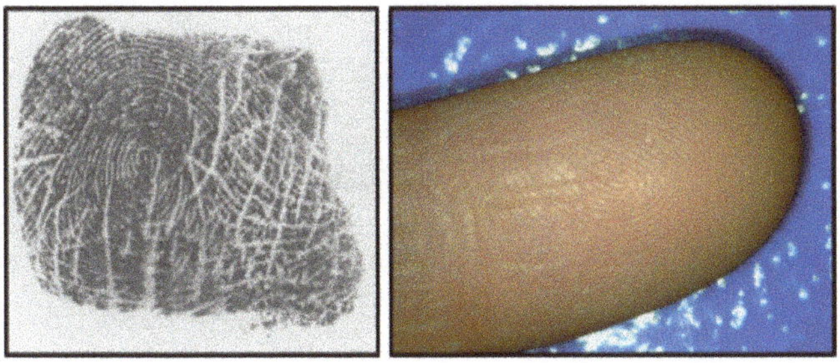

Figure 7.23 Colagenosis. Source: Database

7.4.11 Raynaud's phenomenon

It is a vascular skin disease that often accompanies an associated disease (most often scleroderma). The fingers have sequential discolorations: they first become pale and cold, then white, blue and finally red. This is caused by constrictions of the small arteries and arterioles in fingers [4,13]. As Raynaud's phenomenon causes discoloration only fingerprints in the database are always healthy and undamaged. Disease can be seen in Figure 7.24, left image shows papillary lines on sensor, middle image shows white fingers and right image shows blue color at the end of the fingers.

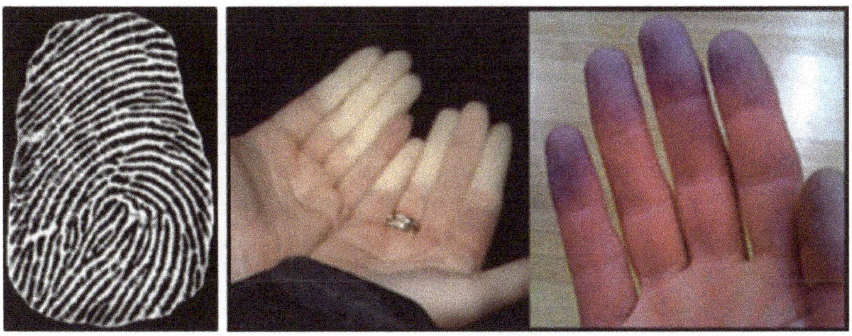

Figure 7.24 Raynaud's phenomenon. Source: Database and [14]

7.4.12 Effusion of fingers

Although being stated as a disease in the database, effusion of fingers is only a syndrome which manifests itself by a strong swelling. It is one of the symptoms of systemic scleroderma, for instance. Papillary lines are typically disrupted in many places, and small to medium spots are present. In general, papillary lines are clearly visible. Sometimes, however, white spots make them unreadable (Figure 7.25(a)).

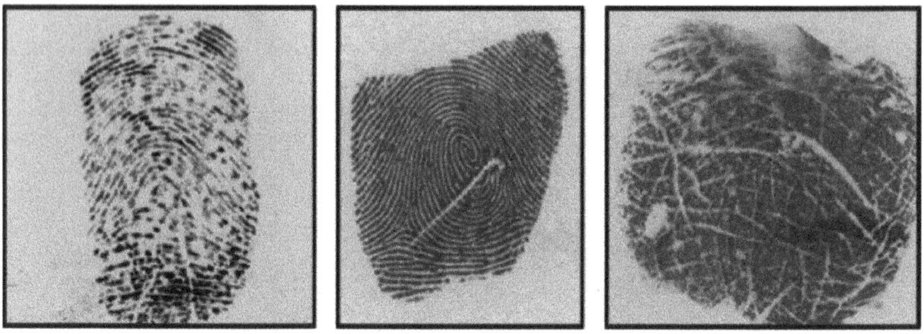

(a) (b) (c)

Figure 7.25 (a) Effusion of fingers, (b) cut wound and (c) "unknown" disease. Source: Database

7.4.13 Cut wounds

A cut wound typically causes either a SL in a fingerprint image or a more blurred white area. The damage is minor and should not be difficult to remove (Figure 7.25(b)).

7.4.14 "Unknown" disease

Fingerprints of this unnamed disease are totally covered with lines of different thickness and length and are therefore unreadable. They are very much alike those with fingertip eczema (Figure 7.25(c)).

7.4.15 Classification of damaged fingerprint images

Based on the analysis of the database, the diseased fingerprint images were classified into five basic feature classes. Such classification is supposed to help access each type of damage individually and facilitate the detection process. For each disease detector, a different combination of features to detect is chosen, which helps differentiating between signs of particular diseases and correctly determining the type of disease present in the fingerprint image.

7.4.15.1 Straight lines and grids

Under these circumstances, representatives are: fingerprint eczema, cut wound, colagenosis, dyshidrotic eczema, "unknown" disease (Figure 7.26).

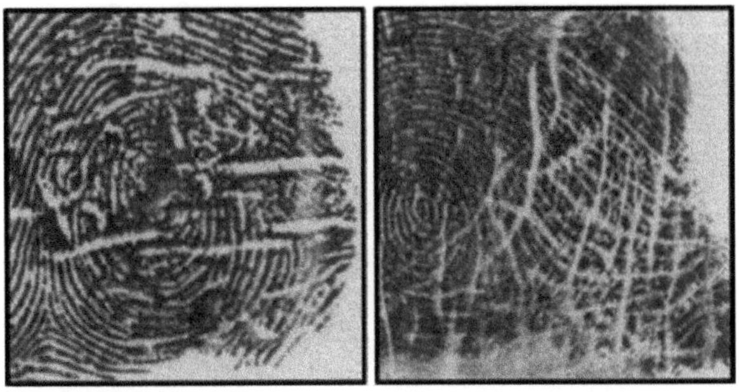

Figure 7.26 Example of fingerprint images with straight lines or grids. Source: Database

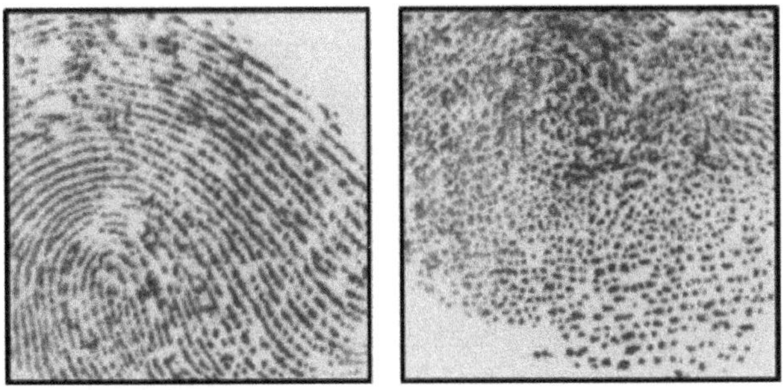

(a) (b)

Figure 7.27 Examples of (a) papillary lines disruptions and (b) "cheetah" spots. Source: Database

7.4.15.2 Small papillary lines disruptions

In this case, papillary lines are disrupted at multiple places but no significant damage is present. Representatives are: dyshidrotic eczema, hyperkeratotic eczema, effusion of fingers and fingertip eczema (Figure 7.27(a)).

7.4.15.3 Small "cheetah" spots

The only representative of this group is acrodermatitis (Figure 7.27(b)).

7.4.15.4 Round/oblong spots

Although ROS occur in most diseases, typical representatives with a significant amount of them are verruca vulgaris, effusion of fingers and psoriasis (Figure 7.28).

7.4.15.5 Large irregular spots

Psoriasis and severe form of fingertip eczema often cause extreme damage to the fingerprint and one of their features are also large spots of irregular shapes (Figure 7.29).

Figure 7.28 Example of fingerprint images with white spots. Source: Database

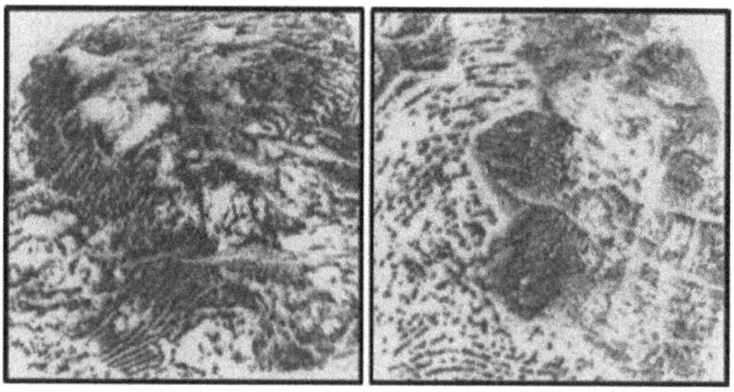

Figure 7.29 Example of fingerprint images with irregular spots. Source: Database

Also, diseases were classified into three categories according to the seriousness of the damage.

1. Minor damage: verruca vulgaris, Raynaud's phenomenon, cut wound, scleroderma.
2. Medium damage: mild form of fingertip eczema, mild form of dyshidrotic eczema, hyperkeratotic eczema, effusion of fingers, colagenosis.
3. Major damage (unrecoverable): acrodermatitis, severe form of fingertip eczema, severe form of dyshidrotic eczema, psoriasis, "unknown" disease.

7.5 Algorithms for dealing with damaged fingerprints

In this section, the specific algorithms used in the disease detector along with their advantages and shortcomings will be discussed, as well as the core methods essential for the program's functionality and data structures used to store and keep important data throughout the process.

7.5.1 Detector

There are three major algorithms that are used for the detection part: *Block orientation field, histogram analysis* and *flood fill*. Their combination provides valuable information about the fingerprint quality and character of the possible disease.

The detector uses a few special data structures. The first of them is cv::Mat, a data type implemented in OpenCV used for storing images [10], in other words a matrix of numerical values. The program makes use of this data type not only for keeping the processed images themselves, but also for storing the intermediate steps and StatusMaps. A StatusMap is a data structure that is used for the visualization of the extent of damage in the fingerprint. It consists of an $n \times m$ matrix (cv::Mat), where n is the number of columns and m is the number of rows. Both n and m are always smaller than the width and height of the input image so that the visualization can capture the global extent of damage in $w \times w$ subfields of the image. The values of this matrix are between -1 and 1. Negative values stand for background, 0 stands for a healthy area and positive values imply a damaged area, with 1 being the most damaged.

$$x = \begin{cases} (-1;0) & \text{background} \\ 0 & \text{healthy area} \\ (0;1) & \text{damaged} \end{cases} \tag{7.1}$$

Another essential data structure is the feature. It is used to store the signs of diseases extracted from the image and consists of a feature type, location of the first pixel, size and the exact pixels belonging to that particular area. It is used both for storing the detection results and visualizing them.

7.5.2 Block orientation field

The computation of block orientation field is commonly used in the fingerprint recognition process for the purposes of estimating the ridges direction and classifying the fingerprint image into one of the several fingerprint classes [1,13]. Because a typical fingerprint pattern consists of alternating dark and white lines, this information can be easily processed by a gradient operator that estimates the image gradient for each pixel. This low-level information is gathered and averaged for each $w \times w$ block in the image [12]. The transformation can result in a relatively smooth and continual image of the ridges direction estimates—for a healthy fingerprint of course—see Figure 7.30 on the left.

If we try to compute the block orientation field for a damaged or a partially damaged fingerprint however, we can easily recognize with the naked eye which areas contain possible damage, because the orientation field in these areas will be discontinuous, as displayed in Figure 7.30 on the right. Exceptions to this are the peripheral areas and deltas and cores.

These discontinuities can be detected by scanning the field for differences in direction angles. In the program's pipeline, a gradient-based method of block orientation field computation is used [12,15,16].

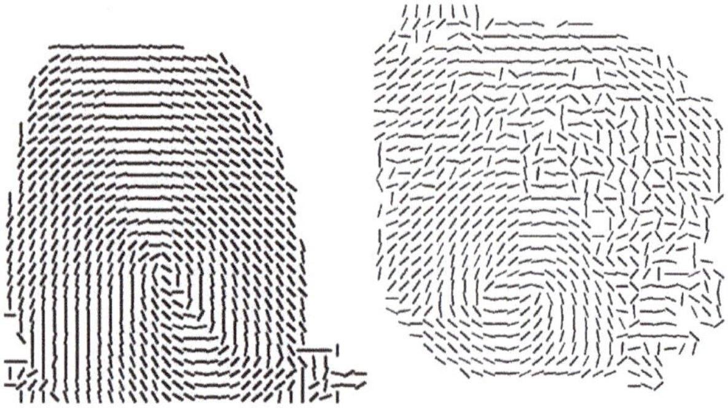

Figure 7.30 Examples of block orientation images (left: healthy fingerprint, right: fingerprint affected by a skin disease)

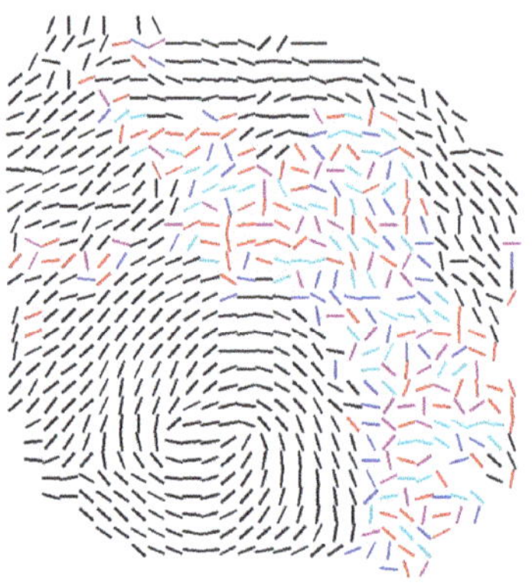

Figure 7.31 Damaged area detected using the orientation field

The resulting block orientation field is afterwards analyzed for any discontinuities that may occur. The analysis is done using a row-wise and column-wise scanning approach that reveals areas of possible damage in the fingerprint. Sometimes, the method detects single discontinuities that may be erroneous, and on the other hand, under different circumstances, one unmarked block may appear in the midst of discontinuous blocks. In order to make the algorithm as accurate as possible, although mistakes never disappear completely, these cases are taken into account. The algorithm handles them by copying the properties of their neighboring blocks (marking the single ones either as alright or as a discontinuity, depending on the neighborhood). Example detection is shown in Figure 7.31.

The advantage of this method is that it is already a part of the standard fingerprint recognition pipeline, so the algorithm can be easily implemented into existing methods. Also, it provides a fairly accurate estimate of the fingerprint damage in the sample. However, it is not always able to detect local damages, such as spots or lines. For this reason, we use the flood fill algorithm. Before that one is

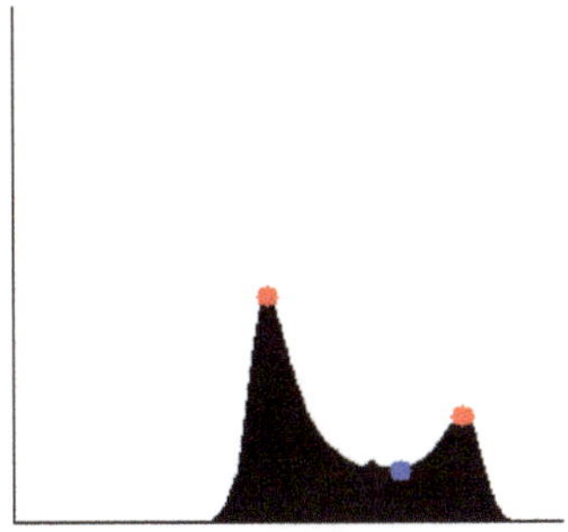

Figure 7.32 A nearly ideal bimodal histogram

explained, however, let's take a look at how analyzing histograms from image subfields can be used to obtain useful information about damage.

7.5.3 Histogram analysis

This method is based on the presumption that a fingerprint image with a high quality consists of equally distributed ridges and valleys. If we assume that ridges are roughly the same dark color while valleys are light-colored, a histogram computed from each subfield of the fingerprint's area should ideally consist of two peaks of approximately the same height and one valley between them. The transition between the peaks and the valley should be smooth, as displayed in Figure 7.32, and the peaks' height difference vary slightly according to the width of ridges in the image. This shape is called bimodal.

On the other hand, the intensity distribution in a fingerprint image part that belongs to a damaged area is not always as equal as in the quality one. Thus, if a histogram is computed for this subfield, it is very likely that it will not have the ideal bimodal appearance as described above. Experiments showed that the majority of damaged areas break the rules of the bimodal histogram. The lower the quality, the less the histogram resembles the ideal one. A nonbimodal histogram always implies a damaged or low-quality area.

At the same time, however, there is a certain percentage of damaged areas whose histograms still fall into the valid category. A damaged subfield therefore does not necessarily imply a nonbimodal histogram. This is due to the fact that a histogram is a measure for the distribution of intensities only and it does not take into account the pattern or neighborhoods of pixels. Figure 7.33 shows examples of invalid histograms.

Figure 7.34 shows an example output of this method, along with the particular histograms that were being analyzed. Red background implies an invalid histogram, green means valid and blue stands for background.

Since the histogram analysis method is an experimental one, its results are not always accurate. The major drawback of this method is its inability to cope with low-quality, especially dark, images. By implementing appropriate preprocessing steps, the method's performance and accuracy can be improved.

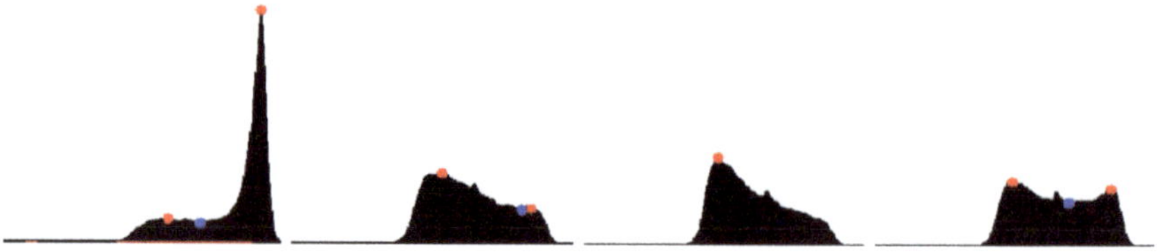

Figure 7.33 Examples of invalid histograms

Figure 7.34 Histogram analysis result with particular histograms

Although the method can never find all damaged areas, histogram analysis is able to detect many areas the block orientation field method might have omitted. The sets of damaged areas detected from these two methods are never identical; therefore, the histogram analysis method is extremely valuable for the final determination of healthy areas.

7.5.4 Flood fill

Flood fill is a very well-known algorithm used for graphical purposes [17] and is especially handy for detecting and filling connected single-colored areas of an image. This characteristic was used in the application in order to find local features of damaged fingerprints, such as SL or spots.

For the detection of such features, the Hough transform [18] was initially used, but it was later rejected for its inaccuracy. Flood fill turned out to be far more exact and appropriate. The flood fill algorithm has three parameters: a target color, a replacement color and a start pixel. It is based on examining the color of all pixels in the 4- or 8-neighborhood of the start pixel and changing the color of those pixels that have the target color to the replacement color. Using either recursion or stack/queue, the colored pixels become the next start pixels and the process is repeated. In the end, the entire single-colored area is filled.

In our case, for better memory management, the scanline flood fill algorithm [17] is used. This one is extended by a stack and differs from its basic version by a

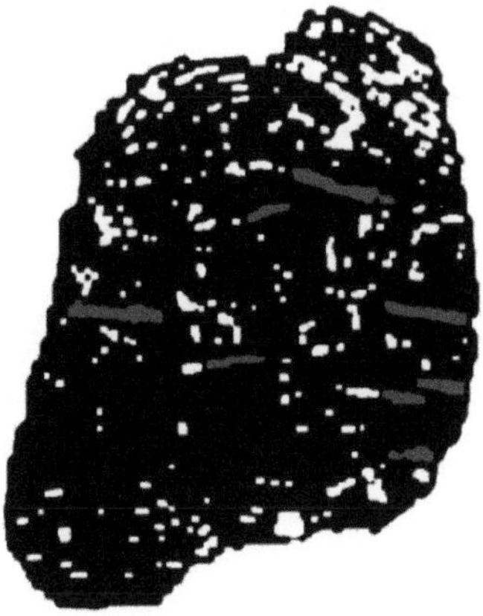

Figure 7.35 Extraction of straight white lines

reduced space and time complexity, which is achieved by filling whole lines instead of single pixels. Also, it is able to retrieve all points belonging to the area and store them later in the Feature class. When flood fill is used for a fingerprint image, the sample first needs to be preprocessed in order to obtain a black and white image that can be used as an input for the algorithm. The preprocessing steps are tricky because they heavily depend on the image quality, as well as the type of sensor used for the acquisition. In this implementation, all algorithms are tailored for our internal fingerprint database, in particular, for the fingerprints from dactyloscopic cards.

The preprocessing steps consist of contrast and brightness adjustment, a series of dilations, erosions, closing and opening operators, combined with fingerprint area detection according to [19], Gaussian blur and thresholding (Figure 7.35).

There are four types of features the flood fill algorithm is programed to detect: large white spots, thick white lines, small CS and PLD (for explanation of the groups)

Apart from the algorithm, the flood fill class also enables us to set parameters that closely specify how big the filled areas should be and what shape they should have. The shape's determination is based on the ratio between the longer and the shorter side of the area's bounding rectangle: if it is below 1.8, it is considered round, and if it is over 2.3, it is considered oblong. Others are not taken into account. Thanks to the parameters, it is possible to tailor the results for different detectors.

7.5.5 Methods merging using a status map

All three of the above-described methods detect a different kind of damage in the image and only flood fill provides logically structured results that can be used in classification.

However, connecting the three methods together results in a surprisingly accurate description of the extent of damage in an entire area of a fingerprint

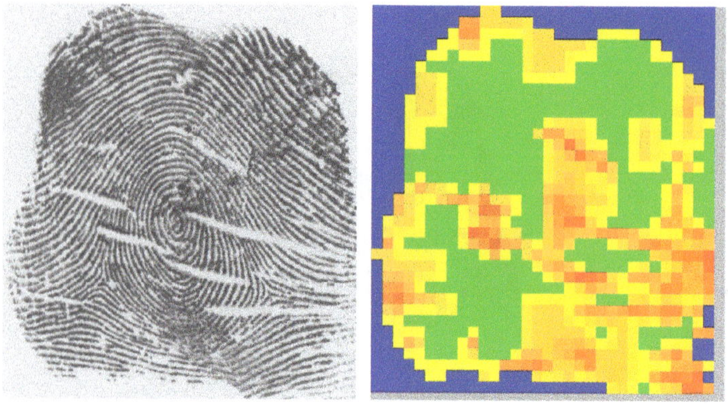

Figure 7.36 Example of a final StatusMap

image. At the end of each detection process, every image pixel is assigned a value between −1 and 1. Negative values stand for background, 0 means a healthy area and positive values indicate damage. The higher the value, the more damaged the area to which the pixel belongs, as explained in the previous parts.

The challenge was to connect these three output matrices together into a so-called StatusMap which would give a good overview of the damage state every $w \times w$ block of pixels.

For the purpose of a correct merge, all methods' output pixel values were limited to nonnegative.

The information about background is stored separately using Background-Extractor. This extraction method marks pixels −1 (background) or 1 (fingerprint area) and produces a fourth matrix of pixel values. The resulting StatusMap gives a very good overview of the damage. The right side image of Figure 7.36 is the StatusMap. The blue surroundings is background, green parts show healthy area and everything else shows damaged area (on yellow to red scale – red color is the most damage area).

7.5.6 Classifier

The classifier decides based on features extracted by the flood fill algorithm and classifies the fingerprint image, according to the features' numbers, sizes and shapes into one of these six categories: acrodermatitis, atopic eczema, psoriasis, verruca vulgaris, unknown disease or healthy.

In order to determine the decision rules, a script that counted the numbers and types of detected features from the whole database for each disease was implemented. Medians and standard deviations of these numerical values were used to support the Classifier's decision—see Table 7.4.

Given the normal probability distribution, it is supposed that the majority of values are going to be one standard deviation away from median. Then, a significant amount of values will lay two standard deviations away from median and almost no values will be farther. These characteristics are used in order to compute an estimated likelihood that a certain set of features belong to a particular disease.

Table 7.4 Statistics of features extracted from each disease

	Acrodermatitis		Atopic eczema		Psoriasis		Verruca vulgaris	
	Med.	**Std. dev.**	**Med.**	**Std. dev.**	**Med.**	**Std. dev.**	**Med.**	**Std. dev.**
White sp.	5	3.97	5	4.31	8	5.35	1	3.02
Lines	2	1.84	3	3.06	4	2.65	1	1.63
Cheetah sp.	47	42.70	29	17.50	21	19.61	18	10.90
Disruptions	7	8.37	17	19.80	8	9.22	15	39.76

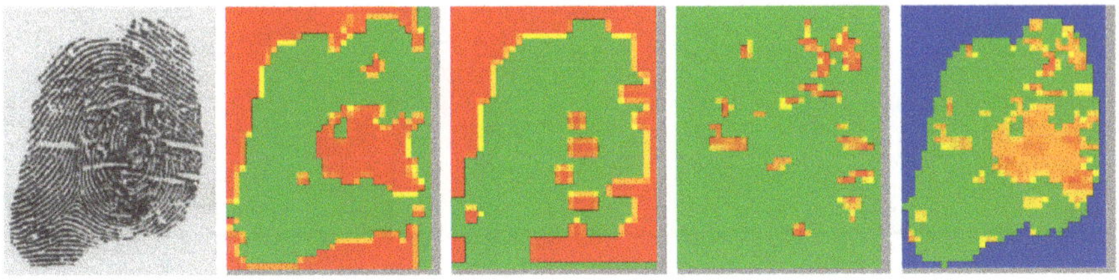

Figure 7.37 Example of the pipeline of StatusMaps and the final distribution of damage in the image (atopic eczema). Green color marks the healthy areas, blue color highlights the background and for the damaged areas, a scale from yellow to red is used. Yellow stands for minor damage, whereas red implies extremely damaged places

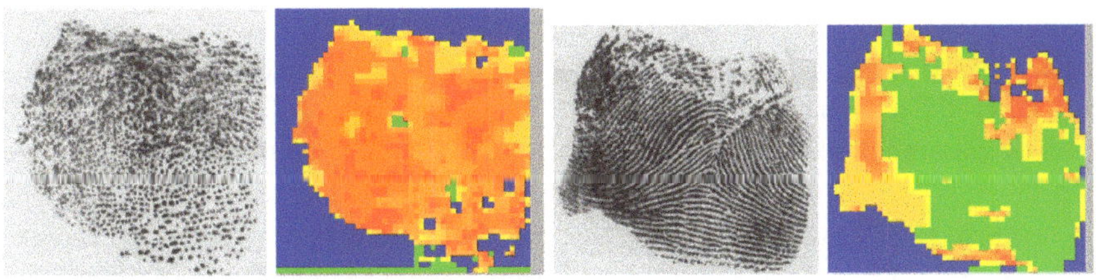

Figure 7.38 Damage detection results. Green color marks the healthy areas, blue color highlights the background and for the damaged areas, a scale from yellow to red is used. Yellow stands for minor damage, whereas red implies extremely damaged places

7.5.7 Damage localizer results

Each of the three detection methods separately provides interesting outputs, but it is their connection that makes the resulting application so notable. Thanks to the connection, very satisfactory results have been achieved for locating the damaged areas (Figures 7.37 and 7.38).

Table 7.5 Rejected and accepted samples

	FAR	FRR	F1	ACC
Acrodermatitis	0.1394	0.6667	0.1655	0.8347
Atopic eczema	0.1968	0.7021	0.4300	0.4533
Psoriasis	0.3408	0.7373	0.1956	0.5827
Verruca vulgaris	0.2329	0.5000	0.2073	0.7496

Table 7.6 Classifier accuracy measures

	TP	FN	FP	TN
Acrodermatitis	12	18	103	478
Atopic eczema	134	289	25	163
Verruca vulgaris	23	17	314	257
Total	611	611	611	611

7.5.8 Classifier accuracy

The classifier itself is ready to be further extended and improved. It relies on the detection results. So far, the following accuracy measures have been computed: FAR (*false accept rate*) and FRR (*false reject rate*) [20], ACC (*accuracy*) and F1 score [20]—see Table 7.5. A total of 611 fingerprint images from dactyloscopic cards from the database were used for the test. Table 7.6 shows the numbers of fingerprint images that were correctly/incorrectly classified. TP (*true positives*) = number of positives that were correctly accepted, FN (*false negatives*) = number of negatives that were incorrectly rejected, FP (*false positives*) = number of positives that were incorrectly accepted and TN (*true negatives*) = number of negatives that were correctly rejected.

The classification accuracy reached high values for acrodermatitis (83.5%) and verruca vulgaris recognition (75.0%), whereas it was lower for atopic eczema (45.3%) and psoriasis (58.3%). Better performance could be gained by improving the classification decision rules, as well as coming up with new types of features detection.

7.6 Conclusion

This chapter includes a description of fingerprints which are common in a population, however, cause big troubles in automatic processing and recognition. In the past, we did a lot of acquisition tests of problematic fingerprints, including acquirement of a database with diseased fingerprints. Furthermore, we invested our effort to development of algorithms for detection of various troubles in fingerprints.

Before the methods were designed, the faculty's database of fingerprint images affected by skin disease was thoroughly analyzed and, based on this foundation, possible algorithms for damaged area detection were tested, including the Hough detection for lines and circles, which was rejected for its inaccuracy. In the end, the following methods were implemented: detection from block orientation field, histogram analysis method and the flood fill method. The best results were achieved by connecting the methods together using a special data structure, StatusMap.

The classifier makes decisions based on statistics that resulted from testing the algorithms on the whole fingerprint database. Using the methods described in this work, the program reached an accuracy of 83.5% for acrodermatitis, 45.3% for atopic eczema, 58.3% for psoriasis and 75.0% for verruca vulgaris.

The resulting program is a graphical user interface (GUI) application that enables the user to load an image, adjust the processing pipeline and view and save the results. It can be used as an analytical tool for future researchers.

There is a great potential for improvements and enhancements, and it is assumed that the research will continue. There are opportunities for the results of this research to be used in real-life applications in the future, such as medical applications or programs for police and security purposes.

At the moment, we work on algorithms for detection of concrete troubles in fingerprints and other skin diseases. Furthermore, we work on development of algorithms for simulation of diseases and damages into synthetic fingerprints.

Acknowledgment

This work was supported by The Ministry of Education, Youth and Sports of the Czech Republic from the National Programme of Sustainability (NPU II); project IT4Innovations excellence in science—LQ1602 and BUT grant "Secure and Reliable Computer Systems" FIT-S-17-4014.

References

[1] Maltoni, D.; Maio, D.; Jain, A.K.; *et al. Handbook of Fingerprint Recognition*, 2nd Edition, Springer, 2009. ISBN 978-1-84882-253-5.

[2] Drahanský, M.; Orság, F. *Biometrie*, 1st Edition, CRC Press, 2011. ISBN 978-80-254-8979-6.

[3] Kanich, O.; Drahanský, M. *Simulation of Synthetic Fingerprint Generation Using Petri Nets*. IET Biometrics, 2017, Vol. 6, No. 6, pp. 402–408. ISSN 2047-4938.

[4] Habif, T.P. *Clinical Dermatology*, 4th Edition, Mosby, China, 2004, p. 1004, ISBN 978-0-323-01319-2.

[5] Wolff, K.; Johnson, R.A.; Suurmond, D. *Fitzpatrick's Color Atlas and Synopsis of Clinical Dermatology*, 5th Edition, McGraw-Hill, USA, 2005, p. 1085, ISBN 0-07-144019-4.

[6] Drahanský, M.; Kanich, O. *Vulnerabilities of Biometric Systems*, Conference Security and Protection of Information, Brno, 2015, pp. 53–60. ISBN 978-80-7231-997-8.

[7] Spurný, J.; Doležel, M.; Kanich, O.; Drahanský, M.; Shinoda, K. *New Materials for Spoofing Touch-based Fingerprint Scanners*, Proceedings of International Conference on Computer Application Technologies, Matsue, 2015, p. 15. ISBN 978-1-4673-8211-3.

[8] Drahanský, M.; Kanich, O.; Březinová, E. *Is fingerprint recognition really so reliable and secure?, Challenges for fingerprint recognition—spoofing, skin diseases and environmental effects*, Handbook of Biometrics for Forensic Science. Springer, p. 21, ISBN 978-3-319-50671-5.

[9] Doležel, M.; Drahanský, M.; Urbánek, J.; Březinová, E.; Kim, T. *Influence of Skin Diseases on Fingerprint Quality and Recognition*. In: New Trends and Developments in Biometrics, London: InTech, 2012, pp. 275–303.

[10] Barotová, Š.; Drahanský, M. *Fingerprint Damage Localizer and Detector of Skin Diseases from Fingerprint Images*, Proceedings of the 16th International Conference of the Biometrics Special Interest Group. Darmstadt: Springer Verlag, 2017, pp. 267–274. ISBN 978-3-88579-664-0.

[11] Drahanský, M.; Pernický, R.; Kanich, O.; Barotová, Š. *Verarbeitung von beschädigten Fingerabdrücken in der polizeilichen Praxis*. DuD—Datenschutz und Datensicherheit, 2017, Vol. 41, No. 7, pp. 407–414. ISSN 1614-0702.

[12] Barotová, Š. *Detector of Skin Diseases by Fingerprint Technology*. Bachelor's thesis. FIT BUT, 2017.

[13] James, W.D.; Berger, T.G.; Elston, D.M. *Andrews' Diseases of the Skin Clinical Dermatology*, 10th Edition, Elsevier Inc., Canada, 2006. ISBN 0-7216-2921-0.

[14] Shiel, Jr., W.C. *Raynaud's Phenomenon*. 2016. Retrieved from on 2017-11-01: http://www.medicinenet.com/raynauds_phenomenon/page4.htm.

[15] Lin, H.; Yifei, W. *Fingerprint Image Enhancement: Algorithm and Performance Evaluation*. IEEE Transactions Pattern Analysis and Machine Intelligence, 1998, Vol. 20, No. 8, pp. 777–789.

[16] Wieclaw, L. *A Review on Fingerprint Estimation Methods*. Journal of Medical Informatics & Technologies, 2012, Vol. 21, pp. 95–102. ISSN 1642-6037.

[17] Godse, D.A. *Computer Graphics*. Technical Publications, 2008. ISBN 9788189411589.

[18] Gary Bradski, A.K. *Learning OpenCV*, O'Reilly Media, Inc., 2008. ISBN 978-0-596-51613-0.

[19] Doležel, M. *Segmentation Procedure for Fingerprint Area Detection in Image Based on Enhanced Gabor Filtering*. International Journal of Bio-Science and Bio-Technology, December 2010, Vol. 2, No. 4, pp. 39–50.

[20] Powers, D. *Evaluation: From Precision, Recall and F-measure to ROC, Informedness & Correlation*. Journal of Machine Learning Technologies, 2011, Vol. 2, pp. 37–63. ISSN 2229-3981. Retrieved from on 2017-09-05: http://www.bioinfo.in/contents.php?id=51.

Chapter 8
Generating synthetic fingerprints

Raffaele Cappelli[1], *Matteo Ferrara*[1], *and Davide Maltoni*[1]

Synthetic fingerprint generation (SFinGe) techniques and associated tools (e.g., SFinGe) were introduced more than 15 years ago [1]. The main aim was to generate large databases for performance evaluation without allocating huge amount of resources for acquisition campaigns and, at the same time, to conform with the privacy directives that in many countries limit the exchange of biometric data. While the original scope remains central today, since the generation of very large synthetic dataset is crucial to predict accuracy on very large scenarios, new security needs (such as detecting altered fingerprints) and algorithms improvements (supervised learning approaches) are continuously renewing interest in the generation of synthetic fingerprints.

8.1 Introduction

SFinGe allows to create, without any privacy issues and at negligible costs, large fingerprint databases, whose characteristics can be tuned according to the specific application goals. In addition, ground-truth data on the main fingerprint features can be automatically made available by the generation software. During the last 15 years, SFinGe has been successfully adopted in several contexts: its main applications are here summarized [2].

- *Performance evaluation*—although real fingerprint databases cannot be completely substituted, especially when performance has to be measured referring to a given real environment/application, in many cases synthetic generation can be an effective tool to overcome the problem of collecting large fingerprint databases. In particular, synthetic fingerprints proved to be well suited for technology evaluations [3]: a comparison of the behavior of several fingerprint comparison algorithms on real and synthetic databases showed that not only the performance is very similar, but the genuine/impostor distributions and the False Match Rate (FMR)/False Non-Match Rate (FNMR) curves are also surprisingly close [4]. For instance, in [5], a synthetic database containing 2

[1]Department of Computer Science and Engineering, University of Bologna, Italy

million fingerprints was used to perform scalability experiments on a new fingerprint indexing method.

- *Training*—many classifiers and pattern recognition techniques require a large training set for an accurate learning stage. This is especially true when considering recent deep learning approaches (e.g., ConvNet) whose efficacy is proportional to the amount of training data and associated labels. Synthetic fingerprint images are very well suited to this purpose: in fact, the generator input parameters allow to explicitly control the type and features of the generated sets (e.g., class, type of noise, distortion), and this can be exploited in conjunction with boosting techniques to drive the learning process. For example, in [6], a large synthetic training set was successfully used to derive optimal parameters for fingerprint indexing.
- *Security evaluation*—synthetic generation can produce large sets of fingerprints whose features (e.g., minutiae distribution) can be varied independently of other fingerprint characteristics (e.g., orientation field). For this reason, synthetic fingerprints can help to test the robustness of fingerprint verification systems against "hill-climbing" attacks and "Trojan horse" attacks [7].
- *Semantic conformance to standards*—algorithms syntactically compliant to standards such as the ISO/IEC 19794-2 [8] are often not semantically compliant (for instance, the location, direction, and type of minutiae extracted from the same finger image by different minutiae extraction algorithms is often different). This leads to interoperability problems, which are difficult to be found in advance, since testing semantic conformance to a minutiae extraction standard would require a lot of data with manually labeled minutiae points (ground-truth); furthermore, in low-quality areas, even the manual labeling of minutiae points is not reliable. The automatic generation of ground-truth data, which is possible with synthetic fingerprints, allows semantic conformance and interoperability studies to be easily carried out. For instance, in [9], a synthetic database has been used to analyze the distribution of minutiae positions and directions of some algorithms with respect to the ground-truth.

The SFinGe approach was first introduced in [1], improved in [10], and comprehensively described in [4]. Although other generation techniques exist (see [4] for a brief literature review), today SFinGe remains the most widely used tool by the research community and public/private organizations.

In Section 8.2, we briefly summarize the SFinGe approach. Then in Section 8.3, we explain how the synthetic generator has been recently improved to better mimic the natural feature distribution. Section 8.4 discusses large-scale accuracy prediction based on synthetic data and reports the results of a large-scale experiment on 10 million synthetic fingerprints. Section 8.5 introduces the generation of altered fingerprints, a type of fingerprint typically not available to researchers and where synthetic generation can play a crucial role. In Section 8.6, we discuss system improvement/tuning based on synthetic data, and finally in Section 8.7, we provide some concluding remarks.

8.2 The SFinGe approach

The typical steps involved in fingerprint feature extraction are summarized in Figure 8.1: (i) the fingerprint is first segmented from the background, (ii) the local frequency and orientation maps are then estimated, (iii) based on this information the ridge–valley pattern is enhanced and the minutiae are detected. In order to generate fingerprint images, SFinGe "inverts" the above steps (see Figure 8.2): a fingerprint area, an orientation image, and a frequency image, randomly generated

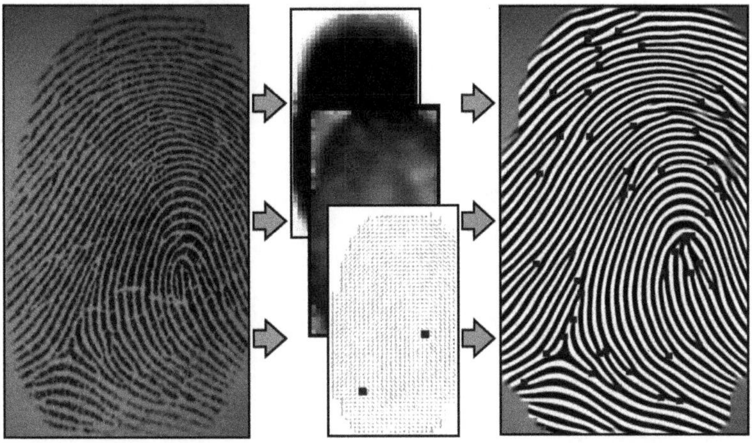

Fingerprint image — Fingerprint area, frequency image, and orientation image — Ridge pattern and minutiae points

Figure 8.1 A typical fingerprint feature extraction process. © 2009 Springer. Reprinted, with permission, from Reference [4]

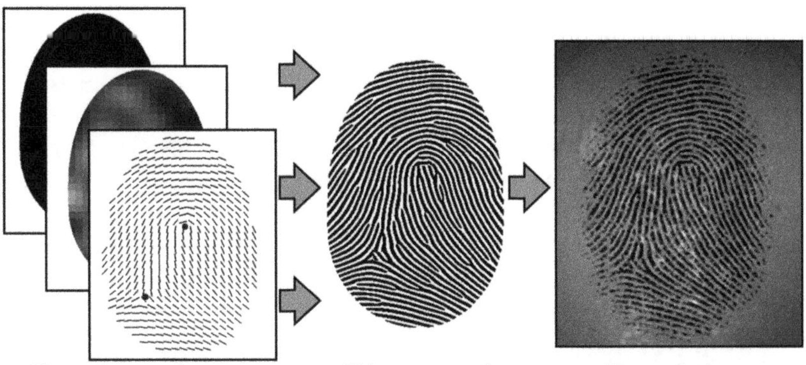

Fingerprint area, frequency image, and orientation image — Ridge pattern and minutiae points — Fingerprint image

Figure 8.2 The basic idea of the fingerprint generation method. © 2009 Springer. Reprinted, with permission, from Reference [4]

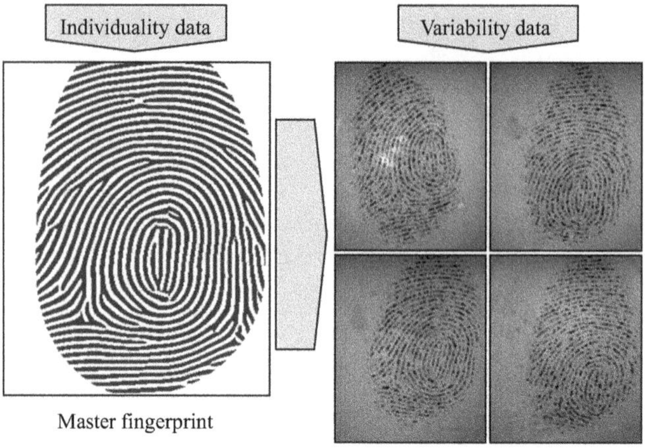

Figure 8.3 Starting from a set of parameters that represent the unique and immutable features of a "synthetic finger" (individuality data), SFinGe creates a master fingerprint; then a number of fingerprint impressions can be generated by changing several parameters that control the fingerprint appearance (variability data). © 2009 Springer. Reprinted, with permission, from Reference [4]

according to ad-hoc models, are the inputs to a ridge generation module; the resulting binary ridge pattern is rendered by adding fingerprint-specific noise.

To produce multiple impressions of the same "synthetic finger," a more complex approach is necessary: a *master fingerprint* (i.e., a ridge pattern that represents the unique and immutable characteristics of a "synthetic finger") is first generated; then, several impressions can be derived from the master fingerprint by explicitly tuning displacement, rotation, distortion, skin condition, and noise (see Figure 8.3).

Figure 8.4 shows the complete generation process: Steps 1–4 lead to the generation of a master fingerprint; Steps 5–10 are repeated for each fingerprint impression derived from the master fingerprint.

The most important step is the generation of the ridge-line pattern from the orientation and frequency images. Here, by iteratively applying contextual filtering (starting from a few random points), we grow regions whose merging leads to the formation of minutiae (terminations and bifurcation) at somewhat random positions. See [4] for more details.

Figure 8.5 shows some synthetic fingerprints generated by SFinGe using different backgrounds. It is worth noting that the fingerprint images appear realistic.

8.3 Tuning SFinGe to mimic natural feature distributions

A number of experiments have been performed in the past to check whether synthetic fingerprints generated by SFinGe are good surrogates of natural fingerprints or not. In general, visual analysis from biometric experts and cross-testing of

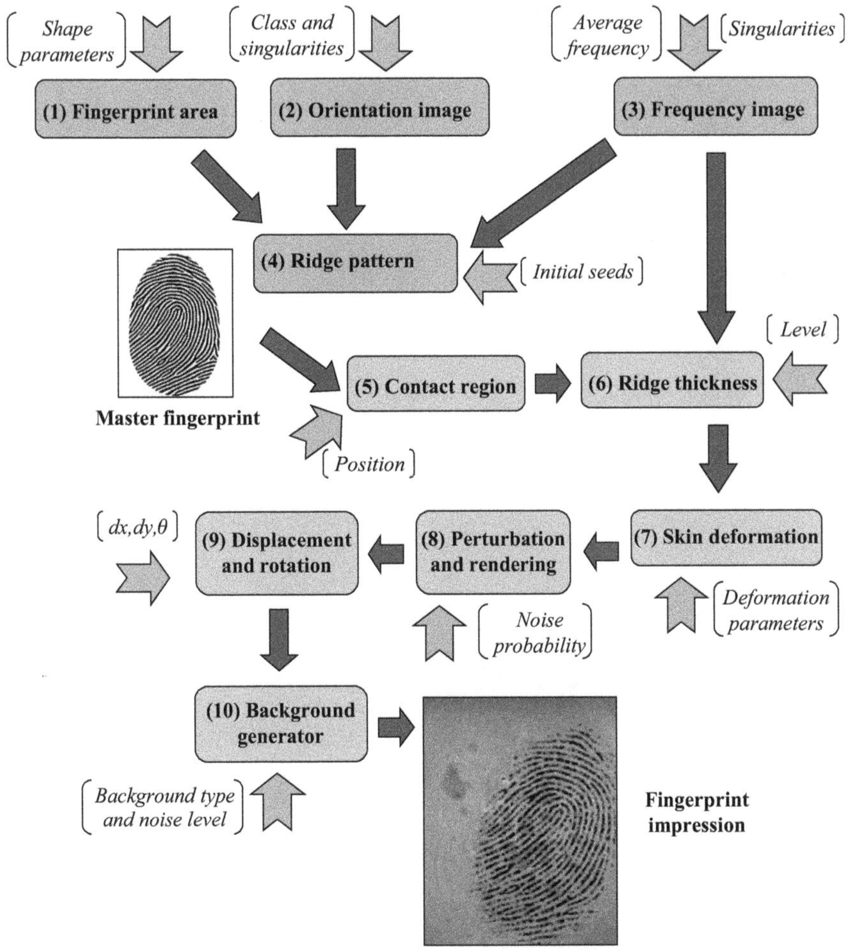

Figure 8.4 A functional schema of the synthetic fingerprint generation approach, SFinGe: each rounded box represents a generation step; the main input parameters that control each step are reported between brackets. Steps 1–4 create a master fingerprint; Steps 5–9 derive a fingerprint impression from the master fingerprint and can be iterated to produce multiple impressions of the same finger. © 2009 Springer. Reprinted, with permission, from Reference [4]

several recognition systems on both real and synthetic datasets confirmed the validity of the generation approach. More recently, in the scope of the Fidelity project [11], a specific study has been carried out to

- define appropriate measures to compare real and synthetic databases;
- improve the synthetic generator to better mimic the characteristic of real databases;

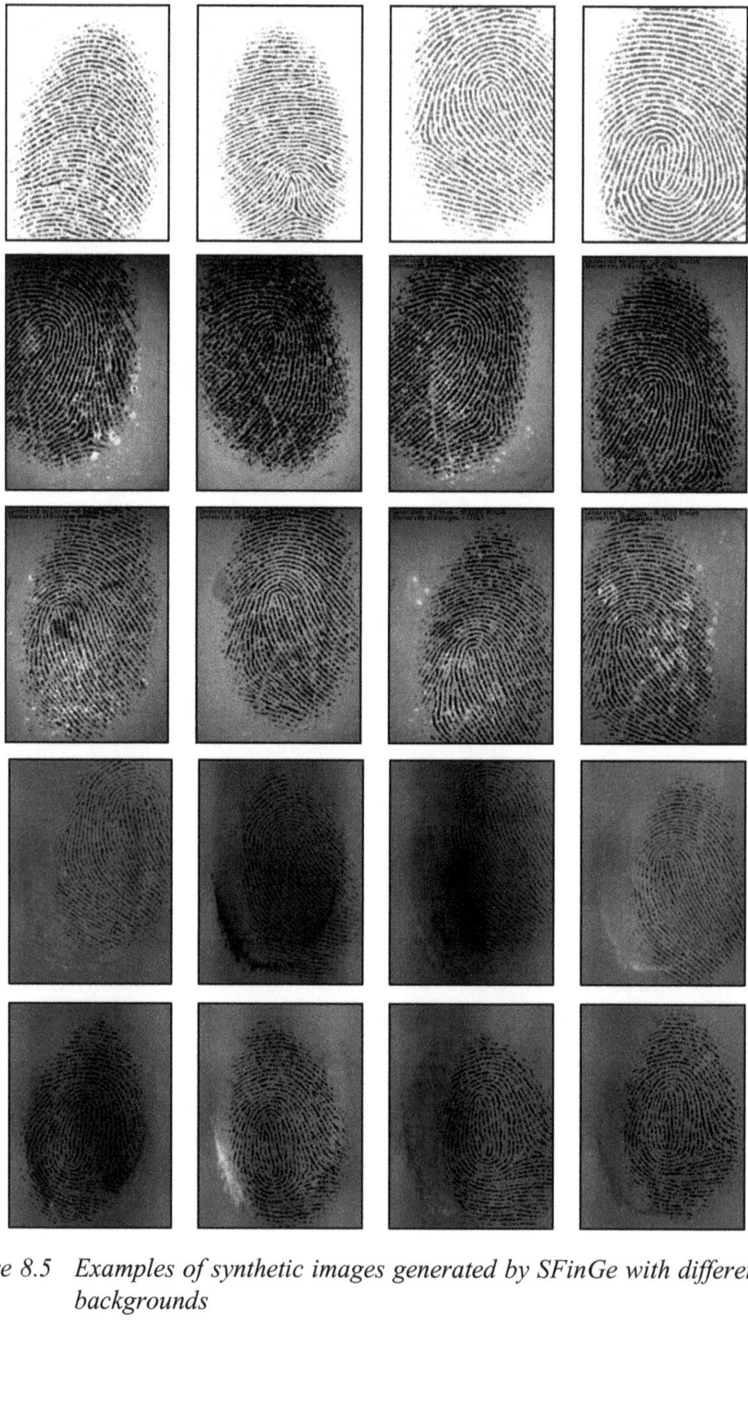

Figure 8.5 Examples of synthetic images generated by SFinGe with different backgrounds

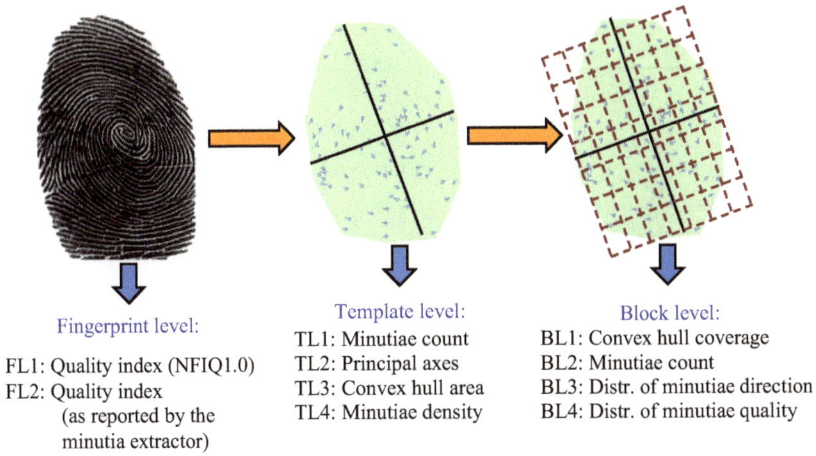

Figure 8.6 Comparison indicators

- generate a large synthetic database and analyze the performance of fingerprint comparison algorithms on both real and synthetic data, as the size of the database grows.

Figure 8.6 shows the indicators that have been used to compare real and synthetic databases: two of them are measured at image-level and describe the quality of the fingerprint, the others are derived from the minutiae, which are the most common feature used in fingerprint recognition. The distribution of each indicator has been calculated over six (from DB1 to DB6) real fingerprint databases (all collected with 500-dpi optical scanners) and over a synthetic database generated by the original version of SFinGe (DB00).

Figure 8.7 shows an example of the result for the NIST Fingerprint Image Quality (NFIQ) 1.0 comparison indicator (FL1) [12]: it is well evident that the distribution of this indicator over the real database is very different from its distribution over DB00.

A similar conclusion can be drawn from Figure 8.8 (distribution of minutiae count), Figure 8.9 (spatial distribution of minutiae over the fingerprint area), and, to a smaller extent, from Figure 8.10 (distribution of minutiae directions).

Table 8.1 reports, for each row, the difference between the distributions of each comparison indicator computed over two real databases; the last row shows the average values. The difference between two distributions is computed as the vector distance between the two corresponding histograms.

Table 8.2 reports the differences calculated between each real database considered and the first synthetic database DB00: by comparing the last row of the two tables, it can be observed that the average difference between synthetic and real databases is almost always larger than that between real databases.

In order to improve the synthetic generator, the workflow depicted in Figure 8.11 has been followed. Starting from DB00, a sequence of synthetic databases has been generated: each time the comparison indicators have been

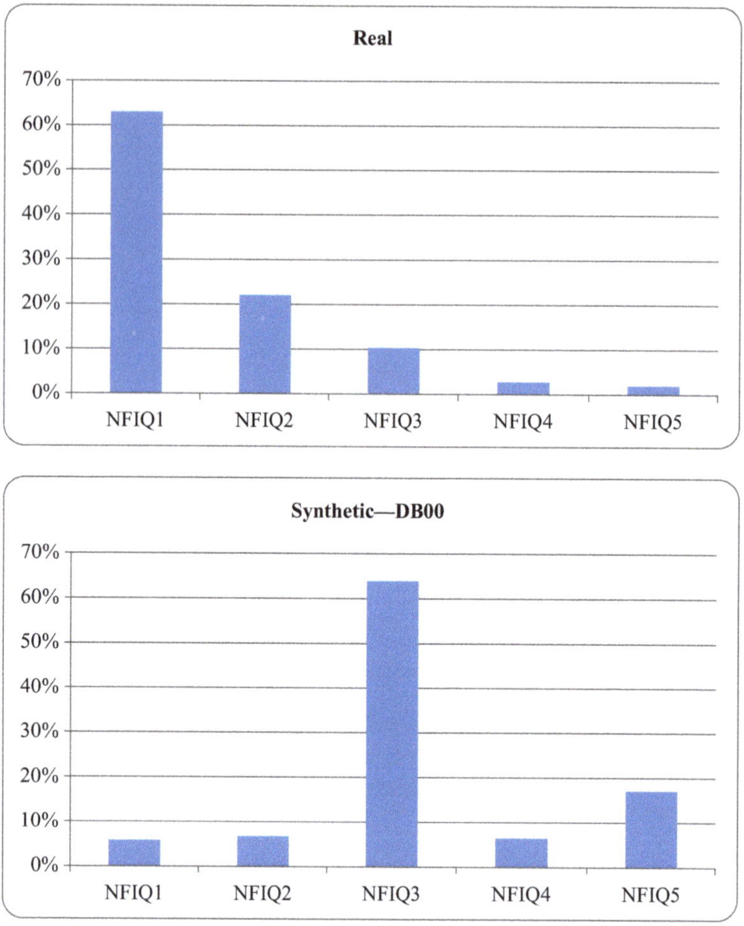

Figure 8.7 Distribution (histogram) of NFIQ 1.0 classes: on the left from a real database, on the right from a synthetic database generated with the original version of SFinGe

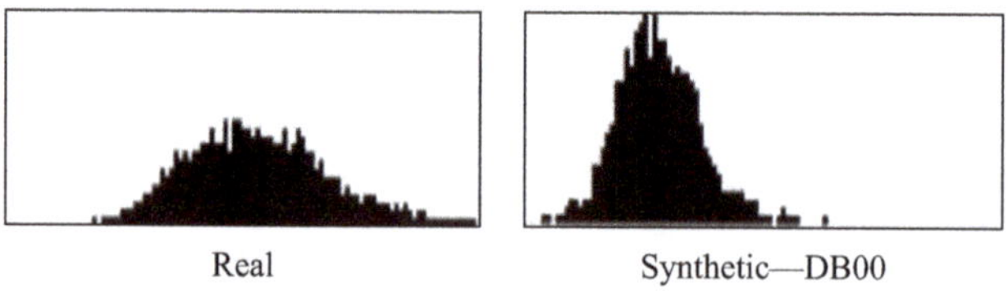

Figure 8.8 Distribution (histogram) of minutiae count: on the left from a real database, on the right from a synthetic database generated with the original version of SFinGe

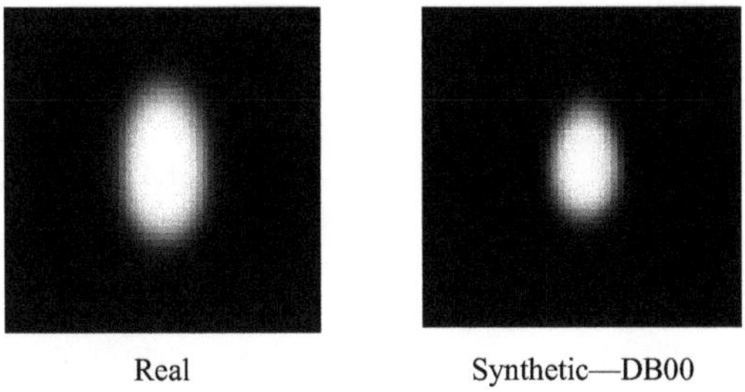

Figure 8.9 *Spatial distribution of minutiae: on the left from a real database, on the right from a synthetic database generated with the original version of SFinGe*

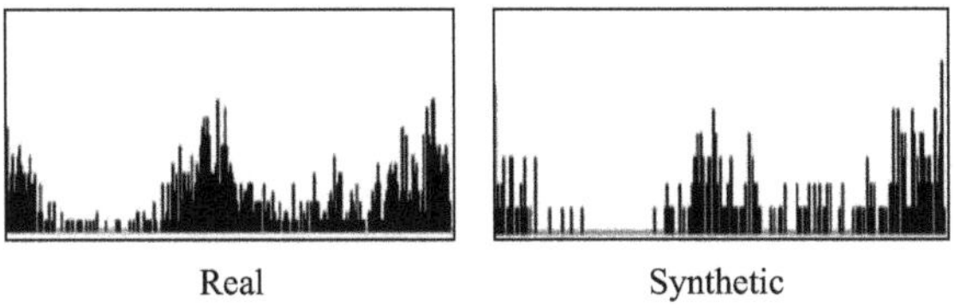

Figure 8.10 *Distribution of minutiae directions in a given block: on the left from a real database, on the right from a synthetic database generated with the original version of SFinGe*

Table 8.1 *Difference of the distribution of each comparison indicator for each possible pair of the real databases considered*

First DB	Second DB	TL1	TL2	TL3	TL4	FL2	BL1	BL2	BL3	BL4
DB1	DB2	0.0331	0.0032	0.0064	0.0139	0.0000	0.7671	0.1887	0.0006	0.0005
DB1	DB3	0.1893	0.0122	0.0196	0.0327	0.5409	5.0188	0.8818	0.0012	0.0055
DB1	DB4	0.1541	0.0097	0.0134	0.0518	0.5561	3.2095	0.6390	0.0010	0.0061
DB1	DB5	0.2865	0.0079	0.0135	0.0868	0.4916	7.6749	0.8815	0.0013	0.0139
DB1	DB6	0.0354	0.0040	0.0080	0.0418	0.4851	1.0093	0.3482	0.0008	0.0030
DB2	DB3	0.1664	0.0117	0.0181	0.0209	0.5409	4.3246	0.7161	0.0010	0.0053
DB2	DB4	0.1246	0.0095	0.0121	0.0394	0.5561	2.5022	0.4815	0.0009	0.0060
DB2	DB5	0.2959	0.0087	0.0142	0.0736	0.4916	8.2118	0.9662	0.0014	0.0139
DB2	DB6	0.0451	0.0045	0.0078	0.0446	0.4851	0.6025	0.4479	0.0010	0.0028
DB3	DB4	0.0969	0.0118	0.0183	0.0217	0.3402	2.4317	0.3053	0.0008	0.0012
DB3	DB5	0.3396	0.0141	0.0230	0.0568	0.2634	11.8394	1.5186	0.0017	0.0148
DB3	DB6	0.1719	0.0122	0.0186	0.0571	0.2720	4.0818	0.9980	0.0014	0.0055
DB4	DB5	0.3361	0.0123	0.0209	0.0402	0.3220	10.0185	1.2759	0.0015	0.0151
DB4	DB6	0.1427	0.0105	0.0134	0.0716	0.2616	2.4757	0.7958	0.0012	0.0062
DB5	DB6	0.2748	0.0084	0.0144	0.0991	0.1849	8.3444	1.0745	0.0016	0.0138
Average difference		**0.1795**	**0.0094**	**0.0148**	**0.0501**	**0.3861**	**4.8342**	**0.7679**	**0.0012**	**0.0076**

Table 8.2 Difference of the distribution of each comparison indicator between the synthetic databases DB00 and real databases

Synth. DB	Real DB	TL1	TL2	TL3	TL4	FL2	BL1	BL2	BL3	BL4
DB00	DB1	0.1949	0.0098	0.0214	0.0470	0.6536	7.2350	1.1715	0.0015	0.0032
DB00	DB2	0.1792	0.0098	0.0206	0.0466	0.6536	6.5980	1.0291	0.0014	0.0030
DB00	DB3	0.0864	0.0102	0.0196	0.0559	0.5409	2.7217	0.5104	0.0010	0.0048
DB00	DB4	0.1549	0.0117	0.0236	0.0686	0.5561	4.7948	0.6943	0.0012	0.0057
DB00	DB5	0.3337	0.0107	0.0222	0.0910	0.4389	13.4679	1.7895	0.0019	0.0136
DB00	DB6	0.1778	0.0097	0.0202	0.0199	0.4861	6.2834	1.1797	0.0015	0.0020
Average difference		**0.1878**	**0.0103**	**0.0213**	**0.0548**	**0.5549**	**6.8501**	**1.0624**	**0.0014**	**0.0054**

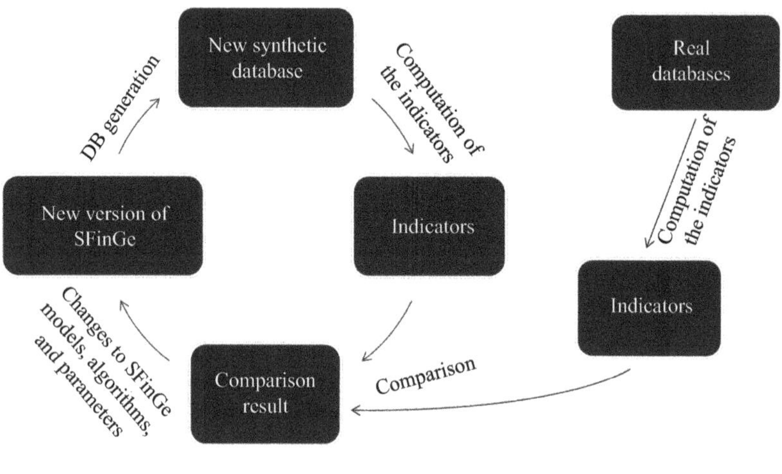

Figure 8.11 The workflow

calculated and their distributions compared to those obtained on the real databases; the results of such comparison guided the adjustment of parameters, the improvement of algorithms and mathematical models, etc., to make the synthetic databases more similar to the real ones. These steps have been iteratively followed generating a total of 27 synthetic databases after DB00: databases from DB00 to DB22 contained 1,000 fingers with two impressions per finger, then from DB23 to DB27, the size of the databases was increased up to 10 million fingers with two impressions per finger in DB27. The generation of DB27 took about 3 days using a cluster of PCs with a total of 180 CPU cores.

The results obtained with DB27 can be considered very satisfactory: the distributions of the various indicators are very similar to real databases, as it can be observed in Figure 8.12 (distribution of NFIQ 1.0 quality indices), Figure 8.13 (distribution of minutiae count), Figure 8.14 (spatial distribution of

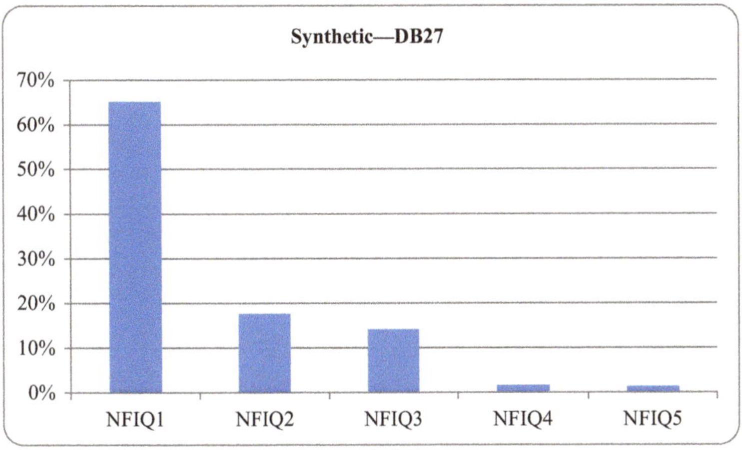

Figure 8.12 Distribution (histogram) of NFIQ 1.0 classes on a synthetic database generated with the improved version of SFinGe

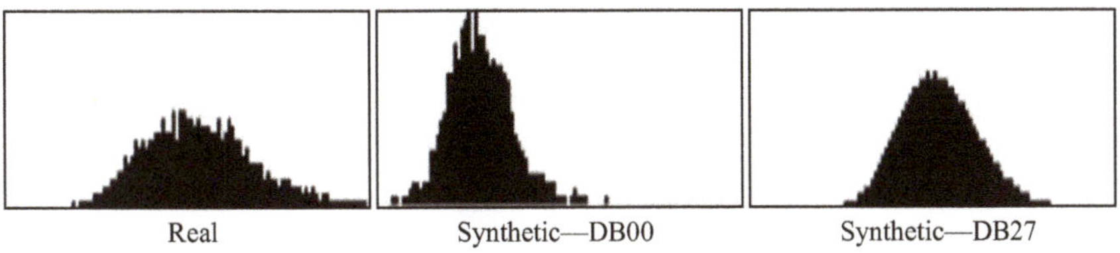

Figure 8.13 Distribution (histogram) of minutiae count: from left to right: a real database, a synthetic database generated with the original version of SFinGe, a synthetic database generated with the improved version of SFinGe

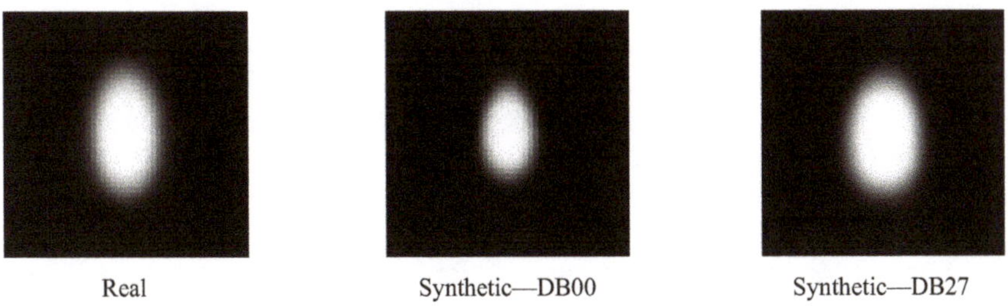

Figure 8.14 Spatial distribution of minutiae: from left to right: a real database, a synthetic database generated with the original version of SFinGe, a synthetic database generated with the improved version of SFinGe

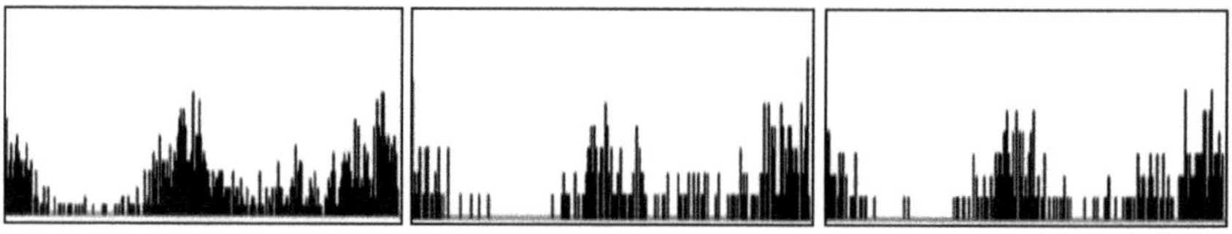

Figure 8.15 Distribution of minutiae direction in a given block: from left to right: a real database, a synthetic database generated with the original version of SFinGe, a synthetic database generated with the improved version of SFinGe

Table 8.3 Difference of the distribution of each comparison indicator between the synthetic databases DB27 and real databases

Synth. DB	Real DB	TL1	TL2	TL3	TL4	FL2	BL1	BL2	BL3	BL4
DB24	DB1	0.0238	0.0035	0.0045	0.0198	0.4841	1.3935	0.3095	0.0010	0.0028
DB24	DB2	0.0371	0.0038	0.0048	0.0144	0.4841	1.2986	0.3419	0.0011	0.0026
DB24	DB3	0.1974	0.0120	0.0186	0.0244	0.2541	4.6956	0.8492	0.0012	0.0054
DB24	DB4	0.1544	0.0083	0.0129	0.0416	0.2542	2.6638	0.5891	0.0011	0.0061
DB24	DB5	0.2971	0.0076	0.0132	0.0742	0.1788	7.8127	0.9377	0.0014	0.0138
DB24	DB6	0.0477	0.0049	0.0069	0.0344	0.0260	1.3231	0.4161	0.0011	0.0011
Average difference		**0.1263**	**0.0067**	**0.0101**	**0.0348**	**0.2802**	**3.1979**	**0.5739**	**0.0012**	**0.0053**

minutiae over the fingerprint area), and Figure 8.15 (distribution of minutiae directions).

Comparing Table 8.3 to Tables 8.1 and 8.2, it can be noted that the average difference between synthetic and real databases is now always lower than or equal to the difference between real databases.

8.4 Predicting performance in large-scale scenarios

In this chapter, we report some results of fingerprint recognition experiments on both real and synthetic fingerprint datasets. In particular, we focus on the accuracy indicators trend as the database size increases. We report experiments both on verification (1:1) and identification (1:*N*) scenarios. In particular, the last identification experiment (see Figure 8.18) involves more than 200 billion fingerprint comparisons and is the largest test with synthetic fingerprint data published so far.

Figure 8.16 shows the results of fingerprint verification experiments (1:1 comparison) on four real databases: the tests have been carried out varying the size of the databases (the horizontal axis refers to the number of different fingers in each database). Each graph shows the results of two fingerprint comparison algorithms in terms of five performance indicators: equal error rate (EER), FMR100

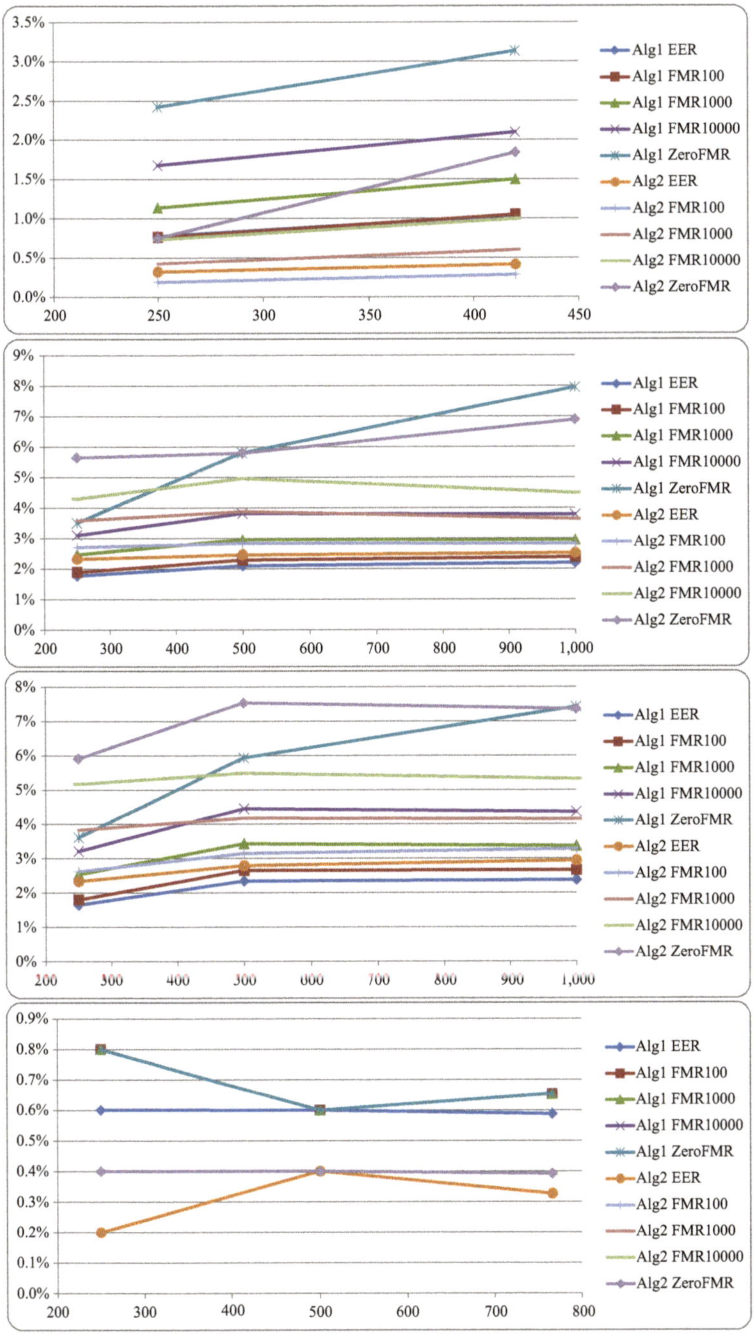

Figure 8.16 Fingerprint verification (1:1 comparison) results on real databases, as a function of the number of fingers in the database

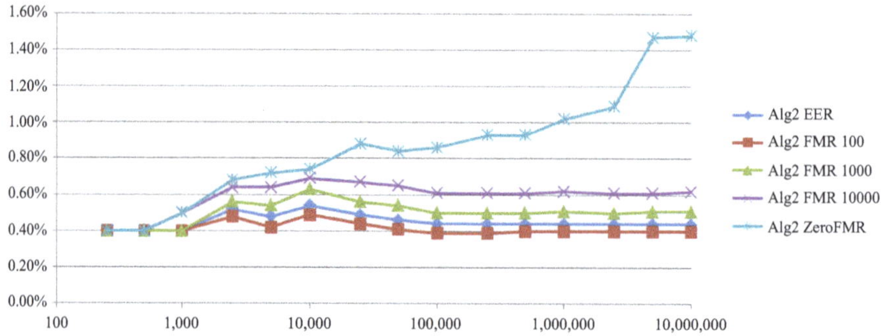

Figure 8.17 Fingerprint verification (1:1 comparison) results on synthetic database DB27, as a function of the number of fingers in the database (note that the horizontal axis uses a logarithmic scale)

(the value of FNMR when FMR = 1%), FMR1000 (the value of FNMR when FMR = 0.1%), FMR10000 (the value of FNMR when FMR = 0.01%), and ZeroFMR (the value of FNMR when FMR = 0%). Although the size of the databases is not very large (this incidentally confirms the difficulty in obtaining large databases of fingerprints and the need for synthetic generation), it can be observed that the trend tends to be constant for all performance indicators, except for ZeroFMR, which grow: this behavior corresponds to what is expected from a theoretical point of view, since ZeroFMR is conditioned by outliers, while the other indicators depend on the characteristics of the population of the database and not on its size; hence, they should remain stable as the number of fingers grows.

Figure 8.17 reports the results of verification experiments on the synthetic database DB27: the constant trend of the performance indicators, except ZeroFMR, is well evident and confirms that the generator has enough "degrees of freedom" to create large databases (there are no "collisions" of synthetic identities due to limitations in the synthetic models).

Finally, Figure 8.18 reports the results of fingerprint identification (1:*N* comparison) experiments on DB27; three performance indicators are reported:

- EER;
- FPIR1K [the value of false-negative identification rate (FNIR) when false-positive identification rate (FPIR) = 0.1%];
- FPIR10K (the value of FNIR when FPIR = 0.01%)

where

- FNIR [13,14] is the percentage of queries with a mate in DB27 that are not correctly identified;
- FPIR [13,14] is the percentage of queries without mate in DB27 that are mistakenly identified.

From the graph, it can be noted that the error rates have a smooth and reasonable trend. The points corresponding to the results of two recent large-scale fingerprint

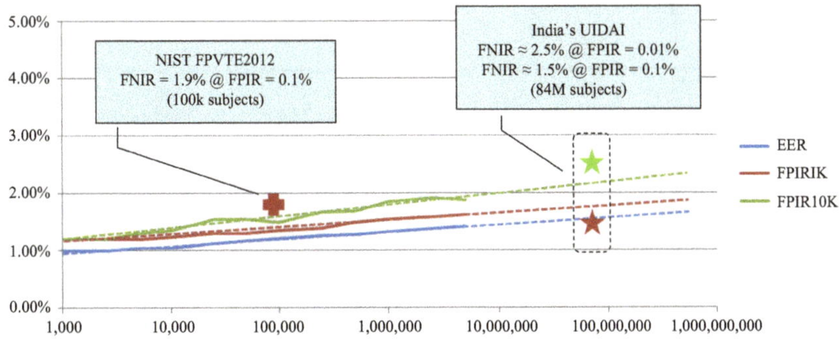

Figure 8.18 *Fingerprint identification (1:N comparison) results on synthetic database DB27 (solid lines), as a function of the number of fingers in the database (note that the horizontal axis uses a logarithmic scale). Three results from recent large-scale fingerprint identification evaluations on real data have been plot on the graph. The trend of the three indicators (dashed lines) is reported to further predict identification performance increasing the DB size*

identification evaluations on real data are also plot on the graph: India's UIDAI [15] and NIST FPVTE2012 [16]: it can be noted that these results are quite in-line with the error rates obtained on synthetic fingerprint data, suggesting the feasibility of predicting fingerprint identification results with synthetic data.

8.5 Generating altered fingerprints

In the recent years, automatic fingerprint recognition systems are being increasingly used for border and immigration control. With the clear intent of deceiving fingerprint-based identification, a subject can produce intentional alterations to his/her fingerprints: if a fingerprint pattern is severely damaged, the chance of finding the corresponding prealtered sample in an Automated Fingerprint Identification System (AFIS) greatly diminishes [17]. The development of specific solutions for detection/comparison of altered fingerprints is very critical due to the lack of public databases containing altered fingerprints and the corresponding prealtered versions. Today, only police agencies or research groups working together with police agencies have access to altered fingerprint datasets under strict confidentiality agreements, thus limiting the number of scientific contributions to this difficult problem.

In this chapter, we summarize an approach for generating synthetic altered fingerprints (see [18] for more details). Three of the most frequent alterations are modeled: namely, obliteration by burning, distortion by central rotation, and by Z-cut. The modeling approaches proposed in [19] have been improved in order to obtain more realistic alterations including scars and transitions between native and altered regions and by designing more realistic alteration shapes.

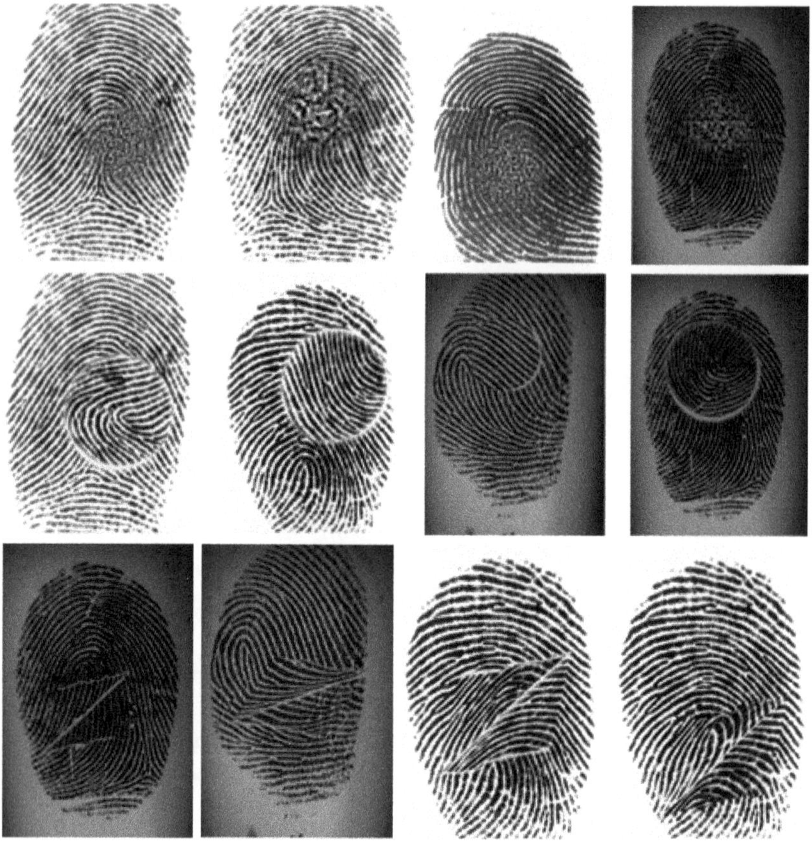

Figure 8.19 Examples of simulated alterations: obliteration (top), central rotation (middle), and Z-cut (bottom). © 2017 IEEE. Reprinted, with permission, from Reference [18]

From a visual analysis of generated altered fingerprints (see Figure 8.19 for some examples) it is quite evident that the generated samples are not visually very realistic; however, at this stage, the scope is not the image realism but the sample utility for the development/tuning of automatic algorithms.

8.5.1 Categories of fingerprint alterations

Usually the alterations are classified into three main categories based on the changes made to the ridge patterns: obliteration, distortion, and imitation [17].

The problem of altered fingerprints was handled by Samischenko [20], where a collection of altered papilla patterns are reported.

The most frequent way to obfuscate the ridge patterns is the obliteration where the pattern is modified by dermal abrasion, burning, or using chemical factors. Some dermal diseases can also cause severe alterations of the fingerprint. This kind of alteration, which is easily achievable without any surgery, produces a "granular effect" which completely deletes the ridge patterns (see examples in [21]).

In surgery-based distortion, parts of the skin are cut and plant out at different (unnatural) positions. Usually the local ridge/valley structure remains unaltered while the global structure is heavily changed. Since this kind of alteration does not change the ridge frequency and width, the image quality doesn't change much and the distorted fingerprints often pass an automatic quality control.

Fingerprint imitation refers to an invasive method to cut out a portion of the fingerprint and (i) stitching up the adjacent skin (as shown in Figure 11(a) of [17]) or (ii) replacing the missing part with skin from other ridged parts (e.g., palms, soles, or fingers) as shown in Figure 11(b) of [17].

Table 2 in [17] reports an exclusive categorization of 4,433 altered fingerprints based on the examination of ridge patterns: 89% are obliterated fingerprints, 10% are distorted, and only 1% are imitated. Based on this statistic, special care must be taken in modeling obliteration and distortion.

8.5.2 Synthetic altered fingerprint generation

The generation focuses on the three most common alterations encountered in real situations: burning obliteration, central rotation, and Z-cut [17].

For each type of alteration, a parametric model is introduced to modify an input (real) fingerprint. After the main pattern modification, some noise (e.g., scars, blurring) is introduced to create a more realistic pattern. The position of the alteration is randomly chosen inside the fingerprint central area to cover the most distinctive fingerprint region, since the alteration intent is to obscure the fingerprint identity. In the following sections, the model of each alteration type is described.

- **Burning obliteration**—By observing real burned fingerprints, we can note that the ridge pattern is completely destroyed and a granular effect is introduced. 2D Perlin noise [19–22], a gradient-type noise often used in computer graphics for texturing, is applied to simulate burning. The alteration is produced by adding to the ridge pattern waves with decreasing frequencies and increasing amplitudes. The obliteration external shape is a four-sector ellipse. This shape is generated starting from a circle of center (x_c, y_c) and radius r and computing the length of the four semi axes (a_1, a_2, b_1, b_2) by randomly varying r of a factor $\pm\Delta$. In the first row of Figure 8.19, some examples of synthetic burning obliterations are shown.
- **Central rotation**—In central rotation, a portion of the fingerprint is cut, rotated, and pasted again. The cut shape is defined by a four-sector ellipse E, and a simple geometric rotation is performed on the alteration mask E obtaining E_θ. In the second row of Figure 8.19, some examples of central rotation are shown.
- **Z-cut**—Z-cut is obtained by a Z-shaped incision of the skin; the lift of the two triangular strip patches and final sewing up in switched mode. The Z-cut geometry is regulated by the following parameters: the center position (x_c, y_c), the length L of the three lines of the Z letter, an angle α (see Figure 8.20) in the

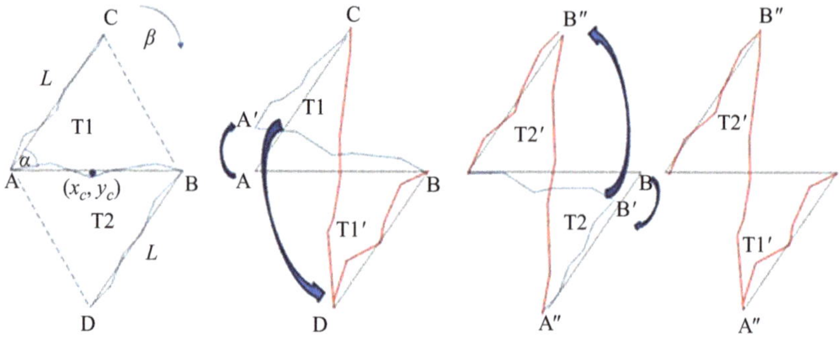

Figure 8.20 Scheme of a Z-cut alteration. © 2017 IEEE. Reprinted, with permission, from Reference [18]

range [40°, 50°], and an angle β representing the cut orientation with respect to the x axis. Once the coordinates of the four points ABCD composing the Z-cut have been determined (see Figure 8.20, left), the three lines defining the Z are replaced by three less regular polygonals (whose deviation from the reference lines is randomly defined) (see Figure 8.20). The two near-triangular shapes T1 and T2 are then moved, as shown in Figure 8.20, obtaining T1′ and T2′, respectively. To this purpose, two affine transforms are used to map the triangle ABC over DBC and the triangle ABD over ACD. In the third row of Figure 8.19, some examples of Z-cut are shown.

8.6 Using synthetic data to improve recognition algorithms

In this chapter, we discuss two applications of synthetic fingerprint data to the tuning/improvement of algorithms. First, we focus on the generation of synthetic pattern and ground-truth labels, which can be useful to train supervised learning algorithms; then, we introduce the optimization of comparison algorithms for altered fingerprints.

8.6.1 Generation of minutiae ground-truth data

Synthetic generation of fingerprint patterns allows us to easily obtain the ground-truth about relevant features, such as the orientation image, the local ridge-line frequency, and the minutiae information. The availability of the ground-truth is very useful for the development, optimization, and evaluation of feature extraction algorithms, especially for deep learning techniques that usually need large amount of labeled data. For instance, creation of minutiae ground-truth can be performed in parallel with the fingerprint generation (see Figure 8.21): the standard minutiae extraction rules defined in ISO/IEC 19794-2 [8] are applied to the master fingerprint, then all the relevant transformations executed on the pattern are applied to the minutiae (e.g., translation, rotation, distortion). This approach has some clear advantages

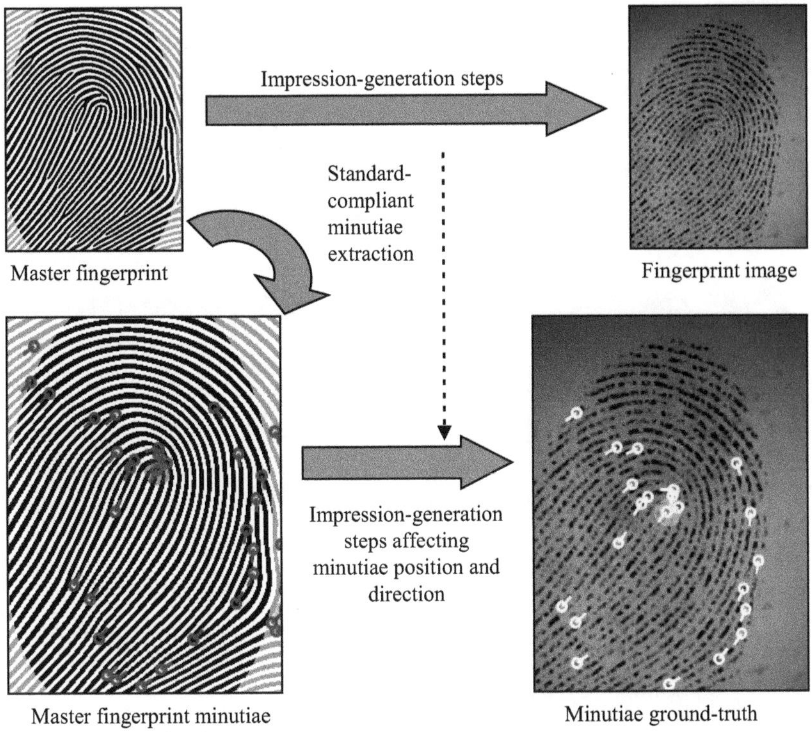

Figure 8.21 Generation of minutiae ground-truth data. © 2009 Springer. Reprinted, with permission, from Reference [4]

- the features can be reliably extracted through simple algorithms, since the extraction occurs in a binary image without any noise;
- the ground-truth is always unique and accurate, even when the quality of the final image is relatively low (see Figure 8.22).

Generation of other ground-truth features can be performed in a similar fashion; for instance, all the relevant transformations can be applied to the orientation image calculated at Step 2 (Section 10.2), thereby obtaining the true orientation image of the final synthetic fingerprint impression.

It is important to underline that this process not only allows to save a great amount of time with respect to manual labeling of fingerprint features but also makes it possible to include very low-quality cases where manual labeling would be totally unreliable.

8.6.2 Optimization of comparison algorithm for altered fingerprints

In this section, we show how the well-known Minutia Cylinder-Code (MCC) approach [23] can be tuned on a synthetic database of altered fingerprints in order to improve its accuracy on a real dataset. Since MCC works directly with minutiae

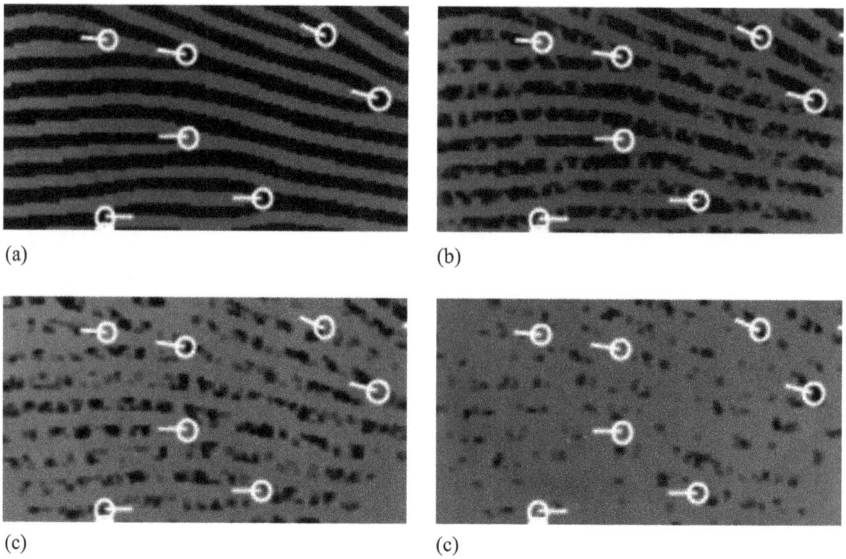

Figure 8.22 Minutiae ground-truth as generated by SFinGe for the same fingerprint portion at different levels of noise: low noise (a), medium noise (b, c), and high noise (d). © 2009 Springer. Reprinted, with permission, from Reference [4]

templates, the minutiae extraction algorithm described in [24] has been used to create minutiae templates. Two datasets have been used for this experiment: UNILDB is a real database of altered fingerprints, while SYNTHDB is a synthetic dataset of altered fingerprints generated according to the method explained in Section 10.5.

- UNILDB—University of Lausanne has access to a database of real altered fingerprints containing 1,206 rolled fingerprints from 1,020 different fingers. The database consists of (i) 1,020 unaltered fingerprints from 1,020 different fingers (subset U) and (ii) 186 altered fingerprints from 141 different fingers with a mate in U (subset A).
- SYNTHDB—SYNTHDB was synthetically generated starting from the NIST special database 14 (NISTDB14) [25]. NISTDB14 has been selected as input database, since it contains rolled fingerprints with an NFIQ 1.0 distribution comparable to UNILDB. NISTDB14 is a public database containing 54,000 rolled fingerprints from 27,000 different fingers (two impressions per finger). The SYNTHDB consists of (i) a subset SU of 1,000 unaltered fingerprints from 1,000 different fingers selected from NISTDB4 to match NFIQ distribution of UNILDB subset U and (ii) a subset SA of 600 synthetically altered fingerprints from 200 different fingers with a mate in SU. In particular, for each prealtered fingerprint, one alteration for each of the three types of alterations has been generated.

Table 8.4 Accuracy of the original tuned parameters on SYNTHDB and UNILDB. © 2017 IEEE. Reprinted, with permission, from Reference [18]

Algorithm parameters		SYNTHDB		UNILDB	
Minutiae extractor	**MCC**	**EER (%)**	**FMR100 (%)**	**EER (%)**	**FMR100 (%)**
Original	Original	12.0	24.2	19.4	33.9
Tuned	Tuned	10.3	19.6	16.1	30.6

Since both minutiae extraction and MCC algorithms were not developed to work with altered fingerprint, their default parameters could be suboptimal for this task, so a number of experiments on SYNTHDB have been performed to better tune the parameters of both algorithms (UNILDB was kept out during this tuning phase) for altered fingerprints comparison.

To evaluate if the improvement achieved on the synthetically altered fingerprints can be transferred to real altered ones, the minutiae extraction and MCC algorithms have been tested on the UNILDB with both original and tuned parameters (see Table 8.4). An improvement of about 3% has been obtained in terms of both EER and FMR100, thus confirming the usefulness of synthetic data for optimization purposes.

8.7 Conclusions

Synthetic data proved to be a useful instrument for optimizing, evaluating, and certifying biometric systems. This chapter introduced the SFinGe approach, its recent improvements, and some case studies.

The outcomes of the large-scale experiments discussed in Section 8.4 prove that SFinGe approaches are nowadays mature and reliable and can be conveniently used for several applications.

We are currently considering to generate a synthetic database with a number of samples comparable to the world population, and run fingerprint identification experiments on it. Such experiments would require, today, a powerful hardware and some months of computing time, but based on coming technologies and expected trends, we believe we could easily run it in the coming years, thus earning new insights on fingerprint individuality and AFIS ultimate scalability.

References

[1] R. Cappelli, A. Erol, D. Maio, and D. Maltoni, "Synthetic Fingerprint-image Generation," in *15th International Conference on Pattern Recognition (ICPR2000)*, Barcelona, 2000, pp. 475–478.

[2] R. Cappelli, "SFinGe," in *Encyclopedia of Biometrics (Second Edition)*. Springer, London, 2015.

[3] R. Cappelli, D. Maio, D. Maltoni, J. L. Wayman, and A. K. Jain, "Performance Evaluation of Fingerprint Verification Systems," *IEEE Transactions on Pattern Analysis and Machine Intelligence*, vol. 28, no.1, pp. 3–18, 2006.
[4] R. Cappelli, "Synthetic Fingerprint Generation," in *Handbook of Fingerprint Recognition (Second Edition)*. Springer, London, 2009, ch. 6.
[5] R. Cappelli, "Fast and Accurate Fingerprint Indexing Based on Ridge Orientation and Frequency," *IEEE Transactions on Systems, Man, and Cybernetics, Part B: Cybernetics*, vol. 41, no. 6, pp. 1511–1521, December 2011.
[6] R. Cappelli, D. Maio, and D. Maltoni, "Indexing Fingerprint Databases for Efficient 1:N Matching," in *Sixth International Conference on Control, Automation, Robotics and Vision (ICARCV2000)*, Singapore, 2000.
[7] D. Maltoni, D. Maio, A. K. Jain, and S. Prabhakar, *Handbook of Fingerprint Recognition (Second Edition)*. Springer-Verlag: New York, NJ, USA, 2009.
[8] ISO/IEC 19794-2, *Information Technology—Biometric Data Interchange Formats—Part 2: Finger Minutiae Data*, 2011.
[9] R. Cappelli, "Use of Synthetic Data for Evaluating the Quality of Minutia Extraction Algorithms," in *2nd NIST Biometric Quality Workshop*, Gaithersburg, Maryland, 2007.
[10] R. Cappelli, D. Maio, and D. Maltoni, "Synthetic Fingerprint-database Generation," in *16th International Conference on Pattern Recognition (ICPR2002)*, Québec City, 2002, pp. 744–747.
[11] The FIDELITY Consortium. FIDELITY European Project Web Site, May 2017. [Online]. http://www .fidelity-project.eu/.
[12] E. Tabassi, C. L. Wilson, and C. Watson, "Fingerprint Image Quality," NIST—NISTIR 7151, 2004. [Online]. http://www.nist.gov/customcf/get_pdf.cfm?pub_id=905710.
[13] Unique Identification Authority of India, *Role of Biometric Technology in Aadhaar Enrollment*. 2012.
[14] P. J. Grother, G. W. Quinn, and P. Jonathon, *Report on the Evaluation of 2D Still-Image Face Recognition Algorithms*. NIST, Gaithersburg, 2010.
[15] Unique Identification Authority of India. India's Aadhaar (UIDAI). [Online]. https://uidai.gov.in/about-uidai/about-uidai.html, February 2018.
[16] NIST. Fingerprint Vendor Technology Evaluation 2012 (FpVTE 2012). [Online]. http://www.nist.gov/itl/iad/ig/fpvte2012.cfm, February 2018.
[17] S. Yoon, J. Feng, and A. K. Jain, "Altered Fingerprints: Analysis and Detection," *IEEE Transactions on Pattern Analysis and Machine Intelligence*, vol. 34, no. 3, pp. 451–464, March 2012.
[18] S. Papi, M. Ferrara, D. Maltoni, and A. Anthonioz, "On the Generation of Synthetic Fingerprint Alterations," in *International Conference of the Biometrics Special Interest Group (BIOSIG)*, Darmstadt, Germany, 2016.
[19] Petrovici, "Simulating Alteration on Fingerprint Images," in *IEEE Workshop on Biometric Measurements and Systems for Security and Medical Applications*, 2012.

[20] S. S. Samischenko, *Atlas of the Unusual Papilla Patterns/Atlas Neobychnykh Papilliarnykh Uzorov*. Urisprudentsiia: Moscow, 2001.

[21] T. R. Anoop, "Detection of Fingerprint Alteration Using Scars, Minutiae Density and Ridge Discontinuity," *International Journal of Scientific & Engineering Research*, vol. 4, no. 8, August 2013.

[22] K. Perlin, "An Image Synthesizer," *Computer Graphics*, vol. 19, no. 3, pp. 287–296, 1985.

[23] R. Cappelli, M. Ferrara, and D. Maltoni, "Minutia Cylinder-code: A New Representation and Matching Technique for Fingerprint Recognition," *IEEE Transactions on Pattern Analysis and Machine Intelligence*, vol. 32, no. 12, pp. 2128–2141, December 2010.

[24] R. Cappelli, M. Ferrara, and D. Maio, "A Fast and Accurate Palmprint Recognition System Based on Minutiae," *IEEE Transactions on Systems, Man and Cybernetics—Part B*, vol. 42, no. 3, pp. 956–962, June 2012.

[25] C. I. Watson, *Nist Special Database 14, Fingerprint Database*. U.S. National Institute of Standards and Technology, Gaithersburg, 1993.

Chapter 9
Robustness of finger-vein recognition

Christof Kauba[1] and Andreas Uhl[1]

9.1 Introduction

One of the big issues in biometric recognition is robustness of recognition accuracy against sample signal quality degradation. The performance of a biometric recognition system is usually heavily affected by sample signal quality. A wide variety of factors potentially influence the quality of acquired biometric samples.

The different types of features that can be extracted from biometric samples influence the impact of quality degradations on recognition performance in various ways. Moreover, there is interplay among different types of feature extraction and acquisition technology/conditions such that it is not clear a priori which type of feature extraction is favourable under which conditions. Therefore, it is essential to provide reliable methodology to comparatively assess biometric recognition robustness under varying conditions.

Fingerprint recognition is a prominent biometric field in which robustness of recognition accuracy against sample image quality degradation is a central issue [1,2]. The quality of fingerprint images is influenced by skin conditions (e.g. dryness, moisture, dirt, age, cuts, and bruises), sensor conditions (e.g. dirt, noise, and size), and other acquisition conditions like user cooperation or crime scene preservation in forensic settings, etc. Some of these factors are inevitable, and some of them change over time. Poor quality images often result in spurious and missed features, therefore decreasing the recognition accuracy of the overall system.

In fingerprint recognition, this issue is classically tackled from two sides: First, benchmarking frameworks have been established, which facilitate a common evaluation basis with standardised protocols for various fingerprint recognition algorithms, see e.g. the fingerprint verification contests (FVC [2]) as well as independent suggestions like [3] and the BioSecure evaluation framework [1]. Second, usually these frameworks rely on the establishment of test data which are used to compare the different algorithms on a common basis. A very good example, specifically focusing onto the robustness issue, is the FVC datasets. FVC2002 (only (i) and (iv)) and FVC2004 data have been acquired in a way to introduce higher

[1]Department of Computer Sciences, University of Salzburg, Austria

intra-class variation by (i) putting the finger at slightly different vertical position, (ii) applying low or high pressure against the sensor, (iii) exaggerating skin distortion and rotation, and (iv) drying or moistening fingers. For FVC2006, the population was chosen to be more heterogeneous, including manual workers and elderly people.

While the availability of these and similar datasets is a significant achievement, the data collection and database establishment are tedious works. Moreover, if additional acquisition conditions should be considered which have not been included into the original dataset, re-enrolment is required, involving complicated procedures for getting the original people back to enrolment. Also, it is hard to compare the different quality degradations from dataset to dataset (e.g. FVC, MCYT, BIOMET, and MSU), since usually, there is no standardised manner to generate the acquisition conditions applied. Therefore, the experimental results of recognition algorithms in case applied to different datasets are hardly comparable, and the results shown in many papers are difficult to interpret.

A strategy to cope with the various problems of generating natural datasets is to generate synthetic fingerprints, the SFinGe [4] being the most well-known tool for doing this. The generated fingerprints have proven to be highly realistic and serve as a sensible tool to generate large datasets for benchmarking. While SFinGe also allows to apply some manipulations to the images, e.g. noise insertion, translations, rotations, and uses a skin deformation model, a simulation of specific sensor types is not foreseen. In related work, we have proposed the watermarking robustness assessment toolkit *StirMark* [5] to be used as standardised tool in fingerprint recognition robustness assessment [6,7], which is able to simulate a wide class of acquisition conditions, applicable to any given (natural) dataset. We have also applied a simulation of sensor ageing to natural image datasets to assess sensor ageing robustness of fingerprint recognition systems [8].

However, for hand- and finger-vein recognition systems, there are neither natural benchmark datasets focussing on robustness issues nor corresponding robustness evaluation results available. Also, robustness evaluations on simulated datasets are limited to our own previous work on

- hand-vein recognition, where we assess
  - the impact of simulated sensor ageing and StirMark-based distortions on the recognition performance of hand-vein recognition [9], and
- finger-vein recognition, where we assess
  - the impact of simulated sensor ageing [10] and
  - the impact of ISO lossy compression standards on the recognition performance of finger-vein recognition [11].

For finger-vein recognition, the data available so far on sensor ageing and compression robustness [10,11] is restricted to a single dataset only (with unknown generalisation to different datasets/sensors), and for compression robustness assessment, results only include equal error rate (EER) data and a single compression scenario (both probe and gallery compressed). StirMark-based distortions, simulating various acquisition conditions, have not yet been investigated for their impact on this modality at all. We will tackle these issues by providing more complete experimental results to provide a better understanding of finger-vein recognition techniques' robustness in this chapter.

In Section 9.2, we describe the finger-vein recognition algorithms applied and the datasets used. Section 9.3 deals with the simulation of varying acquisition conditions using the StirMark toolkit and corresponding impact on recognition accuracy. In Section 9.4, we describe the impact of applying four different lossy compression algorithms on recognition discriminating two different application scenarios. Section 9.5 presents results on sensor-ageing robustness considering artificially aged datasets. Finally, Section 9.6 concludes this chapter.

9.2 Finger-vein recognition and datasets

9.2.1 Finger-vein recognition algorithms

Pre-processing can be grouped into methods to align the finger position and to improve the low contrast and image quality. For finger alignment, we use a method adopted from Lee *et al.* [12] (**LeeRegion**) which simply masks out background pixels (setting them to 0). This is followed by a **normalisation** step, i.e. rotation compensation as done in [13].

For image enhancement, we apply Contrast Limited Adaptive Histogram Equalisation (CLAHE) [14] as the final stage of **high frequency emphasis (HFE) filtering** which was originally proposed for hand vein image enhancement [15]. Also, CLAHE is used to post-process a filtering using a **Circular Gabor Filter (CGF)** as proposed by Zhang and Yang [16].

To foster reproducible research, we have only included *feature extraction* techniques in this study for which basic implementations are available as open-source software. The first scheme discussed aims to extract the vein pattern from the background resulting in a binary image, followed by a comparison of these binary images using a correlation measure.

Maximum curvature (MC [17]) aims to emphasise only the centre lines of the veins and is therefore insensitive to varying vein width. The first step is the extraction of the centre positions of the veins. Therefore, the local MC in the cross-sectional profiles, based on the first and second derivatives, is determined. Afterwards, each profile is classified as being concave or convex where only local maxima in concave profiles indicate valid centre positions of the veins. Then a score according to the width and curvature of the vein region is assigned to each centre position, which is recorded in a matrix called locus space. Due to noise or other distortions, some pixels may not have been classified correctly at the first step; thus the centre positions of the veins are connected using a filtering operation. Finally, binarisation is done by thresholding using the median of the locus space.

For matching the binary feature images, we adopted the approach in [17,18]. As the input images are not registered to each other and only coarsely aligned (rotation is compensated), the correlation between the input image and in x- and y-direction shifted versions of the reference image is calculated. The maximum of these correlation values is normalised and then used as final matching score.

In contrast to MC, the Scale Invariant Feature Transform (SIFT) keypoint [19] based technique uses information from the most discriminative points as well as considering the neighbourhood and context information of these points by extracting key points and assigning a descriptor to each key point. We employ an additional key point filtering as described in [20], where keypoints close to the finger/background boundary are discarded in addition to eventual background keypoints.

SIFT matching is done using the keypoint descriptors – the keypoint with the smallest distance to the reference keypoint is the matched one if the distance is below a threshold, otherwise there is no match. To resolve the problem with ambiguous matches (i.e. one keypoint may have small distances to more than one other point), the classical ratio threshold scheme is used: A match is only valid if the distance of the best point match is at least k (threshold) times smaller than to all other points.

For MC, LeeRegion, Normalisation, and CGF pre-processing has been applied (MC software of B.T. Ton[1] is used), while for SIFT feature extraction, LeeRegion, Resize, CGF, and HFE pre-processing has been used (SIFT feature extraction and matching software is used as provided by VL_Feat SIFT[2]). For more details on the used pre-processing and feature-extraction methods, the interested readers are referred to [20].

9.2.2 Datasets

For experimental evaluations in this chapter, we use the following two publicly available finger-vein datasets:

- **UTFVP**: University of Twente Finger Vascular Pattern (UTFVP) Database [21], consisting of a total of 1,440 images, taken from 60 subjects, 6 fingers per subject and 4 images per finger. The images have a resolution of 672 × 380 pixel with 8-b greyscale depth.
- **SDUMLA-HMT**: This multi-modal dataset was collected during the summer of 2010 at Shandong University, Jinan, China. 106 subjects, including 61 males and 45 females with age between 17 and 31, participated in the data-collecting process, in which all the five biometric traits – face, finger vein, gait, iris, and fingerprint are collected for each subject [22]. SDUMLA-HMT is available at http://mla.sdu.edu.cn/sdumla-hmt.html. The finger-vein dataset consists of 6 fingers per subject, 6 images per finger, 3,816 images, 320 × 240 pixel with 8-b grey level.

The test procedure of the FVC2004 [23] was adopted to determine the EER. For the genuine matches (to determine False Non-Match Rate (FNMR)), each image of each finger is compared with all remaining images of the same finger, no symmetric matches are performed. For the impostor matches (to determine False Match Rate (FMR)), the first image of each finger is compared against the corresponding first image of the same finger of all remaining subjects, again no symmetric matches are performed.

[1]Publicly available on MATLAB Central: http://www.mathworks.nl/matlabcentral/fileexchange/authors/57311.
[2]http://www.vlfeat.org/.

9.3 StirMark distortion robustness: modelling acquisition conditions

Petitcolas *et al.* [24] developed a benchmark test in the context of robustness evaluation for digital image watermarking methods, called StirMark.[3] The basic idea behind their benchmark test is that a digital watermark embedded inside an image can be attacked by introducing different small image manipulations to the watermarked image. The specific types of perturbations are pre-defined, and their intensity can be adjusted via a given set of parameters for each type of perturbation.

Utilising StirMark to generate datasets for robustness evaluations has several advantages. First of all, the tests are reproducible if their parameters are known and the test dataset is available. Moreover, it is possible to systematically simulate different strengths of distortions corresponding to different levels of external influence during the capturing process. This is hardly possible to achieve using real data because one would have to create many different acquisition conditions and ensure stable acquisition conditions during the whole acquisition process across all of the test subjects. In addition, it becomes a tedious and very time-consuming work to establish such datasets for a broad variety of different acquisition conditions in combination with different levels of strength. Finally, it may be impossible to isolate specific external influences from others if there is the need to investigate the impact of a specific type of influence only.

In the following, we describe the StirMark image manipulations which are chosen to be appropriate for finger-vein images and used during the experiments. Not all manipulations provided by StirMark are suitable to simulate natural acquisition conditions. Thus, only a subset of the complete range of image manipulations provided by StirMark is used. For each manipulation, the relation to realistic finger-vein capturing scenarios which could be modelled thereby is outlined. Moreover, the way it is defined is described as well as how it is parametrised to achieve varying strengths. The example images shown have been generated by applying the respective StirMark manipulation.

Median filtering results in a kind of blur to the image, additionally corrupting the clarity of the vein structure. This corresponds to small finger movements during the acquisition of the image (motion blur) or defocus and in general blurry vein structures due to the interaction of the infrared light with different types of tissue inside the finger. The size of the filter mask can be set (height and width of the mask). Figure 9.1(a) shows an example of maximal filter size used.

Remove lines corresponds to errors in finger-vein images resulting either from transmission/processing or errors of the scanner while reading the finger-vein image itself (might not be able to read the entire finger and miss or skip some lines) can be simulated. This manipulation removes lines from the input image the amount of which can be adjusted by a single parameter k which corresponds to the frequency of removing lines, where k means 'remove 1 line in every k lines'.

[3]Currently version 4.0 of the toolkit is available at http: //www.petitcolas.net/fabien/watermarking/stirmark/.

Of course, the output image is reduced in its size while applying this manipulation. An example image with maximal value k can be seen in Figure 9.1(b).

Rotation and cropping corresponds to rotated positioning of the finger on the sensor, with subsequent image cropping to retain image size. Rotation is a very typical, not to say – omnipresent – challenge for finger-vein matching, as in very few cases, a finger will be presented twice in exactly the same orientation to the sensing area during image acquisition. An example image with maximal rotation value used can be seen in Figure 9.2(a).

Embedding watermark Watermarks have been intensively discussed to enhance certain properties of biometric systems [25]. In case robust embedding techniques are being used, impact on recognition accuracy may be expected [26]. StirMark internal watermarks are embedded with increasing strength, resulting in decreasing peak signal-to-noise ratio (PSNR). An example image with maximal watermark embedding strength used can be seen in Figure 9.2(b).

Shearing transformation The application of affine transformations to finger-vein images is intended to simulate distortions of the entire finger-vein image, that can appear in real-life situations during acquisition. These distortions depend on the way the finger is positioned in 3-D space relative to the sensor plane. As special cases, we consider *shearing* in both coordinate directions. An example image with

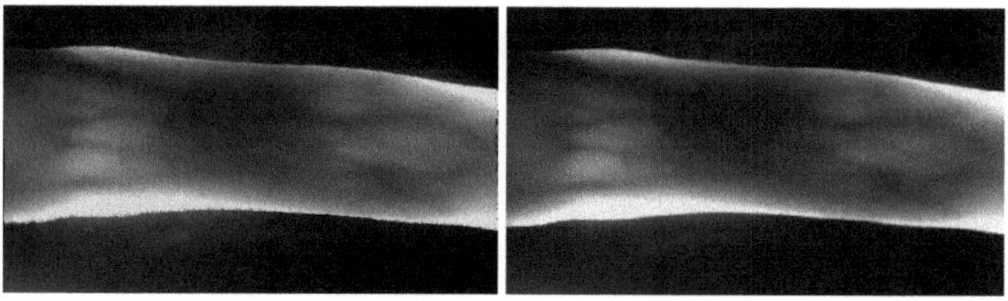

(a) (b)

Figure 9.1 Examples images: (a) median filtering and (b) remove lines

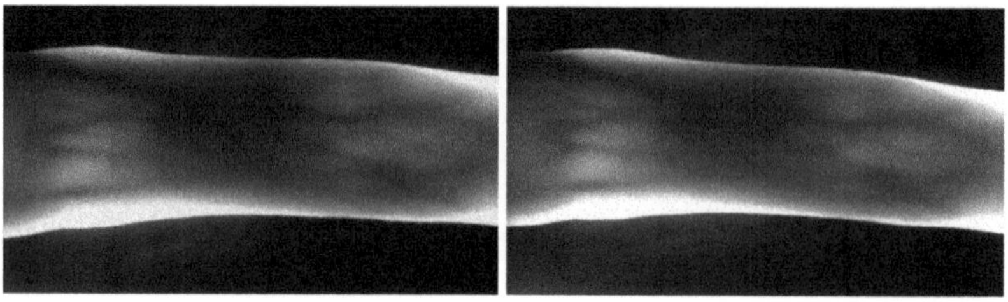

(a) (b)

Figure 9.2 Examples images: (a) rotation and cropping and (b) watermark embedding maximal shearing strength used can be seen in Figure 9.3(a) where a lesser extent of bending of the finger in *y*-direction can be observed due to the shearing.

Small random distortions These are usually termed *The StirMark Test*. Being a combination of several basic manipulations (i.e. random minor geometric distortion followed by resampling and interpolation, a transfer function to emulate analogue/digital converter imperfections, global 'bending', high frequency displacement, and JPEG compression), this test originally aims to simulate a resampling process, i.e. the errors introduced when printing an image and then scanning it again. The involved image warping is performed both on a global, as well as on a very local level. For the finger-vein scenario, this test can be seen as modelling a very generic, diffuse form of distortion. An example image with maximal strength used can be seen in Figure 9.3(b) where contrasting to most other visual example, the impact of the distortion can easily be observed.

9.3.1 Results

For these experiments, MC feature extraction and matching are used on the UTFVP dataset. As shown in Figures 9.1–9.3, most of the distortions do not affect the subjective quality of the images too much (at least at the resolution possible to use in the display). However, impact on recognition performance can be seen as discussed in the following. In Figure 9.4, we display EER, ZeroFMR (i.e. the lowest FNMR for FMR = 0), as well as ZeroFNMR (i.e. the lowest FMR for FNMR = 0). Please note that polygons not extended until distortion grade 10 indicate that the matching process could not be terminated for these settings (e.g. in the EER plot rotcrop grade > 9).

With respect to EER [Figure 9.4(a)], we notice that shearing (in both directions) affects recognition performance clearly, where shearing in *y*-direction is significantly worse. Also, the removal of entire lines leads to a drastic increase of EER, especially for the higher distortion grades. Also, random distortions clearly affect EER. An interesting effect is observed w.r.t. rotation. While we observe good robustness (only a slight EER increase for grade = 9), the matching software does not work anymore for grade = 10. Obviously, the rotation compensation is no longer able to correct the introduced extent of rotation. Median filtering as well as

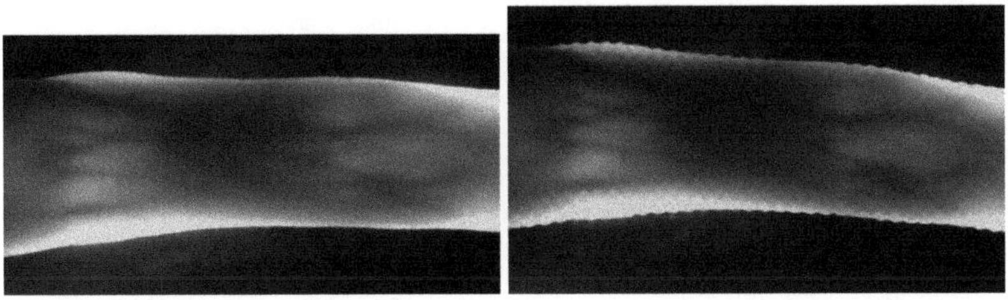

(a) (b)

Figure 9.3 Examples images: (a) shearing in y-direction and (b) small random distortions

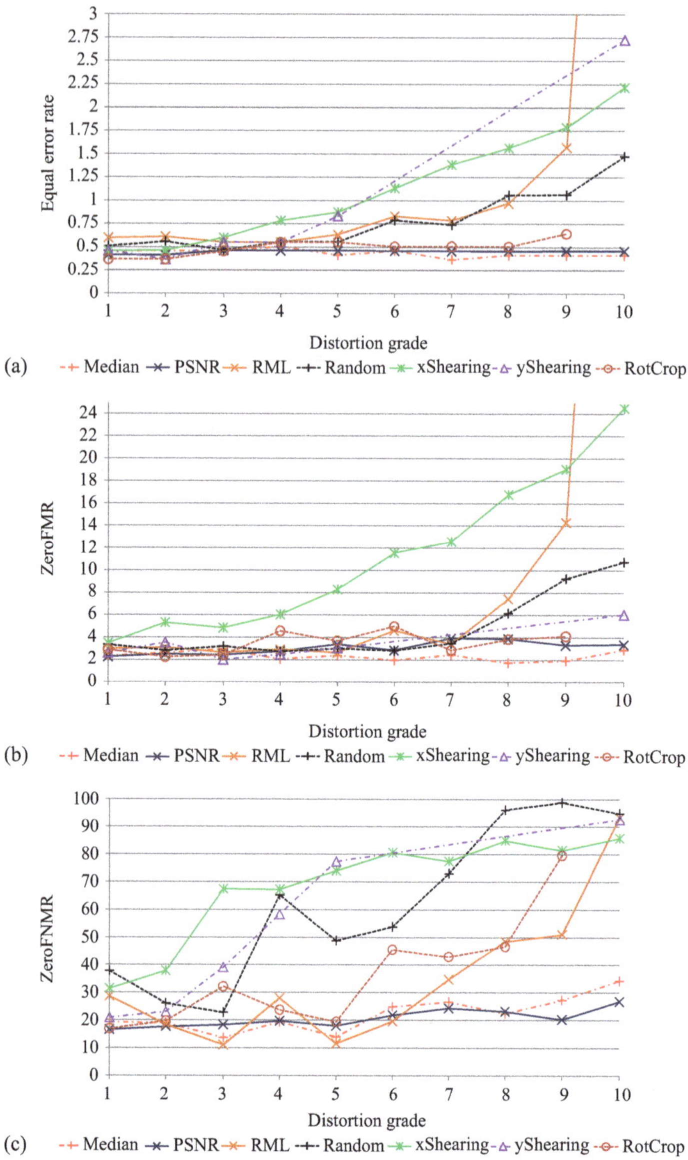

Figure 9.4 Recognition accuracy under StirMark distortions of increasing strengths: (a) EER, (b) ZeroFMR, and (c) ZeroFNMR

watermark insertion does not impact recognition accuracy, at least not up to grade 10. When considering ZeroFMR [Figure 9.4(b)], we observe almost identical relative behaviour of the different distortion types, except for y-shearing which shows less strong impact as compared to EER.

ZeroFNMR as shown in Figure 9.4(c) shows slightly different behaviour. While Median Filtering and Watermark Insertion do not create significantly more false positive matches when increasing their distortion strength, we see small random distortion taking 'the lead' (worst FMR) for high distortion grades and also rotation getting more problematic as observed with EER and ZeroFMR.

Summarising, we may state that different distortion types have a very different impact to MC finger-vein recognition accuracy, and also we have to differentiate between the creation of false positive (higher FMR) or false negative (higher FNMR) matches for different distortion types.

9.4 Compression robustness

Contrasting and in addition to the first and only work on compression robustness of finger-vein recognition [11], we consider an additional non-standardised recent still image compression scheme, we extend the analysis to ZeroFMR (instead of only considering EER), and we consider two datasets (instead of a single one in order to be able to judge generalisation potential of our results). Furthermore, a systematic consideration of compression employment scenarios is conducted.

The certainly most relevant standard for compressing image data relevant in biometric systems is the International Organization for Standardization/International Electrotechnical Commission (ISO/IEC) 19794 standard suite on Biometric Data Interchange Formats where in the most recently published version (ISO/IEC 19794-9:2011 for vascular data), JPEG, JPEG_LS, and JPEG2000 are included for lossy compression (see clause 8.3.13). The ANSI/NIST-ITL 1-2011 standard on 'Data Format for the Interchange of Fingerprint, Facial & Other Biometric Information' (former ANSI/NIST-ITL 1-2007) only supports JPEG2000 for applications tolerating lossy compression.

In this chapter, we consider three different ISO/IEC (lossy) image compression standards and a non-standardised technique for increasing compression rates (i.e. ratio between original file size and file size after compression) up to 110 using the respective default configurations unless stated otherwise:

1. *JPEG (JPG)*: The well-known (ISO/IEC IS 10918-1) Discrete Cosine Transform (DCT)-based image compression method. By adjusting the divisors in the quantisation phase, different compression ratios can be achieved. We adjust the quality parameter iteratively to achieve a file size closest to the desired compression rate. The MATLAB® implementation is used.
2. *JPEG 2000 (J2K)*: The wavelet-based image compression standard (ISO/IEC IS 15444-1) can operate at higher compression ratios as well. J2K is also a part of the Digital Imaging and Communications in Medicine (DICOM) standard where it replaced lossless JPEG compression. Results typically do not generate block-based artefacts as the original DCT-based JPG standard. J2K facilitates explicit rate control, i.e. target bitrates are met with high accuracy. We use JJ2000 version 5.1 available at https://code.google.com/p/jj2000/. For J2K, we additionally employ three variants of region of interest coding (Region of

Interest (ROI), i.e. the pixels corresponding to the finger): first, the classical variant where all the ROI data is coded into the bitstream before the background data; second, where also resolution Level 0 of the background is encoded together with the ROI data; and third, where the first four resolution levels of the background is encoded with the ROI data.

3. *JPEG-XR (JXR)*: This compression standard based on Microsoft's HD Photo is known to produce higher quality than JPEG but provides faster compression than JPEG 2000. In the default configuration, the Photo Overlay/Overlap Transformation is only applied to high pass coefficients prior to the Photo Core Transformation (ISO/IEC IS 29199-2). We adjust quantisation levels iteratively to achieve a target bitrate closest to the desired one. Software available at https://jxrlib.codeplex.com/ is used in experiments.
4. *BPG*: The 'Better Portable Graphics' algorithm is based on a subset of the H.265 (HEVC, ISO/IEC 23008-2) video compression standard. We adjust quantisation levels iteratively to achieve a target bitrate closest to the desired one. The employed software can be downloaded from https://bellard.org/bpg/.

The classical scenario to employ compression in biometric systems is a distributed architecture, where the sensor for authentication sample-data acquisition is dislocated from the biometric matching module. Thus, sample data (the probe) is compressed to facilitate efficient transmission. On the other hand, also enrolment sample data might be kept to enable a change of the template representation without the need for re-enrolment, and if done so, data is certainly compressed (and encrypted) for storage (the gallery). Thus, we have two scenarios how compression may be applied: First, only one sample is compressed (before template generation), i.e. either probe or gallery data (scenario *1-compressed*). And second, both samples involved in matching have been compressed before template extraction (scenario *2-compressed*). Depending on the results we obtain, we may give a recommendation which of the two options should be applied in case we have at least one sample being subjected to compression in mandatory manner (because we may decide if the second sample should be compressed or not depending on the results).

In the experiments, we use both datasets described in Section 9.2, and also the outcome of both recognition schemes (i.e. MC and SIFT) will be compared. Figure 9.5 shows compressed UTFVP images.

It can be clearly seen that JPEG compression is already beyond its limits for sensible operation at a compression ratio of 70 for these type of image data. At compression ratio 110, compression artefacts are clearly visible for JXR and BPG, while J2K subjectively provides the best visual quality.

9.4.1 Results

Subsequent result plots show the compression ratio on the x-axis and EER/ZeroFMR on the y-axis. We first look into the results of the UTFVP data in the 1-compressed scenario. Figure 9.6 shows a comparison of EER for MC and SIFT recognition on the UTFVP data. It is clearly displayed that JPEG is not competitive for compression ratios >30. Also, BPG is clearly worse compared to JXR and J2K for SIFT recognition.

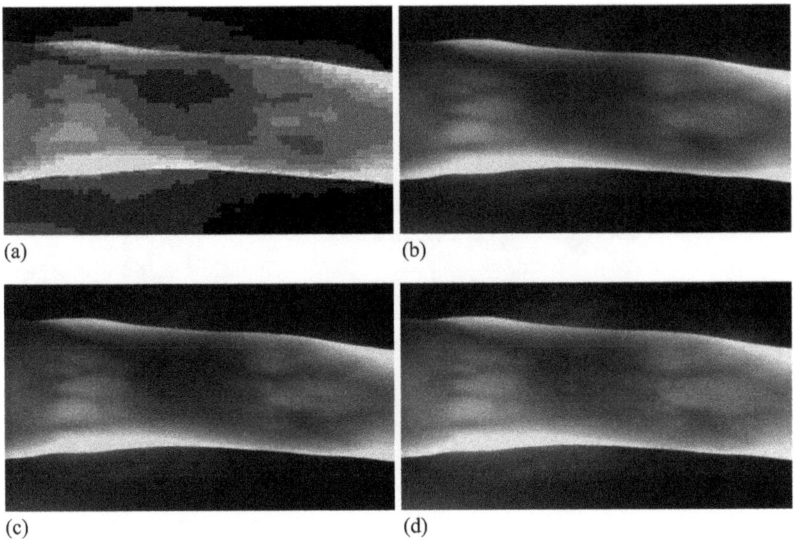

Figure 9.5 Compressed UTFVP images: (a) JPEG, ratio = 70; (b) BPG, ratio = 110; (c) J2K, ratio = 110; and (d) JXR, ratio = 110

More details hidden by the superimposed polygons in Figure 9.6 are shown in Figure 9.7. For MC recognition, the various J2K options are clearly the best, with the 'pure' ROI coding variant being the best option. JXR is clearly inferior, while BPG is the worst option in this detailed view [Figure 9.7(a)]. For SIFT, the different J2K options are pretty close, while JXR is the worst option for most compression ratios.

Figure 9.8 shows results on ZeroFMR. For MC recognition [Figure 9.8(a)], JPEG turns out to lead to very poor results (which are not shown), while the J2K variants are clearly better than BPG and JXR (where again the 'pure' ROI coding is best in most cases). BPG is worst for high compression ratios. For SIFT recognition [Figure 9.8(b)], JPEG and BPG give very poor results, while J2K, J2K ROI, and JXR are rather close in performance.

When comparing the trend for EER and ZeroFMR, we see similar overall trends, while in some details, differences are found: for SIFT, JPEG, and BPG, performance is rather close for ZeroFMR, while in terms of EER BPG is much better. For MC recognition, JXR and BPG exhibit surprisingly good values for compression ratio 70 in terms of ZeroFMR, while this is not visible in terms of EER.

In Figure 9.9, we compare the two scenarios in exemplary plots. Interestingly, in most cases (four out of six), recognition accuracy in terms of EER is superior in the 1-compressed scenario.

When looking more closely to the cases with opposite behaviour, we notice that better results for the 2-compressed scenario are only observed in cases with overall very low EER, i.e. JPEG for high compression ratios (both MC and SIFT

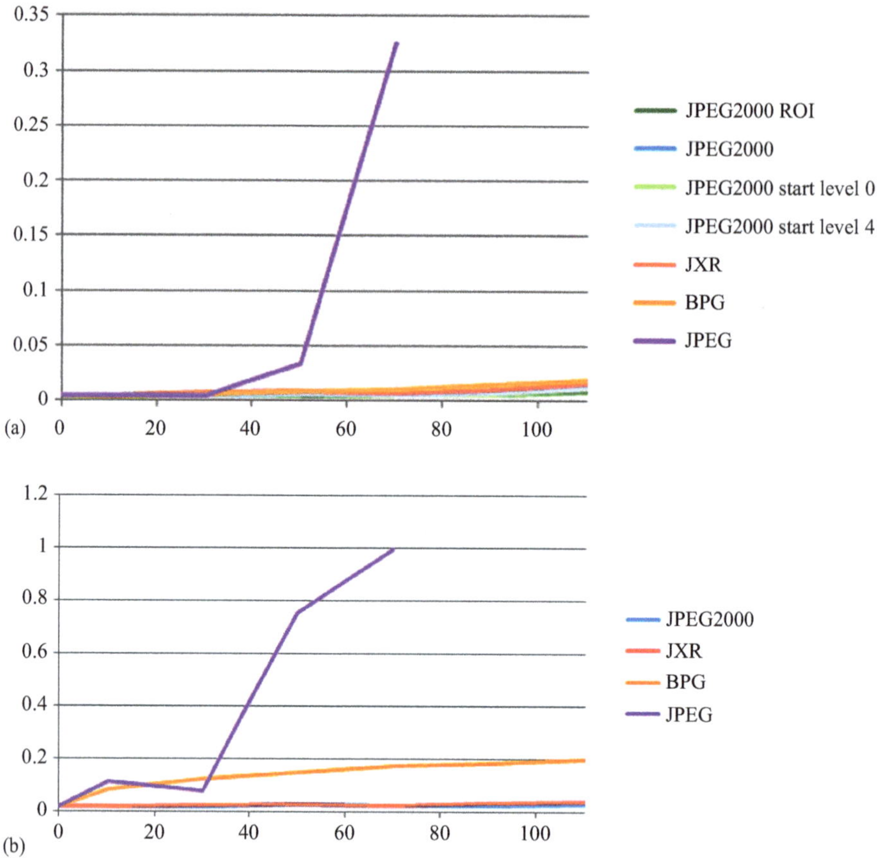

Figure 9.6 Recognition accuracy (EER) under increasing compression strength on UTFVP data (scenario 1-compressed): (a) MC and (b) SIFT

recognition) as well as BPG and SIFT recognition. Thus, under desirable operation conditions, keeping one sample uncompressed is the better option (also saving computational cost for the compression of the second sample).

After the view on isolated single error curves, we additionally provide a comparison of the relation among the error curves of the different compression techniques in Figure 9.10. Figure 9.7(a) (depicting the EER behaviour in scenario 1-compressed) can be directly compared to Figure 9.10(a). While the J2K variants exhibit a very similar behaviour in both scenarios, the superiority of BPG over JXR as seen in scenario 1-compressed can no longer be stated for scenario 2-compressed. The relation of the latter two techniques is actually interchanged when switching scenarios.

By analogy, Figure 9.10(b) displays ZeroFMR behaviour of SIFT recognition in the 2-compressed scenario and can be directly compared to Figure 9.8(b) displaying the 1-compressed case. In this case, we observe very similar relations with

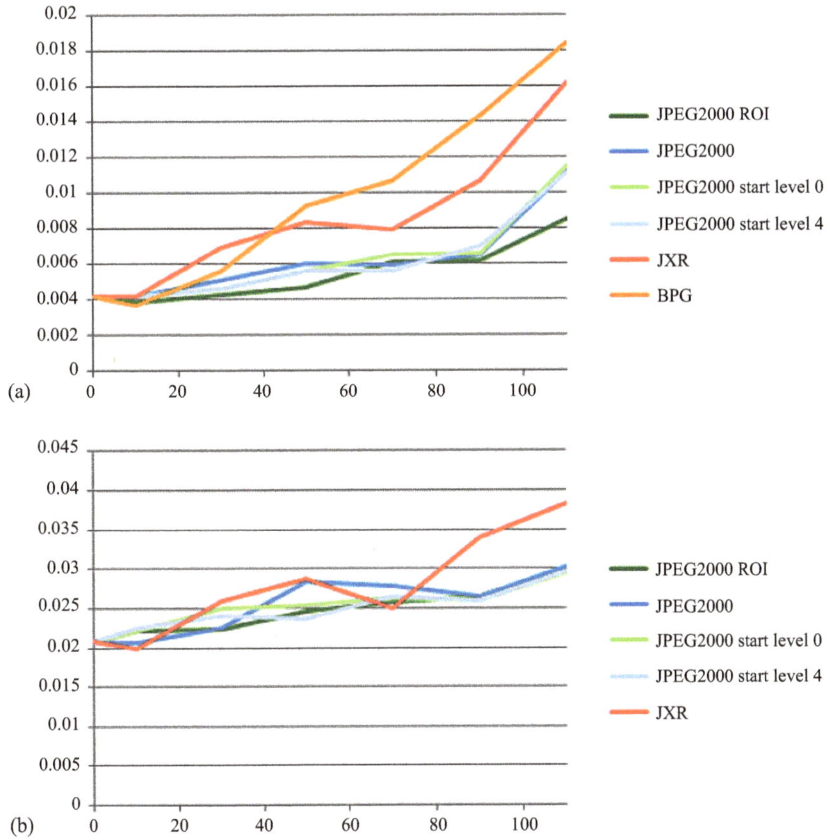

Figure 9.7 Details on recognition accuracy (EER) under increasing compression strength on UTFVP data (scenario 1-compressed): (a) MC and (b) SIFT

the major difference that JXR gets clearly inferior to the J2K variants for higher compression ratios.

In the following, we investigate if the observed results do indeed carry over to the second dataset, i.e. the SDUMLA-HMT data. Again, JPEG results are by far worst, and JPEG is only able to compress the images up to compression ratio 50 (due to the lower resolution of the SDUMLA-HMT data as compared to UTFVP). Figure 9.11 displays EER for the 1-compressed scenario for MC and SIFT recognition, respectively. We again observe for both recognition schemes that the J2K variants are very close and give clearly better behaviour as compared to JXR. BPG behaviour is also quite similar to the UTFVP case – while being close to JXR for MC recognition, the results for SIFT recognition are clearly inferior to J2K and JXR.

ZeroFMR results as shown in Figure 9.12 correspond even better to the expectations. For MC recognition, the clear ranking is identical to the UTFVP data

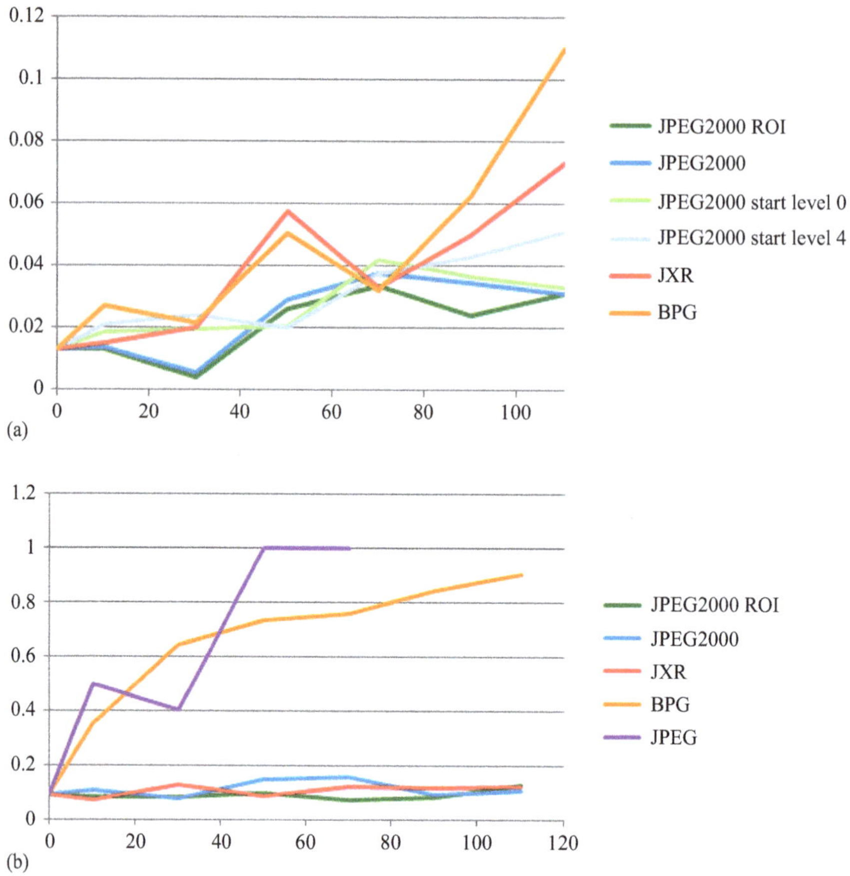

Figure 9.8 Details on recognition accuracy (ZeroFMR) under increasing compression strength on UTFVP data (scenario 1-compressed): (a) MC and (b) SIFT

(i.e. J2K variants best, followed by BPG and JXR), and also the poor BPG results for SIFT recognition do entirely carry over from the UTFVP results.

Overall, results of the two different datasets are in good correspondence. Finally, we aim to verify if the results on comparing 1-compressed to 2-compressed scenario do (also) carry over from the UTFVP data. Thus, in Figure 9.13, we compare the two scenarios in exemplary plots. At first sight, results seem very different from the UTFVP case, as in most cases (four out of six), recognition accuracy in terms of EER is superior in the 2-compressed scenario.

However, when looking at the results more closely, we observe that for the two cases in which scenario 1-compressed is superior, EER behaviour is better as for the other four cases. So overall, again, in the case of good recognition performance, the 1-compressed scenario is advantageous; however, the differences are more subtle as compared to the UTFVP data.

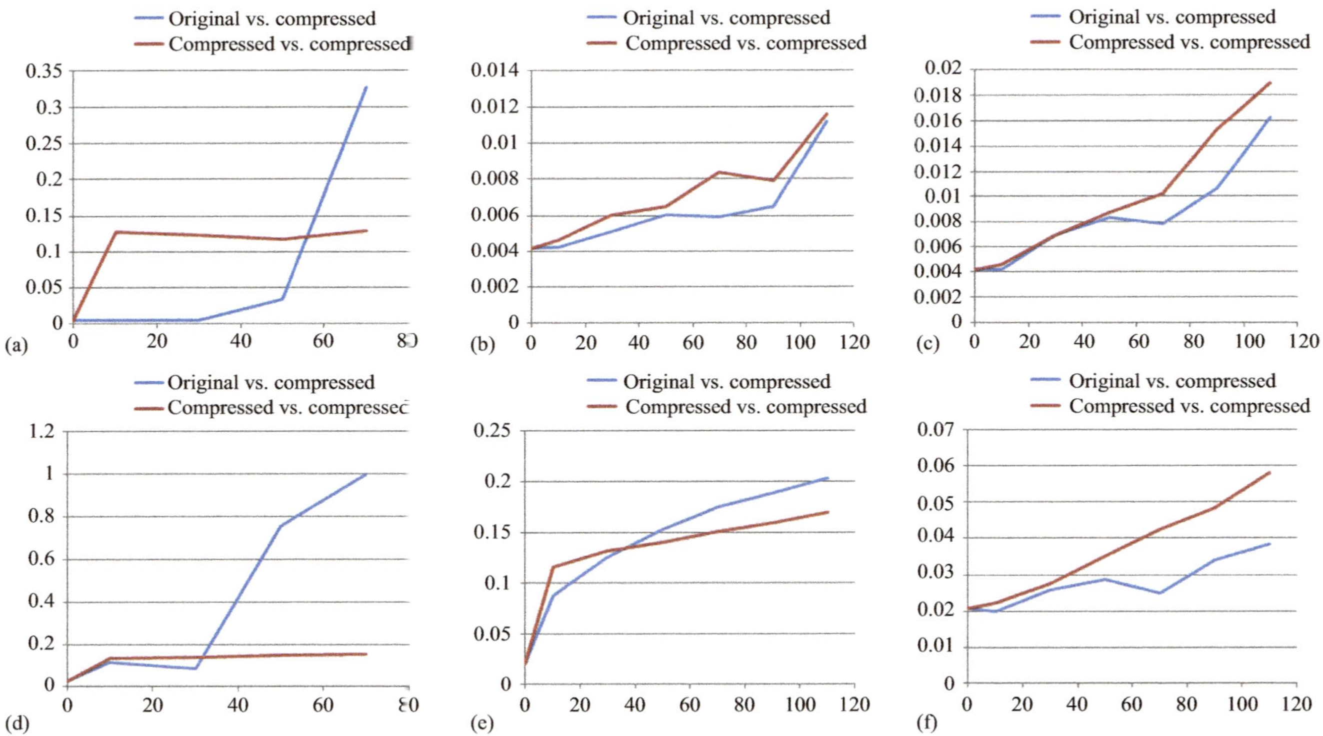

Figure 9.9 Comparing the two scenarios [1-compressed (blue) vs. 2-compressed (red)] on UTFVP data: (a) JPEG, MC; (b) J2K, MC; (c) JXR, MC; (d) JPEG, SIFT; (e) BPG, SIFT; and (f) JXR, SIFT

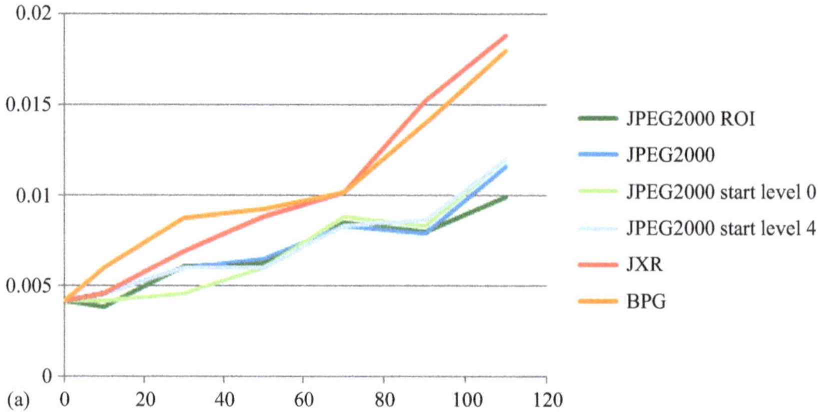

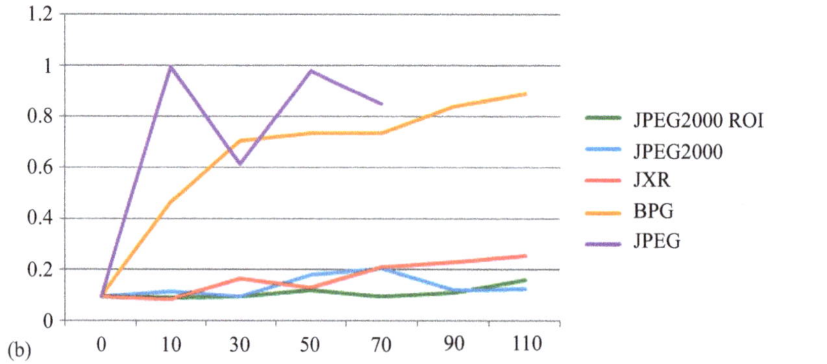

Figure 9.10 Recognition accuracy in the 2-compressed scenario under increasing compression strength on UTFVP data: (a) MC, EER details and (b) SIFT, ZeroFMR

Finally, after the view on isolated single error-curves, we again additionally provide a comparison of the relation among the error curves of the different compression techniques in Figure 9.14. Figure 9.14(a), showing the ZeroFMR with MC recognition in scenario 2-compressed, directly compares to Figure 9.12(a) providing the same data for the 1-compressed scenario. The overall shape of the figures appears to be very similar, but note that the FNMR values for all compression techniques are higher for the 1-compressed scenario. Also, JXR is superior to BPG in the 2-compressed scenario case, which is not true for 1-compressed.

Figure 9.14(b) directly compares to Figure 9.12(b) where we observe very similar behaviour for both scenarios. The only exception is JXR, which is inferior to the J2K variants for a wide range of compression ratios in scenario 2-compressed, but only inferior for the higher ratios in scenario 1-compressed.

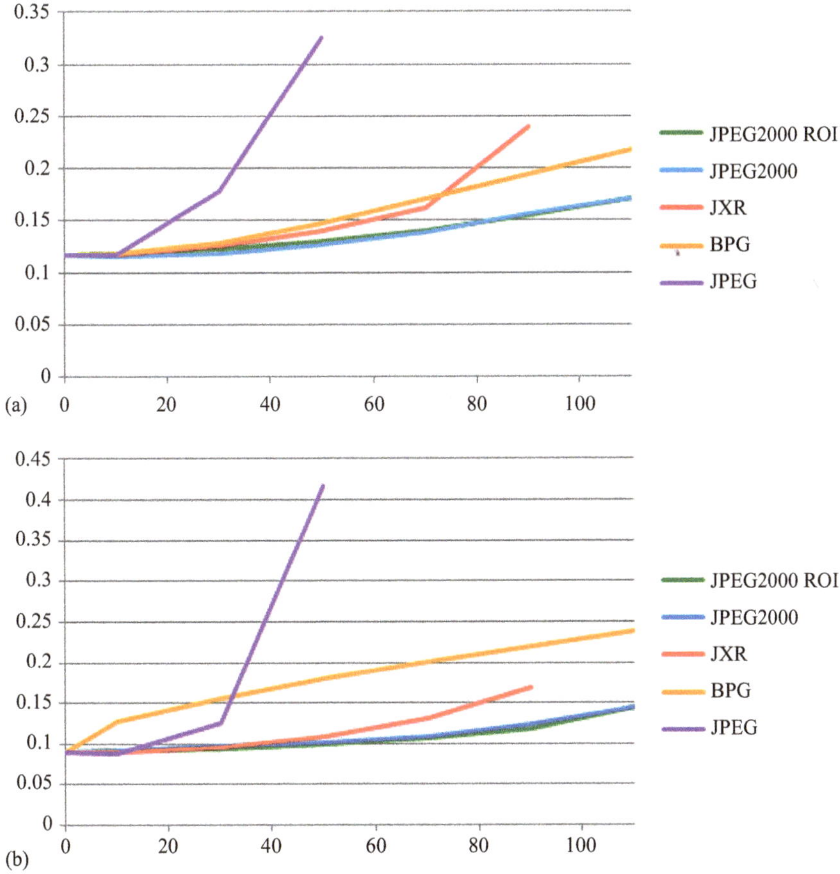

Figure 9.11 Recognition accuracy (EER) under increasing compression strength on SDUMLA-HMT data (scenario 1-compressed): (a) MC and (b) SIFT

9.5 Sensor ageing robustness

Contrasting and in addition to the first and only work on sensor-ageing robustness of finger-vein recognition [10], we consider a different dataset (SDUMLA-HMT instead of UTFVP) to allow to judge generalisation potential of the results and instead of considering EER results only, we also look at the ZeroFMR results to allow more standardised accuracy assessment. To facilitate a fair comparison, we use the identical sensor ageing simulation procedure as in [10] but focusing on the effects of an increasing number of hot and stuck pixels only without looking into combined effects. When being applied to unaltered SDUMLA-HMT data, we find EER for MC recognition to be 0.1045 and ZeroFMR to be equal to 0.2515. For SIFT recognition, we find the EER to be 0.1289, while ZeroFMR is 0.9999.

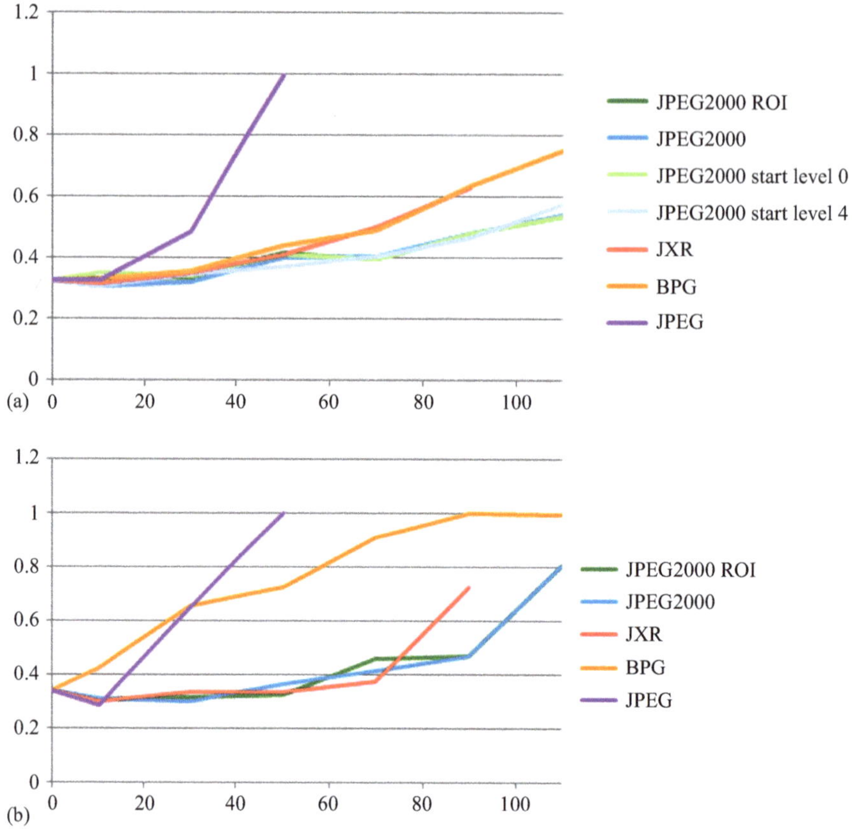

Figure 9.12 Recognition accuracy (ZeroFMR) under increasing compression strength on SDUMLA-HMT data (scenario 1-compressed): (a) MC and (b) SIFT

Thus, comparing EER on unaltered data to the results in [10, Table 2], results for SDUMLA-HMT are significantly worse. In particular, we also note an interesting effect when comparing EER and ZeroFMR: while EER is clearly better for MC on UTFVP as compared to SIFT, on SDUMLA-HMT, the EER exhibits a comparable value for both recognition schemes. Only looking at the ZeroFMR reveals the clearly inferior behaviour of SIFT on this dataset.

9.5.1 Results

The results are evaluated in terms of the absolute EER/ZeroFMR accuracy.

Figure 9.15 displays the results for the SDUMLA-HMT dataset, whereas Figures 1.2 and 1.3 in [10] show the corresponding results for the UTFVP data. Overall, it has to be clearly noted again [10] that under realistic ageing conditions, sensor ageing does not impact finger-vein recognition accuracy.

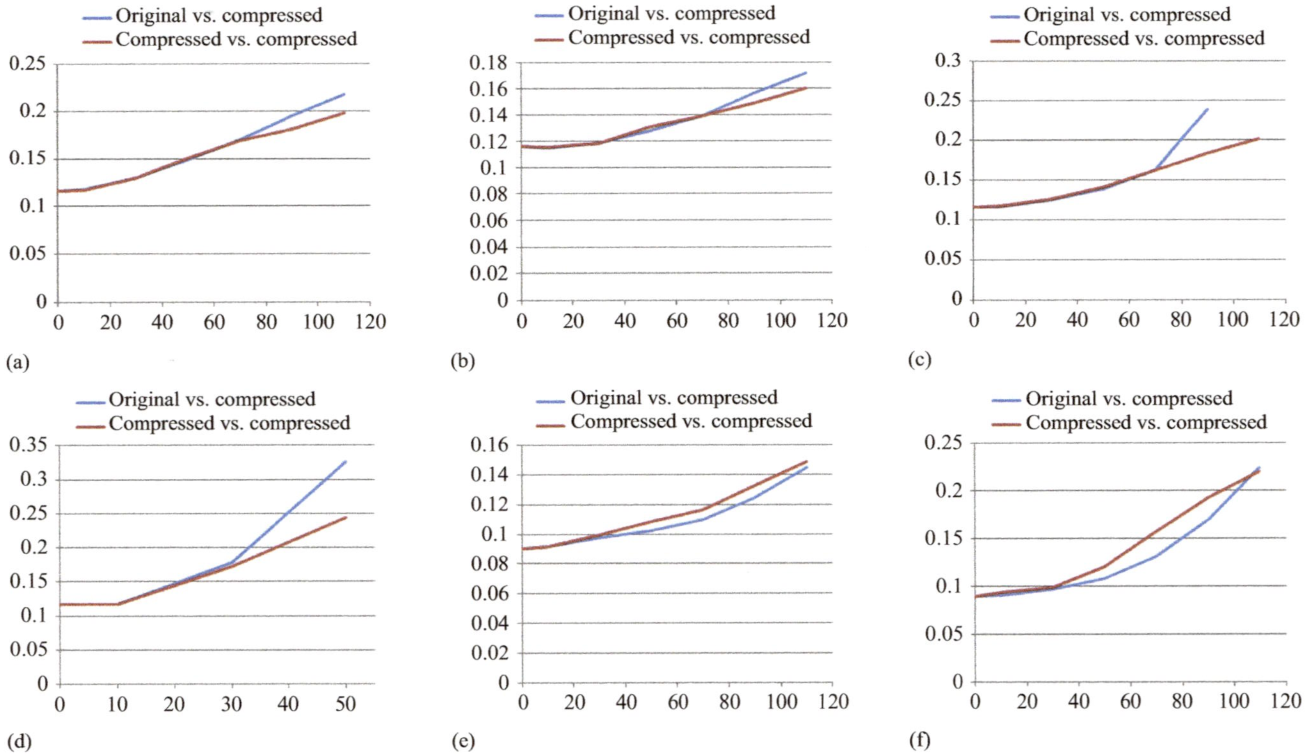

Figure 9.13 *Comparing the two scenarios [1-compressed (blue) vs. 2-compressed (red)] on SDUMLA-HMT data: (a) BPG, MC; (b) J2K, MC; (c) JXR, MC; (d) JPEG, SIFT; (e) J2K, SIFT; and (f) JXR, SIFT*

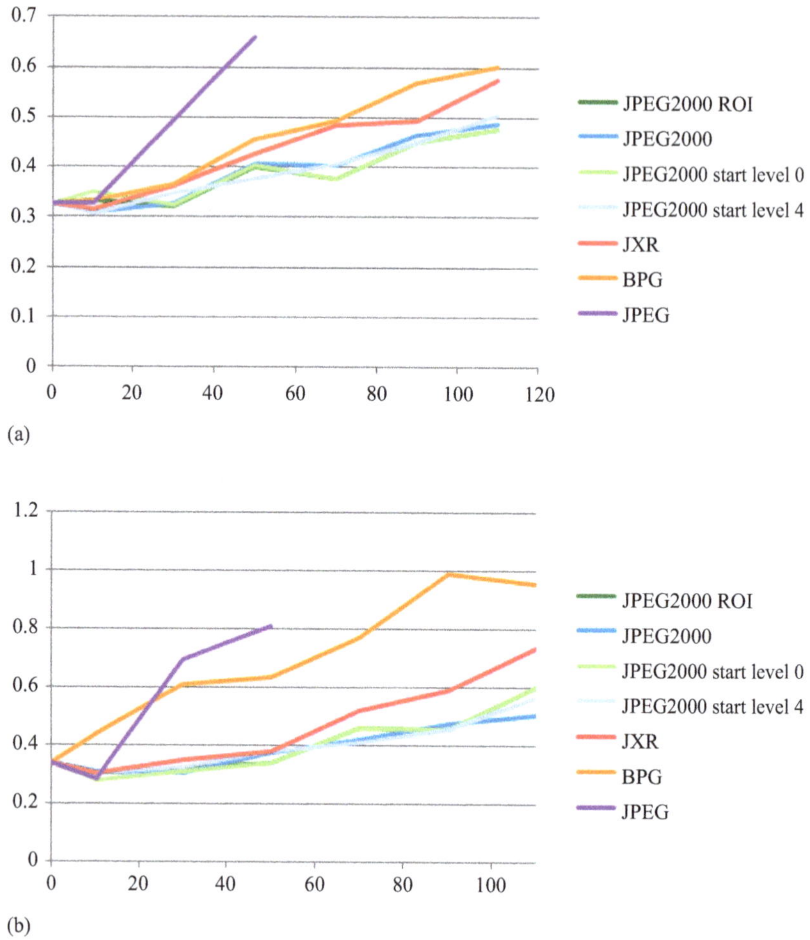

Figure 9.14 Recognition accuracy in the 2-compressed scenario under increasing compression strength on SDUMLA data: (a) MC, ZeroFMR and (b) SIFT, ZeroFMR

However, for high defect rates, interestingly, we notice some differences in the results. For UTFVP data, we do not observe an increase of EER under ageing for hot pixels at all (for both MC and SIFT), and only considering MC we find increasing EER for stuck pixels (EER is constant for SIFT under an increasing number of stuck pixels as well). For SDUMLA-HMT data, different behaviour is seen as follows [Figure 9.15(a)]. For MC recognition, we find increasing EER as long as the number of defects is increased, for both, hot and stuck pixels (for stuck pixels, the situation is worse). For SIFT recognition, EER first increases but then stabilises at values clearly superior to MC in terms of relative difference. Also, the worse behaviour of stuck pixels is only true for a lower number of defects; for higher ones, hot pixels have more impact on SIFT recognition.

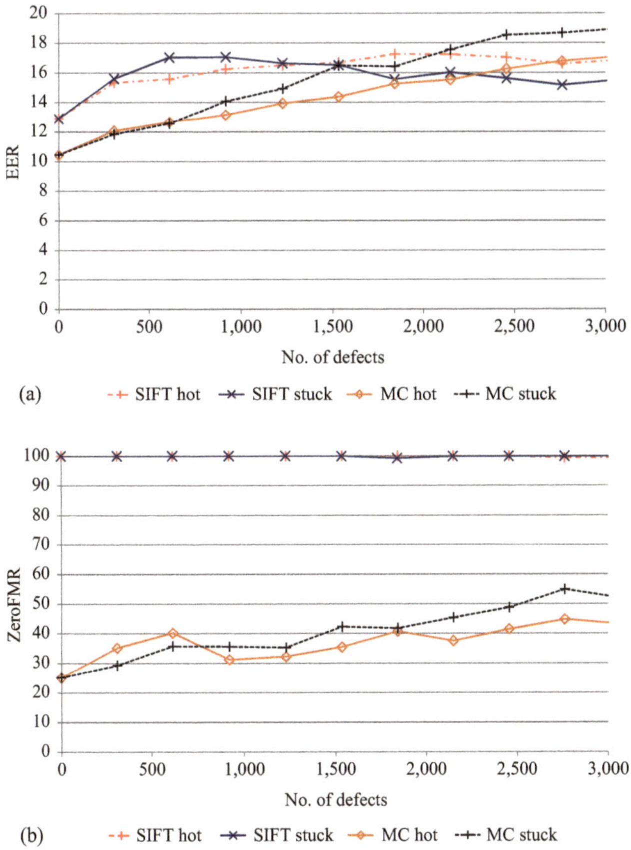

Figure 9.15 Relative recognition accuracy (MC and SIFT algorithms, respectively) under simulated sensor ageing of increasing strength: (a) EER and (b) ZeroFMR

ZeroFMR results, not being considered in [10] [Figure 9.15(b)], do not indicate any change of the SIFT results for an increasing number of defects. This is due to the already catastrophic FNMR values which cannot increase any more. For MC recognition, we find significantly increasing FNMR for increasing defect numbers, more pronounced for stuck pixels at high defect numbers and more pronounced for hot pixels at lower defect numbers.

9.6 Conclusion

For the three different aspects of robustness, we reach the following conclusions:

- **Acquisition conditions:** Incorrect finger placement in 3-D space relative to the sensor plane as modelled by shearing turns out to be highly problematic for recognition performance. Also, an excessive line loss or rotation significantly impacts on recognition accuracy. On the other hand, it turns out that watermark insertion and median filtering (used to model motion blur or defocus) does not affect finger-vein recognition accuracy too much. It is also interesting to observe that different types of distortions tend to generate different amounts of false positives and false negatives, changing the ranking among distortion types clearly.

- **Lossy compression:** Result correspondence between the two datasets considered is very good; thus, we may assume that our findings also do carry over to other finger-vein data (i.e. sensors). Also, results do carry over from EER-based rankings to ZeroFMR ones; however, differences are more pronounced in some cases for ZeroFMR. With respect to compression scenarios, we have found that in the case of higher recognition accuracy, it is beneficial to have one template being derived from uncompressed sample data, while for situations with lower overall recognition accuracy, it is of advantage to compute both templates involved in matching from compressed data. In terms of compression schemes, JPEG is clearly not useful in medium and high compression ratio settings. On the other hand, JPEG2000 robustly delivers the best performance of all compression schemes; however, the ROI coding option hardly improves over the non-ROI baseline compression technique. Obviously, the image background does not play an important role due to its low information content. BPG, being competitive to the ISO still image compression standards for MC recognition, provides much worse behaviour for SIFT recognition. So it seems that the usefulness of BPG is highly dependent on the used recognition scheme.
- **Sensor ageing**: In usual sensor operation conditions, sensor ageing does not pose a problem for finger-vein recognition accuracy. However, when increasing sensor defects to surpass 'natural' levels, we find partially significant dataset dependencies in the results. Thus, we recommend to experimentally identify the most appropriate feature extraction and pre-processing schemes given specific data sets and ageing conditions.

Acknowledgements

We acknowledge and highly appreciate the dedicated work of the students Christian Barthel, Tamara Lipowski, Babak Maser, Simone Oblasser, and Elaheh Yousefiamiri for generating experimental data for this chapter in the context of the Media Data Formats Lab at the University of Salzburg.

References

[1] F. Alonso-Fernandes, J. Bigun, J. Fierrez, H. Fronthaler, K. Kollreider, and J. Ortega-Garcia. Fingerprint recognition. In D. Petrovska-Delacretaz, G. Chollet, and B. Dorizzi, editors, *Guide to Biometric Reference Systems and Performance Evaluation*, pages 51–88. Springer-Verlag, 2009.

[2] D. Maltoni, D. Maio, A. K. Jain, and S. Prabhakar. *Handbook of Fingerprint Recognition (2nd Edition)*. Springer-Verlag, 2009.

[3] A. Noviyanto and R. Pulungan. A comparison framework for fingerprint recognition methods. In *Proceedings of the 6th SEAMS-UGM Conference (Computer, Graph and Combinatorics)*, pages 601–614, 2011.

[4] R. Cappelli. Synthetic fingerprint generation. In D. Maltoni, D. Maio, A.K. Jain, and S. Prabhakar, editors, *Handbook of Fingerprint Recognition (2nd Edition)*, pages 271–302. Springer-Verlag, 2009.

[5] M. Kutter and F. A. P. Petitcolas. Fair evaluation methods for image watermarking systems. *Journal of Electronic Imaging*, 9(4):445–455, October 2000.

[6] J. Hämmerle-Uhl, M. Pober, and A. Uhl. Towards standardised fingerprint matching robustness assessment: The stirmark toolkit – cross-feature type comparisons. In *Proceedings of the 14th IFIP International Conference on Communications and Multimedia Security (CMS'13)*, volume 8099 of *Springer Lecture Notes on Computer Science*, pages 3–17, Magdeburg, Germany, September 2013.

[7] J. Hämmerle-Uhl, M. Pober, and A. Uhl. Towards standardised fingerprint matching robustness assessment: The stirmark toolkit – cross-database comparisons with minutiae-based matching. In *Proceedings of the 1st ACM Workshop on Information Hiding and Multimedia Security (IH&MMSec'13)*, pages 111–116, Montpellier, France, June 2013.

[8] C. Kauba and A. Uhl. Fingerprint recognition under the influence of sensor ageing. In *Proceedings of the 4th International Workshop on Biometrics and Forensics (IWBF'16)*, pages 1–6, Limassol, Cyprus, 2016.

[9] C. Kauba and A. Uhl. Robustness evaluation of hand vein recognition systems. In *Proceedings of the International Conference of the Biometrics Special Interest Group (BIOSIG'15)*, pages 1–8, Darmstadt, Germany, 2015.

[10] C. Kauba and A. Uhl. Sensor ageing impact on finger-vein recognition. In *Proceedings of the 8th IAPR/IEEE International Conference on Biometrics (ICB'15)*, pages 1–8, Phuket, Thailand, May 2015

[11] V. Ablinger, C. Zenz, J. Hämmerle-Uhl, and A. Uhl. Compression standards in fingervein recognition. In *Proceedings of the 9th IAPR/IEEE International Conference on Biometrics (ICB'16)*, pages 1–7, 2016.

[12] E. C. Lee, H. C. Lee, and K. R. Park. Finger vein recognition using minutia-based alignment and local binary pattern-based feature extraction. *International Journal of Imaging Systems and Technology*, 19(3):179–186, 2009.

[13] B. Huang, Y. Dai, R. Li, D. Tang, and W. Li. Finger-vein authentication based on wide line detector and pattern normalization. In *Pattern Recognition (ICPR), 2010 20th International Conference on*, pages 1269–1272. IEEE, 2010.

[14] K. Zuiderveld. Contrast limited adaptive histogram equalization. In P. S. Heckbert, editor, *Graphics Gems IV*, pages 474–485. Morgan Kaufmann, 1994.

[15] J. Zhao, H. Tian, W. Xu, and X. Li. A new approach to hand vein image enhancement. In *Intelligent Computation Technology and Automation, 2009.*

ICICTA'09. Second International Conference on, volume 1, pages 499–501. IEEE, 2009.

[16] J. Zhang and J. Yang. Finger-vein image enhancement based on combination of gray-level grouping and circular gabor filter. In *Information Engineering and Computer Science, 2009. ICIECS 2009. International Conference on*, pages 1–4. IEEE, 2009.

[17] N. Miura, A. Nagasaka, and T. Miyatake. Extraction of finger-vein patterns using maximum curvature points in image profiles. *IEICE Transactions on Information and Systems*, 90(8):1185–1194, 2007.

[18] N. Miura, A. Nagasaka, and T. Miyatake. Feature extraction of finger-vein patterns based on repeated line tracking and its application to personal identification. *Machine Vision and Applications*, 15(4):194–203, 2004.

[19] D. G. Lowe. Object recognition from local scale-invariant features. In *Proceedings of the Seventh IEEE International Conference on Computer Vision (CVPR'99)*, volume 2, pages 1150–1157. IEEE, 1999.

[20] C. Kauba, J. Reissig, and A. Uhl. Pre-processing cascades and fusion in finger vein recognition. In *Proceedings of the International Conference of the Biometrics Special Interest Group (BIOSIG'14)*, Darmstadt, Germany, September 2014.

[21] B. T. Ton and R. N. J. Veldhuis. A high quality finger vascular pattern dataset collected using a custom designed capturing device. In *International Conference on Biometrics, ICB 2013*. IEEE, 2013.

[22] Y. Yin, L. Liu, and X. Sun. SDUMLA-HMT: A multimodal biometric database. In *The 6th Chinese Conference on Biometric Recognition (CCBR 2011)*, volume 7098 of *Springer Lecture Notes on Computer Science*, pages 260–268, 2011.

[23] D. Maio, D. Maltoni, R. Cappelli, J. L. Wayman, and A. K. Jain. FVC2004: Third fingerprint verification competition. In *ICBA, volume 3072 of LNCS*, pages 1–7. Springer Verlag, 2004.

[24] F. A. P. Petitcolas, R. J. Anderson, and M. G. Kuhn. Attacks on copyright marking systems. In D. Aucsmith, editor, *Information Hiding: Second International Workshop*, volume 1525 of *Lecture Notes in Computer Science*, pages 218–238, Portland, OR, USA, April 1998. Springer Verlag, Berlin, Germany.

[25] J. Hämmerle-Uhl, K. Raab, and A. Uhl. Watermarking as a means to enhance biometric systems: A critical survey. In T. Filler, T. Pevny, S. Craver, and A. Ker, editors, *Proceedings of the 2011 Information Hiding Conference (IH'11)*, volume 6958 of Springer LNCS, pages 238–254, Prague, Czech Republic, 2011.

[26] J. Hämmerle-Uhl, K. Raab, and A. Uhl. Experimental study on the impact of robust watermarking on iris recognition accuracy (best paper award, applications track). In *Proceedings of the 25th ACM Symposium on Applied Computing*, pages 1479–1484, 2010.

Chapter 10

An overview of palmvein biometrics

Mohanad Abukmeil[1], Gian Luca Marcialis[1], and Sébastien Marcel[2]

Palmvein recognition (identification or verification) is gaining ground as a biometric system for personal recognition applications, with an intensive investigation during the past few years. Some researchers have devoted attention to this field, but a lot of issues related to palmvein recognition are still open. This paper provides an overview of current palmvein recognition research, describing in particular the current state of biometric palmvein system components. The beginner researchers can find in this paper the more recent approaches at each stage of biometric palmvein recognition. Also, this overview points out current open issues that need to be addressed and investigated.

10.1 Introduction

Automatic biometric recognition systems are concerned with recognizing individuals either by physiological traits such as face, iris, fingerprints, palmvein, and palmprint or by using some behavioral characteristics such as the voice or the signature [1]. Fingerprint personal recognition has received considerable interest over the last 30 years [2]. Voice, face, and iris-based verification techniques have been studied extensively [3]. Palmprint recognition is gaining popularity for several purposes in digital forensic and commercial applications [3–5]. Similar to fingerprints, palmprints suffer from problems such as problematic skin drying/laceration and spoofing [2,6–8], among others. For example, old people and physical workers may not provide fingerprint or palmprint images with sufficient quality.

The vein structure (texture pattern) which resides in the palm has several interesting characteristics. It is complex and rich, simultaneously and automatically assures the liveness in the presented sample; it is also difficult to imitate [9]. The vascular pattern of the palm is used as a biometric data in technology-based palmvein. A palmvein is characterized by the vascular pattern and the surrounding texture [10]. Moreover, it is not susceptible to visible occlusions or changes in

[1]Department of Electrical and Electronic Engineering, University of Cagliari, Italy
[2]Centre du Parc, Idiap Research Institute, Switzerland

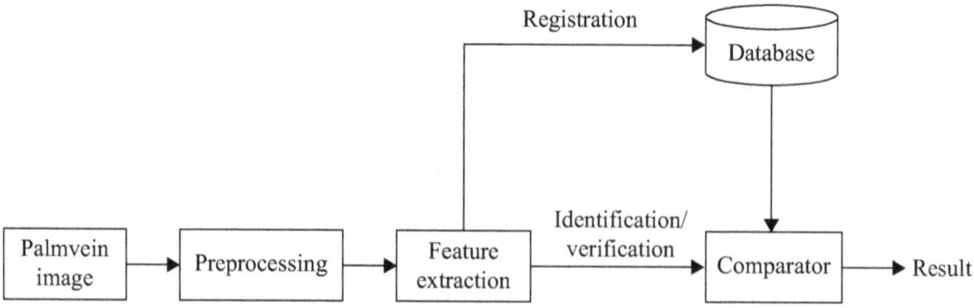

Figure 10.1 Biometric palmvein system components

lighting conditions such as the facial features, or some environmental limitations due to the state of the skin (dryness) as the fingerprint and the palmprint [11].

As any biometric recognition system, a palmvein recognition system consists of five parts: a palmvein-sensing device, a preprocessing module, a feature extraction algorithm, a comparator. The system is completed by a database where the biometric references [12] are stored for each subject. These components are illustrated as in Figure 10.1. The palmvein-sensing device captures images. The preprocessing step sets up the coordinates to get the Region of Interest (ROI). The feature extraction algorithm operates on such ROI. The aim of the comparator[1] is to compute a similarity score (match score) between the palmvein features given as input and the ones stored in the biometric reference. The more the score is, the more the similarity degree is.

Since palmvein biometric research is relatively in its infancy (Figure 10.2), each component can be considered as an independent challenge. The choice of the best configuration between palmvein-sensing device and image clarity to overcome adverse illumination and to achieve high contrast and low blurring [13] is one of such challenges. Using the appropriate image representation and extraction techniques is very important to compute the suitable features of palmvein and ascertain which one will attain the best performance [14]. Dealing with intraclass variations is another problem and must be addressed by intensive investigations [2].

The scientific and industrial trends toward palmvein recognition have remarkably boosted. Recently, the number of scientific articles is dramatically increased according to Scopus and IEEE Xplore.[2] In order to show the importance of research on palmvein recognition, Figure 10.2 compares IEEE and Scopus indexing on advance search on palmvein key terms. The Scopus search includes computer science and mathematics works only, and IEEE includes engineering works.

Moreover, Figure 10.2 shows the incremental publication rate of Scopus and IEEE in palmvein recognition during the period from 2004 to 2015. It may be noted that a strong increment of publications can be located between 2009 and 2010. Before these years, only 13%–15% of works were published. Moreover, the most recent survey on palmvein recognition has been published in 2010 by Sarkar *et al.* [15].

[1] According to the ISO terminology described in [12].
[2] http://www.scopus.com, http://ieeexplore.ieee.org.

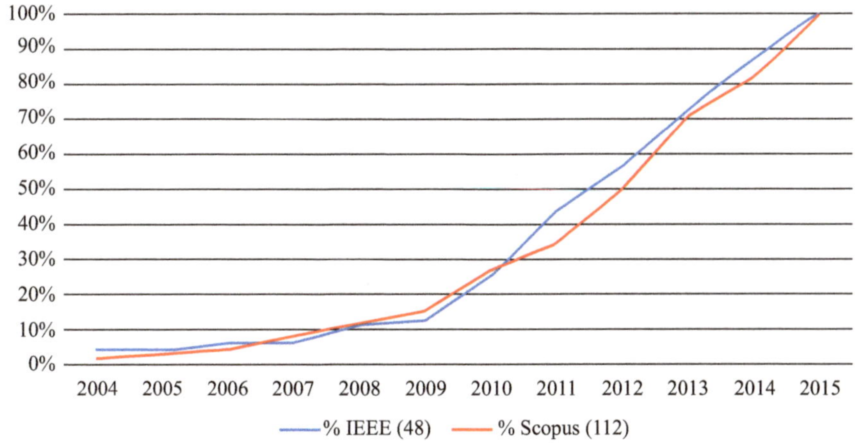

Figure 10.2 Comparison between IEEE and Scopus indexing on advance search on palmvein key terms, where the total number of the works that are published by IEEE and Scopus are 48 and 112, respectively, in the period from 2004 to 2015

As we have shown, significant advances have been done after 2010 especially; thus, the need of a novel assessment of current state of the art is fully motivated.

Accordingly, this paper investigates the current state of biometric palmvein recognition (identification and verification) system components, which are depicted in Figure 10.1. Main works focus on palmvein representation algorithms [15]. We analyzed all stages of palmvein recognition system (sensing, preprocessing, representation and comparison, integration with other biometrics, performance) and also compared pros and cons of current approaches for each stage. Our goal is to help research beginners in being aware on "where we are" in this novel field and also to suggest possible topics to be (better) investigated.

The rest of this paper is organized as follows: Section 10.2 reviews the current state of biometric palmvein system components. Section 10.3 summarizes the state-of-the-art feature extraction techniques. Data sets and experimental protocols are highlighted in Section 10.4. Section 10.5 reports the open issues of biometric palmvein recognition and concludes the paper by drawing possible future directions.

10.2 Components of palmvein recognition systems

To the best of our knowledge, the industrial trends toward biometric recognition have been remarkably boosted.[3] Each biometric mode has its strengths and

[3] The *Research and Markets* rating agency predicted an annual increment rate of 17.6% over the years 2014–2020, similarly to the *MarketsandMarkets* agency, which predicted an annual increment rate of 22.9% over the years 2013–2018.

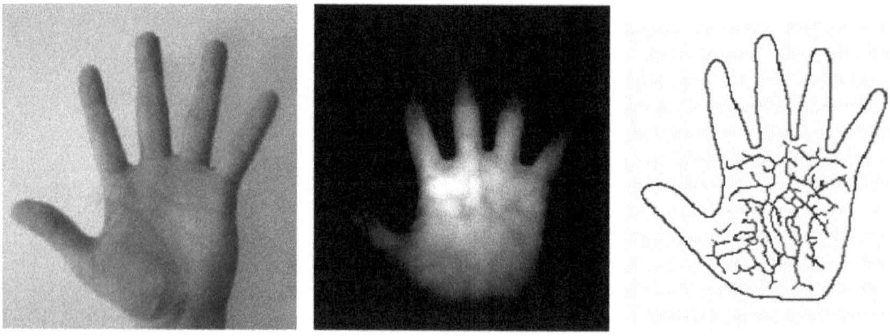

Figure 10.3 Palmprint versus Palmvein: when the palmprint image as in (a) is captured by an infrared sensor, then the vein pattern will be visible as in (b). Based on this feature, the vein recognizer device translates the feature of palmvein, from the conceptual point of view, as in (c). This figure quoted from [10]

weaknesses, and the choice of the biometric trait depends on the application. No single biometric model is expected to effectively satisfy the needs of all biometric recognition applications [16]. In other words, no biometric mode is "optimal" for all applications; the properties of each biometric mode determine the degree of compatibility to the corresponding application. Among others, the palmvein as biometric trait shows interesting and remarkable properties. In the following, we summarize the palmvein recognition system components.

10.2.1 Palmvein sensing

The veins vessels contain deoxidized hemoglobin which can absorb the light having a wavelength of about 760 nm in the infrared area (palmvein image). When the palmvein image is captured [10], differently of the palmprint image seen in Figure 10.3(a), only the vein vessel pattern is visible as a dark texture as shown in Figure 10.3(b). The recognition system aims to translate the texture pattern of the captured image as in Figure 10.3(c) (this figure shows merely the basic concept), and to match it with the previously stored patterns of the subject in the biometric reference database [10].

The vein differs in temperature from the surrounding skin, and the skin possesses a gradient temperature [17,18]. The radiation of actinic infra-red (IR) depth of absorption in palmvein tissue is approximately 3 mm; thus, there is a great possibility to get information only from the surface (skin) by thermal IR radiation. Separating veins from the total palm was proposed by Badawi [19] depending on the vascular tree detection which requires the ROI segmentation (i.e., background elimination and ROI localization). Finally, postprocessing stage is considered to show the image formation and vein structure visualization in the subcutaneous

Table 10.1 Components of several vein image-sensing systems

Literature	Image sensor	IR filter	NIR light source	Light diffuser
Pudzs *et al.* [41]	Aptina MT9V024	Optical filters *(no details)*	LEDs with peak $\lambda = 850$ nm	*No mention*
Zhang *et al.* [13]	JAIAD-080 CL 1/3″ CCD NIR camera	*No mention*	LEDs with peak $\lambda = 850$ nm	Holographic diffusers
Wang *et al.* [24]	Hitachi KP-F2A infrared CCD camera	Hoya RM80 IR filter	LED lamps with peak $\lambda = 850$ nm	*No mention*
Han *et al.* [27], Lee [28]	Digital noise reduction (DNR) CCD camera	Hoya RM72 IR filter [27], Hoya RM80 IR filter [28]	LEDs with peak $\lambda = 750$ nm	*No mention*
Michael *et al.* [42]	IR-sensitive modified webcam	IR filter that passes $\lambda >$ 750 nm	LEDs with λ ranging between 880 and 920 nm	Diffuser paper
Lee *et al.* [43]	Sony XC711-CCD camera	Hoya RM80 IR filter	LEDs with peak $\lambda = 750$ nm	*No mention*
Crisan *et al.* [44]	Philip ToU Cam USB CCD camera	Film that block $\lambda < 720$ nm	Model ELD-740-524 with peak $\lambda = 740$ nm	Diffusion filter

In this table, Literature, the authors of the image acquisition system; Image sensor, the sensor used to capture the vein images; IR filter, the infrared filter adopted; NIR light source, the near-infrared region light source; Light diffuser, the distributor for NIR light. In the table, we also indicated the charge-coupled device (CCD) and the NIR light source wavelength with λ.

layer of the skin, as in the proposed in Bouzida *et al.* [20]. The general requirements of palmvein-sensing device used to acquire palmvein images and shared in several works (as listed in Table 10.1) are as follows [21,22]:

- The near infrared (NIR) light source, used to illuminate and penetrate the palm skin; thus, the blood could absorb more NIR energy, and vein structure could be viewed darker than other areas in the palm.
- The image sensor, to acquire palmvein images illuminated by the NIR light source.
- The IR filter to let NIR sources dispersed into individual wavelength and allow the specific wavelength of NIR light to pass through the image sensor.
- The light diffuser, to ensure that all part of palmvein is illuminated with an evenly distributed NIR light source.

The market is currently lacking of a large amount of capture devices for palmvein acquisition.[4] Therefore, in many papers, researchers adopt a capture device of their own creation: a charge-coupled device (CCD) and light sources are the main parts

[4] We found for example the device by BioSec (http://biosecgroup.com).

of each designed prototype in [7,23,24]; an IR camera was used in [25–31]. Other works exclude this phase depending on publicly offered data sets as in [3,5,32–40]. Other works use publicly offered data sets to test the proposed recognition system's performance.

Table 10.1 shows the basic components of some developed vein-sensing systems. The components choice of the sensing systems varies from work to work, but the developed systems had successfully captured the intended vein images that could be used for recognition tasks in several works [21]. According to [45], transmitting NIR light around 850 nm was suitable to acquire fingervein patterns; thus, this wavelength could help when a multimodal assessment (palmvein and fingervein) must be done. The authors of [23] suggest different spectral bands to capture palmvein images. Moreover, using different spectral bands in image acquisition stage is claimed to be useful to capture clear features of palmvein. In addition, providing scanning systems with controlled pegs overcomes hand displacement and misalignment problems due to the bad cooperation of the subject during the acquisition stage. Therefore, a wide range of choices is possible in current literature; thus, it can be expected that images exhibit different characteristics from work to work.

10.2.2 Preprocessing

The preprocessing stage is used to align different palmvein images in order to remove unwanted areas, such as fingers and external border of the hand, and to segment the center of the palm for the feature extraction step. For palmvein, the same preprocessing approach followed for palmprint is used. Methods for palmprint preprocessing were suggested in [8]. Five common steps are involved:

1. Palmvein image binarization.
2. Contour of hand and fingers extraction.
3. Detection of the key points between fingers. Key points are two points in the boundary of fingers. Figure 10.4 illustrates how these key points are defined.
4. Establishing a coordination system to set up a suitable ROI (Figure 10.4).
5. Extraction of the ROI to be used for features representation, extraction, and matching.

Items 1 and 2 are similar in all the preprocessing algorithms. However, Item 3 differs from work to work. Detecting the key points consists in finding the intersections among boundaries of the fingers (fingers curves) and the tangent of the curves (Figure 10.4). Several implementation techniques for key points detection are summarized in [8] which includes tangent-based, bisector-based, and finger-based techniques.

Figure 10.4 illustrates a typical coordinate system of palmvein image.

Tangent-based approaches [1] estimate the curves (boundaries) from the middle finger and index finger, and the other from the pinky finger and the ring finger as two convex curves; thus, the tangent of them is computed. The two intersections of the curves are regarded as the key points for setting up the coordinate system.

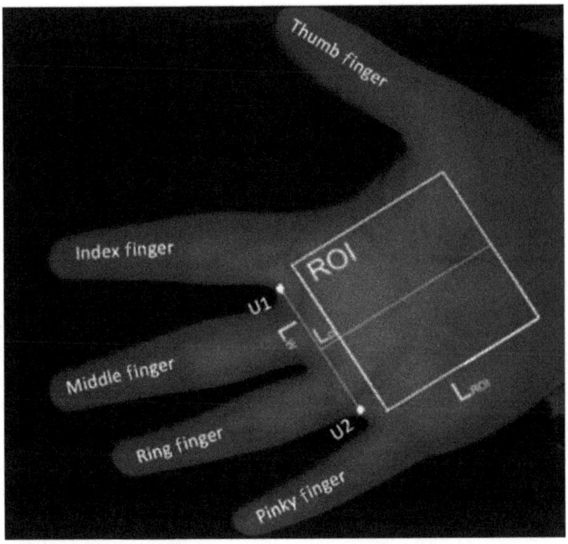

Figure 10.4 *Palmvein coordinate system. The vertices U1 and U2 are considered as key points. The curve between the middle finger and the index finger together with the curve between the pinky finger and the ring finger are utilized as the reference key points/line to build up the coordinate system. L_W is the distance between these curves and determines the size and location of ROI. This picture is quoted and modified from [9]*

Bisector-based approaches [46,47] construct a line based on two key points. The first key point is obtained by considering the midpoint of the start and the end point of the finger boundary, and, the other key point is obtained by the center of the gravity of the whole finger boundary.

Multiple finger approaches [48,49], where three lines are constructed in the image based on a set of predefined boundary points and wavelet on the three visible finger in the image. The position of the coordinate system is associated with the line from the middle finger, and the orientation is determined by the two lines from the ring and point finger.

After obtaining the coordinate system, the central part of the palmvein image (ROI) could be segmented for the feature extraction stage. The majority of pre-processing algorithms segments square regions, but circular [50] or half elliptical regions [51] are considered as well. The square ROI is easier for handling translation variations, whereas handling with rotation variations requires circular and half elliptical regions [8].

10.2.3 Feature extraction

The palmvein trait is described by the ramifications of arterial and venous lines. Once the central part of palmvein is segmented, features could be represented and extracted. High-resolution images (>300 dpi) of palmvein offer accurate features such as smooth curves and textures, while in low-resolution images (<100 dpi), there are ability to obtain line and bifurcations features [8,52].

Several feature extraction techniques have been proposed in the literature [9,23–28,30,32,35]. Palmprint feature extraction techniques have been applied to palmvein in [3,5,29,41,53–56]. The characteristic shapes of veins features of palmvein are similar to principal lines and wrinkles features of palmprint [3,5]. Thus, extracting lines, edges, and smooth curves from the venous structure is potentially useful.

On overall, current approaches are classified to geometry and statistical-based methods [57]. Geometry-based methods extract the local features from the veins such as: curvatures, lines, minutiae, and local statistical features. Statistical-based approaches [8] are either global or local. Both approaches may transform the palmvein image from the spatial domain to another domain (this does not hold for all methods). Global statistical approaches compute features directly from the eventually transformed palmvein image: among others, centers of gravity, density, and moments [58]. Local statistical techniques divide the eventually transformed image into small regions (patches) according to a predefined tessellation. In each region, several calculations for features could be done such as variances and local means, alongside other high-order statistics. For example, Tome *et al.* use histograms of local binary pattern (LBP) as features [59]. Table 10.2 shows the followed techniques for features extraction of recently published works.

Abbreviations in Table 10.2 are described as follows:

- **SIFT:** scale-invariant feature transform.
- **PCA:** principal component analysis.
- **LDA:** linear discriminant analysis.
- **DCT:** discrete cosine transform.
- **SVM:** support vector machine.
- ***k*NN:** *k*-nearest neighbor.
- **Private:** database which is not published publicly for research.
- **CASIA:** data set which is published publicly by National Laboratory of Pattern Recognition (NLPR), Institute of Automation at Chinese Academy of Sciences (CASIA).
- **PolyU hyperspectral:** data set which is published publicly by Biometric Research Centre (UGC/CRC) at the Hong Kong Polytechnic University.
- **PUT Vein:** data set which is published publicly by the Institute of Control, Robotics and Information Engineering, Poznan University of Technology.
- **Vera:** data set which is published publicly by biometrics research group at Idiap Research Institute.

Unfortunately, many data sets are private; thus, the results reported in those works cannot be compared or replicated, whilst the aforementioned publicly offered palmvein data sets are reviewed in Section 10.4.

Table 10.2 Features extraction techniques in recently published works

Ref.	Algorithm	Comparator	Database	Year
[25]	Watershed transform	Boolean function	Private	2004
[7]	Laplacianpalm	Nearest neighbor	Private	2007
[30,57]	SIFT	Key-points matcher	Private	2009, 2015
[9]	Hessian phase	Cosine similarity	CASIA	2010
[23]	$2D^2$ PCA	Not mentioned	PolyU Hyper	2010
[60]	Curvelet	Nearest neighbor	Private	2010
[39,61]	Local texture	Hamming distance	Private	2010
[42]	Directional coding	SVM	Private	2010
[62]	Multiple techniques	*k*NN	Private	2010
[32]	Radon transform	Not mentioned	CASIA	2011
[29]	DualCompCode	DualCompCode	Private	2011
[38,63]	Local binary pattern	Matched pixel ratio	CASIA	2011, 2014
[64]	Affine geometric	Euclidean distance	CASIA	2012
[27,28]	Adaptive Gabor filter	Hamming distance	Private	2012
[3,5]	Multiscale transforms	Neural network	PolyU Hyper	2013
[41,65]	Complex matching filter	Hamming distance	CASIA	2013
[66]	Contourlet transform	Euclidean distance	CASIA	2013
[43]	LDA	LDA correlation	Private	2013
[35]	PCA	Neural network	CASIA	2013
[37]	LDA	Nearest neighbor	PolyU Hyper	2014
[54]	DCT	Euclidean/Canberra distance	Private	2014
[67]	Gaussian filter	Matched filter	Private	2015
[40]	Local tetra pattern	Cosine similarity	PUT VEIN	2015
[59]	Local binary pattern	Histogram intersection	VERA	2015
[68]	Morphological filters	Phase correlation	Private	2015

Ref., the paper's reference in the bibliography; Algorithm, identifier of the features extraction technique; Comparator, the classifier used in each work; Database, the data set adopted; Year, the publication year.

10.2.4 Comparator

Once the ROI is segmented and the palmvein features are extracted, the matching algorithm determines the similarity score (match score) between such features and the ones stored in the system database and named templates [1]. The similarity score is usually a real value in the [0, 1] interval, but this rule is not general.

Two tasks are related to recognition applications, namely, identification and verification [8]. The identification is the one-to-many comparison (1:*N*) of a biometric trait against all other traits in the data set in the attempt to identify an unknown subject. Verification is the one-to-one comparison (1:1) of the given trait for a person for whom the related identity must be confirmed or denied. Current palmvein matching techniques are listed in Section 10.3.3.

10.3 Palmvein images processing

In this section, we first describe and classify the main state-of-the-art approaches to feature extraction from palmvein patterns. The remainder of this section is devoted to the eventual feature reduction step, and the comparator module.

10.3.1 Feature extraction

In palmvein recognition, local features from cardinal veins are computed such as lines, curvatures, local statistical features, and minutiae. Line and texture are the two types features that constitute veins [57]. Similar to fingerprint recognition, minutiae points can be extracted from palmvein to distinguish individuals [3,5,69]. The vein pattern hides under the skin [9], and it is far from fraud and loss; therefore, the palmvein recognition could be dominant in the future as antispoofing biometric mode [33,70].

As pointed out in Section 10.1, many works have dealt with this problem, especially since 2010. Also, some proposals include methods already used for other biometrics. Table 10.2 in Section 10.2 lists some of the recent works with the related approaches. The list is not exhaustive; however, it is representative of the basic methods followed so far. The basic categories of features extraction techniques are listed and summarized in Table 10.3. Some of the feature extraction

Table 10.3 Historical comparison of recently published feature extraction techniques

Technique name	Approach	Year	Reference
Watershed transformation	Other	2004	[25]
Laplacianpalm representation	Statistic	2007	[7]
Shape-based representation	Line	2007	[24]
Scale-invariant feature transform (SIFT)	Statistic	2009	[30]
$2D^2$ PCA	Subspace	2010	[23]
Curvelet	Line	2010	[60]
Local texture	Statistic	2010	[39,61]
Radon transform and Hessian phase	Statistic	2011	[32]
Spectral-means clustering	Subspace	2012	[52]
Adaptive Gabor filter	Statistic	2012	[27,28]
Affine invariant	Statistic	2012	[64]
Directional filter bank	Line	2013	[33]
Multiscale transforms	Line	2013	[3,5]
Principal component analysis	Subspace	2013	[79]
Linear discrimination analysis	Subspace	2014	[37]
Discrete cosine transform	Statistic	2014	[54]
Principal component analysis	Subspace	2014	[35,37]
Local binary pattern	Statistic	2011–2015	[38,59,63]
Local tetra pattern	Statistic	2015	[40]
Morphological filters	Line	2015	[68]

Statistic-, line-, subspace-based, and other approaches are denoted by statistic, line, subspace, and other, respectively.

techniques that are mentioned later could be used for different biometric modes such as fingervein, palmprint, face recognition. We also highlighted some open issues in Section 10.5.2.

To our knowledge, feature extraction techniques can be classified as line-based, statistical-based, subspace-based. Other approaches which cannot be included in none of the above classes are also discussed [8].

10.3.1.1 Line-based techniques

Line-based approaches are aimed to describe the rich line features of the veins of a palmvein image. To this aim, novel or existing edge detectors are used. The extracted vein lines are either matched directly or represented in other formats for matching. Figure 10.5 is quoted from [5] and depicts the line-based features extraction. In the following, we list the basic line-based approaches.

- ***Radon transform and Hessian phase*** [9,32]. This approach utilized a new region-based matching scheme and encoded the orientation in order to compute a roto-translations invariant features set. Two approaches were proposed to extract line-type palmvein features. The first one identified continuous line structures in images by neighborhood matching Radon Transform. The second one was based on the Hessian phase, where the local palmvein image characteristics were observed by using the Taylor's series expansion in the neighborhood of a point.

Figure 10.5 Extracted vein pattern by contourlet transform applied to the ROI. A Laplacian pyramid decomposes the ROI into two subbands or cells, one named high-pass filter and the other low-pass filter. The contourlet coefficients were obtained by directional filter bank. This figure has been taken from [5], where further details can be found

- ***Directional filter bank*** [26,33]. The vein pattern was extracted by involving a directional filter bank with different orientations. The accuracy was increased by neglecting the nonvein pixels through directional filtering magnitude. To extract line features, Wu *et al.* [33] also proposed a directional filter. They exploited the hypothesis that the intensity profile of the vein was similar to Gaussian approximation [26]. The Gaussian envelope was further processed to be like a line by using shape–scale ratio. The features were encoded in a binary code by utilizing minimum directional code, and the texture information in palmvein images were stored in the feature vector.
- ***Multiscale transforms*** [3,5]. Multiscale image transformations such as 2D discrete wavelet, ridgelet, curvelet, and contourlet were applied in these works. The proposed techniques attempted to extract different types of palmprint and palmvein features in different scales under fine resolutions. The lines and edges that constitute the veins were extracted by 2D discrete wavelet and ridgelet transforms, whilst the curves and smoothed curves were extracted by curvelet and contourlet transforms. The size of each transformed image was reduced to one feature vector by applying a projection techniques 2D-LDA and 2D-PCA.
- ***Shape-based representation*** [24,50]. Wang *et al.* [24] investigated two imaging methods: NIR and far-infrared (FIR). Palm, wrist, and back of the hand were involved in image acquisition. The result shows that the data collected from FIR was insufficient at palm–wrist vein acquisition. Separating the vein pattern from other the hand was considered by Wang *et al.* [24]. To segment the vein patterns, a local adaptive thresholding algorithm was introduced. The subjects were identified based on the shape of their vein pattern. A better representation of the vein pattern shape was done by extracting its skeleton after applying the thinning algorithm used by Kumar *et al.* [50].

Further line-based features extraction techniques for palmvein were applied in [42,66,67,71,72], by following algorithms similar to those adopted for other biometric modes (e.g., palmprint, fingerprint, face, voice, gait).

10.3.1.2 Local statistical techniques

Statistical methods are either local or global [52]. Local statistical approaches transform a palmvein image into another domain and then divide it into a set of small regions according to a predefined tessellation. Features such as variance and mean and other statistics are extracted. The statistical features in global approaches are obtained by computing the density, centers of gravity, and moments on the whole transformed palmvein images. Figure 10.6 quoted from [27] and illustrates an example of statistical-based features extraction.

- ***SIFT*** [30,73]. One originality of the proposed work by Ladoux *et al.* [30] was to use SIFT descriptors. The vein pattern was extracted from the background by using local thresholding based on statistical mean values of each pixel neighborhood. To facilitate the comparison of vein patterns, a local vein

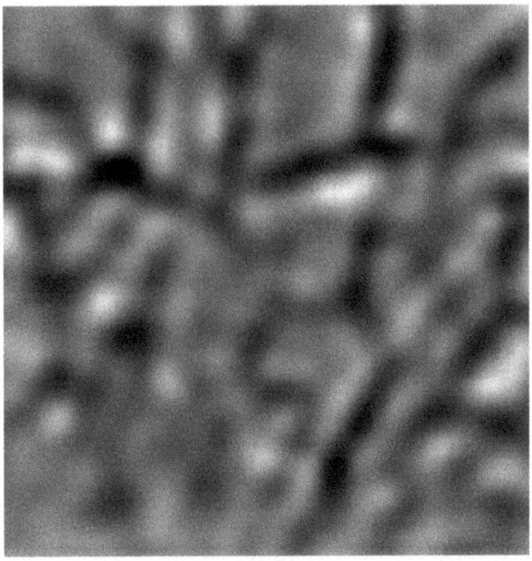

Figure 10.6 Extracted vein pattern by 2D Gabor filter transform, the real parts of the extracted is obtained using a set of parameters and coding procedure. This figure is quoted from [27]

descriptor was chosen: the so-called SIFT, described in [73]. The descriptor was created by sampling orientations and magnitudes of the image gradients in each key-point neighborhood, and building smoothed orientation histograms that included the important aspects of the neighborhood. The potential key points are identified by using the difference of Gaussian function, such points are invariant to scale and orientation. The size of each local descriptor was 4×4 array (histogram) and associated with eight orientation vector. Finally, a vector comprising 128 elements was built for each key point.

- ***Adaptive Gabor filter*** [27,28,74–77]. A robust personal recognition approach using palmvein patterns was presented by Han *et al.* [27]. The palmvein was regarded as a set of textures. In particular, the texture-based feature extraction techniques proposed by Dunn *et al.* [74] were applied. A Gabor filter [75–77] gave fine resolution in both the frequency and spatial domains. Han *et al.* [27] obtained an effective pattern of palm vascular by robust adaptive Gabor filters. Also, they made ensure of the performance of an automatic palmvein identification system by dividing the ROI of palmvein image into nonoverlap blocks; accordingly, the adaptive Gabor filters could be applied to each block in parallel. The overall aim of the Han *et al.*'s work was to discuss the optimization algorithms that determine the best parameters values of a single Gabor filter, thus making it adaptive for all blocks.
- ***Local texture patterns*** [38,40,63]. The spatial structure of local dense descriptor (texture) based on direction of central gray pixel were described by using Local Phase Quantization (LPQ) [63], LBP [38], and Local Tetra Pattern

(LTrb) [40]. LBP is a grayscale and rotation invariant dense descriptor. For each target pixel, x, it takes P neighbors sampled uniformly on a circle or square (called window or patch) of radius R centered on x. The gray level value of x is compared with all neighbor pixels in each patch according to linear filters and threshold. LPQ uses a binary encoding scheme similar to LBP, but the patch surrounding the target pixel is analyzed in the Fourier domain. The LTrb feature is coded in four directions to give most discriminative information.

- ***Laplacianpalm representation*** [7]. A multimodal personal identification system based on palmvein and palmprint images was developed with fusion scenario applied at the image level. Wang *et al.* [7] proposed a fusion method based on wavelet image, whereas contrast-enhancement and line-preservation were integrated. The fusion was implemented by combining the score of multiscale edges come from any two palmvein images. Additionally, they proposed a different feature, "Laplacianpalm," which was obtained by applying Locality Preserving Projection on the fused images.

Other statistical approaches were proposed [1,54,64,78], but the ones discussed here constitute the basis of all methods of this category applied to palmvein recognition.

10.3.1.3 Subspace-based techniques

Subspace- or appearance-based approaches are well-known in face recognition applications. PCA, LDA are the main techniques of this category. The subspace coefficients are considered as features. Different classifiers and distance measures are used to compare such features. Some works employ the subspace-based approaches after the transformations from the image domain to another in order to reduce the feature space size as in [3,5,53]. An example of subspace-based features extraction is showed in Figure 10.7 (from [37]). The list of subspace-based methods for palmvein recognition follows.

- ***Linear discrimination analysis*** [37,43]. Elnasir *et al.* [37] proposed an improved scheme of palmvein recognition model based on the LDA to extract the discriminative features. Nearest neighbor classifier with cosine distance is then followed to perform matching task. From Elnasir *et al.* experiments, LDA was showed to be efficient in class separation.
- ***Principal component analysis*** [79]. Bayoumi *et al.* used PCA to obtain vector of features that contains the most variant information extracted from palmvein. The concept of PCA, which is converting any correlated variables into uncorrelated ones, was the motivation for Bayoumi *et al.* to exploit PCA in feature extraction. The covariance matrix was obtained from normalized palmvein images; therefore, eigenvalues and eigenvectors, called eigenveins, were calculated from the covariance matrix. For each palmvein image, k eigenveins were selected from the eigenvalues weights. The overall accuracy in Bayoumi *et al.* work was 85% in the context of identification, computed on the basis of their private database.

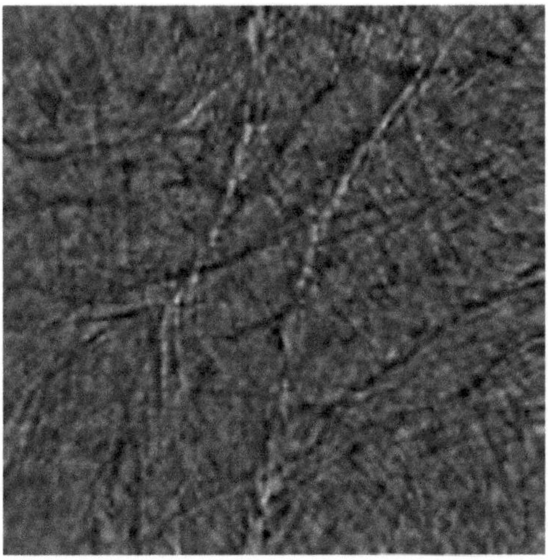

Figure 10.7 Vector basis generated by Linear Discriminant Analysis. To overcome nonsingular issue, the data matrix was projected into a lower dimension space (by using PCA) before applying LDA. This picture taken from [37]

- ***Spectral mean clustering*** [23,52]. Guo *et al.* [23] introduced palmprint hyperspectral data set which contains palmvein features and proposed a spectral-means clustering technique to cluster hyperspectral palmprint/palmvein cubes. Clustering has been used to give an optimal number of feature bands. Guo *et al.* showed that three features bands might be enough to represent. Also, they reported that veins considered as kind of special palmprint features could be captured together with palmprint. This method is better detailed in the Guo's thesis [52].

Further subspace-based features extraction techniques are followed in [35,80].

10.3.1.4 Other techniques

Some techniques are difficult to include in the previous categories, because of the adopted image processing algorithms [81,82]. Therefore, we refer to such methods without a specific reference category. Figure 10.8 was quoted from [25] and depicts an example of such techniques.

- ***Watershed transformation and multiresolution representations*** [25,83–86]. Lin *et al.* [25] proposed a new method applied to palm-dorsal vein images (where the thermal image was considered). Two of the fingers webs (the finger webs are the gaps or the boundaries between any two fingers which takes the form of a curve) constituted the boundary of two fingers. For example, the boundary between point and middle finger were automatically selected as the "Datum Points" [87] to define the ROI of the palm images. Depending on properties of acquired images, the feature points of the vein patterns (FPVPs)

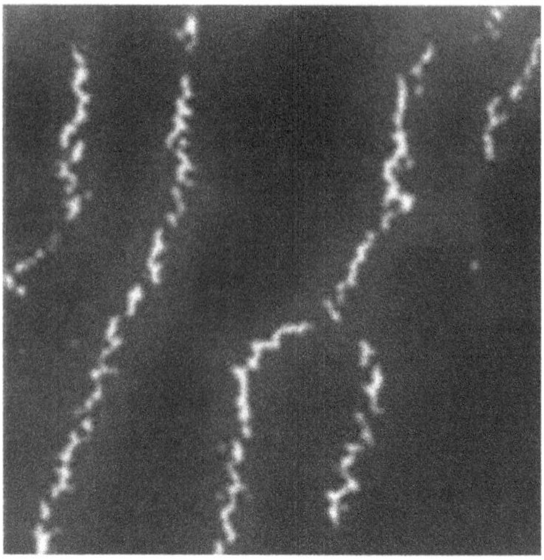

Figure 10.8 Feature points extracted from palm-dorsal vein by Watershed transformation and multiresolution representation. The white points in this figure represent the feature point of vein pattern, and the feature point extracted by wavelet transformation superimposed on the ROI. This figure quoted from [25]

were extracted by using the watershed transformation [83]. Shinagawa *et al.* [85] proposed multiresolution filters to obtain FPVPs with multiresolution representations.

- ***Minutiae triangulation*** [26,86,88–90]. Kumar *et al.* [26] proposed a method to verify candidates based on the triangulation of hand vein images and knuckle shape extraction. The suggested approach was considered for palm dorsal vein pattern. The adaptive histogram equalization was applied by local gray level information (intensity from 0 to 255) for noise reduction and image enhancement. The Laplacian of Gaussian algorithm was also applied in a further process to identify which regions with rapid intensity change while curb the noise. A final thinning step was used to generate the vein pattern structure from the resulting image. The features were extracted after thinning step by using meaningful minutiae points (i.e., triangles or triplets) by using Delaunay triangulation which first used for fingerprint identification [90]. Kumar *et al.* used such method, in the term of their belief, to obtain rotation and translation invariant representation of local information out from the vein pattern structure. The final step in this method is the calculation and indexing of the three lengths formed out triangle minutiae. It's important to point out this feature extraction method is different from the one listed under line-based techniques and offered by Wang *et al.* [24] where it directly matches the shape after thinning step.

10.3.1.5 Discussion on the reported feature extraction methods

The feature extraction stage depends on the application of image processing techniques. Also, each feature category (line, statistical, and subspace) extracts different types of features, among others: line, curve, points, and texture features. Accordingly, the nature of selected features and the related features space size affect the biometric system's speed and accuracy, as we will point out in Section 10.4.2.

Tables 10.2 and 10.3 list the feature extraction techniques in ascending chronological order during the period from 2004 to 2015. Table 10.2 contains feature extraction techniques together with the comparator (classifier), whilst Table 10.3 includes the category of each technique. It can be observed that the techniques in both Tables 10.2 and 10.3 constitute the majority of the methods applied to palmvein as well as to other biometrics (face, fingerprint, palmprint, fingervein, voice, gait).

Figure 10.9 illustrates the distribution of feature extraction categories during the period from 2010 to 2015 according to Tables 10.2 and 10.3. As it can be noticed from Figure 10.9, the earlier trials were focused on subspace-based feature extraction techniques, but now the majority is based on statistical and line-based techniques. The reduction of interest on subspace-based feature extraction approaches might be explained according to what reported by Gu *et al.* [91]: in such methods, the learned projection is a linear combination of all the original features; thus, it is difficult to interpret the results and improve the performance, while, we add, line-based and statistical-based approaches are more directly connected to the enhancement and features extraction from the vein flow variations and ramifications.

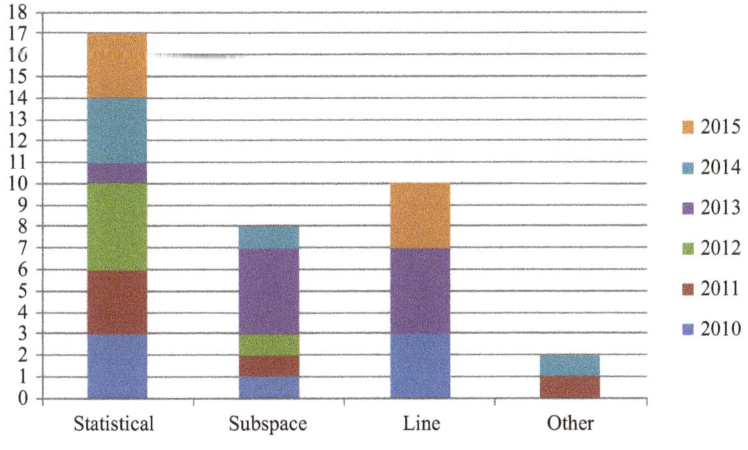

Figure 10.9 Historical distribution of feature extraction categories during the period from 2010 to 2015

10.3.2 Dimensionality reduction

As mentioned in previous section, some methods rely on features reduction methods explicitly [3,5,35,53,63]. As a result of the reduction, low computational cost and reliability could be achieved.

To perform this task, several techniques can be adopted [92]. Linear and nonlinear are the two categories of dimensionality reduction techniques. According to the carried out experiments by Maaten *et al.* in [93], although the satisfactory results of nonlinear techniques in pattern recognition applications (and artificial tasks), they do not outperform the traditional PCA on real-world tasks. Therefore, the authors recommend to employ linear dimensionality reduction techniques in biometric palmvein recognition systems. PCA and LDA were used in [3,5,53,71,72,82,94,95].

The detailed description of these techniques is out of the scope of this paper, since they are general purpose algorithms. Further information on linear and nonlinear techniques is reported in [93].

10.3.3 Comparator

The comparison algorithm determines the degree of similarity between two biometric traits.

The similarity score is usually a real value in the [0, 1] interval, but there is no specific agreement in the literature. What it is usually intended is that the more the similarity score, the more two biometric traits represented by the related features can be considered as similar [8,12].

The taxonomy of biometric comparators according to [96] is as follows:

- **Template based.** Template matching is a high-level technique that identifies the feature sets that match a predefined template. Advanced template matching algorithms allow to find occurrences of the template regardless of their orientation and local brightness. In other words, it relies on the definition of a distance between two palmvein samples independently of their representations. Among others, Euclidean, Mahalanobis, Hamming distances can be used for example if the palmvein is represented by a feature sets of real values or bits. As the majority of approaches represent the palmvein in this way, these distances are the most widely used. Once computed, distance is converted into a match score by linear transformation or using appropriate normalization methods [27,64].
- **Density estimation based.** The prior probabilities and the class-conditional densities depending on the whole features set are estimated. Their values are then interpreted as a match score. Two categories belong to density estimation: (i) parametric approach to model the class-conditional densities as multivariate normals (e.g., quadratic discriminant analysis, LDA). (ii) Nonparametric density estimation in which no fixed parametrically defined form for the estimated density is assumed [e.g., Kernel or Parzen estimates, *k*NN (*k large*)]. The maximum likelihood or a posteriori probability is then followed, on some hypothesis verification test [97].

- **Regression based.** In such methods, the class posterior probabilities are directly estimated using some regression techniques. Two categories associated with regression: (i) parametric regression (e.g., logistic regression, multilayer perceptron) and (ii) nonparametric regression (e.g., additive models, neural networks with flexible size) [3,96].

With regard to palmvein recognition, Wang *et al.* [24] measured the Line segment Hausdorff Distance (LHD) between a pair of vein patterns. LHD [98,99] is a natural measure for comparing similarity of shapes. The Nearest Neighbor classifier was adopted to identify the Laplacianpalm feature space in [7]. In Lin *et al.* [25], the distinction or similarity threshold was applied. Kumar *et al.* [26] computed the matching score for knuckle-shape feature by two ways: (i) by considering the binarized vein structures and use hierarchical matching, and (ii) perimeter distances of knuckle which belong to geometrical features, while the normalized Hamming distance was applied to measure the similarity in [27,28,33]. Template correlation algorithm [100] was adopted in [45]. Table 10.4 shows the recently used matchers according to the above classification along the period from 2010 to 2015.

The neighborhood matching Radon transform, which can effectively account for more translational variations and frequent rotational, also deformations in the

Table 10.4 Recently used biometric comparators

Ref.	Comparator	Comparator class	Year
[100]	Template correlation	Template-based	2004
[24]	Hausdorff distance	Template-based	2007
[33]	Nearest neighbor	Nonparametric density	2008
[9]	Cosine similarity	Template-based	2010
[39,61]	Template-based	Other	2010
[42]	SVM	Template-based	2010
[60,62]	*k*NN	Nonparametric density	2010
[29]	DualCompCode	Template-based	2011
[38,63]	Matched pixel ratio	Template-based	2011, 2014
[64]	Euclidean distance	Template-based	2012
[27,28,33]	Hamming distance	Template-based	2012
[3,5]	Neural network	Nonparametric regression	2013
[41,65]	Hamming distance	Template-based	2013
[66]	Euclidean distance	Template-based	2013
[43]	LDA correlation	Parametric density	2013
[35]	Neural network	Nonparametric regression	2013
[37]	Nearest neighbor	Nonparametric density	2014
[54]	Euclidean/Canberra distance	Template-based	2014
[67]	Matched filter	Template-based	2015
[40]	Cosine similarity	Template-based	2015
[59]	Histogram intersection	Template-based	2015
[68]	Phase correlation	Template-based	2015

Ref., related reference in the bibliography; Comparator, classifier or similarity metric; Comparator class, related matcher's class; Year, year of publication.

acquired image was applied in [32] to find continuous line structures in the images. Neighborhood matching is regarded as a special type of template-based comparator.

10.4 Data sets and experimental protocols

10.4.1 Palmvein data sets

According to the literature [3,5,7,24,27,28,30,53], a few of data sets are publicly available for research on biometric palmvein recognition. Usually, the performance of different works is assessed on a small data set collected by each research group. Other works used public data sets [3,5,32–40].

In this section, we briefly describe the publicly available palmvein data sets. It is worth mentioning that Guo *et al.* [23,52] introduced multispectral palmprint data sets: NIR bands (blue, green, red) were used to acquire palmprint images. Guo *et al.* observed that red NIR band can show palmvein structures as well as partial line information. Thus, in their opinion, palmveins can be regarded as special palmprint features. For this reason, multispectral and hyperspectral palmprint data sets are included in this section.

- ***PUT vein data set***[5]*:* PUT can be applied as a common platform for evaluation and comparison of new segmentation and classification algorithms. This data set consists of 2,400 images presenting human vein patterns of 50 students. Half of images contain a palmvein pattern (1,200 images) and another half contains a wrist vein pattern (another 1,200 images). PUT data set contains 100 different patterns for palm and wrist region.
- ***VERA Palmvein data set***[6]*:* The Biometrics Research Group at Idiap Research Institute and Haute Ecole Spécialisée de Suisse Occidentale in Switzerland have built VERA palmvein database [59]. Haute Ecole spécialisée de Suisse Occidentale developed the used sensor to acquire all palmvein images. A total number of 110 clients have been presented both palms (left and right) and recorded five images for each palm. The database contains 2,200 palmvein images. Location and normal lighting condition were changed in recording images. The average age is 33 years, also 70 men and 40 women whose age between 18 and 60 were involved to collect the database.
- ***CASIA multispectral palmprint data set***[7]*:* CASIA Multi-spectral Palmprint [101,102] (version 1) database is released by the Biometrics Ideal Test website to conduct cutting edge research in palmprint and palmvein recognition based on multispectral sensing. A self-designing sensing device has been used to acquire 7,200 palm images from 100 different individual. All images recorded with no pegs to restrict the position of the palm, also stored in JPEG files at 8-bit gray-level. Evenly illumination source and CCD camera were involved to capture palm images.

[5] http://biometrics.put.poznan.pl/vein-dataseta.
[6] https://www.idiap.ch/dataset/vera-palmvein.
[7] http://biometrics.idealtest.org/dbDetailForUser.do?id=6.

Table 10.5 Comparison of various biometric palmvein data sets

Data set name	Subjects	Images	Sessions	Samples	Image size	Related works
PUT Vein	50	2,400	3	48	1,280 × 960	[103]
CASIA Multispectral	100	7,200	2	72	768 × 576	[101,102]
PolyU Multispectral	250	6,000	2	72	128 × 128 ROI	[1,55]
PolyU Hyperspectral	190	5,240	2	72	128 × 128 ROI	[23,56]
VERA	110	2,200	2	20	480 × 680	[59]

Subjects, total number of volunteers in data set; Images, total number of image in the data set; Samples, total number of each candidate's images; Image size, the size of image pixels; Related work, some published works based on the data sets.

- ***PolyU multispectral palmprint data set***[8]*:* A multispectral palmprint-sensing device has been built by the Biometric Research Centre at the Hong Kong Polytechnic University. It can acquire images from the palmprint under NIR and red, blue, green illuminations [1]. A large-scale multispectral palmprint data set was built. Its purposes are to perform advance research in the field of multispectral palmprint and to provide researchers who are working in the area of multispectral image recognition with a platform to compare the effectiveness of various multispectral palmprint recognition algorithms. This data set is freely available for academic, noncommercial purpose. A database comprising 6,000 palm images was collected from 250 candidates by involving both palms in sensing acquisition process.
- ***PolyU hyperspectral palmprint data set***[9]*:* A hyperspectral palmprint-sensing device has been developed by the Biometric Research Centre at the Hong Kong Polytechnic University, which can acquire images from palmprint with different bands (wavelength) [23]. The wavelength start from 420 to 1,100 nm and has been used to build a large-scale hyperspectral palmprint data set. The total number of volunteers involved in image-sensing process was 190, and the age distribution starts from 20 to 60 years. For each wavelength (from 420 to 1,100 nm with step length of 10 nm), both palms were involved and seven images taken from each one. Overall, the database comprises 5,240 images from 380 different palms for each wavelength.

Several comparison factors of aforementioned palmvein data sets are summarized in Table 10.5. Table 10.2 in Section 10.2 reports recently used feature extraction techniques and data sets in ascending chronological order.

Figure 10.10 shows some samples from each palmvein data set.

For any future work, large and competitive data sets are necessary. Moreover, it is important to focus on palmvein sensing to get wealthy differentiation information and enhancement techniques; the response time could be reduced by using

[8] http://www4.comp.polyu.edu.hk/biometrics/MultispectralPalmprint/MSP.htm.
[9] http://www4.comp.polyu.edu.hk/biometrics/HyperspectralPalmprint/HSP.htm.

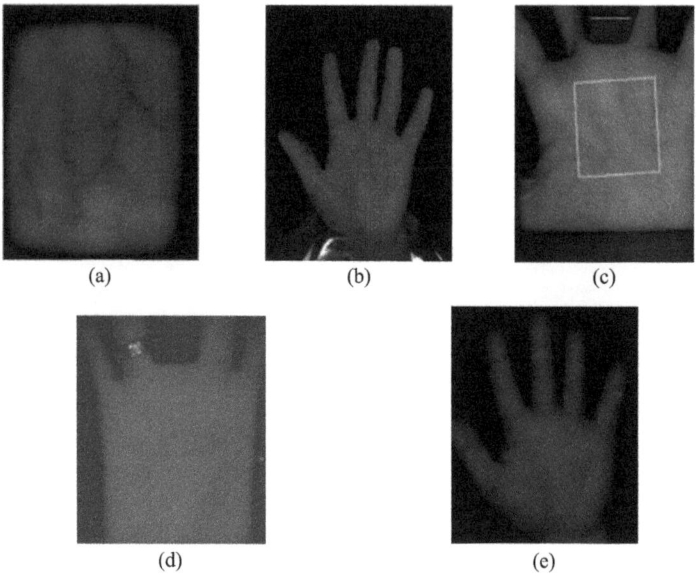

Figure 10.10 Image sample from the following data sets. (a) PUT vein, (b) CASIA multispectral, (c) PolyU multispectral palmprint, (d) PolyU hyperspectral palmprint, (e) VERA palmvein

image clarity techniques in preprocessing stages. We are aware that it is not a trivial task, especially because a few of capture devices is available on the market. Therefore, the data sets collection for palmvein recognition could be greatly boosted by the production and widespread of more capture devices.

10.4.2 Experimental protocols

Biometrics identification is a process of searching versus a biometric enrollment database to return and find the reference biometric trait of a single subject (individual), whilst the identity claim is confirmed through biometric comparison in the verification process.[10] In evaluating biometrics systems, including the one which adopted the palmvein, two stages are considered [104]:

- **Training stage:** The biometric reference database with correctly labeled biometric samples is made available at the start. To achieve optimal performance, the system would be trained on the reference database. Training means the estimation of the classifier/comparator parameters. The benefit of training are maximized if the reference database is split into two parts, the second one used to preliminary validate the system's performance according to the computed parameters. It is also useful to retune and adjust them. For this reason, it is

[10] The term "authentication," which has the same meaning of "verification" in many papers, should be avoided according to the ISO terminology [12].

called "development" set, whilst it is called "validation" set in the majority of the pattern recognition literature.

- **Evaluation stage:** Also called testing stage, performance statistics are measured and defined. When the test samples are submitted, a set of nonmatch (impostor) scores and a set of match (genuine users) scores are computed (Figure 10.11). These scores are used to estimate the false nonmatch rate (FNMR) and false match rate (FMR), namely, the classification errors depending of a match misclassification or a nonmatch misclassification, respectively, as a function of decision threshold R_{TH}. The threshold value tunes the related operational point. The usual rule is that the match score is over such threshold, the test sample is classified as a genuine user; otherwise, it is classified as an impostor in the case of the verification task. Similar rules is followed in case of maximum likelihoods, a posteriori probability or other hypothesis verification tests are followed (e.g., the one based on the Likelihood Ratio). The threshold value impacts on the system's performance. See also R'_{TH} and R''_{TH} in Figure 10.11: R'_{TH} reduces false acceptance rate (FAR) at the expense of the false rejection rate (FRR) increase, whilst the opposite is done by R''_{TH}.

The quality of biometrics samples has a significant influence on the performance of the biometric recognition system. The literature has also taken into account impersonation or evasion attempts from different subjects, i.e., zero-effort (impostor) attacks, and the possibility of presentation attacks [106].

In the following, we list the evaluation parameters in palmvein recognition. It can be noticed that, so far, palmvein has been investigated for verification purposes. As a consequence, all the parameters below are adopted to this aim

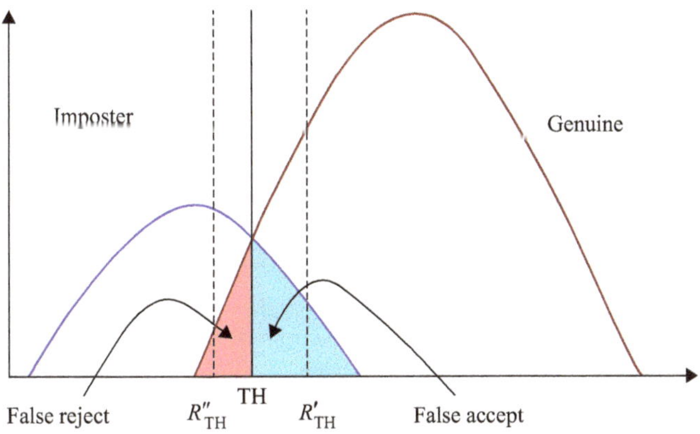

Figure 10.11 Genuine and impostor matching score distributions. If R'_{TH} is chosen as reference threshold, then the person earlier as genuine users become impostors. This means that the person is falsely accepted due to change in reference threshold value. This figure quoted from [105]

[107,108]. Parameters used for identification such as Cumulative Match Characteristic (curves) and the identification rate will be not listed. Further details on performance parameters for the identification task can be found in [109].

FMR, also called FAR [12,107]. It indicates the probability that the system falsely matches the input pattern to a nonmatching pattern (from the template set). In other words, it measures the rate of incorrectly accepted pattern, classified as genuine users.

FNMR, also called FRR. It measures the probability that the system be unsuccessful to detect matches between matching template already stored in the biometric reference database and the input pattern. The FNMR and FMR behavior is opposite and their gradient, as function of the acceptance threshold, depends on the effectiveness of the features set and the ability of the comparator to exploit it for providing high match scores when two patterns come from the same subject and low match scores when they come from different subjects.

Genuine acceptance rate (GAR) is derived from FNMR by the formula: GAR = 1 − FNMR.

Receiver operating characteristic (ROC) is the curve used for depicting the relationship between GAR (or FNMR) and FMR. Figure 10.12 shows an example of ROC curves obtained in [32].

Detection error trade-off (DET) plots the FMR against the FNMR depending on a logarithmic scale for any axis (e.g., FMR axis). The number of match errors is showed at *Y*-axis, the best biometric performance is taken from the curve that is closest to the bottom of the plot. Often ROC and DET plots are used with the same meaning, but the recent International Organization for Standardization (ISO) paper on biometrics standards [12] intends ROC curves in a more general sense, by including eventual failure rates during the image acquisition (FTC – failure to capture) or the feature extraction phases (FTE or FER – failure to enroll). See below for the definition of these two terms.

Crossover error rate, also called equal error rate (EER) [12,107]. It is used for accuracy comparison and refers to the error rate at which FNMR = FMR. This corresponds to the intersection point between the plane's bisector and the ROC curve when both axes are in linear scale.

Figure 10.13 shows the DET curve and related EER of the matched pixel ratio (MPR) verification matching [38].

FTC and failure to enroll rate (FTE or FER). It is used for biometric performance comparison (including the palmvein ones) beside FMR, FNMR, and ROC curve. The FTE is often raised by insufficient input. Its rate refers to the percentage of unsuccessful attempts to create a template from an input image, namely, to process it and extract features. The FTC is introduced to measure the probability that the biometric input is not detected.

Finally, the *Response Time* is taken into account in practical applications and rarely measured. It jointly depends by two factors: the first one is the processing platform capability including the memory size, the CPU speed and characteristics, and other software- and hardware-based factors. The second factor is the computational complexity of the palmvein recognition algorithm.

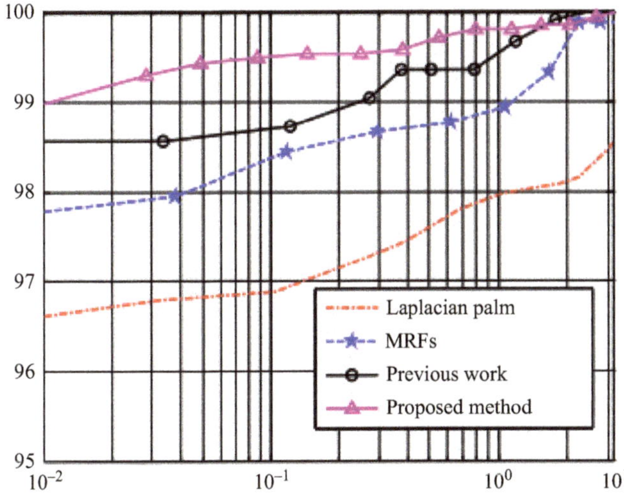

Figure 10.12 Receiver operating characteristics curve (ROC) based on adaptive Gabor filters for features extraction and normalized Hamming distance for matching. Results are obtained on a private data set [27]. To validate the accuracy of the proposed work, the authors compared their method with other existing methods such as Multiple Multiresolution Filters (MRFs), Laplacianpalm, and the previous works presented by Hsu et al. in [63]. The red curve in the figure represents the performance of Laplacianpalm feature, the blue curve represents the MRFs performance, the black curve shows the performance of the previous work, and the proposed work performance appears in the magenta curve. This figure quoted from [27]

The aforementioned evaluation indicators are not all mentioned in some works, especially when neural networks are used. In [5,35,53], for example, the percentage of the overall accuracy in the testing stage is shown. It is pointed that, in the previous works where the evaluation parameters were not included, the authors aimed to show rank-1 averaging in the bases of palmvein identification [3,12]. Several ROC curves have been evaluated in the literature to show the relationship between FAR (FMR) and GAR [9,27,29,32,38,54,61–63]. ROC curves are often coupled with EER [9,40,57,63,64,67]. GAR indicators are mentioned in [23,27,60,66], also the response times are measured in [7,27,28,30]. FTE or FER is not pointed out in recent works.

Table 10.6 contains a list of recent methods and related performance parameters. Although this list is not exhaustive, representative works are reported about the recent efforts in this field. By taking a look at the table, the lower EER and FAR are the more accurate the system is. Lin *et al.* [17] reported EER about 0.23; this work is considered as the oldest work in our survey dealing with vein pattern

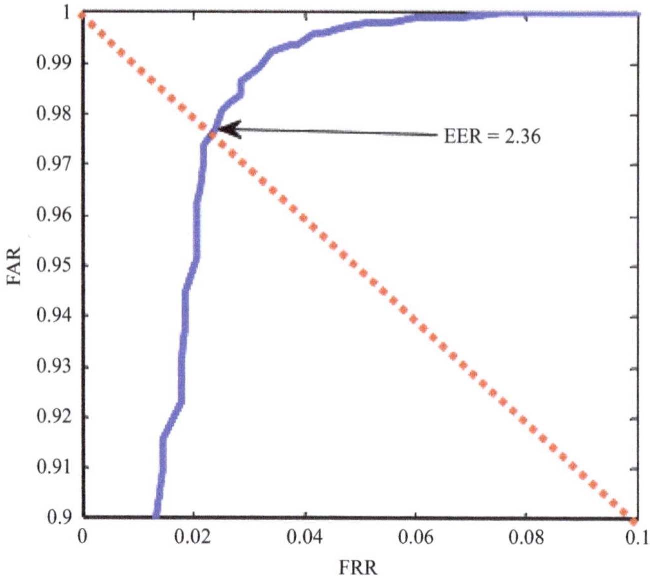

Figure 10.13 Detection Error Trade-off (DET) curve of the MPR verification algorithm. The authors of this figure evaluated the performance of their proposed method, which is mutual foreground LBP method for feature extraction and matched pixel ration for matching. They obtained multiple sets of FMR and FNMR according to different thresholds to plot the ROC curve. Since a trade-off exists between FMR and FNMR, the EER was used as a comparison metric: in this case, its value is 2.36%. In other works, it is represented by an absolute ratio (e.g., 0.236 would be its corresponding value). This figure taken from [38]

recognition and published in year 2004. The performance has developed from 2004 to 2015 in [3,5,23,29,32,37,40,62,64,67,68] with EER near to 0, even in [54], the EER is equal to 0. Some works also report FAR and EER near to 0 as Refs. [62,68]. However, due to the small size of the data set adopted in those papers, where only 28 and 50 subjects are collected, these results might not reflect the real-world conditions. Therefore, the reliability of these values could be not sufficient to some extent.

10.5 Open issues of biometric palmvein recognition

In this section, we try to raise some of those that, in our view, constitute the open issues of biometric palmvein recognition. Observations below have been moved from the analysis of the state of the art done in this paper.

Table 10.6 Performance comparison of some recent works

Ref.	Feature category	Comparator	Indicator
[57]	Statistic	Key-points matcher	$EER = 0.16$
[9]	Statistic	Cosine similarity	$EER = 0.28$
[29]	Statistic	Angular distance	$EER = 0.0757$, $FAR = 0.0005$, $FRR = 0.0289$
[32]	Statistic	Not mentioned	Several results
[64]	Statistic	Euclidean distance	$EER = 0.005$
[27]	Statistic	Hamming distance	$GAR = 0.993$
[61]	Statistic	Hamming distance	$EER = 0.3091$, $FRR = 0.555$
[63]	Statistic	Matched pixel ratio	$EER = 0.004$
[54]	Statistic	Euclidean/Canberra distance	$GAR = 0.99$, $FRR = 0.01$
[40]	Statistic	Cosine similarity	$EER = 0.00$
[60]	Line	Nearest neighbor	$GAR = 0.996$
[66]	Line	Euclidean distance	$GAR = 0.974$
[67]	Line	Matched filter	$EER = 0.08$
[68]	Line	Phase correlation	$GAR = 0.988$, $FAR = 0.055$
[37]	Subspace	Nearest neighbor	$GAR = 1$, $EER = 0.000$
[62]	Other	*k*NN	$GAR = 0.996$, $FRR = 0.004$

The oldest work, which can be considered as a sort of "ground truth" on palmvein verification, is [17]. In this table, Ref., the related reference in the bibliography; Feature category, the taxonomy of feature extraction techniques (see Section 10.3.1); Comparator, the matching approach adopted; Indicator, evaluation parameters used in experimental protocols.

In particular, we believe that some challenges in image acquisition, feature extraction, and matching need to be addressed. Some of them require a systematic cross-disciplinary effort, while others are well-known open problems in the allied areas (e.g., pattern recognition and image processing) [2].

10.5.1 Designing of palmvein-sensing device and data sets availability

Biometric-based palmvein addresses the recognition issue using high and low-resolution images. It has been pointed out in Section 10.2.1 the presence of a few commercial capture devices for palmvein, beside the existence of some homemade sensors. Thus, this causes a limitation of the number of publicly available data sets. As a matter of fact, according to the literature review [7,13,23,24,27,28,30,32–44], there are currently few data sets publicly available. Obviously, the more the data sets, the more the possibility to develop common/shared experimental protocols and results.

A biometric palmvein recognition system operates in two distinct stages: enrollment and verification (or identification). During the enrollment stage, palmvein images are captured from candidates, relevant information from the palmvein are gleaned by the feature extractor, and (feature, image) information are stored in a data set.

Acquiring relevant palmvein data for the recognition purpose is one of the critical processes which has not received adequate attention. The overall system performance is increased as much as the amount of care taken in acquiring palmvein images. There are two associated tasks according to Jain *et al.* [2] which are defined in pattern recognition and image process research: (i) quality assessment: automatically assessing the suitability of the input palmvein for automatic processing (see Section 10.4.2) and (ii) segmentation: separation of the palmvein images into foreground (ROI which could be achieved from the preprocessing stage, see Section 10.2.2) and background (irrelevant information such as the fingers of the hands).

In the *palmvein-sensing stage*, the control of palmvein image acquisition will improve both preprocessing and feature extraction stages; accordingly, it is important to exert a suitable effort to configure the palmvein capture device. The subject's cooperation during this stage can be helped by pegs that may help to control the hand displacement and by modifying the scanning region to be near to required ROI.

The clarity of the vein patterns that are extracted from ROI varies from image to image. Therefore, the quality of the ROI needs to be enhanced before further processing (feature extraction and matching) [24]. It is recommended to apply image clarity and denoising techniques during the enrollment stage of palmvein to enhance the performance and facilitate the recognition.

Therefore, by considering the system's performance in terms of speed, accuracy, and cost, the following issues must be addressed: how to acquire palmvein images? Does the acquired image suitable for recognition? Is it possible to integrate sensing and preprocessing stages together? Is there a possibility to integrate a literature clue such as the presence of blood to make the system actually robust to spoofing? [59].

Addressing the aforementioned issues is extremely important, although it requires environment and conscious efforts.

10.5.2 Feature extraction

Biometrics-based palmvein recognizes subjects depending on local features from cardinal veins. We pointed out in Section 10.3.1 that line-based, statistical-based, subspace-based, and other approaches have been proposed. We also discussed them in Section 10.3.1.5.

One of the properties of extracted features is that they should be as more invariant as possible for a given subject. At the same time, they must differ from those belonging to a different subject. Generally, the features for the same candidate must meet high interclass variation and low intraclass variation. Although state-of-the-art systems appeared to be quite good, in our opinion, there is very likely room for improvements. This motivated by the fact that most of the works in palmvein recognition undertaken on private databases and are not reproducible.

In palmvein recognition, there are two associated tasks defined in pattern recognition to feature extraction stage: (i) image denoising and (ii) dimensionality reduction. Whilst the existence of many image denoising algorithm helps for task (i), we believe that task (ii) has been pursued using general-purpose algorithms, as

PCA. In our opinion, the particular appearance of the palmvein structure should suggest specific feature reduction algorithms to study which ones affect the performance in terms of intraclass variations and interclass variations.

Detecting the reliable features could be achieved right now, perhaps the reasons are increasing the availability of inexpensive of computing and sensing technologies. Once the features are determined then it must be extracted by somewhat ad hoc manner (we summarizes some of state-of-the-art feature extraction in Section 10.3.1). The efficiency of each feature extraction techniques is limited especially when the palmvein images are noisy. Rigorous techniques of feature representations positively affect the feature extraction stage of corresponding palmvein images especially in noisy situations.

In order to act on the denoising and space reduction issues, thus improving the possible interclass/intraclass variations ratio, multiscale and multiresolution extraction techniques could be adopted (Section 10.3.1.1), whereas, in multi (scale or resolution) techniques, the system is described by passband filter with single parameter to control the spatial scale. For example, the wavelength is used in the context of linear filters to describe small-sized object with fine resolution as short wavelengths are needed to describe small-sized objects associated with fine resolution [110]. The combination of feature extraction and dimensionality reduction techniques could help achieving a better performance. Therefore, the availability in data sets published recently increase the productivity and increase the quality of the performance, and the developments of features extraction techniques which extract the most discriminative information, for example, texture techniques as LBP, and LPQ to extract local features form palmvein.

10.5.3 Recognition and comparison

The core point of the matching stage is the computation of the similarity function (see Section 10.3.3), which leads to obtaining the matching score. Determining an appropriate similarity technique helps in improving system quality, since it must be able to discriminate between the representations of two different palmveins despite noise, aging, and structural and statistical variations of the input images.

As mentioned in the previous section, a few number of feature vectors that represent the most discriminative information should be adopted, achieved by dimensionality reduction stage. For example, [3,5,53] used one feature vector as input to neural network classifiers and achieved a good performance reported in Section 10.4.2.

Different types of approaches can perform the matching task, but the aim that must be taken into account is also to reduce the response time of biometric palmvein recognizer. Among others, we noticed that generative approaches which gained a popularity in face [111] and speaker recognition [112] have not yet been investigated, although they are claimed to manage the intra- and interclass variations problem. The overall idea of generative approaches is to extract local features from the image of a subject's palmvein before learning the distribution of these features using a probabilistic model.

The most used generative model is the one based on Gaussian Mixtures (GMM). This consists of K multivariate Gaussian components [113]. Some improvements to the GMM generative approach have been also proposed in the literature, such as joint factor analysis (JFA) [114] and intersection variability (ISV) modeling [115]. Both techniques ISV and JFA propose to explicitly model the variations between different sessions (image modalities) of the same subject and compensate them during the enrollment and testing. This is obtained by excluding the variation of sources of session when generating the subject model using a low-dimensional subspace. Therefore, these approaches could be coupled with multiscale image transforms (curvelet and contourlet) as local features, as well as more traditional ones like 2D DCT and LBP for palmvein recognition, thus offering a novel way of match score computation.

Some works compare the EER of palmvein and fingerprint modes based on several published works generally. For example, the EER for fingerprint recognition can be obtained as 0.2%–0.4% according to Luo *et al.* [107]. However, in [63,64], the reported EER is in the $(0.005, 0.004)\%$ interval, respectively, and in [61,100] FRR is in the $(0.376, 0.555)\%$ interval, respectively. Thus, the performance of palmvein verification can be claimed to be better than that of fingerprint verification, on the basis of current data sets. Obviously, the same considerations made in the previous paragraph about the size of data sets adopted holds when doing such comparisons. Zhang *et al.* [61] reported that the EER of palmvein and palmprint verification were about 0.3091% and 0.0352%, respectively, thus palmvein appeared to be not accurate as palmprint verification. This is largely due to the relatively low quality of palmvein images according to Zhang *et al.* Improving the acquisition of palmvein images may affect positively the system performance.

Another point of comparison between palmvein and existing modes is the cost to accuracy ratio. Jain *et al.* [16] report such relationships for different biometric modes. We rebuilt the cost to accuracy ratio as depicted in Figure 10.14, hereby adding the palmvein according to the best of our knowledge. It is important to refer that palmvein term that appears in Figure 10.14 is a model of newborn biometrics, and the cost to accuracy ratio could be swinging (increase or decrease). Therefore, the stability of the cost and the accuracy of the system will depend on the degree of investigation of palmvein in the future.

Cost and accuracy vary by considering that, although unimodal recognition of palmvein, palmprint, and other biometric modes have been investigated, they are not robust to noise and spoofing attacks [7,17,62,116]. A multimodal solution, that is, the fusion of palmvein [14,54,65,66,101] with other modes could be taken into account when the system's design must be aimed to the increase of robustness and antispoofing ability, thus the cost could be increased.

The vulnerability of palmvein recognition to spoofing attacks was studied by Tome *et al.* [59]. A collection of spoofing palmvein images have been created from real palmvein samples. Palmvein images were printed using a commercial printer, and then presented at a contact-less palmvein sensor. Tome *et al.* used printed palmvein images, and they made their experiments to spoof palmvein regardless detecting the liveness of palmvein. Accordingly, liveness detection is extremely

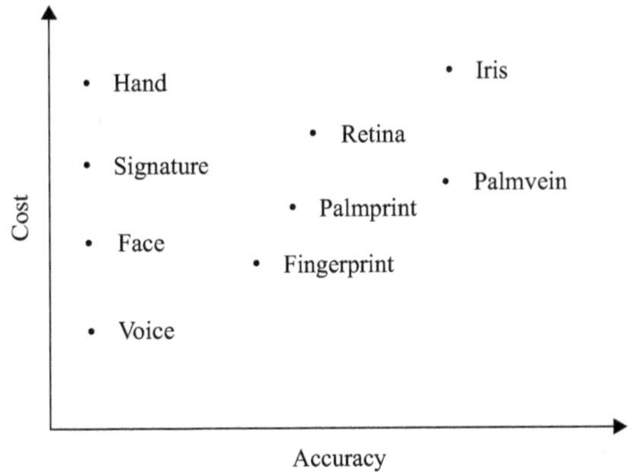

Figure 10.14 *Cost to accuracy ration of the most popular biometric modes. Each biometric mode has specific limitations and strengths; therefore, no biometric mode can outperform others in all operational environment. We believe that palmvein could be set in the top rank in the future, but it requires deep investigation since it can be considered as a "newborn" trait. This figure is driven according to our background and forestation on biometrics, also the practical issues which highlighted in [2]*

important to be applied in palmvein recognition [61], in order to find appropriate countermeasures against presentation attacks.

10.6 Conclusions

In this paper, we have reported the state of the art on biometric palmvein recognition. Many works in the literature have been presented about palmvein recognition, especially the ones published recently. We also analyzed all stages of a palmvein recognition system (sensing, preprocessing, representation and matching, integration with other biometrics, performance) and compared pros and cons of current approaches for each stage.

The scientific and industrial trends toward palmvein recognition have remarkably boosted. Last 5 years showed a noticeable increase of works around this topic. A strong increment of publications can be located between 2009 and 2010. Before these years, only 13%–15% of the overall works were published. Thus, the need of a novel assessment of current state of the art was fully motivated.

Our goal was to help research beginners in being aware on "where we are" in this novel field, also to suggest possible topics to be (better) investigated. In particular, we dealt with the question of: when the biometric palmvein recognition has launched? And how? It is important to highlight on feature extraction techniques and describe how it could be applied in biometric palmvein recognition.

At this point of research (where are we?), we can say that palmvein is a promising biometric, and the majority of analyzed works clearly indicate that very high recognition rates can be potentially attained. The most impressive contradiction is that, despite the large number of publications, feature extraction methods, experiments are still carried out on a few data sets, and no specific capture device is fully available to obtain standardized performance rates, that are necessary to fairly compare different methods. Thus, the main open issues that can be pointed out after our analysis are as follows: how to acquire palmvein images? Is the acquired image suitable for representation? Is it possible to integrate sensing and preprocessing stages together? Is the performance rate fully reliable? Is this system robust to well-known vulnerabilities such as presentation attacks? Is it possible to integrate a literature clue such as the presence of blood to make the system actually robust to them? Which other features captured from palmvein images may assure this robustness?

We believe that answers to above questions will come in the future, and hope that this paper has shed some lights on the possible efforts which may bring to them.

References

[1] Zhang, D.; Kong, W.; You, J.; Wong, M., "Online palmprint identification," IEEE Transactions on Pattern Analysis and Machine Intelligence, vol. 25, no. 9, pp. 1041–1050, Sep. 2003, doi: 10.1109/TPAMI.2003.1227981.

[2] Jain, A.; Bolle, R.; Pankanti, S., "Biometrics: personal identification in networked society," New York, NY: Springer Science & Business Media, 2006.

[3] Elaydi, H.; Alhanjouri, M.; Abukmeil, M., "Palmprint recognition using 2-D wavelet, ridgelet, curvelet and contourlet," i-Manager's Journal on Electrical Engineering, vol. 7, no. 1, p. 9, Sep. 2013.

[4] Morales, A.; Ferrer, M.A.; Kumar, A., "Improved palmprint authentication using contactless imaging," in 2010 Fourth IEEE International Conference on Biometrics: Theory Applications and Systems (BTAS), pp. 1–6, 27–29 Sep. 2010, doi: 10.1109/BTAS.2010.5634472.

[5] Elaydi, H.; Abukmeil, M.; Alhanjouri, M., "Palmprint recognition using multiscale transform, linear discriminate analysis, and neural network," Science Journal of Circuits, Systems and Signal Processing, vol. 2, no. 5, pp. 112–118, 2013, doi: 10.11648/j.cssp.20130205.13.

[6] Kong, W.; Zhang, D.; Li, W., "Palmprint feature extraction using 2-D Gabor filters," Pattern Recognition, vol. 36, no. 10, pp. 2339–2347, Oct. 2003, doi: 10.1016/S0031-3203(03)00121-3.

[7] Wang, J.; Yau, W.; Suwandy, A.; Sung, E., "Person recognition by fusing palmprint and palm vein images based on 'Laplacianpalm' representation," Pattern Recognition, vol. 41, no. 5, pp. 1514–1527, ISSN 0031-3203, May 2008, doi: 10.1016/j.patcog.2007.10.021.

[8] Kong, A.; Zhang, D.; Kamel, M., "A survey of palmprint recognition," Pattern Recognition, vol. 42, no. 7, pp. 1408–1418, ISSN 0031-3203, Jul. 2009, doi: 10.1016/j.patcog.2009.01.018.

[9] Zhou, Y.; Kumar, A., "Contactless palm vein identification using multiple representations," in 2010 Fourth IEEE International Conference on Biometrics: Theory Applications and Systems (BTAS), pp. 1–6, 27–29 Sep. 2010, doi: 10.1109/BTAS.2010.5634470.

[10] Watanabe, M.; Endoh, T.; Shiohara, M.; Sasaki, S., "Palm vein authentication technology and its applications," in Proceedings of The Biometric Consortium Conference, Arlington, USA, 2005.

[11] Miura, N.; Nagasaka, A.; Miyatake, T., "Extraction of finger-vein patterns using maximum curvature points in image profiles," IEICE Transactions on Information and Systems, vol. E90-D, no. 8, pp. 1185–1194, Aug. 2007, doi: 10.1093/ietisy/e90-d.8.1185.

[12] ISO and IEC, "Information technology—vocabulary—Part 37: Biometrics," ISO/IEC 2382-37, Joint Technical Committee ISO/IEC JTC 1, Information Technology, Subcommittee SC 37, Biometrics, 2012.

[13] Zhang, H.; Hu, D. "A palm vein recognition system," in 2010 International Conference on Intelligent Computation Technology and Automation (ICICTA), pp. 285–288, 11–12 May 2010, doi: 10.1109/ICICTA.2010.425.

[14] Wang, J.; Yau, W.; Suwandy, A.; Sung, E., "Fusion of palmprint and palm vein images for person recognition based on 'Laplacianpalm' feature," in IEEE Conference in Computer Vision and Pattern Recognition CVPR'07, pp. 1–8, 17–22 Jun. 2007, doi: 10.1109/CVPR.2007.383386.

[15] Sarkar, I.; Alisherov, F.; Kim, T.; Bhattacharyya, D., "Palm vein authentication system: a review," International Journal of Control and Automation, vol. 3, no. 1, Mar. 2010.

[16] Jain, A.; Ross, A.; Prabhakar, S., "An introduction to biometric recognition," IEEE Transactions on Circuits and Systems for Video Technology, vol. 14, no. 1, pp. 4–20, Jan. 2004, doi: 10.1109/TCSVT.2003.818349.

[17] Lin, C.; Fan, K., "Biometric verification using thermal images of palm-dorsa vein patterns," IEEE Transactions on Circuits and Systems for Video Technology, vol. 14, no. 2, pp. 199–213, Feb. 2004, doi: 10.1109/TCSVT.2003.821975.

[18] Wang, J.; Chen, T., "Building palm vein capturing system for extraction," in 2011 21st International Conference on Systems Engineering (ICSEng), pp. 311–314, 16–18 Aug. 2011, doi: 10.1109/ICSEng.2011.61.

[19] Badawi, A.M., "Hand vein biometric verification prototype: a testing performance and patterns similarity," Idiopathic Polypoidal Choroidal Vasculopathy, vol. 14, pp. 3–9, 2006.

[20] Bouzida, N.; Bendada, A.; Maldague, X., "Near-infrared image formation and processing for the extraction of hand veins," Journal of Modern Optics, vol. 57, no. 18, pp. 1731–1737, Oct. 2010, doi: 10.1080/09500341003725763.

[21] Noh, Z.M.; Ramli, A.R.; Hanafi, M.; Saripan, M.I., "Review on a palm vein infrared image acquisition system," in 2013 IEEE Student Conference on

Research and Development (SCOReD), pp. 334–337, 16–17 Dec. 2013, doi: 10.1109/SCOReD.2013.7002602.

[22] Noh, Z.M.; Ramli, A.R.; Hanafi, M.; Saripan, M.I., "Acquiring palm vein patterns for visual interpretation," in 2015 2nd International Conference on Biomedical Engineering (ICoBE), pp. 1–5, 30–31 Mar. 2015, doi: 10.1109/ICoBE.2015.7235884.

[23] Guo, Z.; Zhang, L.; Zhang, D., "Feature band selection for multispectral palmprint Recognition," in 2010 20th International Conference on Pattern Recognition (ICPR), pp. 1136–1139, 23–26 Aug. 2010, doi: 10.1109/ICPR.2010.284.

[24] Wang, L.; Leedham, G.; Cho, S.-Y., "Infrared imaging of hand vein patterns for biometric purposes," IET Computer Vision, vol. 1, no. 3–4, pp. 113–122, Dec. 2007, doi: 10.1049/iet-cvi:20070009.

[25] Lin, C.; Fan, K., "Biometric verification using thermal images of palm-dorsa vein patterns," IEEE Transactions on Circuits and Systems for Video Technology, vol. 12, no. 2, pp. 199–213, Feb. 2004, doi: 10.1109/TCSVT.2003.821975.

[26] Kumar, A.; Prathyusha, K.V., "Personal authentication using hand vein triangulation and knuckle shape," IEEE Transactions on Image Processing, vol. 18, no. 9, pp. 2127–2136, Sep. 2009, doi: 10.1109/TIP.2009.2023153.

[27] Han, W.; Lee, J., "Palm vein recognition using adaptive Gabor filter," Expert Systems with Applications, vol. 39, no. 18, pp. 13225–13234, ISSN 0957-4174, Dec. 2012, doi: 10.1016/j.eswa.2012.05.079.

[28] Lee, J., "A novel biometric system based on palm vein image," Pattern Recognition Letters, vol. 33, no. 12, pp. 1520–1528, ISSN, Sep. 2012, doi: 10.1016/j.patrec.2012.04.007.

[29] Luo, N.; Guo, Z.; Wu, G.; Song, C., "Joint palmprint and palmvein verification by dual competitive coding," in 2011 3rd International Conference on Advanced Computer Control (ICACC), pp. 538–542, 18–20 Jan. 2011, doi: 10.1109/ICACC.2011.6016471.

[30] Ladoux, P.; Rosenberger, C.; Dorizzi, B., "Palm vein verification system based on SIFT matching," ICB 2009. Lecture Notes in Computer Science, vol. 5558. Springer, Berlin, Heidelberg, pp. 1290–1298, 2009, doi: 10.1007/978-3-642-01793-3-130.

[31] Jiaqiang, W.; Ming, Y.; Hanbing, Q.; Bin, L. "Analysis of palm vein image quality and recognition with different distance," in 2013 Fourth International Conference on Digital Manufacturing and Automation (ICDMA), pp. 215–218, 29–30 Jun. 2013, doi: 10.1109/ICDMA.2013.50.

[32] Zhou, Y.; Kumar, A., "Human identification using palm-vein images," IEEE Transactions on Information Forensics and Security, vol. 6, no. 4, pp. 1259–1274, Dec. 2011, doi: 10.1109/TIFS.2011.2158423.

[33] Wu, K.; Lee, J.; Lo, T.; Chang, K.; Chang, C., "A secure palm vein recognition system," Journal of Systems and Software, vol. 86, no. 11, pp. 2870–2876, ISSN 0164-1212, Nov. 2013, doi: 10.1016/j.jss.2013.06.065.

[34] Vaid, S.; Mishra, D., "Comparative analysis of palm-vein recognition system using basic transforms," in 2015 IEEE International Advance Computing Conference (IACC), pp. 1105–1110, 12–13 Jun. 2015, doi: 10.1109/IADCC.2015.7154875.

[35] Perwira, D.Y.; Agung, B.W.T.; Sulistiyo, M.D., "Personal palm vein identification using principal component analysis and probabilistic neural network," in 2014 International Conference on Information Technology Systems and Innovation (ICITSI), pp. 99–104, 24–27 Nov. 2014, doi: 10.1109/ICITSI.2014.7048245.

[36] Yan, X.; Deng, F.; Kang, W., "Palm vein recognition based on multi-algorithm and score-level fusion," in 2014 Seventh International Symposium on Computational Intelligence and Design (ISCID), pp. 441–444, 13–14 Dec. 2014, doi: 10.1109/ISCID.2014.93.

[37] Elnasir, S.; Shamsuddin, S.M., "Proposed scheme for palm vein recognition based on linear discrimination analysis and nearest neighbour classifier," in 2014 International Symposium on Biometrics and Security Technologies (ISBAST), pp. 67–72, 26–27 Aug. 2014, doi: 10.1109/ISBAST.2014.7013096.

[38] Kang, W.; Wu, Q., "Contactless palm vein recognition using a mutual foreground-based local binary pattern," IEEE Transactions on Information Forensics and Security, vol. 9, no. 11, pp. 1974–1985, Nov. 2014, doi: 10.1109/TIFS.2014.2361020.

[39] Mirmohamadsadeghi, L.; Drygajlo, A., "Palm vein recognition with local texture patterns,"IET Biometrics, vol. 3, no. 4, pp. 198–206, Dec. 2014, doi: 10.1049/iet-bmt.2013.0041.

[40] Mohan, M.; Saxena, J.; Teckchandani, K., *et al.*, "Palm vein recognition using local tetra patterns," in 2015 4th International Work Conference on Bioinspired Intelligence (IWOBI), pp. 151–156, 10–12 Jun. 2015, doi: 10.1109/IWOBI.2015.7160159.

[41] Pudzs, M.; Fuksis, R.; Ruskuls, R.; Eglitis, T.; Kadikis, A.; Greitans, M., "FPGA based palmprint and palm vein biometric system," in 2013 International Conference of the Biometrics Special Interest Group (BIOSIG), pp. 1–4, 5–6 Sep. 2013.

[42] Michael, G.; Connie, T.; Jin, A., "Design and implementation of a contactless palm print and palm vein sensor," in 2010 11th International Conference on Control Automation Robotics & Vision (ICARCV), pp. 1268–1273, 7–10 Dec. 2010, doi: 10.1109/ICARCV.2010.5707951.

[43] Lee, Y., "Palm vein recognition based on a modified text $2D^2$ LDA," Signal, Image and Video Processing, vol. 9, no. 1, pp. 229–242, Jan. 2013, doi: 10.1007/s11760-013-0425-6.

[44] Crisan, S.; Tarnovan, J.G.; Crisan, T.E., "A low cost vein detection system using near infrared radiation," in Sensors Applications Symposium, 2007. SAS '07. IEEE, pp. 1–6, 6–8 Feb. 2007, doi: 10.1109/SAS.2007.374359.

[45] Vanoni, M.; Tome, P.; El Shafey, L.; Marcel, S., "Cross-data set evaluation using an open finger vein sensor," in 2014 IEEE Workshop on Biometric

Measurements and Systems for Security and Medical Applications (BIOMS) Proceedings, pp. 30–35, 17–17 Oct. 2014, doi: 10.1109/BIOMS.2014.6951532.

[46] Li, W.; Zhang, D.; Xu, Z., "Palmprint identification by Fourier transform," International Journal of Pattern Recognition and Artificial Intelligence, vol. 16, no. 04, pp. 417–432, Jun. 2002, doi: 10.1142/S0218001402001757.

[47] Wu, X.; Wang, K.; Zhang, D., "HMMs based palmprint identification," First International Conference, ICBA, Hong Kong, China, vol. 3072, pp. 775–781, 2004, doi: 10.1007/978-3-540-25948-0-103.

[48] Han, C.; Cheng, H.; Lin, C.; Fan, K., "Personal authentication using palmprint features," Pattern Recognition, vol. 36, no. 2, pp. 371–381, ISSN 0031-3203, Feb. 2003, doi: 10.1016/S0031-3203(02)00037-7.

[49] Han, C., "A hand-based personal authentication using a coarse-to-fine strategy," Image and Vision Computing, vol. 22, no. 11, pp. 909–918, ISSN 0262-8856, Sep. 2004, doi: 10.1016/j.imavis.2004.05.008.

[50] Kumar, A.; Zhang, D., "Integrating shape and texture for hand verification," in Third International Conference on Image and Graphics (ICIG'04), pp. 222–225, 18–20 Dec. 2004, doi: 10.1109/ICIG.2004.87.

[51] Poon, C.; Wong, D.C.M.; Shen, H.C., "A new method in locating and segmenting palmprint into region-of-interest," in Proceedings of the 17th International Conference on Pattern Recognition, 2004. ICPR 2004. vol. 4, pp. 533–536, 23–26 Aug. 2004, doi: 10.1109/ICPR.2004.1333827.

[52] Guo, Z., "Online Multispectral Palmprint Recognition," Ph.D. Dissertation, Hong Kong Polytechnic University (People's Republic of China), Advisor(s) David Zhang. AAI3470132, 2010, dl.acm.org/citation.cfm?id=1970660.

[53] Abukmeil, M.A.M.; Elaydi, H.; Alhanjouri, M., "Palmprint recognition via bandlet, ridgelet, wavelet and neural network," Journal of Computer Sciences and Applications, vol. 3, no. 2, pp. 23–28, 2015, doi: 10.12691/jcsa-3-2-1.

[54] Gupta, A.; Malage, A.; More, D.; Hemane, P.; Bhautmage, P.; Dhandekar, D., "Feature level fusion of face, palm vein and palm print modalities using discrete cosine transform," in 2014 International Conference on Advances in Engineering and Technology Research (ICAETR), pp. 1–5, 1–2 Aug. 2014, doi: 10.1109/ICAETR.2014.7012805.

[55] Zhang, D.; Guo, Z.; Lu, G.; Zhang, L.; Zuo, W., "An online system of multispectral palmprint verification," IEEE Transactions on Instrumentation and Measurement, vol. 59, no. 2, pp. 480–490, Feb. 2010, doi: 10.1109/TIM.2009.2028772.

[56] Guo, Z.; Zhang, D.; Zhang, L.; Liu, W., "Feature band selection for online multispectral palmprint recognition," IEEE Transactions on Information Forensics and Security, vol. 7, no. 3, pp. 1094–1099, Jun. 2012, doi: 10.1109/TIFS.2012.2189206.

[57] Yan, X.; Kang, W.; Deng, F.; Wu, Q., "Palm vein recognition based on multi-sampling and feature-level fusion," Neurocomputing, vol. 151, no. 2, pp. 798–807, ISSN 0925, Mar. 2015, doi: 10.1016/j.neucom.2014.10.019.

[58] Qian, X.; Guo, S.; Li, X.; Zhong, F.; Shao, X., "Finger-vein recognition based on the score level moment invariants fusion," in International Conference on Computational Intelligence and Software Engineering, 2009. pp. 1–4, 11–13 Dec. 2009, doi: 10.1109/CISE.2009.5366021.

[59] Tome, P.; Marcel, S., "On the vulnerability of palm vein recognition to spoofing attacks," in 2015 International Conference on Biometrics (ICB), pp. 319–325, 19–22 May 2015, doi: 10.1109/ICB.2015.7139056.

[60] Li, Q.; Zeng, Y.; Peng, X.; Yang, K., "Curvelet-based palm vein biometric recognition," Chinese Optics Letters, vol. 8, pp. 577–579, 2010.

[61] Zhang, D.; Guo, Z.; Lu, G.; Zhang, L.; Liu, Y.; Zuo, W., "Online joint palmprint and palmvein verification," Expert Systems with Applications, vol. 38, no. 3, pp. 2621–2631, Mar. 2011, doi: 10.1016/j.eswa.2010.08.052.

[62] Deepamalar, M.; Madheswaran, M., "An enhanced palm vein recognition system using multi-level fusion of multimodal features and adaptive resonance theory," International Journal of Computer Applications, vol. 1, no. 20, pp. 95–101, Feb. 2010, doi: 10.5120/414-612.

[63] Mirmohamadsadeghi, L.; Drygajlo, A., "Palm vein recognition with local binary patterns and local derivative patterns," in 2011 International Joint Conference on Biometrics (IJCB), pp. 1–6, 11–13 Oct. 2011, doi: 10.1109/IJCB.2011.611.

[64] Yuan, W.; Li, W., "A palm vein feature extraction method based on affine invariant," in 2012 IEEE International Conference on Robotics and Biomimetics (ROBIO), pp. 2323–2326, 11–14 Dec. 2012, doi: 10.1109/ROBIO.2012.6491316.

[65] Peidan, L.; Zhenjiang, M.; Zhiqun, M., "Fusion of palmprint and palm vein images for person recognition," in 2014 12th International Conference on Signal Processing (ICSP), pp. 2407–2410, 19–23 Oct. 2014, doi: 10.1109/ICOSP.2014.7015426.

[66] Gaikwad, D.; Narote, S., "Multi-modal biometric system using palm print and palm vein features," in 2013 Annual IEEE India Conference (INDICON), pp. 1–5, 13–15 Dec. 2013, doi: 10.1109/INDCON.2013.6726010.

[67] Li, Q.; Li, X.; Guo, Z.; You, J., "Online personal verification by palmvein image through palmprint-like and palmvein information," Neurocomputing, vol. 147, pp. 364–371, ISSN 0925-2312, Jan. 2015, doi: 10.1016/j.neucom.2014.06.050.

[68] Zhang, Y.; Li, Q.; You, J.; Bhattacharya, P., "Palm vein extraction and matching for personal authentication," 9th International Conference, Visual Shanghai, China, vol. 4781, pp. 154–164, Jun. 2007, doi: 10.1007/978-3-540-76414-4-16.

[69] Hartung, D.; Aastrup Olsen, M.; Xu, H.; Thanh Nguyen, H.; Busch, C., "Comprehensive analysis of spectral minutiae for vein pattern recognition," IET Biometrics, vol. 1, no. 1, pp. 25–36, Mar. 2012, doi: 10.1049/iet-bmt.2011.0013.

[70] Zhang, Y.; Li, Q.; You, J.; Bhattacharya, P., "Palm vein extraction and matching for personal authentication," Advances in Visual Information Systems, pp. 154–164, 2007. https://doi.org/10.1007/978-3-540-76414-4_16

[71] Jenkins, O.; Mataric, M., "Deriving action and behavior primitives from human motion data," in IEEE/RSJ International Conference on Intelligent Robots and Systems, 2002. vol. 3, pp. 2551–2556, 2002, doi: 10.1109/IRDS.2002.1041654.

[72] Zheng, W.; Lai, J.; Li, S., "1D-LDA vs. 2D-LDA: when is vector-based linear discriminant analysis better than matrix-based?," Pattern Recognition, vol. 41, no. 7, pp. 2156–2172, Jul. 2008, doi: 10.1016/j.patcog.2007.11.025.

[73] Lowe, D., "Distinctive image features from scale-invariant keypoints," International Journal of Computer Vision, vol. 60, no. 2, pp. 91–110, doi: 10.1023/B:VISI.0000029664.99615.94.

[74] Dunn, D.; Higgins, W.E.; Wakeley, J., "Texture segmentation using 2-D Gabor elementary functions," IEEE Transactions on Pattern Analysis and Machine Intelligence, vol. 16, no. 2, pp. 130–149, Feb. 1994, doi: 10.1109/34.273736.

[75] Daugman, G., "Uncertainty relation for resolution in space, spatial frequency, and orientation optimized by two-dimensional visual cortical filters," Journal of the Optical Society of America A, vol. 2, no. 7, pp. 1160–1169, 1985, doi: 10.1364/JOSAA.2.001160.

[76] Daugman, J., "High confidence visual recognition of persons by a test of statistical independence," IEEE Transactions on Pattern Analysis and Machine Intelligence, vol. 15, no. 11, pp. 1148–1161, Nov. 1993, doi: 10.1109/34.244676.

[77] Lee, T., "Image representation using 2D Gabor wavelets," IEEE Transactions on Pattern Analysis and Machine Intelligence, vol. 18, no. 10, pp. 959–971, Oct. 1996, doi: 10.1109/34.541406.

[78] Fuehr, H.; Demaret, L.; Friedrich, F., "Beyond wavelets: new image representation paradigms," Document and Image Compression, vol. 7, pp. 179–206, 2006.

[79] Bayoumi, S.; Al-Zahrani, S.; Sheikh, A.; Al-Sebayel, G.; Al-Magooshi, S.; Al-Sayigh, S., "PCA-based palm vein authentication system," International Conference on Information Science and Applications (ICISA), ISBN: 978-1-4799-0604-8, Aug. 2013, doi: 10.1109/ICISA.2013.6579422.

[80] Hsu, C.; Hao, S.; Lee, J.-C., "Personal authentication through dorsal hand vein patterns," Optical Engineering, vol. 50, no. 8, pp. 087201–087201, Jul. 2011, doi: 10.1117/1.3607413.

[81] Bamberger, R.; Smith, M., "A filter bank for the directional decomposition of images: theory and design," IEEE Transactions on Signal Processing, vol. 40, no. 4, pp. 882–893, Apr. 1992, doi: 10.1109/78.127960.

[82] Yang, J.; Zhang, D.; Frangi, A.; Yang, J., "Two-dimensional PCA: a new approach to appearance-based face representation and recognition," IEEE Transactions on Pattern Analysis and Machine Intelligence, vol. 26, no. 1, pp. 131–137, Jan. 2004, doi: 10.1109/TPAMI.2004.1261097.

[83] Beucher, S., "The watershed transformation applied to image segmentation," in Proceedings of the 10th Pfefferkorn Conference on Signal and Image Processing in Microscopy and Microanalysis, Cambridge, UK, pp. 299–314, Sep. 16–19, 1991.

[84] You, J.; Li, W.; Zhang, D., "Hierarchical palmprint identification via multiple feature extraction," Pattern Recognition, vol. 35, no. 4, pp. 847–859, ISSN 0031-3203, Apr. 2002, doi: 10.1016/S0031-3203(01)00100-5.

[85] Shinagawa, Y.; Kunii, T.L., "Unconstrained automatic image matching using multiresolutional critical-point filters," IEEE Transactions on Pattern Analysis and Machine Intelligence, vol. 20, no. 9, pp. 994–1010, Sep. 1998, doi: 10.1109/34.713364.

[86] You, J.; Bhattacharya, P., "A wavelet-based coarse-to-fine image matching scheme in a parallel virtual machine environment," IEEE Transactions on Image Processing, vol. 9, no. 9, pp. 1547–1559, Sep. 2000, doi: 10.1109/83.862635.

[87] Zhang, D.; Shu, W., "Two novel characteristics in palmprint verification: datum point invariance and line feature matching," Pattern Recognition, vol. 32, no. 4, pp. 691–702, ISSN 0031-3203, Apr. 1999, doi: 10.1016/S0031-3203(98)00117-4.

[88] Oden, C.; Ercil, A.; Buke, B., "Combining implicit polynomials and geometric features for hand recognition," Pattern Recognition Letters, vol. 24, no. 13, pp. 2145–2152, ISSN 0167-8655, Sep. 2003, doi: 10.1016/S0167-8655(03)00087-4.

[89] Kumar, A.; Zhang, D., "Personal recognition using hand shape and texture," IEEE Transactions on Image Processing, vol. 15, no. 8, pp. 2454–2461, Aug. 2006, doi: 10.1109/TIP.2006.875214.

[90] Bebis, G.; Deaconu, T.; Georgiopoulos, M., "Fingerprint identification using Delaunay triangulation," in 1999 International Conference on Information Intelligence and Systems, 1999. Proceedings. pp. 452–459, 1999, doi: 10.1109/ICIIS.1999.810315.

[91] Gu, Q.; Li, Z.; Han, J., "Joint feature selection and subspace learning," IJCAI Proceedings—International Joint Conference on Artificial Intelligence, vol. 22, no. 1, Jul. 2011, doi: 10.5591/978-1-57735-516-8/IJCAI11-219.

[92] Roweis, S.; Saul, L., "Nonlinear dimensionality reduction by locally linear embedding," Science, vol. 290, no. 5500, pp. 2323–2326, 2000.

[93] Maaten, V.; JP, L.; Postma, E.; Herik, H., "Dimensionality reduction: a comparative review," Journal of Machine Learning Research, vol. 10.1, no. 41, pp. 66–71, 2009.

[94] Hughes, N.; Tarassenko, L., "Novel signal shape descriptors through wavelet transforms and dimensionality reduction," Optical Science and Technology, SPIE's 48th Annual Meeting, International Society for Optics and Photonics, vol. 2, pp. 763–773, Aug. 2003, doi: 10.1117/12.506045.

[95] Wang, L.; Wang, X.; Zhang, X.; Feng, J., "The equivalence of two-dimensional PCA to line-based PCA," Pattern Recognition Letters, vol. 26, no. 1, pp. 57–60, ISSN 0167-8655, Jan. 2005, doi: 10.1016/j.patrec.2004.08.016.

[96] Hölmstrom, L.; Koistinen, P.; Laaksonen, J.; Oja, E., "Neural and statistical classifiers-taxonomy and two case studies," IEEE Transactions on Neural Networks, vol. 8, no. 1, pp. 5–17, Jan. 1997, doi: 10.1109/72.554187.

[97] Jain, A.; Flynn, P.; Ross, A., "Handbook of biometrics," Springer, US, ISBN: 978-0-387-71040-2, 2008, doi: 10.1007/978-0-387-71041-9.

[98] Rucklidge, W.J., "Efficiently locating objects using the Hausdorff distance," International Journal of Computer Vision, vol. 24, no. 3, pp. 251–270, Sep. 1997, doi: 10.1023/A:1007975324482.

[99] Gao, Y.; Leung, M., "Line segment Hausdorff distance on face matching," Pattern Recognition, vol. 35, no. 2, pp. 361–371, ISSN 0031-3203, Feb. 2002, doi: 10.1016/S0031-3203(01)00049-8.

[100] Miura, N.; Nagasaka, A.; Miyatake, T., "Feature extraction of finger-vein patterns based on repeated line tracking and its application to personal identification," Machine Vision and Applications, vol. 15, no. 4, pp. 194–203, Oct. 2004, doi: 10.1007/s00138-004-0149-2.

[101] Hao, Y.; Sun, Z.; Tan, T.; Ren, C., "Multispectral palm image fusion for accurate contact-free palmprint recognition," in 15th IEEE International Conference on Image Processing, 2008. ICIP 2008. pp. 281–284, 12–15 Oct. 2008, doi: 10.1109/ICIP.2008.4711.

[102] Hao, Y.; Sun, Z.; Tan, T., "Comparative studies on multispectral palm image fusion for biometrics," 8th Asian Conference on Computer Vision, Tokyo, Japan, vol. 4844, pp. 12–21, doi: 10.1007/978-3-540-76390-1-2.

[103] Kabaciński, R.; Kowalski, M., "Vein pattern data set and benchmark results," Electronics Letters, vol. 47, no. 20, pp. 1127–1128, 29 Sep. 2011, doi: 10.1049/el.2011.1441.

[104] Bolle, R.M.; Connell, J.; Pankanti, S.; Ratha, N.K.; Senior, A.W., "Guide to biometrics," Springer Science & Business Media, ISBN: 978-1-4419-2305-9, Jan. 2013, doi: 10.1007/798-1-4757-4036-3.

[105] Malik, J.; Girdhar, D.; Dahiya, R.; Sainarayanan, G., "Reference threshold calculation for biometric authentication," International Journal of Image, Graphics and Signal Processing, vol. 6, no. 2, pp. 46, Jan. 2014, doi: 10.5815/ijigsp.2014.02.06.

[106] Rattani, A.; Poh, N., "Biometric system design under zero and non-zero effort attacks," International Conference on Biometrics (ICB), pp. 1–8, ISSN 2376-4201, Jun. 2013, doi: 10.1109/ICB.2013.6612999.

[107] Luo, H.; Yu, F.; Pan, J.; Chu, S.; Tsai, P., "A survey of vein recognition techniques," Information Technology Journal, vol. 9, no. 6, pp. 1142–1149, 2010, doi: 10.3923/itj.2010.1142.1149.

[108] Biometric Precise, "Understanding biometric performance evaluation," Precise Biometrics AB-SPA 133 1000-4160/wpbpe RA.

[109] Bolle, R.M.; Connell, J.H.; Pankanti, S.; Ratha, N.K.; Senior, A.W., "The relation between the ROC curve and the CMC," Fourth IEEE Workshop on Automatic Identification Advanced Technologies, pp. 15–20, doi: 10.1109/AUTOID.2005.48.

[110] Bangham, J.; Aldridge, R., "Multiscale decomposition using median and morphological filters," in IEEE Winter Workshop on Nonlinear Digital Signal Processing, 1993. pp. 6.1-1.1–6.1-1.4, 1993, doi: 10.1109/NDSP.1993.767731.

[111] Cardinaux, F.; Sanderson, C.; Marcel, S., "Comparison of MLP and GMM classifiers for face verification on xm2vts," in Audio- and Video-Based Biometric Person Authentication, vol. 2688 of LNCS, pp. 911–920, Springer, 2003, doi: 10.1007/3-540-44887-X_106.

[112] Reynolds, D.A.; Quatieri, T.F.; Dunn, R.B., "Speaker verification using adapted Gaussian mixture models," Digital Signal Processing, vol. 10, no. 1–3, 2000, doi: 10.1006/dspr.1999.0361.

[113] Reynolds, D.; Quatieri, T.; Dunn, R., "Speaker verification using adapted Gaussian mixture models," Digital Signal Processing, vol. 10, no. 1–3, pp. 19–41, Jan. 2000, doi: 10.1006/dspr.1999.0361.

[114] Kenny, P.; Boulianne, G.; Ouellet, P.; Dumouchel, P., "Joint factor analysis versus eigenchannels in speaker recognition," IEEE Transactions on Audio, Speech, and Language Processing, vol. 15, no. 4, pp. 1435–1447, May 2007, doi: 10.1109/TASL.2006.881693.

[115] Vogt, R.; Sridharan, S., "Explicit modelling of session variability for speaker verification," Computer Speech and Language, vol. 22, no. 1, pp. 17–38, Jan. 2008, doi: 10.1016/j.csl.2007.05.003.

[116] Kong, A.; Zhang, D., "Feature-level fusion for effective palmprint authentication," First International Conference, ICBA, Hong Kong, China, vol. 3072, pp. 761–767, 2004, doi: 10.1007/978-3-540-25948-0-103.

Chapter 11

Finger vein recognition and new processing approaches

Lu Yang[1] and Gongping Yang[2]

In this chapter, we review the state of the arts and discuss the main challenges in finger vein recognition. First, the brief introduction of finger vein recognition is given, and then the finger vein image acquisition methods are presented, including imaging principle, imaging devices and open databases. Additionally, we give all kinds of methods in image preprocessing, feature extraction and matching, listing the recognition performance of the state of the arts. Some finger vein involved multifeature and multitrait recognition systems, which are also given. Lastly, the open research issues in finger vein recognition are analyzed.

11.1 Introduction

Biometrics employs personal physiological characteristics or behavioral characteristics to perform identity recognition, which has higher safety over traditional passwords, ID cards and so on. Among all biometric traits, finger vein is a relatively new trait, which was first proposed by Kono in 2000 [1]. In finger-vein-based individual identification system, superficial vein vessels in finger palmar side were captured and used.

As a biometric trait, finger vein satisfies all basis requirements:

- **Universality:** Like other traits, finger vein is highly universal among persons. It can be easily understood that, as part of the blood circulation system, superficial vein vessels exist in each finger of a person. But, it is inevitable that physiological changes and some diseases may be adverse to the usage of finger vein [2]. As there are five fingers in each hand of one person, finger vein vessels in multiple fingers are alternative in recognition.
- **Distinctiveness:** It is believed that finger vein in each finger of a person, even among identical twins, is unique [2,3]. Two statistical studies have been conducted to prove the uniqueness of finger vein. Yanagawa *et al.* experimentally proved the diversity of human finger vein patterns and the usability of finger

[1]School of Computer Science and Technology, Shandong University of Finance and Economics, PR China
[2]School of Software Engineering, Shandong University, PR China

vein for identity recognition on 2,024 fingers of 506 persons [4]. This work reported that two finger vein patterns are identical only if they are from the same finger in the same hand of the same person. Recently, Ye *et al.* performed recognition experiments on 710,399 images from 363,703 fingers and concluded that finger vein was sufficiently unique to distinguish one person from another in such a large-scale population [5].

- **Permanence:** Although finger vein can be altered with physiological growth during the childhood and shrink with strength decline of bones and the muscles in agedness, it is quite stable at the age of 20–50 years [6].
- **Collectability:** The superficial vein vessels in finger can be captured by the near-infrared light with wavelengths between 700 and 1,000 nm [3,7]. The near-infrared light can be intensively absorbed by the hemoglobin in the blood of vein, but easily transmits other tissues of finger. Therefore, vein vessels can be captured as shadows in the image, but other part of the finger is captured as relatively bright area.

In addition, finger vein has some unique advantages over other traits [3]:

- **Living body identification.** It means that only vein vessels in the living finger can be captured and further used in human identification. So finger vein has higher security and antispoofing capabilities than face, fingerprint and so on.
- **Internal characteristic.** Unlike surface traits (e.g., face and fingerprint), finger vein is hidden under skin. Copying or stealing finger vein is very difficult, which guarantees the security of the trait. At same time, the condition of finger surface (e.g., finger humidity, dirt and so on) has little influence on the image capturing of finger vein.
- **Uncontrolled image acquisition.** Finger vein imaging is always performed in a contactless and noninvasive way, which is more acceptable for the users.

Similar to all biometric systems, finger vein recognition system mainly includes four steps, i.e., image acquisition, preprocessing, feature extraction and matching, which is shown in Figure 11.1. Lots of efforts from research communities were given to all involved steps, and some promising achievements have been achieved. And some industry groups (e.g., Hitachi and Fujisoft) have also promoted the development of finger vein largely. Based on the efforts of research communities

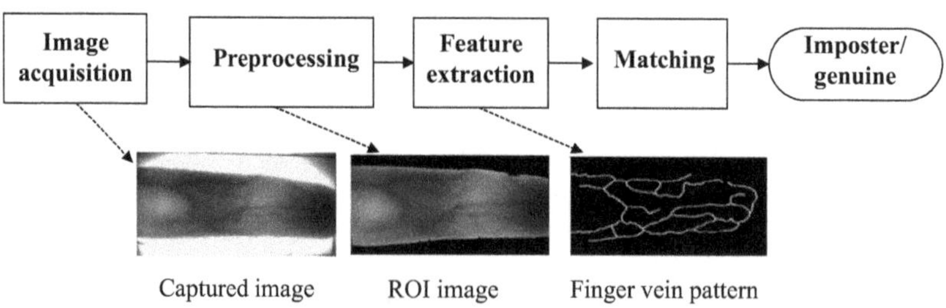

Figure 11.1 Flow chart of finger vein recognition system

and industry groups, finger vein recognition was gradually applied in the daily life. For example, finger vein recognition system has been embedded into Automatic Teller Machine (ATM) and steering wheel of car. In first case, the traditional passwords were replaced by finger vein, which overcomes forgetting and stealing, and highly improves the property security. In second one, the recognition system in car steering wheel can capture finger vein and perform human identification, so only the person with successfully matched vein pattern can drive the car. Recently, finger vein is also used in electronic payment, access control of computer and so on.

This chapter will give an overview of the current state of the arts in finger vein recognition. Section 11.2 introduces the acquisition of finger vein images and the public finger vein databases. Section 11.3 mainly gives all kinds of finger area localization methods. In addition, finger vein feature extraction and matching methods are reviewed in Section 11.4. Finger-vein-involved multifeature/trait recognition systems are also discussed in Section 11.5. Major challenges in finger vein recognition are analyzed in Section 11.6. Finally, Section 11.7 concludes this chapter.

11.2 Finger vein image acquisition

Imaging principle. As mentioned above, finger vein can be captured by near-infrared light. There are totally three ways of finger vein image acquisition, i.e., light reflection method, light transmission method and light scattering method, which are shown in Figure 11.2. The main difference between these methods is the relative position of near-infrared light to finger. We can see from the figure, the light and image sensor are both under finger in light reflection method, in which finger vein is captured by the reflected light. The light and the sensor are separately placed above and under the finger in light transmission method. In this method, the light will penetrate through the finger, and the captured vein pattern is relatively clear. Keeping the sensor fixed, the light is settled on two sides of finger in the third

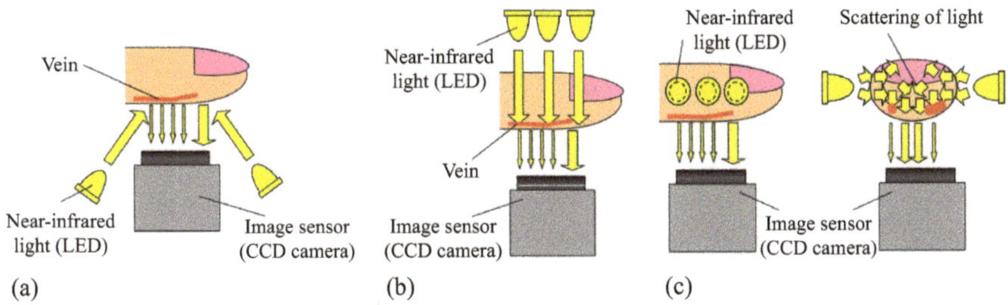

Figure 11.2 Three ways of finger vein image acquisition [3]: (a) light reflection method; (b) light transmission method; and (c) light scattering method (the first subfigure is on the side view and the second on is on the front view)

(a) (b)

Figure 11.3 One typical device and its schematic diagram [9]: (a) imaging device using light transmission method and (b) schematic diagram of the device

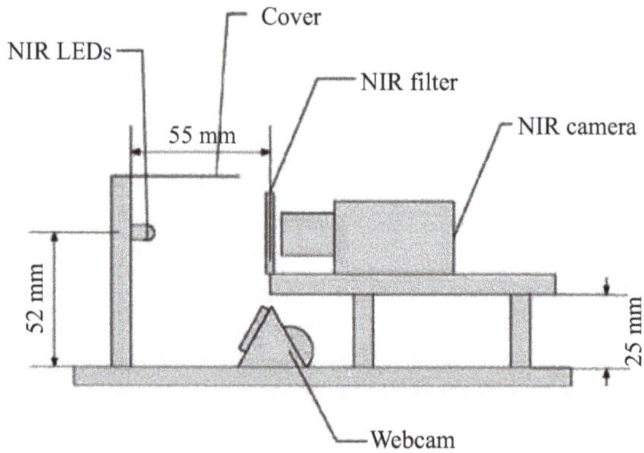

Figure 11.4 The schematic diagram of one finger vein and finger palmar texture imaging device [2]

method. The scattering light passes through finger palmar side and is captured by the sensor.

Imaging devices. In practice, most of image acquisition devices employ light transmission method [2,8–11]. A typical device based on light transmission method and its schematic diagram is shown in Figure 11.3. In the device, there are mainly three components, i.e., the near-infrared light source in the top plate, the camera on the right side and the mirror. In detail, the mirror is installed to place the camera in the horizontal plane. A USB LightBox is used to regulate the intensity of LED light. In image acquisition, the subject places his/her finger in the groove of the device below the top plate. (The captured image is shown in Figure 11.5(c) of the following subsection named "Open databases".)

In addition, there are some other devices, which can simultaneously capture finger vein and other biometric traits, for example, finger vein and fingerprint imaging device [12], finger vein and finger palmar texture imaging device [2] and finger vein and finger dorsal texture imaging device [13]. The schematic diagram of finger vein and finger palmar texture imaging device is shown in Figure 11.4 as

Table 11.1 Some public finger vein databases

Database	**SDUD [14]**	**HKPUD [2]**	**UTD [9]**	**MMCBNU_ 6000 [15]**	**USMD [16]**	**THUD [17]**
Subject number	106	156	60	100	123	610
Finger number/subject	6	2	6	6	4	1
Image number/finger	6	6/12	4	10	12	8
Image size (pixels)	320 × 240	513 × 256	672 × 380	640 × 480	640 × 480	720 × 576

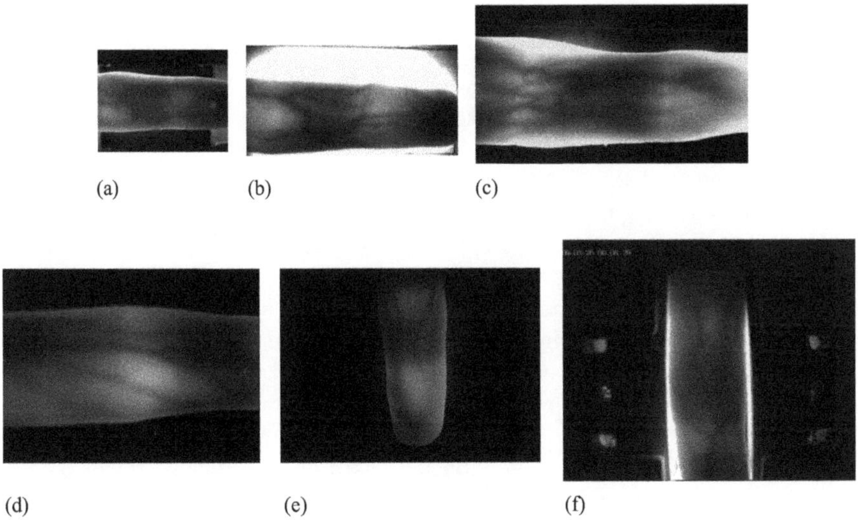

Figure 11.5 Typical images from different open databases: (a) SDUD [14], (b) HKPUD [2], (c) UTD [9], (d) MMCBNU_6000 [15], (e) USMD [16], (f) THUD [17]

one example. In the schematic diagram, the webcam is along with NIR components. In image acquisition, finger palmar side is in front of two cameras, and finger dorsal side is near the NIR LEDs. The webcam and NIR camera are separately used to capture finger palmar texture image and finger vein image. With this kind of device, user conveniences can be guaranteed in multitrait recognition system as there is no need to repeatedly capture images for different traits.

Open databases. In recent years, several finger vein databases have been opened. Some open databases are given in Table 11.1, and the typical images from these databases are given in Figure 11.5. Each of the first three databases is a part of the homologous multimodal database, and others are single finger vein databases. All images from these databases are captured by light transmission method but with different devices. At the same time, we can see that other factors, like the subject number, finger number, image size, image quality and so on, are all varied from one database to another.

11.3 Image preprocessing

Image preprocessing in finger vein recognition mainly includes image enhancement/restoration and ROI (region of interest) extraction. As image quality is a very tough problem in finger vein recognition, image enhancement/restoration will be introduced as a major challenge in Section 11.6. So, here only ROI extraction methods are given.

In captured finger vein image, there is not only finger region but also image background. ROI of a finger vein image refers to the finger region, in which the finger vein network exists. In ROI extraction, finger region is located, and image background is set as zero or removed. Therefore, only finger region will be involved in the subsequent feature extraction and matching, saving recognition time and keeping the discrimination of finger vein feature. In the beginning, a predefined window with a fixed size was employed to crop ROI image from captured finger vein image. Two typical methods are listed. Considering big gray difference between finger region and image background, Rosdi *et al.* [18] built a gray threshold-based binarization to obtain image foreground (i.e., finger region), and then the predefined window was used on image foreground to crop the ROI image. Furthermore, a keypoint, which is located on finger distal interphalangeal joint, was taken as a reference in cropping ROI image by the predefined window [19,20]. Obviously, this kind of methods cannot adapt the variation of finger size and image size.

Next, ROI image is generally extracted based on the detected finger boundaries, which is adaptive to finger size and image size. In detail, Kumar and Zhou [2] performed the binarization to coarsely localize the finger shape on the captured image with a fixed threshold value (i.e., 230), and the Sobel edge detector was applied to the binarized image to obtain the finger boundaries. Different from the combination of binarization and edge detector, Yang *et al.* [21] only employed the edge detector on ROI candidate region, which is obtained by removing the noise region around the captured image. Two masks, instead of edge detectors, were designed and used in finger boundary detection by Lu *et al.* [22], and finger boundary restoration was further conducted to remove the false boundary and connect the broken boundary. Additionally, all these three methods corrected finger direction to overcome finger displacement by the orientation of binarized finger mask [2] or the inclination of the middle line of two finger boundaries [21,22], which was ignored in the predefined window-based ROI extraction method. Moreover, like the predefined window-based method, ROI image was cropped from the captured image based on the internal tangents of the finger boundaries, and normalized into the fixed size in the latter two finger-boundary-based methods. But, the image background outside two finger boundaries was set as 0 in the first finger-boundary-based method.

All above-mentioned methods were designed for images captured by one particular imaging device. However, the imaging devices used in recognition system, especially in practical application, may be various. We can see from Section 11.2 that there are large variations about image size, gray level and background noise in the open images from different sensors. These variations will degrade the performance of the above-mentioned ROI extraction methods. For example, the gray value in image background is different in images captured by different devices, and the gray-value-based binarization [2] is sensitive to gray variation. To overcome these image variations, Yang *et al.* proposed a cross-sensor ROI extraction method [23] in which a robust superpixel segmentation method was conducted on finger vein image, and finger boundaries were tracked from the segmented image. The finger boundary postprocessing and finger direction correction were also considered in this work.

11.4 Feature extraction and matching

Various features have been proposed to represent finger vein image, and these features can be grouped into two classes according to if vein pattern is segmented from the preprocessed image, i.e., vein-pattern-based method and nonvein-pattern-based method. As only vein pattern in image, without nonvein region, is used in recognition, vein-pattern-based method explores the intrinsic characteristics of finger vein image. So, in this section, we discuss vein-pattern-based method and their matching method in detail and others relatively roughly.

11.4.1 Vein-pattern-based method

In this kind of method, the segmented vein pattern is used as vein feature to represent finger vein image. As we know, vein pattern has lower gray value than nonvein region in finger vein image, as near-infrared light is absorbed by the hemoglobin in vein vessel. Generally, the gray relationship between the current point and its surrounding points is exploited in vein pattern segmentation. The points on cross-sectional profile, or in a rectangle or circle neighborhood region, even or in one images are considered. According to the considered surrounding points, these vein-pattern-based methods are classified into three main categories, i.e., cross-sectional profile-based method, neighborhood region-based method and whole image-based method.

Cross-sectional profile-based method: The cross-sectional profile(s) on one or multiple direction(s) of the current point will be analyzed in vein pattern detection. On the cross-sectional profile of a vein point, a valley shape can always be detected in the distribution of gray values as the points on two sides of the profile have higher gray value than the center point (i.e., the finger vein point). But the profile of nonvein point does not have the valley shape. Based on the discovery, two growth methods, (i.e., repeated line tracking [8] and region growth [24]), two curvature methods (i.e., maximum curvature point [25] and anatomy structure

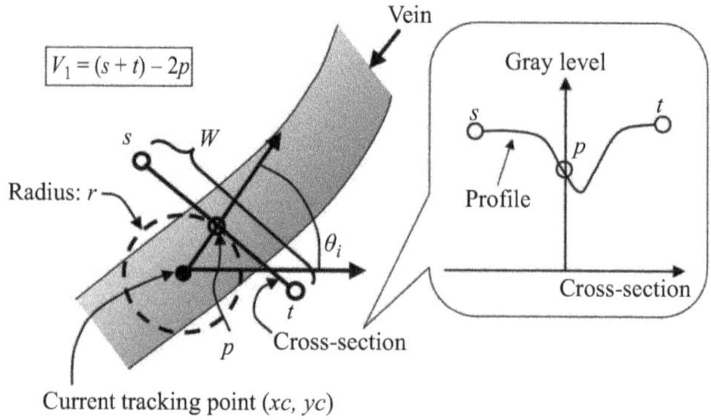

Figure 11.6 Cross-sectional profile of one finger vein point [8]

analysis-based vein extraction method [26]) and a spatial curve filter method [27] were proposed.

In each tracking of the first method, the initial point was randomly selected from all points in the image, and the neighbor vein point of the current point was detected by the depth of the valley-shaped cross-sectional profile and tracked. The current point p and its cross-sectional profile spt are shown in Figure 11.6. From the figure, we can see the cross-sectional profile of the vein point p has a valley shape. If there is no valley-shaped profile for all neighbor points, the tracking begins again. Finally, vein pattern was obtained by binarization of the tracked times, which was based on the fact that the tracked number of vein point was larger than it of nonvein point. In the binarized image, vein points were labeled by 1, and non-vein points were labeled by 0. In this method, the total tracking number was pre-defined, and it has big relationship with image size. And in general, if a larger tracking number is defined, more time will be taken and more abundant vein pattern can be extracted. In region growth method, there are two primary improvements to the former method. First, the initial point was randomly selected in the rough segmented vein pattern, not the whole image, which can largely reduce the number of initial points and further reduce the growth number. Second, in vein point detection, this method also considered the symmetry of the cross-sectional profile to avoid the false acceptance of noises.

In the above two growth methods, the gray values of cross-sectional profile were directly used in vein point detection, but in maximum curvature method, the curvature values of the profile were computed and used. All points with larger tracked times over the threshold are labeled as vein points in growth methods, but not all points with positive curvature will be seen as vein points in maximum curvature methods. Only the center line of vein with local maximum curvature on the cross-sectional profile is saved as image feature [25], which was believed to be robust to the variation of vein width caused by the fluctuations in the amount of blood in the vein. In addition to the center vein line, the vein points with larger curvature values were mined as vein backbone and fused with the center line in

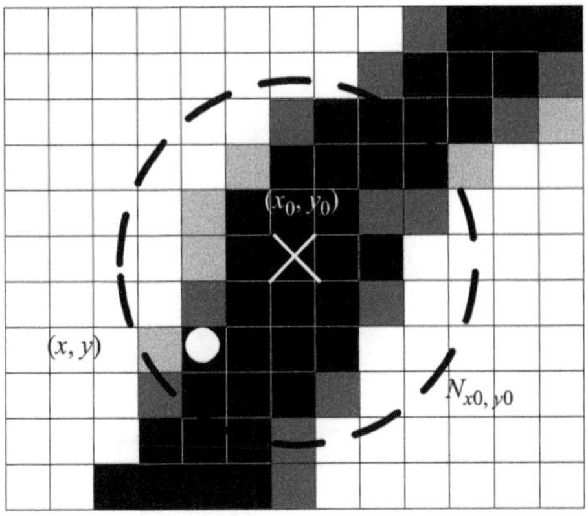

Figure 11.7 Circle neighborhood region of one finger vein point [28]

matching [26]. Besides, a weighted spatial curve filter was designed on the cross-sectional profile to extracted finger vein pattern in [27].

Neighborhood region-based method: Here, the neighborhood region of the current point was employed in vein point detection. For a vein point, the distribution of gray value in its neighborhood is close to a reverse bell, as the center current point has lower gray value than others in the neighborhood. The circle neighborhood region and the rectangle one were separately used in wide line detection [28] and Gabor filters [2]. In the first method, a circle neighborhood region was first located for the current point, and then the current point was labeled as vein point if the number of its neighbors, with higher gray value over it of the current point and in the located region, was bigger than a predefined threshold. If not, the point was seen as nonvein point. In Figure 11.7, the circle neighborhood region of point (x_0, y_0) is labeled. In the second method, a set of self-similar Gabor templates was built and convoluted with finger vein image to detect vein pattern. In addition, the morphological operations were conducted to enhance the structure of the detected vein pattern in this method. Apparently, the latter method is much more robust to noises than the former one. But, the setting of parameters, like template size, filter wavelength, filter length–width ratio and so on, in this method is relatively difficult.

Whole image-based method: In this kind, there are two typical methods, i.e., mean curvature [29] and convolutional neural network (CNN)-based method [30]. In first method, the finger vein image was seen as a three-dimensional geometric shape. Theoretically, vein pattern exists in valley shape and nonvein pattern looks like the ridge. In this method, the valley-like vein pattern was detected by negative mean curvature. An image segment and its mean curvature values are shown in Figure 11.8. In latter method, the deep learning method was used in finger vein pattern extraction. Specifically, a combination of the handcrafted methods was first employed to extract vein pattern, and the resulting images were used to train a CNN. The probability of each testing pixel of being vein point was predicted by the trained CNN.

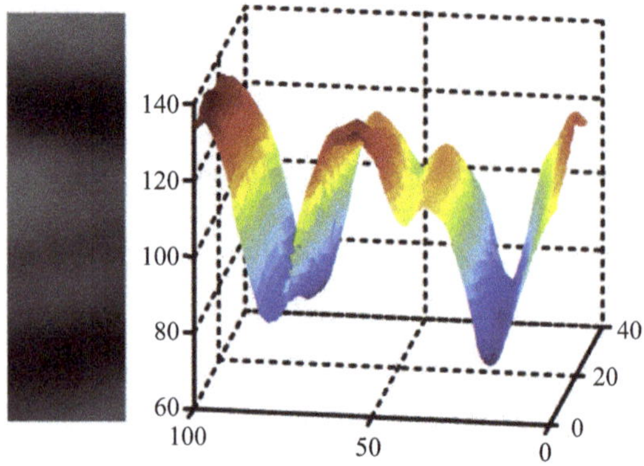

Figure 11.8 An image segment and its mean curvature [29]

Generally, in extracted vein pattern map, the vein points are labeled by 1 or 255, and the nonvein points are labeled by 0. The template matching was popularly used in similarity measurement of two vein patterns, in which the number of the overlapped vein points between two vein patterns was seen as the similarity score. Owing to large variations of genuine images caused by finger displacement in image acquisition, the traditional template matching was improved by incorporating the relative image translation [2,8,24], relative image rotation [29] and adding normalization to the extracted vein pattern prior to matching [28].

In the improved template matching, the multiple matches were performed to the enrolled feature by translating the input feature in both horizontal and vertical directions or/and rotating the input feature in both clockwise and anticlockwise directions, and the maximum of the matching scores in multiple matching was taken as the final score. The translation scale and rotation angle were both predefined according to the relative translation and rotation of genuine images in the used database. This improved method can enhance recognition performance to some extent, but it spends more time than the original method owing to larger matching times. For example, if the translation radius on both horizontal and vertical directions is 5 pixel and the translation interval is 1 pixel, the matching number in the improved matching method with only relative translation is 121 [i.e., $(2 \times 5 + 1)^2$] times of it in original method.

Another improved matching method conducted the ellipse projective coefficients-based vein pattern normalization ahead of the template matching. In vein pattern normalization, the cross-section of finger was approximated to an ellipse with fixed diameter. The approximated ellipse, the finger boundaries and the middle line of finger were all used in the computation of projective coefficients, and based on the computed coefficients, the vein pattern was normalized to overcome image variations. Lastly, the normalized vein pattern will be used in template matching.

Furthermore, the robustness and reliability of all detected vein points may be different. Considering this problem, some methods only use the robust part of the extracted vein pattern in matching. In repeated line-tacking method, the downsampling and relabeling of vein pattern were performed before similarity measurement. In detail, the vein pattern was reduced to one-third of its original size, in which the mean gray of points in nonoverlapped image patch with 3×3 pixel was set as the gray value of downsampled point. In the following vein pattern relabeling, the downsampled image was transformed as a ternary image by two thresholds, i.e., 86 and 170. The downsampled points, with gray value less than 85, and these, with gray value bigger than 170, were separately seen as nonvein points and vein points, which were used in similarity measurement. But, other points, with gray value between 86 and 170, will not considered in matching. And in Gabor filters method, finger mask was used in matching to locate finger region, which can prevent the influence of noises outside finger mask.

11.4.2 Nonvein pattern-based method

In this kind of methods, the features are directly extracted from the preprocessed finger vein images, without finger vein extraction. And these features can be classified into four main subgroups, i.e., local binary-based method, keypoints-based method, dimensionality reduction-based method, high-level feature. In the following, we give the analysis of these four kinds of nonvein-pattern-based methods.

Local binary-based method. The local binary was encoded by comparison between gray value of the current point and the values of its neighbors [10,31], or learned by support vector machine (SVM) [32]. In comparison-based method, the bit of one neighbor, with larger gray value than the current point, was encoded as 1 and 0 if otherwise. The considered neighbors were, respectively, in a rectangular area and on a line, with the current point as the center, in local binary pattern (LBP) [10] and local line binary pattern (LLBP) [31]. The rectangular area and the computation of binary codes of LBP are given in Figure 11.9. This kind of methods focused on capturing representative information of subject, but ignored the discriminative information among different subjects. In order to overcome this problem, the discriminative binary code was learned by considering the subject relation graph [32]. For the binary feature, the Hamming distance was always used in similarity measurement.

In addition, some improvements have been made for local binary feature. Considering that the values of some binary bits, with same location in multiple genuine images, may be inconsistent, personalized best bit map (PBBM) [11] was learned to label the consistent bits of each subject, and only the learned consistent

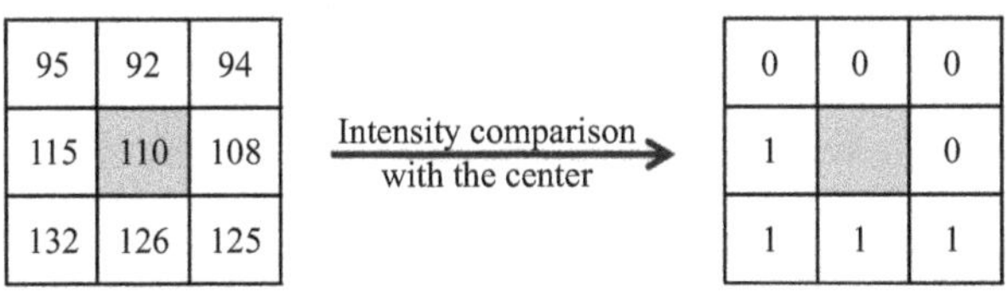

Figure 11.9 One example of LBP operator [10]

bits were considered in Hamming distance-based matching. In other words, the method qualitatively classified the binary bits into consistent set and inconsistent set, and discarded the bits in latter set. However, the inconsistent bits may also have discrimination to some extent. For example, 1 b is one in five of six genuine images, and it in the rest image is 0. In this case, the discrimination may be discounted if the bit is discarded. In view of this problem, in personalized weight map (PWM) [33], the stability of each bit in binary feature was measured quantificationally and used as a weight for the bit in matching.

The discrimination of each bit was explored in PBBM and PWM, and each image patch of it was also measured in optimal partitioning uniform rotation invariant LBP [34] and weighted LBP [35]. The motivation is that in finger vein image, some patches may be more discriminative than others. For example, the patch with vein pattern is believed to be more discriminative than the one without vein pattern. Two selection methods of discriminative patch were proposed in the former method, and the LBP histograms in only selected patches were used in histogram intersection-based similarity measurement. In the latter method, all image patches were classified in three kinds according to the amount of their vein points, and LBP in all patched was matched by weighted Hamming distance, in which the weight was given based on the kinds of two matched image patches.

Keypoints-based method. Minutiae (i.e., bifurcation points and ending points in finger vein pattern) [36,37] and keypoints in scale-invariant feature transform (SIFT) [38] were two kinds of popular point features in finger vein recognition. The modified Hausdorff distance and the number of matching pairs were used for similarity measurements of minutiae and SIFT. But affected by image quality and image variation, it is difficult to build the correspondence of minutia pair between two images. To overcome this problem, a singular value decomposition (SVD)-based minutiae correspondence building method [39] was proposed. More importantly, limited minutiae in finger vein pattern leads to the unsatisfied recognition performance. The local extensive binary pattern of each minutia [39] and the relationship between different minutiae [40] were used in fusion with minutiae to enhance recognition performance.

Dimensionality reduction-based method. In this kind of methods, finger vein image was transformed into a low-dimensional space by dimensionality reduction methods, like unsupervised principal component analysis (PCA) [41] and supervised linear discriminant analysis [42]. In the transformation, the main image components were saved and noises were removed. For the saved main image components, the neural networks (e.g., back-propagation neural network and adaptive neuro-fuzzy inference system) [41] and SVM [42] were used in classification. In addition, some improvements have been made to PCA, resulting in two-dimensional PCA (2DPCA) [43] and two-directional and two-dimensional PCA $(2D)^2$PCA [44], which were also used in finger vein recognition. The Euclidean distance and a metric learning method (i.e., large margin nearest neighbor) were separately employed for similarity measurement in two improved methods. Furthermore, a manifold learning method, called orthogonal neighborhood preserving projection (ONPP) [45], and the corresponding point-manifold distance were involved in feature extraction and similarity measurement of finger vein recognition, which was more robust to image variation than other dimensionality reduction-based methods as the neighboring relation of finger vein images were preserved in dimensionality reduction.

High-level feature. In all former methods, the features are extracted from finger vein image. However, in this kind of methods, the number and the distribution of the extracted features in one image are further explored. In detail, the high-level features were acquired by the idea of bag of word [46,47], sparse representation [48] and a tree model [49,50]. In bag of word-based methods, LBP represented superpixels or image patches were clustered by *K*-means algorithm to learn visual words (i.e., finger vein texton), and the superpixel or image patch was represented by the nearest visual word. The image patch-based finger vein texton learning is illustrated in Figure 11.10. The best image patches, which are the consistent visual word between the multiple genuine images, were learned and used in similarity measurement by matched best patches ratio. In superpixel-based method, the spatial pyramid histogram of the visual words was used as image feature, and a weighted histogram intersection-based matching was employed.

In sparse representation-based method, the image was represented by a sparsely linear representation of all learned attributes. In detail, the pyramid histograms of gray and orientation gradients was used as base feature, and *K*-means clustering was performed to learn the image attributes. Furthermore, in tree-model-based feature representation, the distribution of the detected Harris–Laplacian feature point and the extracted vein pattern were separately represented by a vocabulary tree and a binary tree. A point set matching method was used in similarity measurement of the former tree model. In another tree model, the similarities of all vein segment pairs were measured first, and then combined to obtain the final score.

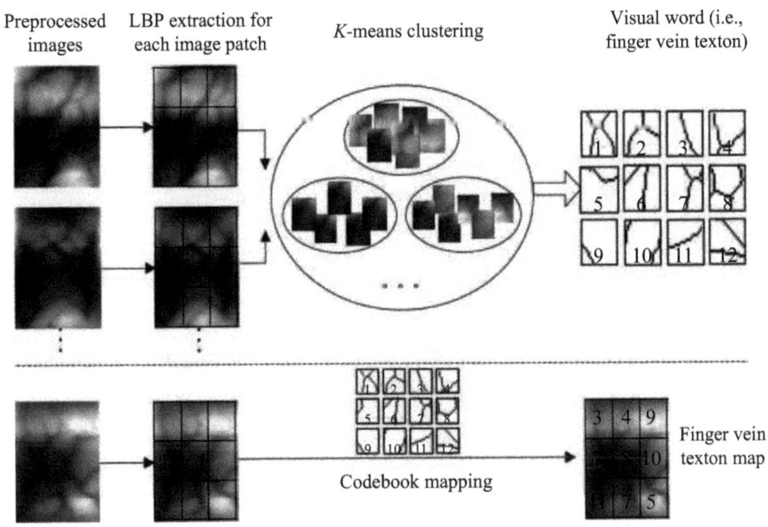

Figure 11.10 Image patch-based finger vein texton learning [47]

Table 11.2 Comparative summary of typical methods in finger vein recognition

Reference	Method	Matching	Database, finger number	Performance (EER%)
[8]	Repeated line tracking	Template matching	Private, 339	0.145
[24]	Region growth	Template matching	Private, 320	0.0369
[25]	Maximum curvature	Template matching	Private, 339	0.0009
[28]	Wide-line detector	Template matching	Private, 10140	0.87
[2]	Gabor filter	Template matching	Open, 312	0.65
[29]	Mean curvature	Template matching	Private, 125	0.25
[10]	LBP	Hamming distance	Private, 240	0.21
[11]	PBBM	Hamming distance	Private, 106	0.38
[38]	SIFT	Number of matched point pairs	Private, 95	1.71
[39]	SVD-based minutiae	Euclidean distance	Open, 312	2.45
[43]	2DPCA	Euclidean distance	Open, 636	1.59
[45]	ONPP	Point-manifold distance	Private, 328	0.8
[46]	Superpixel feature	Weighted histogram intersection	Open, 312	1.47
[49]	Tree model	Point set matching	Open, 492	0.96

11.4.3 Performance reported in existing work

As we introduced above, all kinds of methods have been proposed for improving the performance of finger vein recognition. Now, a summary of some typical methods and their recognition performance are presented in Table 11.2, which can give an intuitive comparison of the existing methods.

11.5 Finger vein involved multifeature/trait system

To improve the recognition performance, the fusion of multiple finger vein features and the fusion of multiple biometric traits have been proposed. About multifeature fusion, there was one image level fusion method, in which images captured by near-infrared lights with different wavelengths were fused by an improved regional energy fusion strategy [51]. In feature level fusion, local feature extracted by Gabor filter and global feature extracted by moment invariant were fused by canonical correlation analysis and a weight feature fusion strategy [52]. Two vein patterns, extracted by multiscale-matched filtering and repeated line tracking, were fused in [53]. In addition, some features were fused in score level. In one case, two similarity scores from LLBP and global Wavelet transform were combined by the SVM [54]. In another case, pixel level feature and superpixel level feature were also fused in score level by weighted average strategy [55]. There was also one decision level fusion related work, in which the classification decisions, from local moment

Table 11.3 Typical finger vein involved multitrait recognition systems

Involved traits	Reference	Fusion level	Database, subject number	Performance (EER%)
Finger shape, finger vein	[58]	Score level fusion	Real, 816	0.075
	[59]	Score level fusion	Real, 120	About 4
	[16]	Score level fusion	Real, 492	1.78
Fingerprint, finger vein	[60]	Feature level fusion	Real, 64	1.35% FAR at FRR = 0%
Finger palmar texture, finger vein	[2]	Score level fusion	Real, 312	0.08
Finger dorsal texture, finger vein	[17]	Feature level fusion	Real, 220	0.435
	[61]	Feature level fusion	Real, 220	0.450
Finger palmar texture, fingerprint, finger vein	[62]	Score level fusion	Real, 378	0.109
Finger dorsal texture, fingerprint, finger vein	[63]	Feature level fusion	Real, 300	0.415
Fingerprint, face, finger vein	[64]	Score level fusion	Synthetic, 510	99.8% GAR at FAR = 0.01%
Finger dorsal texture, fingerprint, finger shape, finger vein	[65]	Feature level fusion	Synthetic, 100	7.9631e−03

feature, topological feature and vein shape feature, were fused by Dempster–Shafer algorithm [56].

In the above methods, all kinds of finger vein features are directly fused in different level. Different from these methods, personalized feature was selected from the extracted multiple features, and only the selected feature was used in similarity measurement [57]. In detail, three histogram features, i.e., pyramid histogram of gray, pyramid histogram of gray and pyramid histogram of orientation gradient, were first extracted from finger vein image. Next, personalized feature was selected by LASSO from these histograms for each subject.

Moreover, multiple biometric traits were also fused in some recognition systems, for example, finger shape and finger vein fused system, fingerprint and finger vein fused system and so on. Some finger vein involved multitrait recognition systems are summarized in Table 11.3, mainly including the used traits, the fusion level, the used database and recognition performance. The table shows that the trait number varies from 2 to 4, and the traits were fused mostly at feature level and score level. Multiple finger-traits-based recognition system were generally built on real homologous, as some image devices were designed for simultaneously capturing multiple finger traits. However, face involved system and four traits involved system were built on synthetic databases. The reported recognition performance was all comparatively more promising than it of finger vein recognition.

11.6 Major challenges

11.6.1 Image quality

The quality of finger vein image can be affected by some factors, such as the imaging device performance, the temperature and humidity of imaging environment and the individual difference (e.g., gender and fat thickness). So, low-quality image is inevitable if it is captured by a device with unsatisfactory imaging performance in cold weather especially from a female. In order to overcome image quality problem, some works focused on image quality evaluation and others aimed at image enhancement and image restoration.

In image quality evaluation, some quality features were artificially designed according to the domain knowledge, for example, image gradient, image contrast, information capacity [66], information entropy [67]. The deep feature, learned by deep neural network from finger vein image, was also used in quality evaluation [68]. All three methods need training images for learning the classifier model, but the way of labeling training images was varied among these methods. In designed feature-based methods, the training images were labeled by the expert according to their domain knowledge. However, in deep feature-based method, the training images were classified by a verification system. In detail, the images, false rejected by the used verification system, were labeled as low quality, and other images were labeled as high quality. There is another difference in these methods. For one testing image, the first method gave its quality label (i.e., low quality or high quality), but the other two methods outputted a quality score (i.e., from 0 to 1). For the predicted low-quality images, filtering them and recapturing images, or using robust image processing methods is helpful to avoid its harmfulness to the recognition performance.

In addition, some image enhancement or restoration methods have been proposed to improve image quality. The optical blurring and the scattering of finger biological tissue can both damage the image quality. To restore the blurred finger vein image, the depth-dependent point spread function [69,70], the biological optical model (BOM) [71] and the weighted BOM [72] were proposed. In image enhancement, there are Gabor filter [19], fuzzy-based fusion method of Gabor and Retinex filtering [73] and guided filter-based intensity variation normalization method [74].

11.6.2 Finger displacement

As we have introduced, finger vein image capturing is always conducted in a contactless and unrestrained way, in which the random finger placement in the big space between the near-infrared light and the camera causes the relative variations (i.e., translation, rotation, scaling and so on) between genuine images. What's worse, these relative image variations will enlarge the difference of genuine images, and further lead to the serious degeneration of recognition performance. Fortunately, some methods have been proposed to deal with the problem.

To overcome in-plane finger rotation, finger alignment was conducted in most of finger vein image preprocessing methods [2,21]. In [75], one minutia-based image alignment was executed ahead of feature extraction. Different from these three methods, in which image alignment were performed in image preprocessing, there are some finger displacement robust feature extraction methods, for example, ONPPs [45], SIFT [38]. In addition, there are some robust matching methods. One example is the nonrigid registration-based matching [76], in which keypoint-based nonrigid image registration was performed before the similarity measurement. The others are vein pattern normalization-based template matching [28] and maximal matched pixel ratio-based matching [2,29], which are introduced in Section 11.4.1.

11.6.3 Essence of finger vein recognition

Vein pattern is the essential characteristic of finger vein image. Hence, in finger vein recognition, the vein pattern-based methods may be superior to the nonvein pattern-based ones. In most of vein pattern-based methods, the number of overlapped vein points between two images was seen as the similarity score [2,25,29]. It is a pity that some discriminative information, like the topology of vein pattern and the vein direction, were not exploited in these matching methods. Luckily, some works have made the preliminary attempt to exploit the discriminative features from the extracted finger vein pattern. There were three typical examples: (1) eight-neighbors-based method [77], in which the local structure of the edge of binary vein pattern was encoded by the neighbor pattern coding method. (2) Vein segment-based methods [78,79]. In two methods, the extracted vein pattern was both divided into multiple vein segments based on the detected keypoints (i.e., junction points in [78] and singular points in [79]). The divided vein segment was then encoded by a modified angle chain in the former method, and the segment-wise matching was performed by the optimized generalized Hough transform in the latter method. (3) Image patch-based method [80], in which the local shape of vein pattern was characterized by the patch-based Histogram of Oriented Gradient (HOG).

11.7 Conclusions

We present the state-of-the-art methods in image acquisition, image processing and multifeature/trait fusion of finger vein recognition and point out the main challenges. Based on the analysis in this chapter, we make some conclusions about the current situation of finger vein recognition:

1. It is urgent to improve the imaging performance of the device and build large-scale open database. The low image quality is one of the primary reasons for the unsatisfactory recognition performance, as some vein segments, especially the thin subsegments, are too vague to be extracted and used in recognition. In addition, the size of the existing open database is very limited, and lots of methods can achieve good recognition performance, which causes that the comparison of these methods cannot be evaluated effectively.

2. Vein pattern-based methods explore the intrinsic characteristic of vein vessel in the palmar side of finger and have achieved very promising recognition performance. So, these methods can be seen as the mainstream recognition algorithm. However, there is a large space for the exploitation of the discriminative information in vein pattern, which has potential to further promote the improvement of finger vein recognition.
3. The fusion of multiple finger traits may be an extremely practicable way to improve the recognition performance in practical application. Some imaging devices can simultaneously capture fingerprint image and finger vein image, or finger palmar/dorsal texture image and finger vein image, which provides large convenient for the users in multitrait recognition system.
4. More powerful methods are needed for dealing with the relative variations in genuine images. In the worst case, part of vein pattern, which can be captured in normal condition, is missing in the image owing to the large finger displacement. The existing methods are weak to this case. So, how to deal with this terrible case is a valuable study.

Acknowledgments

The chapter is supported in part by National Science Foundation of China under Grant No. 61703235, 61472226 and 61573219.

References

[1] Miyuki Kono, Hironori Ueki, and Shin-ichiro Umemura. A new method for the identification of individuals by using of vein pattern matching of a finger. In *Proceedings of the 5th Symposium on Pattern Measurement*, pages 9–12, 2000.

[2] Ajay Kumar and Yingbo Zhou. Human identification using finger images. *IEEE Transactions on Image Processing*, 21(4):2228–2244, 2012.

[3] Junichi Hashimoto. Finger vein authentication technology and its future. In *2006 Symposium on VLSI Circuits, Digest of Technical Papers*, pages 5–8. IEEE, 2006.

[4] Takashi Yanagawa, Satoshi Aoki, and Tetsuji Ohyama. Human finger vein images are diverse and its patterns are useful for personal identification. *MHF Preprint Series*, 12:1–7, 2007.

[5] Yapeng Ye, He Zheng, Liao Ni, Shilei Liu, and Wenxin Li. A study on the individuality of finger vein based on statistical analysis. In *2016 International Conference on Biometrics (ICB)*, pages 1–5. IEEE, 2016.

[6] Amioy Kumar, Tanvir Singh Mundra, and Ajay Kumar. Anatomy of the hand. In Stan Z. Li and Anil K. Jain (eds), *Encyclopedia of Biometrics*. Boston, MA: Springer, pages 19–28, 2015.

[7] Miyuki Kono, Hironori Ueki, and Shin-ichiro Umemura. Near-infrared finger vein patterns for personal identification. *Applied Optics*, 41(35):7429–7436, 2002.

[8] Naoto Miura, Akio Nagasaka, and Takafumi Miyatake. Feature extraction of finger-vein patterns based on repeated line tracking and its application to personal identification. *Journal of Machine Vision and Applications*, 15(4): 194–203, 2004.

[9] Bram T Ton and Raymond NJ Veldhuis. A high quality finger vascular pattern dataset collected using a custom designed capturing device. In *2013 International Conference on Biometrics (ICB)*, pages 1–5. IEEE, 2013.

[10] Eui Chul Lee, Hyunwoo Jung, and Daeyeoul Kim. New finger biometric method using near infrared imaging. *Sensors*, 11(3):2319–2333, 2011.

[11] Gongping Yang, Xiaoming Xi, and Yilong Yin. Finger vein recognition based on a personalized best bit map. *Sensors*, 12(2):1738–1757, 2012.

[12] Ramachandra Raghavendra, Kiran B Raja, Jayachander Surbiryala, and Christoph Busch. A low-cost multimodal biometric sensor to capture finger vein and fingerprint. In *2014 IEEE International Joint Conference on Biometrics (IJCB)*, pages 1–7. IEEE, 2014.

[13] Wenming Yang, Xiaola Huang, Fei Zhou, and Qingmin Liao. Comparative competitive coding for personal identification by using finger vein and finger dorsal texture fusion. *Information Sciences*, 268:20–32, 2014.

[14] Yilong Yin, Lili Liu, and Xiwei Sun. SDUMLA-HMT: a multimodal biometric database. In *Chinese Conference on Biometric Recognition*, pages 260–268. Springer, 2011.

[15] Yu Lu, Shan Juan Xie, Sook Yoon, Zhihui Wang, and Dong Sun Park. An available database for the research of finger vein recognition. In *Image and Signal Processing (CISP), 2013 6th International Congress on*, volume 1, pages 410–415. IEEE, 2013.

[16] Mohd Shahrimie Mohd Asaari, Shahrel A Suandi, and Bakhtiar Affendi Rosdi. Fusion of band limited phase only correlation and width centroid contour distance for finger based biometrics. *Expert Systems with Applications*, 41(7):3367–3382, 2014.

[17] Wenming Yang, Xiaola Huang, Fei Zhou, and Qingmin Liao. Comparative competitive coding for personal identification by using finger vein and finger dorsal texture fusion. *Information Sciences*, 268:20–32, 2014.

[18] Bakhtiar Affendi Rosdi, Chai Wuh Shing, and Shahrel Azmin Suandi. Finger vein recognition using local line binary pattern. *Sensors*, 11(12):11357–11371, 2011.

[19] Jinfeng Yang and Yihua Shi. Finger–vein ROI localization and vein ridge enhancement. *Pattern Recognition Letters*, 33(12):1569–1579, 2012.

[20] Jinfeng Yang and Xu Li. Efficient finger vein localization and recognition. In *2010 20th International Conference on Pattern Recognition (ICPR)*, pages 1148–1151. IEEE, 2010.

[21] Lu Yang, Gongping Yang, Yilong Yin, and Rongyang Xiao. Sliding window-based region of interest extraction for finger vein images. *Sensors*, 13(3):3799–3815, 2013.

[22] Yu Lu, Shan Juan Xie, Sook Yoon, Jucheng Yang, and Dong Sun Park. Robust finger vein ROI localization based on flexible segmentation. *Sensors*, 13(11):14339–14366, 2013.

[23] Lu Yang, Gongping Yang, Lizhen Zhou, and Yilong Yin. Superpixel based finger vein ROI extraction with sensor interoperability. In *2015 International Conference on Biometrics (ICB)*, pages 444–451. IEEE, 2015.

[24] Huafeng Qin, Lan Qin, and Chengbo Yu. Region growth-based feature extraction method for finger-vein recognition. *Optical Engineering*, 50(5):057208-1–057208-8, 2011.

[25] Naoto Miura, Akio Nagasaka, and Takafumi Miyatake. Extraction of finger-vein patterns using maximum curvature points in image profiles. *IEICE Transactions on Information and Systems*, 90(8):1185–1194, 2007.

[26] Lu Yang, Gongping Yang, Yilong Yin, and Xiaoming Xi. Finger vein recognition with anatomy structure analysis. *IEEE Transactions on Circuits and Systems for Video Technology*, 2017.

[27] Jinfeng Yang, Yihua Shi, and Guimin Jia. Finger-vein image matching based on adaptive curve transformation. *Pattern Recognition*, 2017.

[28] Beining Huang, Yanggang Dai, Rongfeng Li, Darun Tang, and Wenxin Li. Finger-vein authentication based on wide line detector and pattern normalization. In *2010 20th International Conference on Pattern Recognition (ICPR)*, pages 1269–1272. IEEE, 2010.

[29] Wonseok Song, Taejeong Kim, Hee Chan Kim, Joon Hwan Choi, Hyoun-Joong Kong, and Seung-Rae Lee. A finger-vein verification system using mean curvature. *Pattern Recognition Letters*, 32(11):1541–1547, 2011.

[30] Huafeng Qin and Mounim A El-Yacoubi. Deep representation based feature extraction and recovering for finger-vein verification. *IEEE Transactions on Information Forensics and Security*, 12(8): 1816–1829, 2017.

[31] Yu Lu, Shanjuan Xie, and Dong Sun Park. Finger vein identification using polydirectional local line binary pattern. In *International Conference on ICT Convergence*, pages 61–65, IEEE, 2013.

[32] Xiaoming Xi, Lu Yang, and Yilong Yin. Learning discriminative binary codes for finger vein recognition. *Pattern Recognition*, 66:26–33, 2017.

[33] Gongping Yang, Rongyang Xiao, Yilong Yin, and Lu Yang. Finger vein recognition based on personalized weight maps. *Sensors*, 13(9):12093–12112, 2013.

[34] Bang Chao Liu, Shan Juan Xie, and Dong Sun Park. Finger vein recognition using optimal partitioning uniform rotation invariant LBP descriptor. *Journal of Electrical and Computer Engineering*, 2016:1–10, 2016.

[35] Hyeon Chang Lee, Byung Jun Kang, Eui Chul Lee, and Kang Ryoung Park. Finger vein recognition using weighted local binary pattern code based on a support vector machine. *Journal of Zhejiang University Science C*, 11(7):514–524, 2010.

[36] Cheng-Bo Yu, Hua-Feng Qin, Yan-Zhe Cui, and Xiao-Qian Hu. Finger-vein image recognition combining modified Hausdorff distance with minutiae feature matching. *Interdisciplinary Sciences: Computational Life Sciences*, 1(4):280–289, 2009.

[37] Kejun Wang and Hun Ma. A finger vein recognition method using improved oriented filter and modified Hausdorff distance. *Journal of Computer-Aided Design and Computer Graphics*, 23(3):385–391, 2011.

[38] Shaohua Pang, Yilong Yin, Gongping Yang, and Yanan Li. Rotation invariant finger vein recognition. In *Chinese Conference on Biometric Recognition*, pages 151–156. Springer, 2012.

[39] Fei Liu, Gongping Yang, Yilong Yin, and Shuaiqiang Wang. Singular value decomposition based minutiae matching method for finger vein recognition. *Neurocomputing*, 145:75–89, 2014.

[40] Yu-Chun Cheng, Huan Chen, and Bo-Chao Cheng. Special point representations for reducing data space requirements of finger-vein recognition applications. *Multimedia Tools and Applications*, 76(9): 2016.

[41] Jian-Da Wu and Chiung-Tsiung Liu. Finger-vein pattern identification using principal component analysis and the neural network technique. *Expert Systems with Applications*, 38(5):5423–5427, 2011.

[42] Jian-Da Wu and Chiung-Tsiung Liu. Finger-vein pattern identification using SVM and neural network technique. *Expert Systems with Applications*, 38(11):14284–14289, 2011.

[43] Shirong Qiu, Yaqin Liu, Yujia Zhou, Jing Huang, and Yixiao Nie. Finger-vein recognition based on dual-sliding window localization and pseudo-elliptical transformer. *Expert Systems with Applications*, 64:618–632, 2016.

[44] Gongping Yang, Xiaoming Xi, and Yilong Yin. Finger vein recognition based on $(2D)^2$ PCA and metric learning. *BioMed Research International*, 2012:1–9, 2012.

[45] Zhi Liu, Yilong Yin, Hongjun Wang, Shangling Song, and Qingli Li. Finger vein recognition with manifold learning. *Journal of Network and Computer Applications*, 33(3):275–282, 2010.

[46] Fei Liu, Yilong Yin, Gongping Yang, Lumei Dong, and Xiaoming Xi. Finger vein recognition with superpixel-based features. In *2014 IEEE International Joint Conference on Biometrics (IJCB)*, pages 1–8. IEEE, 2014.

[47] Lumei Dong, Gongping Yang, Yilong Yin, Fei Liu, and Xiaoming Xi. Finger vein verification based on a personalized best patches map. In *2014 IEEE International Joint Conference on Biometrics (IJCB)*, pages 1–8. IEEE, 2014.

[48] Xiaoming Xi, Gongping Yang, Yilong Yin, and Lu Yang. Finger vein recognition based on the hyperinformation feature. *Optical Engineering*, 53(1):013108-1–013108-10, 2014.

[49] Jinjun Wang, Jing Xiao, Weiyao Lin, and Chuanfei Luo. Discriminative and generative vocabulary tree: with application to vein image authentication and recognition. *Image and Vision Computing*, 34:51–62, 2015.

[50] Tong Liu, Jianbin Xie, Wei Yan, Peiqin Li, and Huanzhang Lu. Finger-vein recognition with modified binary tree model. *Neural Computing and Applications*, 26(4):969–977, 2015.

[51] Jinfeng Yang and Yongshuang Jia. A method of multispectral finger-vein image fusion. In *2012 IEEE 11th International Conference on Signal Processing (ICSP)*, volume 1, pages 753–756. IEEE, 2012.

[52] Jinfeng Yang and Xu Zhang. Feature-level fusion of global and local features for finger-vein recognition. In *IEEE 10th International Conference on Signal Processing Proceedings*, pages 1702–1705. IEEE, 2010.

[53] Puneet Gupta and Phalguni Gupta. An accurate finger vein based verification system. *Digital Signal Processing*, 38:43–52, 2015.

[54] Kang Ryoung Park. Finger vein recognition by combining global and local features based on SVM. *Computing and Informatics*, 30(2):295–309, 2012.

[55] Fei Liu, Gongping Yang, Yilong Yin, and Xiaoming Xi. Finger-vein recognition based on fusion of pixel level feature and super-pixel level feature. In *Biometric Recognition*, pages 274–281. Springer, 2013.

[56] Jinfeng Yang, Yihua Shi, Jinli Yang, and Lihui Jiang. A novel finger-vein recognition method with feature combination. In *2009 16th IEEE International Conference on Image Processing (ICIP)*, pages 2709–2712. IEEE, 2009.

[57] Xiaoming Xi, Gongping Yang, Yilong Yin, and Xianjing Meng. Finger vein recognition with personalized feature selection. *Sensors*, 13(9):11243–11259, 2013.

[58] Beyun J Kang and Kang R Park. Multimodal biometric method based on vein and geometry of a single finger. *IET Computer Vision*, 4(3):209–217, 2010.

[59] Xiaoming Xi, Yilong Yin, Gongping Yang, and Xianjing Meng. Personalized fusion method based on finger vein and finger contour. *Journal of Computer Research and Development*, 50(9):1914–1923, 2013.

[60] Jinfeng Yang and Xu Zhang. Feature-level fusion of fingerprint and finger-vein for personal identification. *Pattern Recognition Letters*, 33(5):623–628, 2012.

[61] Wenming Yang, Xiaola Huang, and Qingmin Liao. Fusion of finger vein and finger dorsal texture for personal identification based on comparative competitive coding. In *2012 19th IEEE International Conference on Image Processing*, pages 1141–1144. IEEE, 2012.

[62] Wenxiong Kang, Xiaopeng Chen, and Qiuxia Wu. The biometric recognition on contactless multi-spectrum finger images. *Infrared Physics & Technology*, 68:19–27, 2015.

[63] Jinfeng Yang, Zhen Zhong, Guimin Jia, and Yanan Li. Spatial circular granulation method based on multimodal finger feature. *Journal of Electrical and Computer Engineering*, 2016:1–7, 2016.

[64] Mingxing He, Shi-Jinn Horng, Pingzhi Fan, *et al*. Performance evaluation of score level fusion in multimodal biometric systems. *Pattern Recognition*, 43(5):1789–1800, 2010.

[65] Jialiang Peng, Qiong Li, Ahmed A Abd El-Latif, and Xiamu Niu. Linear discriminant multi-set canonical correlations analysis (LDMCCA): an efficient approach for feature fusion of finger biometrics. *Multimedia Tools and Applications*, 74(13):4469–4486, 2015.

[66] Lu Yang, Gongping Yang, Yilong Yin, and Rongyang Xiao. Finger vein image quality evaluation using support vector machines. *Optical Engineering*, 52(2): 027003-1–027003-9, 2013.

[67] Lizhen Zhou, Gongping Yang, Lu Yang, Yilong Yin, and Ying Li. Finger vein image quality evaluation based on support vector regression. *International Journal of Signal Processing, Image Processing and Pattern Recognition*, 8(8):211–222, 2015.

[68] Huafeng Qin and Mounim A El Yacoubi. Deep representation for finger-vein image quality assessment. *IEEE Transactions on Circuits and Systems for Video Technology*, 2017.

[69] Eui Chul Lee and Kang Ryoung Park. Restoration method of skin scattering blurred vein image for finger vein recognition. *Electronics Letters*, 45(21):1, 2009.

[70] Eui Chul Lee and Kang Ryoung Park. Image restoration of skin scattering and optical blurring for finger vein recognition. *Optics and Lasers in Engineering*, 49(7):816–828, 2011.

[71] Jinfeng Yang, Ben Zhang, and Yihua Shi. Scattering removal for finger-vein image restoration. *Sensors*, 12(3):3627–3640, 2012.

[72] Jinfeng Yang and Yihua Shi. Towards finger-vein image restoration and enhancement for finger-vein recognition. *Information Sciences*, 268:33–52, 2014.

[73] Kwang Yong Shin, Young Ho Park, Dat Tien Nguyen, and Kang Ryoung Park. Finger-vein image enhancement using a fuzzy-based fusion method with Gabor and Retinex filtering. *Sensors*, 14(2):3095–3129, 2014.

[74] Shan Juan Xie, Yu Lu, Sook Yoon, Jucheng Yang, and Dong Sun Park. Intensity variation normalization for finger vein recognition using guided filter based singe scale Retinex. *Sensors*, 15(7):17089–17105, 2015.

[75] Eui Chul Lee, Hyeon Chang Lee, and Kang Ryoung Park. Finger vein recognition using minutia-based alignment and local binary pattern-based feature extraction. *International Journal of Imaging Systems and Technology*, 19(3):179–186, 2009.

[76] Yusuke Matsuda, Naoto Miura, Akio Nagasaka, Harumi Kiyomizu, and Takafumi Miyatake. Finger-vein authentication based on deformation-tolerant feature-point matching. *Machine Vision and Applications*, 27(2): 237–250, 2016.

[77] Yang Wenming, Ma Guoli, and Li Weifeng. Finger vein verification based on neighbor pattern coding. *IEICE Transactions on Information and Systems*, 96(5):1227–1229, 2013.

[78] Di Cao, Jinfeng Yang, Yihua Shi, and Chenghua Xu. Structure feature extraction for finger-vein recognition. In *2013 2nd IAPR Asian Conference on Pattern Recognition*, pages 567–571. IEEE, 2013.

[79] Li Peiqin, Xie Jianbin, Liu Tong, and Yan Wei. Finger vein recognition algorithm based on optimized GHT. *Optik—International Journal for Light and Electron Optics*, 125(6):1780–1783, 2014.

[80] Munalih Ahmad Syarif, Thian Song Ong, Andrew BJ Teoh, and Connie Tee. Enhanced maximum curvature descriptors for finger vein verification. *Multimedia Tools and Applications*, 76(5):1–29, 2016.

Chapter 12

Hand shape recognition and palm-print recognition using 2D and 3D features

Michal Dvořák[1] and Martin Drahanský[1]

12.1 Introduction

Many security systems today use various biometrics to identify or verify identity of a given person; among most used ones, we often name fingerprint recognition, facial markers, iris or retina scans. These systems, however, are not applicable in every situation either due to environmental conditions, costs or other requirements arising from place or userbase. In these cases, other methods may be required. Hand geometry represents an alternative approach to human identification and verification. Due to its acquisition technology, it may be used even in areas where other technologies fail. Such as dust-heavy environments or the clean rooms that require glove use. Due to lower technological requirements, it may also prove to be a more affordable solution. Palm-print recognition on the other hand offers a technology that is more preferable to users, as it does not have the bias the fingerprint technology has despite the entropy level of extracted features being similar.

The principle and basis for hand-based geometry and palm-print-based recognition utilizes the assumption that every individual possesses a unique hand, which upon analysis can be used to positively identify the user. In other words, if a reference shape of user's hand is on record, this person's identity can be established by comparing his immediate hand shape or features of his palm-print with the one's on the record. It was this assumption that gave birth to the first systems designed to do so.

Identification of the characteristics that can be efficiently extracted without compromising the uniqueness requirement is the first necessary step, for chosen biometrics. 2D hand geometry and palm-print biometrics serve of an abstraction of a model of human hand and define the array of features to be extracted. For hand geometry, the features such as silhouette or various geometrical measurements, of width and/or thickness, are recorded [1], whereas in palm-print recognition friction ridges (papillary lines) found on palm are used.

[1]Faculty of Information Technology, Centre of Excellence IT4Innovations, Brno University of Technology, Czech Republic

To increase the possible database sizes, and to improve security, the technology around these biometrics is still being improved on. One such direction of research and development is a modification of system from purely 2D acquisition and recognition to 3D. In order to expand into a 3D system, a measurement of depth must be included, this then in turn allows parameters such as curvatures of individual fingers and 3D position of palm-print features to be measured and recorded.

12.2 2D hand geometry acquisition

2D hand geometry biometric systems rank among oldest foray into identifications based on human measurements. Due to robustness and space for further development, the systems based on this technology can be found in many industrial productions and are still being developed. To understand the issues and advantages of hand biometrics, let's first introduce the process of 2D acquisition.

Since for the identification only a silhouette of a hand is needed, the process of acquisition can be very simple. For this task, all that is required is camera sensor with adequate resolution. While for fingerprint sensor this resolution needs to be high due to the dimensions of papillary lines, e.g. 500 DPI, the hand geometry systems require much lower resolution; in commercial systems, 100–200 DPI sensors are used, but it has been demonstrated that even resolution as low as 100 DPI can be sufficient [2].

For 2D acquisition, the back of a hand is usually captured on an image; from this image, chosen features are then extracted. Some systems utilize multiple sensors to acquire images from different angles and thus making it possible to extract additional features [1]. As can be seen on Figure 12.1, majority of the features are easily extractable when the camera is placed directly above the scanned hand.

The acquisition process may be further enhanced by performing material modification to the device. Such as limiting the entrance of daylight, using artificial illumination and adding nonreflective background to limit the impact of shadows, all these modifications will make the preprocessing simpler. Figure 12.4 shows

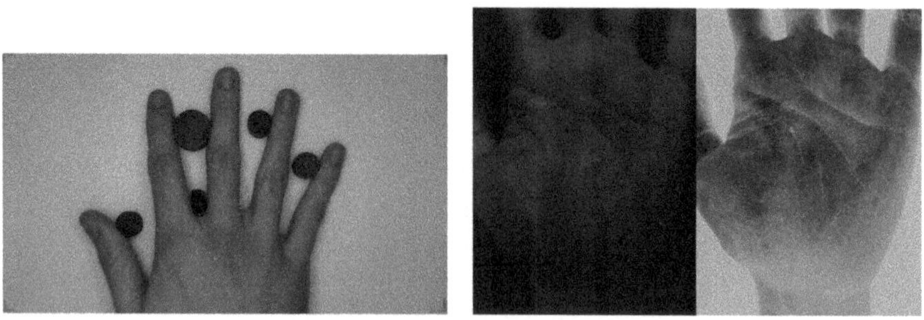

(a) (b)

Figure 12.1 (a) Gathered image of a hand for 2D hand geometry extraction of features and (b) image of acquired palm (left) and after enhancement (right) possible set of features that can be extracted from the top view of the hand. All commercially available solutions today are built around this approach.

12.3 Palm-print acquisition

Palm-print acquisition can be realized using any technology for fingerprint acquisition. Due to advantages offered and the cost effectiveness, the optical approach is now used almost exclusively to the knowledge of the authors.

In the optical approach, a camera of sufficient parameters is used to record an image. The parameters of this camera need to have sufficient resolution, to extract the papillary lines. While the sensor of 500 DPI is recommended, the mobile phone cameras with resolution as low as 640 × 480 pixel have been demonstrated to be sufficient for this acquisition [3].

12.4 Existing commercial solutions for hand geometry biometrics

To the best knowledge of the authors, there are only two commercially available systems being currently developed, both by SCHLAGE. Line intended for use as a security admittance mechanism HandPunch® and line for time logging mechanism HandKey®.

At the time of writing this chapter, the newest model of the HandKey series was HandKey II [4]. HandPunch GT-400 [5] and HandPunch 4000 then belong to the HandPunch line [6]. HandKey II, as can be seen on Figure 12.2, is primarily

Figure 12.2 Image of HandKey II biometrics hand geometry reader [4]

Figure 12.3 Image of HandPunch GT-400 biometrics hand geometry reader [5]

designed as an access device; it is of sturdy construction and offers security features such as duress code. The base user memory of 512 user fields can be further expanded. As per available information, the hand geometry features are stored as a 9-byte large compressed template.

The HandPunch series is primarily designed as an attendance and time logging terminal. HandPunch 4000 also utilizes 9-byte large template, with default database size of 530 users, which can be expanded to 3,498. The HandPunch GT line is also intended as an attendance terminal. The main difference is in the size of the saved template, which is increased to 20 bytes from original 9. The device does not offer a database expansion, but the default size of 1,000 users is larger than the base model of 4000 line. On Figure 12.3, the appearance of HandPunch device can be seen.

12.5 Commercially available solution for palm-print biometrics

To the best knowledge of the authors, there is currently no ready-to-use solution commercially available on the market.

There is, however, a software solution available from Redrock Biometrics—PalmID—that offers software development kit (SDK) for all major platforms [7].

12.6 Identification methods of hand geometry biometrics

Unlike fingerprint-based biometrics, the hand geometry uses multiple different approaches to recognition. These approaches need to consider the fact that there are multiple joints inside ones hand, which allows for different position of individual

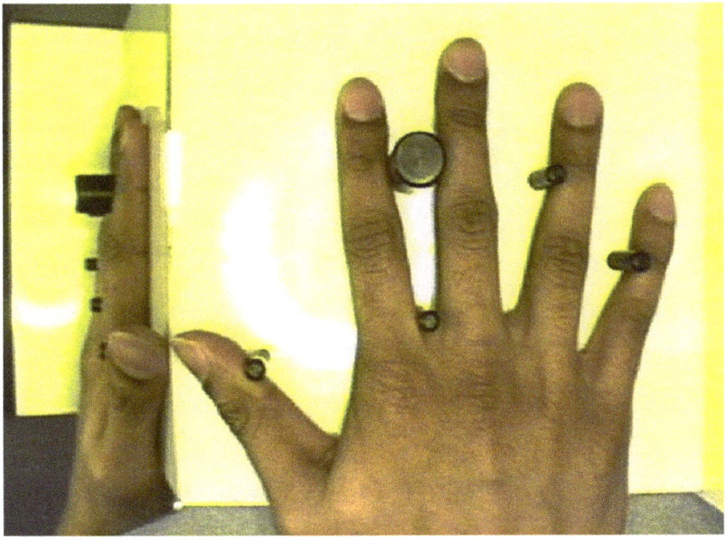

Figure 12.4 Viable hand 2D hand geometry features [2]

fingers and their phalanxes on every image capture; this combined with fact that different pressure, placement or even physical state of the hand will make each reading unique and chosen method needs to take this under consideration. In general, we can split the today's methods in two categories: methods based on direct measurements of various hand parts and hand shape alignment comparison.

12.6.1 Methods based on direct measurement

This method compiles a vector of dimensions of individual fingers and other parts of hand gained directly from the acquired image of a person's hand. The example on Figure 12.4 shows configuration where 14 features are being inferred from a downward facing charge coupled device (CCD) sensor. Side facing camera is used to collect two additional dimensions. To the best knowledge of the authors, all commercially available devices today use only the top-view camera.

The extraction of features can be then done either by thresholding the image and gaining the dimensions straight from the binary image, alternatively, array of pixel values is taken from an unprocessed image, and the features are determined individually. The second approach may prove advantageous, especially in case of localized skin discolorations, which can be incorrectly processed, if global or even local transformations are used. It is worth noting that if utilizing 2D imaging, increasing the number of features above certain threshold adds very little additional useful information for purposes of recognition (Figure 12.5).

During enrollment, one or multiple measurements are performed and saved as a model vector, which we then use during identification/verification phase where the model vector is matched against the newly acquired vector. Difference can be calculated using any metrics such as sum of absolute difference (1), weighted sum of

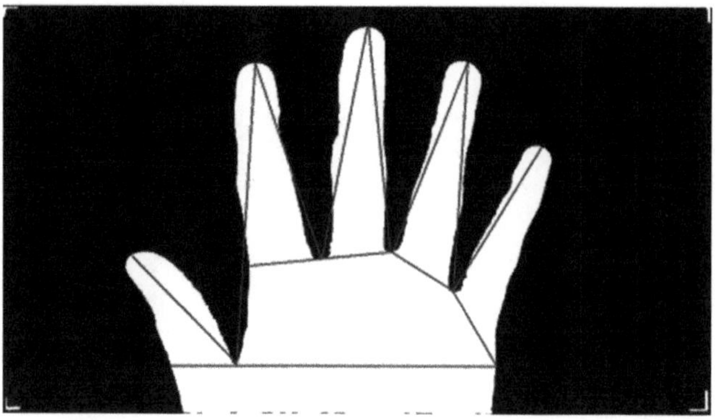

Figure 12.5 Other possible set of extractable features

absolute difference (2), Euclidian distance (3) or weighted Euclidian distance (4) [1]. The measured features $x_i = [x_1, x_2, \ldots, x_z]$ are then compared to template features $y_i = [y_1, y_2, \ldots, y_z]$ with σ_i being the weight of a given feature. While the usage of weights is not required, given the differences in absolute sizes of these feature, and difference in mobility of individual segments, it has proven to be beneficial.

$$\sum_{i=1}^{z} |x_i - y_i| \tag{12.1}$$

$$\sum_{i=1}^{z} \frac{|x_i - y_i|}{\sigma_i} \tag{12.2}$$

$$\sqrt{\sum_{i=1}^{z} (x_i - y_i)^2} \tag{12.3}$$

$$\sqrt{\sum_{i=1}^{z} \frac{(x_i - y_i)^2}{\sigma_i^2}} \tag{12.4}$$

When using this method, the device usually includes positioning pins. These pins guide the hand into a position and simplify the feature extraction, as the position will be, to an extent, guaranteed. If we attempt to locate chosen features, on an image of an unfixed hand, the pixel ranges for acquiring these features would need to be larger or dynamic, and in some cases, the uncertainty of features would increase as additional information would be determined from the image, such as knuckle and joints positions and then used for feature extraction.

The pins however prohibit the system from being touchless and dynamic, as the hand needs to come in contact with the pins and stay still for the duration of acquisition. This may be problematic for hygienic and medical reasons.

Alternative approach that allows unguided hand position is often based on silhouette matching.

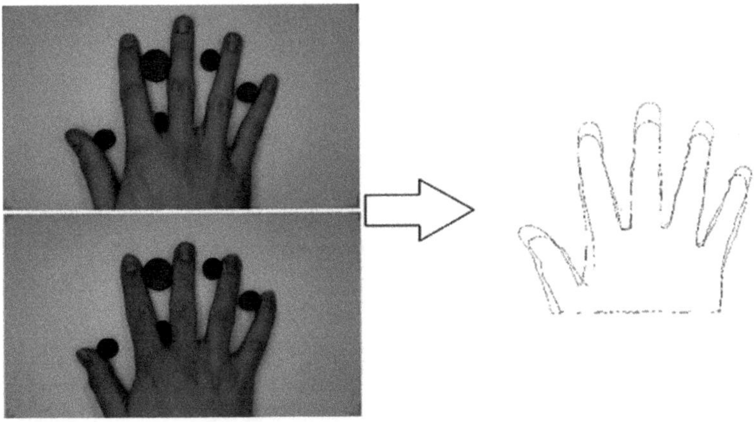

Figure 12.6 Hand image used as a source and other hand image (left), outline overlap after finger positioning (right)

12.6.2 Methods based on shape matching

Unlike the direct measurement approach, this method utilizes whole outline of the hand. The template is an outline itself or a normalized set of information that may be used for reconstruction.

The following steps describe the general approach to this method:

1. *Preprocessing*. In this step, the input image is normalized, if necessary. Noise removing filters may be applied.
2. *Outline extraction*. If the device is well designed, only thresholding is needed for object separation and then applying morphological operations to gain an outline of the hand.
3. *Extraction of individual fingers*. To compensate for variable finger orientation, the outlines of individual fingers need to be extracted. The tips and valleys of the fingers can be searched for, and the range of pixels between them is considered the finger. The orientation normalization can be performed here to prepare for the next step.
4. *Perform alignment operations with the template*. To do this, any algorithms that measures distance of two metric subspaces can be used, for example Modified Hausdorff Distance [8], where the orientation and position of one outline is gradually changed to achieve closest alignment with the template.
5. *Calculate the distance of the pair*. Mean alignment error can be calculated, as an average distance difference between corresponding points of the outlines.
6. If *mean alignment error* (MEA) is less then threshold, it is declared that the measured hand is of the same person as the one saved in template.

Figure 12.6 shows the principle of the comparison outline extracted from two hands: one during the enrollment process and other during the identification process. The template of the first is stored. During the identification, we prepare a maximum possible overlap of individual fingers and calculate the MEA. Based on the resulting score, we decide whether the hand belongs to the same user. On Figure 12.6 (right), we can see that while the thickness of fingers displays a high degree of similarity, the length is different. Our threshold value should determine that these two hands are not of the same individual.

Both of the described methods serve the purpose they are designed for, both, however, are plagued by limitations we will look at now.

12.7 Identification methods of palm-print biometrics

Palm-print recognition is very similar to fingerprint recognition, in that it uses the friction ridges for feature extraction. The larger total area can be seen both as an advantage and a disadvantage. It does offer greater number of identifiable features, which makes it potentially safer; at the same time, the device needs to be larger and is often more expensive, hindering the adoption rate. The features we can extract depend on the design of the device, lowest resolution, can only record the ridge flow and general patterns, also known as Level 1 details. By increasing resolution, smaller details can be used for features so-called Level 2 details, such as ridge ending, bifurcation, dot, island, lake, hook and variations. The further increase in resolution allows for detecting path variation of ridges width and pores, so-called Level 3 details.

The detail level we acquire during the acquisition determines the palm matching method we use. For Level 1 details, simple template matching or correlation method will be used.

Minutiae matching uses the Level 2 details; the orientation and position of detected occurrences is noted and during verification; the matching score is determined by comparing the recoded features with the newly acquired. If the match rate is high enough based on predetermined score system, the hand is identified as a match. This is the currently most widely used method. Level 3 details can be used in ridge-based matching, but is frequently used as complimentary source of features to the minutiae matching and for liveness detection.

12.8 Limitations of currently employed 2D-based biometrics

While the lower requirements on the resolution of the sensor can make the system more cost effective and research shows that the resolution required for hand geometry can be further decreased [9], it also is a cause of several short comings that need to be addressed. We can split them into those that can be solved or mitigated by modifying or improving on existing concept, and those that originate from the 2D acquisition itself, which makes them unsolvable without changing the basic concept.

The database size is a large issue for the hand geometry-based technologies, as we can see from examination of commercially available solutions such as HandKey II [4]; the focus on increasing viable database size is necessary for further spread of this technology. As has been discussed earlier in the chapter, increasing the number of features extracted from the 2D images does not yield an adequate increase in the uniqueness. However, it has been demonstrated that, by including the 3D features of the hand, the entropy is increased, which in turn allows for larger user databases.

Just as the requirements for resolution are low for these biometrics, so is the difficulty of developing a spoof that can be used to bypass this technology. The approaches to mitigating this risk will be elaborated on here. The palm-print recognition faces the similar issue, especially if the simple optical approach is used.

To increase the security of these systems, two approaches are being utilized and further researched. The first approach is to expand the unimodal system into multimodal system, where instead of recognition based solely on one biometrics, we use one or more of others, alternatively we use these additional biometrics simply as means of liveness detection and focus solely on detecting their presence.

The liveness detection aims at measuring parameters, we expect to be present in a human hand, but not in a spoof. In case the system is developed as touch based, we commonly encounter checks based on electric properties, such as conductivity and capacity. Among other used, we can encounter systems detecting heat, utilizing skin spectroscopy, odor analysis and high-resolution skin analysis to name a few.

For the system to remain touchless, the vein detection can be utilized. The vein detection can either be performed as a surface detection, whereby utilizing a near-infrared light source (NIR) and performing an acquisition with a sensor without an IR filter, i.e. see Figure 12.7. We can detect veins on palm or dorsal part of hand as the optical absorption of NIR by hemoglobin is higher than that of other tissue and will appear darker on the resulting image.

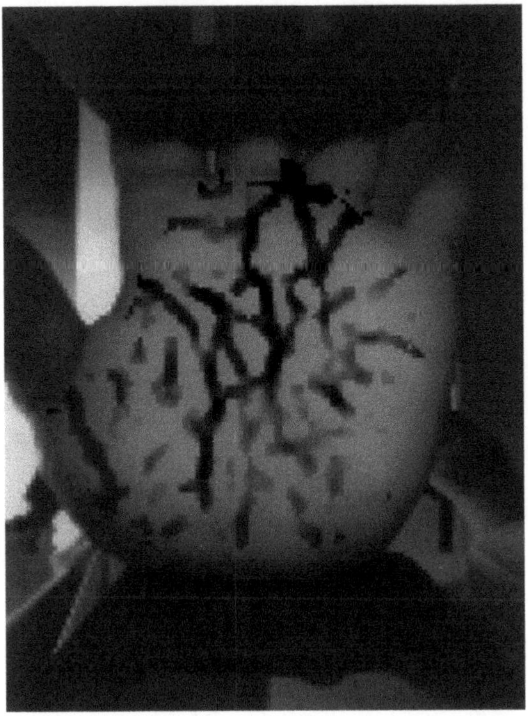

Figure 12.7 Surface vein detection using NIR

Instead of capturing the reflection, system can instead capture the dispersed light passing through the hand as in Figure 12.8. Due to the hand structure, systems using this approach can detect the passing light on finger area only; otherwise, a very strong light source would have to be used. As with the reflection approach, the NIR absorption by hemoglobin is used and the veins on the resultant image can be detected. Both these methods can be easily implemented into biometric system, as the NIR can be used as the sole source of the artificial lighting, and both CCD and complementary metal-oxide-semiconductor (CMOS) sensors have high sensitivity in the infared (IR) range of the electromagnetic spectrum. As stated earlier, the vein detection can be used solely as a liveness detection or a source of additional unique features.

The other approach is to record the 3D as well as 2D features during enrollment and verification.

Both the transition from 2D to 3D and the multimodal approach can serve as a spoof protection, as the data required for construction of spoof are both less accessible and make the model reconstruction more complex. To the knowledge of the authors, no study has demonstrated a 3D hand model or hand geometry with bloodstream model reconstruction from an uncooperative user with accuracy high enough to create a viable spoof.

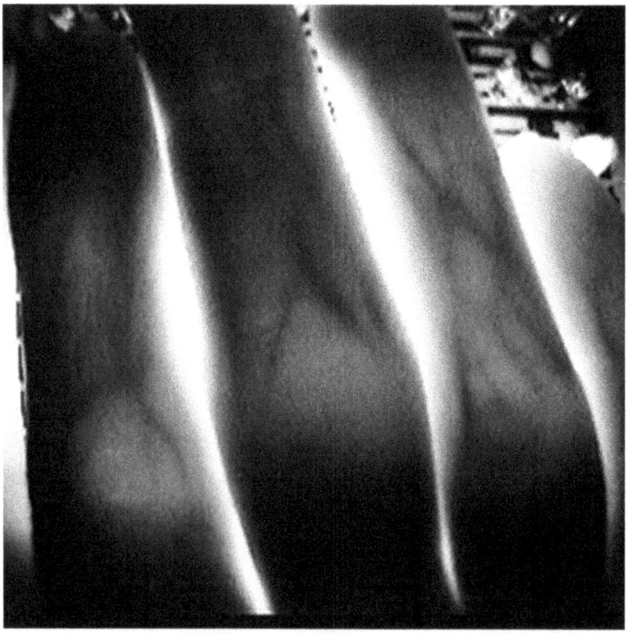

Figure 12.8 Vein detection using NIR light source behind object

As will be further discussed in the chapter, the transition from 2D to 3D biometrics has been experimentally accomplished and supports the assumption that 3D features will lead to increased entropy of the model.

To the extent of knowledge of authors, there are no existing pinless and, by extension, contactless commercial devices for hand geometry recognition now available on the market. While it has been demonstrated that pure 2D approach can be used to remove the pin requirement for hand geometry recognition, the chapter presented one of such approaches. There is research into utilization of depth sensing devices as a method of tackling this problem [10] as well as the need for the physical contact with the sensing device.

12.9 3D acquisition of hand biometrics

The principle of 3D acquisition in biometrics has been successfully tested and researched. However, the price of hardware and high computational demands for processing this data limited the application to academic research. With the advance of low-cost 3D acquisition devices, and increasing computational power of even the mobile devices, the 3D acquisition has become a viable alternative to 2D-based systems. For hand geometry, the expansion into 3D offers additional source and type of features that can be extracted. For palm-print recognition, the advantage lays in the feature description. Whereas in 2D-based systems, the features location was defined using only two coordinates, the extension into 3D allows for three coordinates description, which serves both as a spoof protection and source of increased entropy.

Creative VF0800 low-cost camera [11] has been used to acquire the image at Figure 12.9. As we can observe that the depth map not only offers data for 3D feature acquisition, but due to the nature of the image, it allows for extraction of 2D features and even assists at the segmentation stages of the analysis, as the background is separated more clearly than on the illumination map. At the same time, the depth map allows us to determine the absolute dimensions of the recorded object, thus allowing for the freedom of position within the frame without losing any information regarding size.

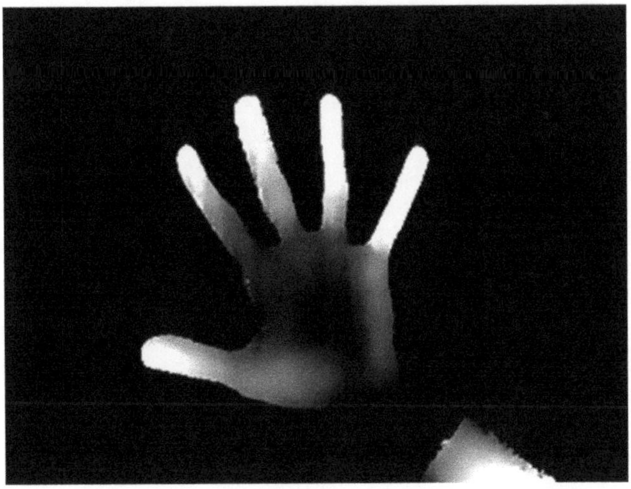

Figure 12.9 3D image of hand using low cost 3D camera

While any 3D mapping method may be used, it can be clearly observed that in current research, most utilized method is that of active triangulation. Whether the 3D laser digitizers are used for this purpose [12,13], which offer superior accuracy, or systems based on structured light projection [14] that are now being mass produced and are therefore affordable.

12.9.1 Using 3D scanner

In [12], high-accuracy 3D scanner is employed, and in this research, Minolta Vivid 910 has been used for compilation of a large database of right-hand scans (3,540 items). Per documentation, the device offers an accuracy up to 0.22 mm in x-axis, 0.16 mm in y-axis and 0.10 mm in z-axis [15].

Research done in [12] performs a detection of four fingers on intensity map; this information is then used for extraction of corresponding data from the depth map. Each finger is then split into chosen amount of segments along the length of the finger. At these points, the features such as curvature and normals along the finger are calculated based on the profile of the cross-section. Figure 12.10 shows a depth profile of a cross section, calculated curvature and normal features.

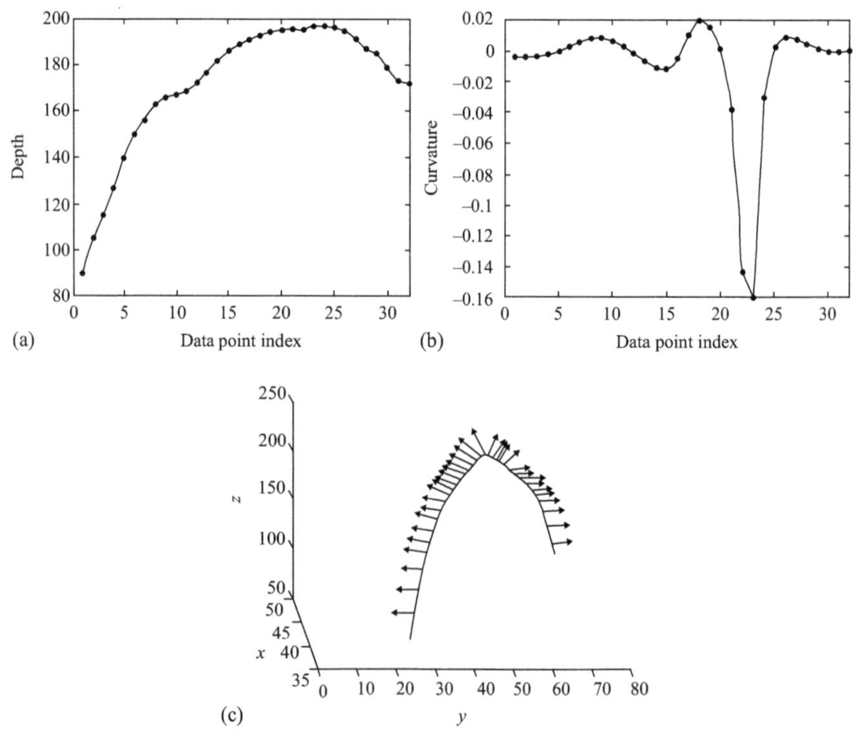

Figure 12.10 (a) Cross-sectional finger segment, (b) its computed curvature features and (c) normal features computed for a finger segment [13]

Table 12.1 EER and AUC of 2D, 3D and a combined matcher [13]

Matcher	EER (%)	AUC
3D hand geometry	3.5	0.9655
2D hand geometry	6.3	0.9722
(2D + 3D) hand geometry	2.3	0.9888

In the paper, the 2D and 3D data are experimentally matched and EER (Equal Error Rate) and AUC (area under receiver operating characteristic (ROC) curve) are calculated (Table 12.1).

From the data, it can be observed that the combination of 2D and 3D features offers an improved performance.

12.9.2 Using structured light

When a light pattern of an observable frequency is projected onto a deformed surface such as human hand or palm, the pattern itself becomes apparently deformed. By detecting and measuring this deformation relative to the non-deformed pattern, a 3D mesh can be reconstructed that serves as a model of an observed object. This approach using active triangulation has been used in several publications, varying both in the type of light and the pattern used.

12.9.2.1 IR pattern projection

Utilizing IR camera and IR projector became viable, thanks to the mass production of affordable 3D sensing technologies for purposes of entertainment, such as Microsoft Kinect and Intel RealSense. Both aforementioned devices include not only the depth sensor, but also standard RGB imagining sensor, with software libraries, that map the coordinates of one onto the other.

In [13], 2D and 3D features of human hand are extracted using Intel RealSense camera. After normalizing the image during preprocessing, markers such as fingertips, finger valleys and wrist lines are located. Using these markers, feature vectors containing 41 2D features and 137 3D features are constructed.

- Finger length (2D)
- Finger valley distance (2D)
- Finger width (2D)
- Wrist-to-valley distance (2D)
- Finger axis surface distance (3D)
- Finger width (3D)

Figures 12.11 and 12.12 show the respective features on the depth images and intensity images.

The matching of the features vectors is performed using Mahalanobis distance [16], where the weights for individual features are calculated using the large-margin nearest neighbors method (LMNN) [17].

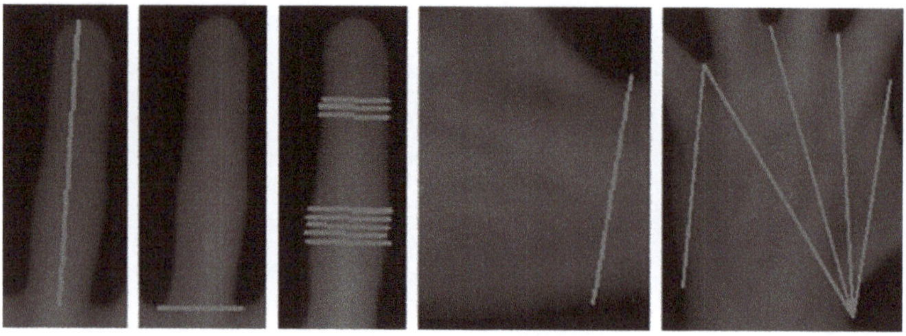

Figure 12.11 Extracted 2D features [14]

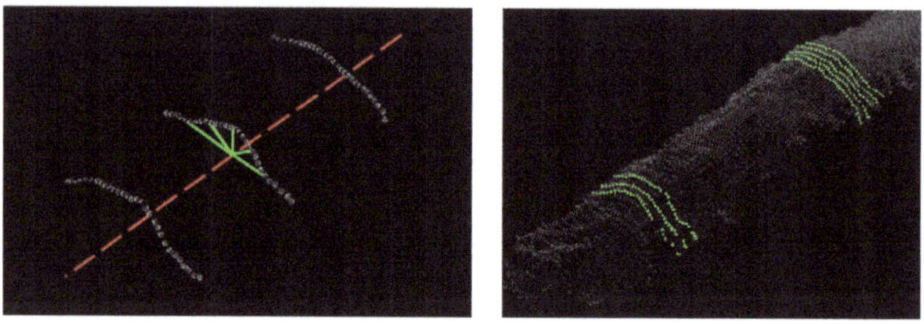

Figure 12.12 Extracted 3D features [14]

Table 12.2 Overall performance of Svoboda et al. *system using LMNN metric learning approach [14]*

Features	**ERR (%)**	**FRR (%)**	
		@FAR = 0.5%	**@FAR = 1%**
2D	2.76	6.50	1.63
3D	2.92	10.70	6.42
2D + 3D	1.61	4.81	2.23

In the paper, 2D and 3D feature vectors are calculated separately; this allowed the authors to compare both the performance of individual vectors and the combination of two. The results can be seen in Table 12.2.

Unlike in the experiment with the high-precision 3D scanner, the performance of only 3D features in recognition algorithm is inferior to the 2D hand-based

Table 12.3 Qualitative comparison of hand biometrics-based matchers [13]

Method	Features	Templates	Database	FAR (%)	FRR (%)	EER (%)
Jain and Duta	2D	1–14	53	2.00	3.50	N/A
Jain *et al.*	2D	1 (avg.)	50	2.00	15.00	N/A
Woodard and Flynn	2D + 3D	1 (avg.)	177	5.50	5.50	5.50
Malassiotis *et al.*	2D + 3D	4	73	3.60	3.60	3.60
Kumar *et al.*	2D + 3D	5	100	5.30	8.20	N/A
Kanhangad *et al.*	2D + 3D	5	177	2.60	2.60	2.60
Svoboda *et al.*	2D + 3D	1 (avg.)	88	1.61	1.61	1.61

geometry biometric. However, the performance of combined features is superior to a 2D only approach by a large margin, despite using the low-cost technology. It is also worth noting that the performance is significantly higher when one source of features is burdened by heavy noise, as one technology can be immune to source of noise, that the other is burdened by, i.e. glare in RGB sensor does not influence the depth mapping. Overall the combination of 2D and 3D feature vectors in the algorithm has demonstrated performance comparable to the state of the art approaches (Table 12.3).

12.9.2.2 Lasers and diffraction grating

While the utilization of commercial low-cost cameras has demonstrated acceptable results, the fact that these devices are designed to operate at wider range of distances and greater depth of field means that the principle can be optimized for the specific application. Since we also have the information about the range of colors we can expect from human hand, a specific wavelength can be chosen to better match the intended application.

In [18], the array of 532-nm lasers is being used, with optics that changes the dot projector into the line projector, thus producing number of parallel lines. For the image acquisition, two MS Lifecam HD 3000 camcorders are used. From the acquired images, the model of hand can be reconstructed. Figure 12.13 shows the process of segmentation and feature extraction during the experiment.

Figure 12.14 presents a model reconstructed from the acquired data, demonstrating the concept.

The paper serves as the proof of concept and viability study of the method for the purposes of 3D hand geometry scanning, as such, database has not been created and method cannot be compared with other available approaches. The accuracy of method has been demonstrated by scanning calibrated object of known dimensions. For this purpose, a cube with a side of length 25.5 mm has been used and the root-mean-squared, and normalized root-mean-squared has been calculated as can be seen in Table 12.4.

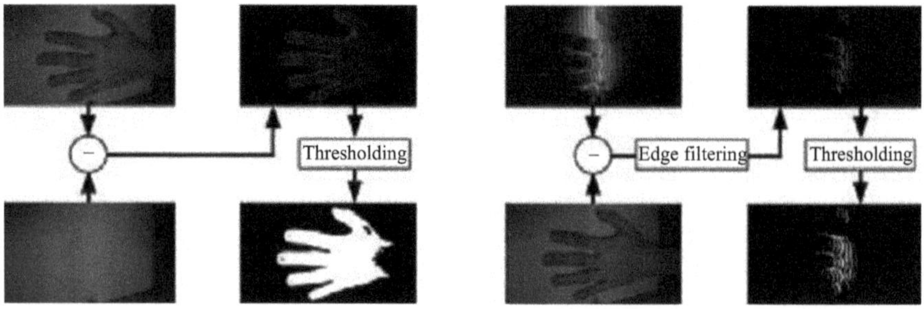

Figure 12.13 *The process of extracting the deformation markers, created by diffracting line projection [18]*

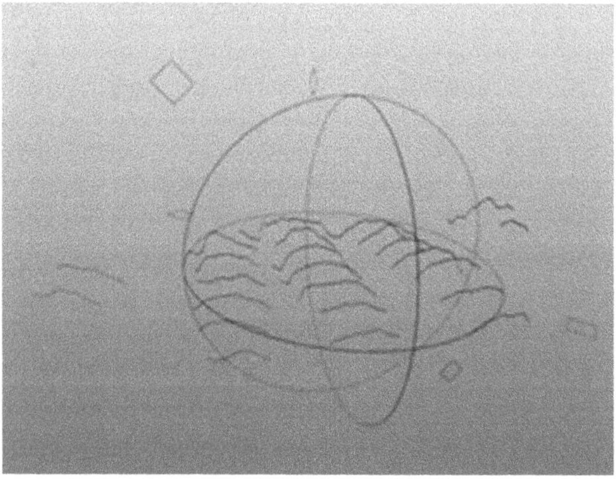

Figure 12.14 *Resultant surface reconstruction from the lines projection [18]*

Table 12.4 *Accuracy of the proposed method based on calibration [18]*

Model	NRMSE	RMSE
Laser 0	0.2190	1.4558
Laser 1	0.1993	0.7977
Laser 2	0.1764	0.4317
Merged	0.1764	1.0474

12.9.3 Time-of-flight (TOF) scanning

Time-of-flight (TOF) cameras due to the principle of their operation and the costs were for a long time more suitable for industrial applications. However, as low-cost

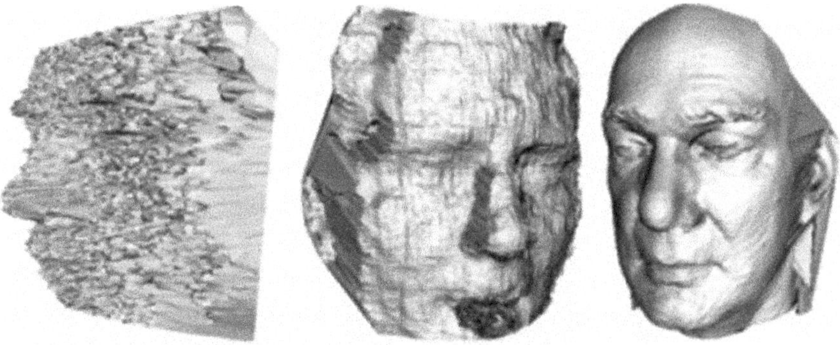

Figure 12.15 Example 3D acquisition using TOF cameras SoftKinetic (left), Kinect (middle) and professional Minolta Vivid scanner (right) [19]

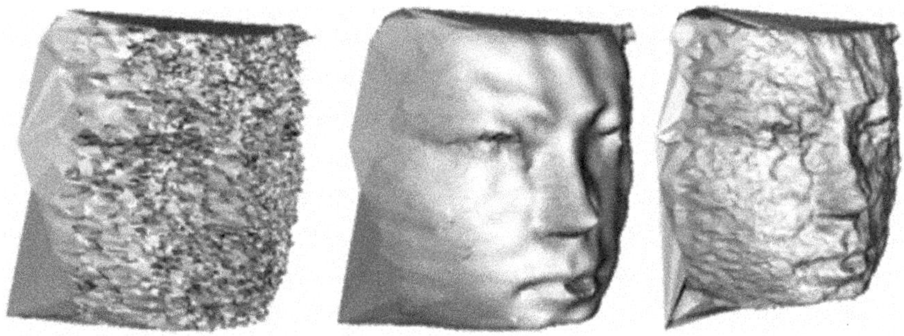

Figure 12.16 High noise DS325 image (left), after application of feature preserving mesh denoising (middle) and Gaussian smoothing (right) [19]

systems such as SoftKinetic's DS325 and Kinect v2 RGB-D have been introduced, their utility became much wider, despite the limitations.

While the research in hand biometry using the TOF has been limited to gesture detection. It has been demonstrated that the technology is viable for biometric recognition uses as well, especially with regards to face recognition. In [19], the TOF cameras have been used to capture a depth map of user's facial features and compared to results from other technologies, as can be seen on Figure 12.15. As expected, the feature details are lost to high degree of noise.

The research however demonstrated, that using feature preserving mesh denoising methods, can be used to suppress the noise and improve the signal-to-noise ratio enough, for an extraction of biometric features and can be used for biometric identification, as can be seen on Figure 12.16.

12.9.4 Stereovision approach

Stereovision is a process where, by combining information from two stationary cameras, we can determine a depth in an image via the method of passive

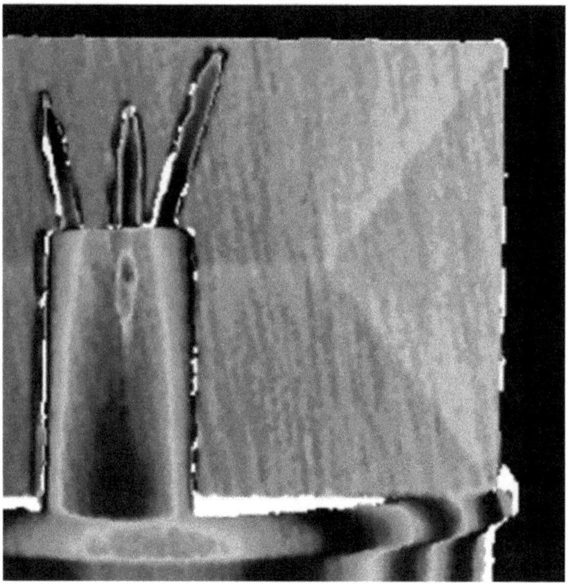

Figure 12.17 Example of 3D image created using stereovision [21]

triangulation. The passive triangulation uses the position difference of significant points on the pair of images, to determine the distance to sensors. If the resolution is high enough and therefore the number of point pairs is sufficient, the shape of the observed object, in our case the human hand, can be reconstructed.

At the moment, there are multiple commercially available 3D cameras, based on the stereoscopy principle, with resolution high enough for the biometric purposes. For example, ENSENSO N10-304-18 can per documentation achieve the resolution in z-axis of up to 0.109 mm [20]. Unlike TOF-based technology, the acquisition time can be as low as a time to capture one image and under proper conditions can offer higher precision than active triangulation via the subpixel methods. The largest source of error for this technology is a lack of detectable significant point pairs on homogenous or smooth surfaces; this however does not apply to human hand, as the papillary lines themselves offer the necessary irregularities for significant point pair identification (Figure 12.17).

12.10 Utilization of line scanners in biometric acquisition

Direction in research currently explored by the authors is the utilization of line scanners for 2D and 3D acquisition of multimodal hand biometrics, where the main advantage comes in the form of ability to performing dynamic acquisition. The principle of operation of line scanner, which records many one-line wide segments at high speed, allows for recoding high-resolution images while the scanned hand is in motion. This in turn allows for development of high-throughput biometric gates, without compromising the image quality. The high-quality image then can be used to extract additional biometrics, such as fingerprints and palm prints at the same time.

12.10.1 Camera system

The proposed parameters of this system are to allow an image acquisition of such quality, so that 2D hand geometry, fingerprints and palm prints can be extracted from acquired image. Minimum speed of the moving hand that must be supported by the solution has been decided as 0.5 m s^{-1}.

To meet these requirements, a line-scanner resolution must provide at least 500 DPI on the cross section of at least 200 mm; this translates to minimum resolution of at least 4,000 pixels. To meet the vertical resolution requirement, the sensor must have a minimum frame rate of 10 kHz.

The raL6144-16gm—Basler racer [22] provides resolution of 6,144 pixel, sufficient for the designed purpose and 17-kHz frame rate, which allows for maximum hand speed of 0.85 m s^{-1}. Secondary reason for choosing this model is the sensor size of 43 mm, allowing for utilization of full-frame optics.

AF Nikkor 50 mm f/1.8D has been chosen for the camera lens, as it has an acceptable minimum focal length of 45 cm and *f*-number that supports depth field necessary for touchless solution.

12.10.2 Proof of concept

In Figure 12.18(a), an image acquired using fixed-line scanner with a hand on a moving platform is shown. The hand on the image is moving at the speed of 0.5 m s^{-1}. Figure 12.18(b) is just an extracted area of interest from Figure 12.18(a) showing a detail of a ring finger. It can be observed that papillary lines are clearly

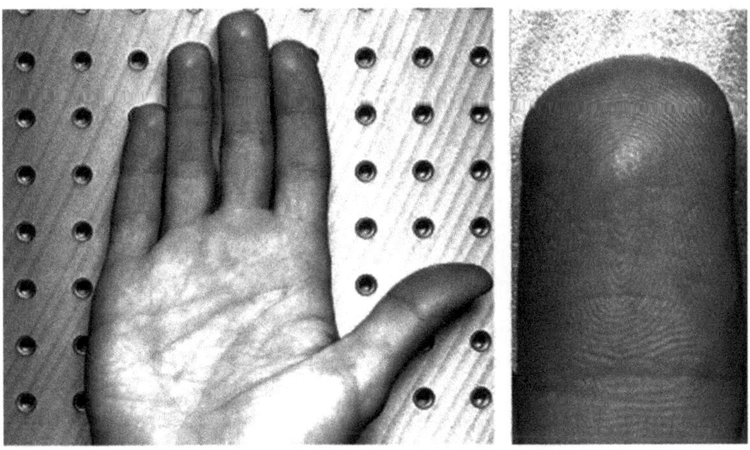

(a) (b)

Figure 12.18 (a) Palm print using line scanner mounted on moving platform and (b) detail of ring finger

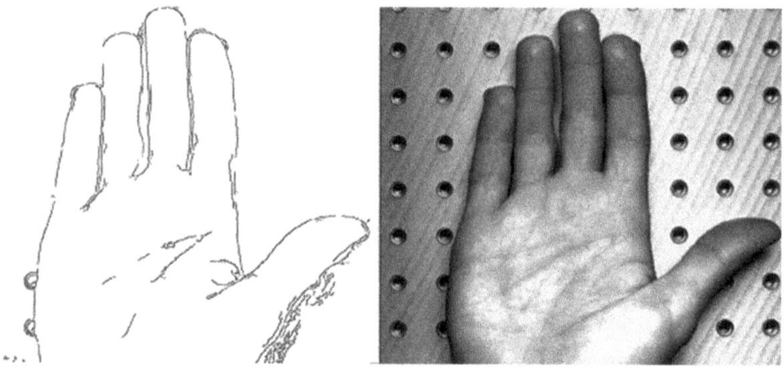

Figure 12.19 Mask filtering and edge detection used to extract outline (left) from an image of hand acquired via line-scanner (right)

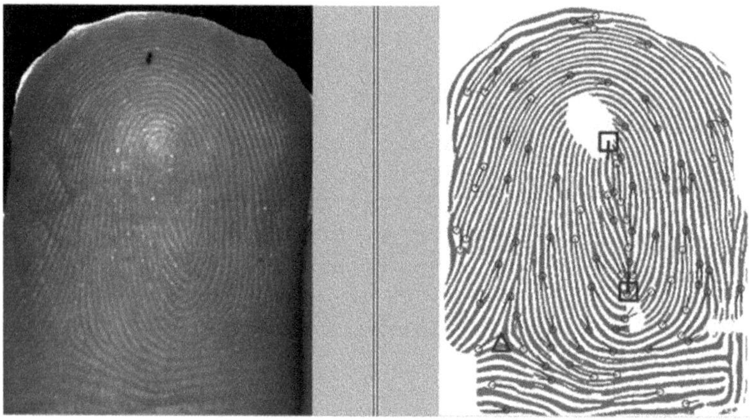

Figure 12.20 Fingerprint image using line scanner (right), extracted fingerprint using VeriFinger SDK (right)

visible, meeting the requirement for possible fingerprint feature extraction and 2D hand geometry features. The latter is demonstrated on Figure 12.19.

For fingerprint extraction, a commercial tool VeriFinger [23] has been used to verify the concept. A detail of finger has been taken from the image and processed using VeriFinger. As can be seen on Figure 12.20 VeriFinger was able to successfully perform segmentation and extract large number of unique features that can be used for identification purposes. As the setup used was designed as a proof of concept only, voids can be observed on the segmented image, where the light on the original image caused a glare. The usability of this setup for fingerprint extraction has been successfully demonstrated.

Figure 12.21 presents a detail of palm and a preprocessed image with enhanced papillary lines. While a level of noise can be noted, the features for palm-print recognition are also visible, meeting the requirements placed on the POC.

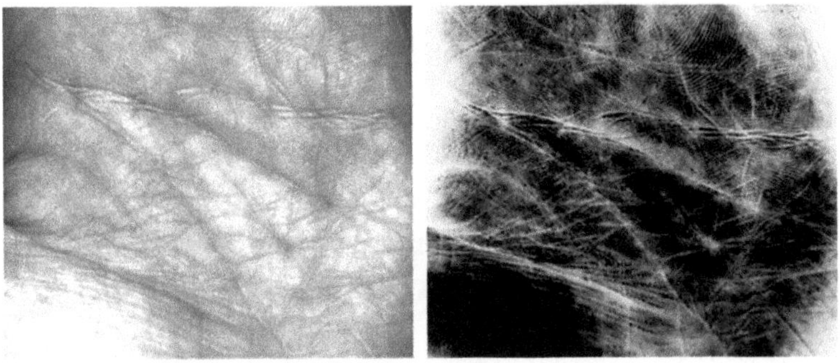

Figure 12.21 The POC palm-print extraction from the image, source image (left) and preprocessed palm print (right)

Figure 12.22 Setup of the system with free hand movement

12.10.3 Reconstruction of hand image in touchless system

As the proof of concept with the line scanner was successful, the current direction of development lies in the image acquisition of hand that performs an uncalibrated movement. Based on the proof of concept, a camera system that allows a free hand movement has been constructed. The setup is presented on Figure 12.22.

Initial experimentation with this setup has been performed and is presented in Figure 12.23; the deformations are caused by nonperfect reconstruction of acquired data; the details on Figure 12.23(a) and (b) demonstrate that the detail level has been retained for required feature extraction.

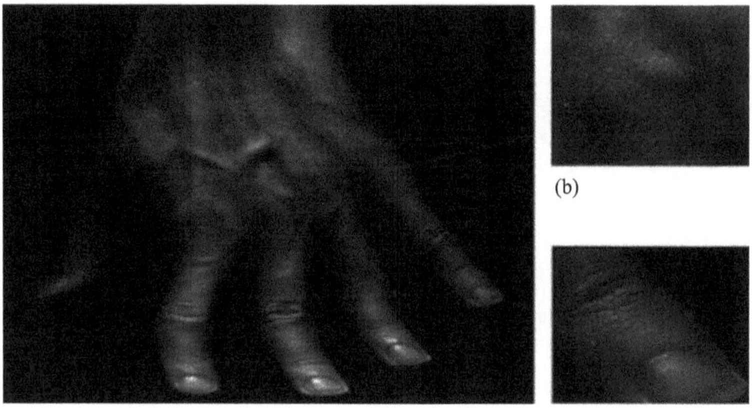

(a) (b) (c)

Figure 12.23 (a) Handprint using a fixed-line scanner scanning a moving hand; (b) detail of middle knuckle and (c) detail of little finger

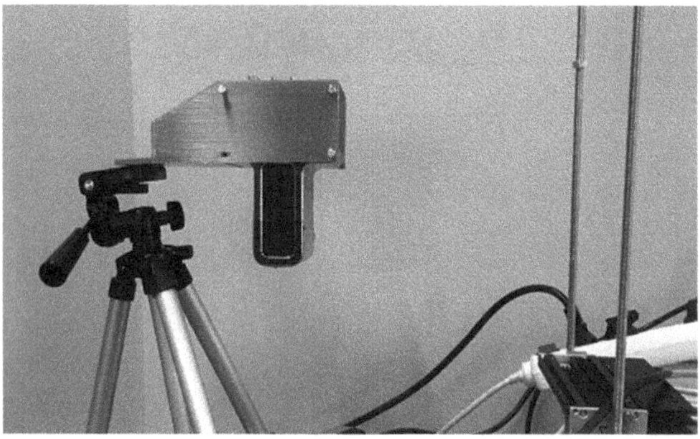

Figure 12.24 Preview of 3DPIXA setup for 3D line scan acquisition

12.10.4 Utilization of 3D line scanner in 3D biometric feature extraction

Future direction of research lies in expanding the above-described technologies into 3D line scanning. Combination of stereovision and line-scanner technology for biometric application is currently being explored and developed. For the proof of concept development, 3D PIXIA [24] 3D line scanner has been purchased and is being tested. With the resolution of up to 10 μm in *z*-axis and high frame rate of up to 21 kHz, the 3D mapping resolution is high enough to capture a high-quality model of human hand that will be usable for 3D hand geometry and palm-print extraction. As of now, only preliminary preparation has been made (Figure 12.24).

12.11 Conclusion

In this chapter, the current state of palm-print and hand-geometry-based biometrics has been presented. The most used methods of acquisition, feature extraction and subsequent identification have been described as well as the current state of the commercially available solutions based on these technologies. The limitations of 2D biometrics have been discussed along with possible approaches to solving them.

3D scanning techniques utilized in biometrics have been introduced and described along with their complimentary nature to the 2D methods. The current development in this area has been outlined.

Novel multimodal method of biometric image acquisition utilizing the line scanner has been introduced, with the experimental results from the proof of concept system demonstrating the capability of hand geometry, fingerprint and palm-print features extraction. The future development direction has been outlined by introducing a 3D line scanner and justifying its specification with the required task in mid.

The palm-print and hand geometry are two technologies that are often being overlooked in favor of alternatives, despite their advantages. With the advance of cheaper sensing technologies and higher available computing power, some of the shortcomings are being overcome. By expanding these systems into 3D, in a fast and affordable way, these biometrics may very well soon gain in popularity.

Acknowledgments

This work was supported by The Ministry of Education, Youth and Sports of the Czech Republic from the National Programme of Sustainability (NPU II); project IT4Innovations excellence in science—LQ1602.

References

[1] M. Drahanský, F. Orság, M. Doležel, *et al.*, Biometrie. Brno: Computer Press, 2011. ISBN 978-80-254-8979-6.

[2] A. K. Jain, A. Ross, S. Pankanti, "A prototype hand geometry-based verification system," Proc. AVBPA, 1999, pp. 166–171.

[3] M. Choraś, R. Kozik, (2012). "Contactless palmprint and knuckle biometrics for mobile devices," Pattern Analysis and Applications [online]. 15(1): 73–85 [cit. 2017-11-02]. doi: 10.1007/s10044-011-0248-4. ISSN 1433-7541. Available at: http://link.springer.com/10.1007/s10044-011-0248-4.

[4] HandKey II. Allegion [online]. Dublin, Ireland, 2014 [cit. 2017-04-25]. Available at: http://us.allegion.com/Products/biometrics/handkey2/Pages/default.aspx.

[5] HandPunch GT-400. Allegion [online]. Dublin, Ireland, 2014 [cit. 2017-04-25]. Available at: http://us.allegion.com/Products/time_attendance/g_series/gt400/Pages/default.aspx.

[6] HandPunch 4000. Allegion [online]. Dublin, Ireland, 2014 [cit. 2017-04-25]. Available at: http://us.allegion.com/Products/time_attendance/f_series/hp_4000/Pages/default.aspx.

[7] PalmID. PalmID [online]. San Francisco, CA, 2017 [cit. 2017-11-02]. Available at: http://redrockbiometrics.com/.

[8] M.-P. Dubuisson, A. K. Jain, "A modified Hausdorff distance for object matching," Pattern Recognition, 1994. Vol. 1-Conference A: Computer Vision & Image Processing, Proceedings of the 12th IAPR International Conference on, Vol. 1, IEEE, 1994.

[9] M. A. Ferrer, J. Fabregas, M. Faundez, J. B. Alonso, C. Travieso, "Hand geometry identification system performance," 43rd Annual 2009 International Carnahan Conference on Security Technology, Zurich, 2009, pp. 167–171.

[10] S.-I. Joo, S.-H. Weon, H.-I. Choi, (2014). "Real-time depth-based hand detection and tracking," The Scientific World Journal. 2014: 17 pages, Article ID 284827. doi: 10.1155/2014/284827.

[11] Intel RealSense™ Camera [online]. 2015 [cit. 2017-04-17]. Available at: https://software.intel.com/en-us/RealSense/SR300Camera.

[12] V. Kanhangad, A. Kumar, D. Zhang, "Combining 2D and 3D hand geometry features for biometric verification," 2009 IEEE Computer Society Conference on Computer Vision and Pattern Recognition Workshops, Miami, FL, 2009, pp. 39–44.

[13] V. Kanhangad, A. Kumar, D. Zhang, (Sep. 2011). "A unified framework for contactless hand verification," IEEE Transactions on Information Forensics and Security. 6(3): 1014–1027. doi: 10.1109/TIFS.2011.2121062.

[14] J. Svoboda, M. Bronstein, M. Drahanský. "Contactless biometric hand geometry recognition using a low-cost 3D camera," Proceedings 2015 International Conference on Biometrics, Phuket, IEEE Biometric Council, 2015, pp. 452–457. ISBN 978-1-4799-7824-3.

[15] Vivid 910: Non-contact 3D digitizer [online]. 2002, p. 6 [cit. 2017-04-17]. Available at: http://www.upc.edu/sct/documents_equipament/d_288_id-715.pdf.

[16] P. C. Mahalanobis, (1936). "On the generalised distance in statistics" (PDF), Proceedings of the National Institute of Sciences of India. 2(1): 49–55. Retrieved 2016-09-27.

[17] K. Q. Weinberger, J. C. Blitzer, L. K. Saul, (2006). "Distance metric learning for large margin nearest neighbor classification," Advances in Neural Information Processing Systems. 18: 1473–1480.

[18] J. Svoboda, O. Klubal, M. Drahanský, "Biometric recognition of people by 3D hand geometry," The International Conference on Digital Technologies 2013, Zilina, 2013, pp. 137–141.

[19] Š. Mráček, M. Drahanský, R. Dvořák, I. Provazník, J. Váňa, "3D face recognition on low-cost depth sensors," Biometrics Special Interest Group (BIOSIG), 2014 International Conference of the, IEEE, 2014.

[20] Ensenso N10. Imaging development systems [online]. Obersulm, Germany, 2014 [cit. 2017-04-25]. Available at: https://en.ids-imaging.com/ensenso-n10.html.

[21] Nové modely 3D kamer Ensenso. Analyza obrazu (Now models of 3D cameras Ensenso. Image analysis) [online]. 2015 [cit. 2017-04-25]. Available at: http://www.analyza-obrazu.cz/aktuality/nove-modely-3d-kamer-ensenso/.

[22] RaL6144-16gm—Basler racer. Basler: The power of sight [online]. Germany, 2013 [cit. 2017-04-25]. Available at: https://www.baslerweb.com/en/products/cameras/line-scan-cameras/racer/ral6144-16gm/.

[23] VeriFinger SDK. Neurotechnology [online]. 2012 [cit. 2017-04-25]. Available at: http://www.neurotechnology.com/verifinger.html.

[24] 3D Line Scan Camera 3DPIXA compact 30μm. Chromasens [online]. 2015 [cit. 2017-04-30]. Available at: https://www.chromasens.de/en/product/3d-line-scan-camera-3dpixa-compact-30um.

Chapter 13

New achievements in 3D hand shape recognition

Jan Svoboda[1] *and Michael Bronstein*[1]

The dawn of research on 3D hand shape as a biometric identifier dates back to 1988, with the invention of apparatus for 3D hand profile identification [1]. To the best of our knowledge, this was the first attempt that motivated further research and later on gave rise to devices such as HandKey II mentioned already in Chapter 12 of this book. Those, however, are not in fact the true 3D-based devices. The third dimension is simulated by obtaining multiple 2D images and preprocessing them separately.

The truly 3D hand shape recognition appears in early 2000, with the increased interest in 3D reconstruction methods and devices. The rapid development of 3D scanning brought new possibilities in biometrics. Researches could have bought a 3D scanner or build their own DIY (do it yourself) setup. The use of 3D data was first evaluated for face recognition showing encouraging results and motivating the use of 3D also with other biometric modalities.

Searching for the research on hand recognition, we come across the work of Woodard and Flynn from 2005 [2], which focuses on using 3D range images of the dorsal side of a hand. The finger profiles are extracted and used as a biometric identifier. Their work was followed by Malassiotis *et al.* [3], who in 2006 presented their work on 3D finger geometry. These pioneering works were later succeeded by others, who, in contrast to the previous, mainly focused on the palm side of the hand as it carries more valuable information to extract. Namely, from 2009, Kanhangad *et al.* [4] were the first ones to exploit the full 3D hand shape. By using precise but bulky and costly 3D scanner, they have collected a dataset of hand 3D range scans, which is now publicly available for other researchers.

Others have followed their path. In 2012, Mathivanan *et al.* [5] used a 3D camera to obtain range images of the human hands and perform biometric recognition. Later, in 2014, Genovese *et al.* [6] have developed a DYI 3D scanning system and algorithms for 3D hand recognition based on analysis of the palmprint surface. Most recently, in 2015, Svoboda *et al.* [7] proposed a contactless system based on an off-the-shelf structured light (SL) 3D camera.

[1]Institute of Computational Science, Facutly of Informatics, Università della Svizzera Italiana, Switzerland

In the following sections, we will introduce the relevant approaches to 3D data acquisition for the purpose of 3D hand recognition. Afterward, we will discuss the current state of the art, the most recent methods and possible future directions will be introduced.

13.1 Data acquisition

With the beginning of the twenty-first century, 3D reconstruction of real-world scenes started getting more and more attention, as the first methods for 3D reconstruction appeared and showed that there is plenty of space for further development.

13.1.1 Methods

There are many 3D reconstruction approaches, such as passive stereo analysis, photometric stereo, SL, time-of-flight and several others (e.g., ultrasound, computed tomography, magnetic resonance imaging, etc.). Each method introduces particular constraints in order to work properly and, therefore, not all of them are suitable for our scenario. The most relevant methods are shortly explained in the following sections. Their comparison is depicted in Table 13.1.

13.1.1.1 Passive stereo

Passive stereo analysis is based on principles closely related to the human vision [8]. The human vision is fundamentally a binocular process that takes two images obtained from a slightly different viewpoint and estimates the depth from the parallax between the two images. One big advantage of the passive stereo methods is that no other devices are needed for the reconstruction process (e.g., no additional light sources, no pattern projectors, etc.).

A passive stereo system uses at least two (possibly more) cameras that capture the scene at the same time. The whole reconstruction process is based on principles of so-called epipolar geometry [8]. Input from all the cameras [Figure 13.1(a)] is used to produce so-called disparity map [Figure 13.1(b)]. Consequently, the corresponding 3D coordinate is computed for each point of the disparity map [Figure 13.1(c)]. Interested reader can find the detailed description of the typical passive stereo pipeline in [8].

Table 13.1 Comparison of the most relevant 3D scanning technologies

Technology	Environment	Noise level	Distance (m)	Issues
Passive stereo	Indoor/outdoor	Medium	>0.4	Homogenous scenes Specular reflections
Structured light	Indoor	Low	>0.15	Environmental light Specular reflections
Time-of-flight	Indoor/outdoor	High	>0.2	Noise

13.1.1.2 Structured light

The SL approach is in principle a modification of the static binocular stereo [8], where one of the cameras is replaced by a light source that projects the light pattern into the scene. The depth information is triangulated by intersecting the projection ray casted from the camera into the scene and the light ray (or light plane) that is coming into the scene from the light source.

Diverse light patterns can be used. The methods can be divided on the light pattern basis into two groups. First group uses simple geometric patterns such as dots, stripes, etc. [Figure 13.2(a)]. The second group includes methods that are based on a spatial or temporal coding of the light patterns, such as binary encoding or phase shifting [Figure 13.2(b)].

13.1.1.3 Time-of-flight

Last of the listed methods uses a laser range finder to capture the surface of an object based on the speed of light. It sends multiple pulses of light toward the object and calculates the time that it takes to the pulses to bounce on the surface of the object and come back as illustrated in Figure 13.3.

13.1.2 Devices

Back in the days, first 3D scanners were costly, bulky and not available to everybody. As the years have passed and the technologies improved, more and more companies were getting interested in manufacturing 3D scanners. One particularly appealing area is low-cost 3D scanning, which makes it possible to employ 3D

(a)

(b) (c)

Figure 13.1 (a) Input images from both cameras; (b) Disparity map that is the original output of the static stereo analysis; and (c) 3D model of the scene reconstructed using the disparity map [9]

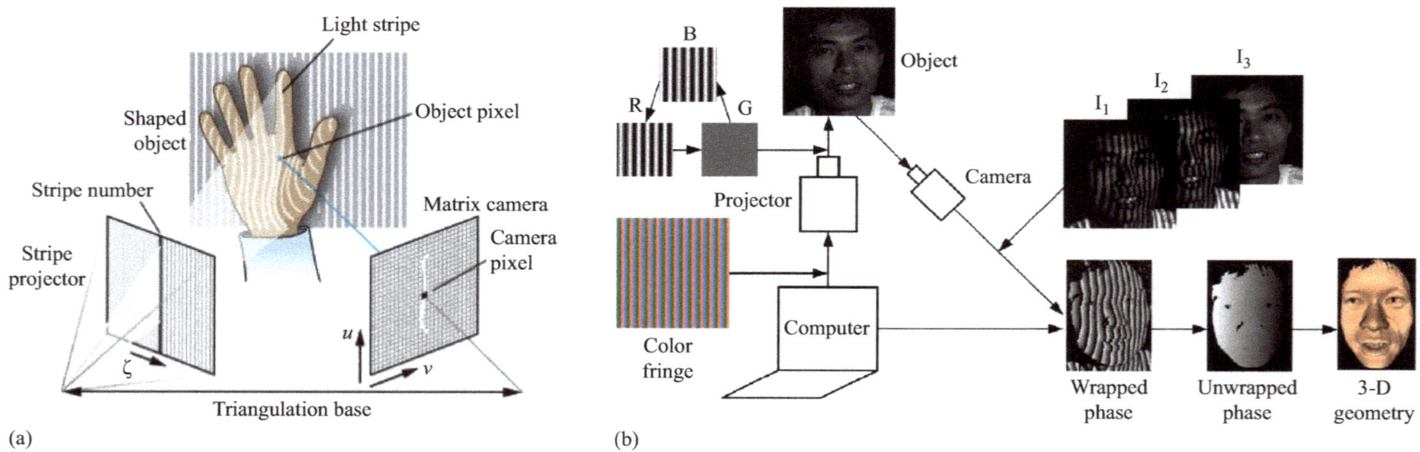

Figure 13.2 (a) Multiple parallel stripes pattern [8] and (b) two plus one phase shifting method processing pipeline [10]

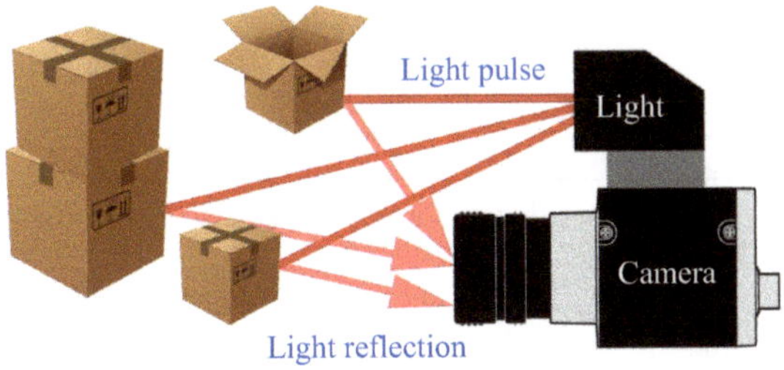

Figure 13.3 Illustration of the time-of-flight imaging principle [11]

vision in everyday tasks. Up to the date, variety of the 3D scanning devices is available on the market.

The selection of the acquisition device (sensor) is one of the key decisions in the design of a biometric system. There are several criteria to be considered. The cost and the form factor of the sensor are of a big importance, since they have a crucial impact on the cost of the material and the size of the final device. On the other hand, we desire the sensor to produce a high-quality data in order to minimize or avoid the postprocessing of the acquired data. The two criteria are typically in conflict, and one has to seek a good trade-off.

For those who prefer precision over the cost and convenience, devices like Minolta Vivid [12] or Artec Eva [13] might be a good choice. Such devices are usually based on the SL technology and provide high precision 3D models. The downside is their price ranging in thousands $ as well as size and possibly cumbersome usage.

Researchers and companies seeking an affordable solution that could be possibly embedded into existing biometric devices in place of a normal camera usually take advantage of the low-cost solutions. Starting with probably the most popular one, which could be rightfully considered the father of the other low-cost commercial 3D scanners, we have Microsoft Kinect [14]. There are currently two versions (v1 and v2). The main difference is in the used 3D reconstruction technology. Kinect v1 uses SL, whereas Kinect v2 time-of-flight. Another possibility could be SoftKinetic DS325 [15] sensor, which uses time-of-flight technology. It has smaller form factor compared to MS Kinect; it can capture from a shorter distance, and it is cheaper. Unfortunately, the output of DS325 is very noisy and therefore practically useless for our purpose without heavy postprocessing. As an alternative with similar properties, Occipital Structure Sensor [16] based on the structured-light technology could do the trick. It however lacks the RGB camera and therefore provides only depth output. That would be compensated by Orbbec Astra sensors [17] that have similar properties, but unlike Occipital, they provide the RGB information up to 720p as well. Taking all the options into consideration, our experiments have shown that currently the best-suited low-cost 3D sensor for hand biometric recognition is the last one we would like to mention, Intel RealSense RS300 [18]. By using SL technology, it combines good properties of the

Table 13.2 Comparison of low-cost 3D sensors. Technology is SL (structured light) or TOF (time-of-flight). Noise level is listed relatively to the other compared sensors

Sensor	Tech.	Depth/color stream	Range (m)	Noise level	Price ($)
RealSense SR300	SL	VGA/1080p	0.2–1.5	Low	150
MS Kinect v1	SL	VGA/VGA	0.4–4.5	Low	150
Orbbec Astra Pro	SL	VGA/720p	0.4–8	Medium	150
Structure Sensor	SL	VGA/N/A	0.4–3.5	Medium	380
SoftKinetic DS325	TOF	QVGA/720p	0.2–1	High	250
MS Kinect v2	TOF	512×424/1080p	0.5–4.5	High	100

previously mentioned sensors and adds something on the top. It has the smallest form factor; it provides depth information as well as RGB up to 1,080p. Opposed to its competitors, it allows one to scan objects starting from distance of about 20 cm and therefore captures closer model of the hand surface. It also comes for the price of only 150$, which is probably the cheapest of all commercially available 3D sensors. More detailed comparison of all the sensors is provided in Table 13.2.

13.1.3 RGB-D data refinement

As a direct consequence of the compromise between accuracy and lower cost, the consumer devices provide significantly less detailed 3D surface. Shortly after the rapid development of low-cost 3D cameras, the idea of using the RGB shading information to improve the accuracy of the coarse initial depth information started attracting interest. This is due to the fact that RGB-D is the typical output of a low-cost 3D camera.

One of such ideas was presented in 2008 by Bohme *et al.* [19], where the range maps measured by time-of-flight cameras were improved by imposing certain shading constraints using the intensity image. Later in 2013, more sophisticated approaches to RGB-D data refinement were proposed. Han *et al.* [20] proposed a solution for uncalibrated natural illumination. They solved the problem by defining global and local lighting model together with surface normal constraints in form of several energies that are jointly optimized. Their approach is able to recover fine details starting from a coarse 3D model. In 2015, Or-El *et al.* [21] proposed a similar approach with some improvements and obtained encouraging results.

Unfortunately, from our experience, these methods typically have several parameters to be set, which influence the quality of the output heavily. Moreover, in order to run efficiently, they usually require optimized implementation on a graphics processing unit (GPU). Nevertheless, it could be interesting to use existing works in biometrics for 3D model improvement, and we believe that in the following years, they will become mature enough to be used commercially as well.

13.1.4 Acquisition setup

The 3D hand data acquisition can be generally divided into two groups—constrained and unconstrained acquisition. Majority of research nowadays uses constrained acquisition. It should not be mistaken for 2D constrained acquisition, where placement of the hand on some surface and even so-called distance pegs are usually required [22]. 3D constrained acquisition is typically done in a contact-free manner, but in a predefined laboratory environment, with constant lighting conditions and sometimes even predefined background. On the other side, we have unconstrained 3D acquisition, where ideally no constraints are imposed. However, this is very challenging, and one would say almost impossible to reach in the case of hand geometry recognition, since human hand is highly nonrigid and in some poses it is impossible to scan a useful 3D model. Therefore, researchers started introducing so-called less-constrained acquisition in the last years, where some of the constraints are dropped in favor of convenience.

13.1.4.1 Constrained acquisition

One of the first works on 3D hand shape recognition proposed in 2005 by Woodard and Flynn [2] is a perfect example of constrained acquisition. They have used Minolta 900/910 sensor [12], which is bulky and heavy (approx. 11 kg) and therefore limits the use of the final system significantly. The sensor is in a fixed position, facing a flat wall that is covered with a piece of black cloth. The subjects are instructed to remove all jewelry and place their hand flat against the wall with the fingers naturally spread.

In 2006, Malassiotis *et al.* [3] proposed a system based on DYI 3D reconstruction setup. He used a light projector to project a color-striped pattern into the scene observed by a CCTV color camera. As the visible part of the color spectrum is used, use of the device with variable lighting conditions is highly limited. Recording of each subject was supervised to ensure good positioning of the hand, and the users were asked to place their hand in front of their face. Since the acquisition was contact-free in this case, slight rotations w.r.t. the image plane of the camera are naturally allowed.

13.1.4.2 Less-constrained acquisition

Kanhangad *et al.* [4] proposed approach that already belongs to the group of less-constrained acquisition. It does not impose any constraints on the environmental lighting; however, the capture has to happen indoor. Also, to simplify the hand-segmentation task, there was a black background placed behind the hand every time. Their system however does not constraint the position much. The users are only required to hold their hand approximately parallel to the image plane of the camera and some rotations are allowed. It neither requires the user to take off the jewelry.

More recently, in 2014, Genovese *et al.* [6] proposed novel system based on stereo reconstruction system consisting of two CCD color cameras that are mounted facing in the upward direction and LED illumination. A user has to place his hand into an acquisition volume, which is defined by the filled of view of the cameras, their depth of focus and the region with uniform illumination. Behind the acquisition volume, there has to be a uniformly colored surface. There are no requirements on spreading the fingers apart, and users can therefore keep the hand in a relaxed position to perform the recognition.

Another system by Wang *et al.* [23] does not impose almost any constraints at all and is based on Microsoft Kinect sensor, which allows simple capture without any background or illumination constraints. The user is requested to show the palm in front of the Kinect sensor and stretch the fingers casually apart. There are no guide pegs as the capture happens in a completely contact free manner.

Svoboda *et al.* [7] proposed novel system with automatic positioning in 2015. It is based on the Intel RealSense F200 3D sensor and does not impose any environmental or illumination constraints. As well as in the case of Wang, there are no special requirements on the background. The user is expected to show his hand in a relaxed position with the fingers spread in front of the sensor, palm facing the sensor. Slight rotations and finger bending are allowed, and the bigger ones are intrinsically eliminated by giving a visual feedback to the user that the hand is positioned wrongly. This automatic system is called the positioning loop, and it removes the necessity of another person supervising the capture process as it keeps the actual capture process from triggering as long as the hand is in a wrong position (some of the automatically checked constraints are not met).

13.1.4.3 Toward unconstrained acquisition

Latest research on hand tracking [24–26] suggests that the challenge of unconstrained acquisition for the purpose of hand biometric recognition could be solvable in the future.

Hand tracking developed at MS Research in Cambridge [26] is able to learn a hand model from examples and afterward fit the learnt model to a newly obtained hand scan. Resulting parameters of the model after fitting could be treated as descriptors of a human hand. The model can be fit to a partial hand scan in any pose and therefore allow completely unconstrained acquisition. Their latest work on hand shape personalization [27], which focuses on fine-tuning the model parameters for a particular person suggests the potential of using their model not only for hand tracking but also for person authentication as well.

13.2 Preprocessing

Although it might not seem like it at first, preprocessing is one of the key stages. Without it, we would not be able to understand where is the hand in the input data. It usually involves depth map smoothing, intensity image denoising, segmentation of the hand and annotation of the important parts of the hand—typically the center of the palm, fingertips and finger valleys (Figure 13.4). Last but not least, since the data are captured in a contact-free manner, three-dimensional pose normalization is essential to improve the stability and therefore performance of the system.

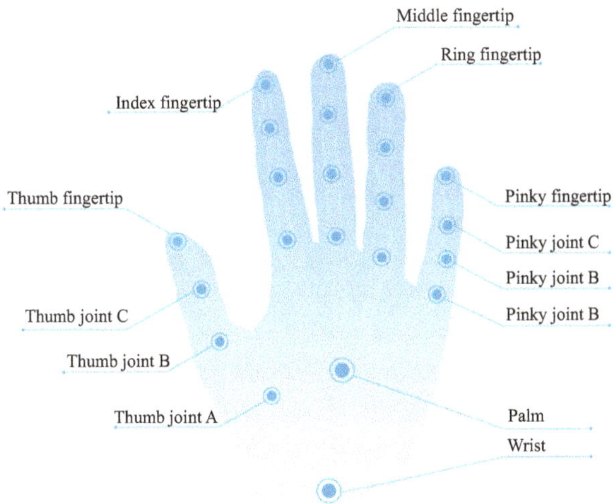

(a)

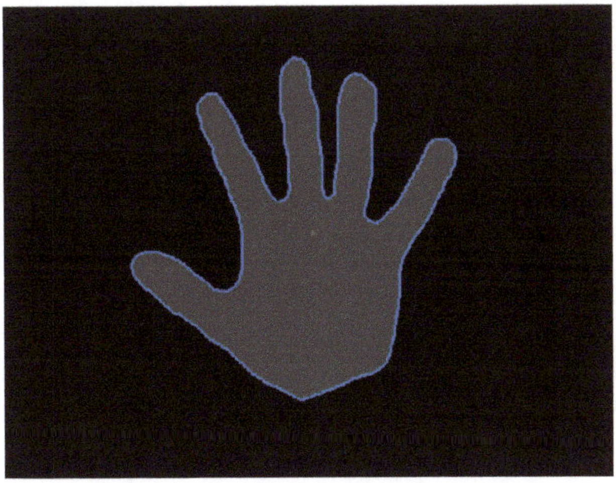

(b)

Figure 13.4 (a) The points of interest on a human hand [18] and (b) contour of the hand detected in the depth map

13.2.1 Input smoothing

In the case of intensity images, it is often enough to apply the traditional median filter [28] in order to remove noise from the image, as it is done for example in [5]. More sophisticated methods are usually not necessary as a good RGB camera is present in most of the low-cost 3D sensors.

As mentioned previously in Section 13.1.2, using a low-cost 3D sensor implies lower precision of the 3D reconstruction that goes hand in hand with higher amount

(a) (b)

Figure 13.5 (a) Noisy human hand 3D scan and (b) the same 3D scan after denoising

of noise. Our desire is to get a smoother model while sacrificing as few details as possible. We therefore usually stipulate to more sophisticated filtering methods such as bilateral filtering [28] or other edge-preserving techniques. Considering the data represented by 3D mesh, feature preserving mesh denoising methods [29,30] provide good solution as previously demonstrated for 3D face recognition [31]. Example of application of such mesh denoising algorithm on hand 3D model is shown in Figure 13.5.

It is worth mentioning that even after denoising, the input data are not guaranteed to be noise-free and therefore any successive processing should be robust to a certain amount of noise.

13.2.2 Hand segmentation

Before performing any hand-specific processing of the input data, we need to determine which portion of the input represents the hand. As the 3D sensors typically provide a one-to-one correspondence between intensity and depth images, information extracted from the intensity image can be easily transferred to the depth map and vice versa. The segmentation methods can therefore be, on a higher level, separated into two classes. First class groups the methods based on the intensity image analysis, whereas the second group represents the methods that operate on the depth information directly. Independently of which method is used, an output of the hand segmentation is typically a binary image that separates the background from the foreground.

Several approaches have been proposed in the past for segmentation of the hand based on the intensity image. Many works [4,6] rely on the fact that there is a constant contrasting background behind the hand, which allows one to perform the segmentation by simply thresholding the intensity image using Otsu's method [32], which is followed by morphological opening and closing in order to remove noisy regions and eventually fill small holes in the resulting hand blob. The method can be even more robust to outliers by further selecting only the largest connected component as the final hand blob as done in [6]. Another approach by Woodard and

Flynn [2] is a combination of edge and skin detection. They use the RGB color space skin detection rules introduced by Kovac *et al.* [33] together with an implementation of a Canny edge detector, which corrects the errors of the skin-based detection due to varying illumination conditions.

The available depth information presents itself as a good candidate for hand segmentation since one would expect the hand to be the object closest to the camera since it is what we are trying to capture. This assumption, however, is not always true and becomes valid only if certain conditions are met. Positioning subsystem designed by Svoboda *et al.* [7] ensures the conditions are met and therefore simple depth map thresholding is enough in their case. In order to improve the separation of the hand from the rest of the arm, so-called weighted-depth-histogram based first-peak hand segmentation [23] can be employed. It performs computation of the depth histogram followed by its reweighting. The first peak of the reweighted depth histogram is assumed to be the hand region. In case the conditions cannot be assumed, we can perform statistical modeling of the hand in the 3D space as shown in [3], which will increase the robustness to noise in the data as well.

13.2.3 Hand annotation

Once we have the hand region segmented, we need to understand the blob as a human hand. That means we need to mark the fingertips, finger valleys and the center of the palm [see Figure 13.4(a)], which are the crucial information for any feature extraction method. We should strive for precision here as any inaccuracy during the hand annotation will be propagated through the feature extraction all the way to the matching subsystem and subvert the result.

The annotation process usually starts by detection of the palm center. Most commonly, a distance transform [28] of the binary image is obtained, which provides the shortest distance between a point and the boundary of the hand. We expect the palm center to be the most distant point from the boundary, and the corresponding distance is taken as an approximate radius of the palm.

In order to detect the fingertips and finger valleys, practically every approach first extracts the contour of the hand by analyzing the boundary pixels of the binary image [Figure 13.4(b)]. That can be carried out for example using simple 8-connected contour tracing algorithm. The contour is smoothed, and curvature is computed at each point. Fingertips and finger valleys can be subsequently detected by finding local minima and maxima of the curvature along the contour as done by [4,7]. For more robust estimate, one can compute the curvature for several down-sampled versions of the contour and subsequently detect the local maxima and minima using curvature histogram [23]. Additionally, the finger orientations can be estimated by analyzing neighboring points on both sides of the fingertip in order to automatically find the finger axis as mentioned in [4]. Another possibility to estimate the orientation of the fingers is to extract the separate finger blobs by means of morphological operations [34] and employ eigendecomposition to compute the principal axis, which is then giving the orientation of the finger.

13.2.4 Three-dimensional pose normalization

Some of the datasets and actually most of the real-world scenarios will contain samples of hands with out-of-plane rotations. As we capture three-dimensional data, we can use that information in order to compensate for this type of rotations. Solution to this problem proposed by [35] and later revised in [23] is depicted in Figure 13.6 and explained below.

The method assumes that the palm of the hand is a rigid planar surface. It is based on fitting a plane to the previously located center of the palm. Considering the set of 3D data points $[x_i, y_i, z_i]^T$, $i = 1, \ldots, n$, where n is the number of points, the problem can be solved by means of Iterative Reweighted Least Squares [35], formulating the weighted least squares optimization at iteration p as

$$\alpha^p = \underset{\alpha}{\operatorname{argmin}} \sum_{i=1}^{n} w_i^{(p-1)} \left(z_i - X_i \alpha^{(p-1)} \right)^2, \tag{13.1}$$

where $\alpha = [\alpha_1, \alpha_2, \alpha_3]^T$ are the plane parameters and $X_i = [1, x_i, y_i]$. The weight w_i depends on how close is the data point i to the fitted plane. A bisquare weighting function can be employed to assign the weights when the least squares residual $r_i = (z_i - X_i \alpha)$ is smaller than a threshold and is defined by

$$w_i = \left(1 - (r_i)^2 \right)^2. \tag{13.2}$$

The weight of the points farther than the threshold is set to 0. Defining the plane fitting problem this way makes the algorithm more robust against outliers in the data. If one would like to avoid solving an optimization problem, a simpler approach based on the finger valley points [23], which is however more sensitive to the noise in the data, can be employed.

Given the hand 3D point cloud data is represented by a $3 \times n$ matrix

$$H_{3D} = \begin{pmatrix} x_1 & x_2 & \cdots & x_n \\ y_1 & y_2 & \cdots & y_n \\ z_1 & z_2 & \cdots & z_n \end{pmatrix}, \tag{13.3}$$

where x, y and z are the 3D data point coordinates, and having the point cloud orientation represented by the plane normal vector $n = [n_x, n_y, n_z]$, the pose corrected point cloud H'_{3D} is defined as

$$H'_{3D} = RH_{3D}, \tag{13.4}$$

where R is the pose correction transformation matrix. Considering the rotation angles about x and y axis are given as

$$\phi_x = -\arctan\left(\frac{n_y}{n_x}\right), \quad \phi_y = \arctan\left(\frac{n_x}{n_z}\right), \tag{13.5}$$

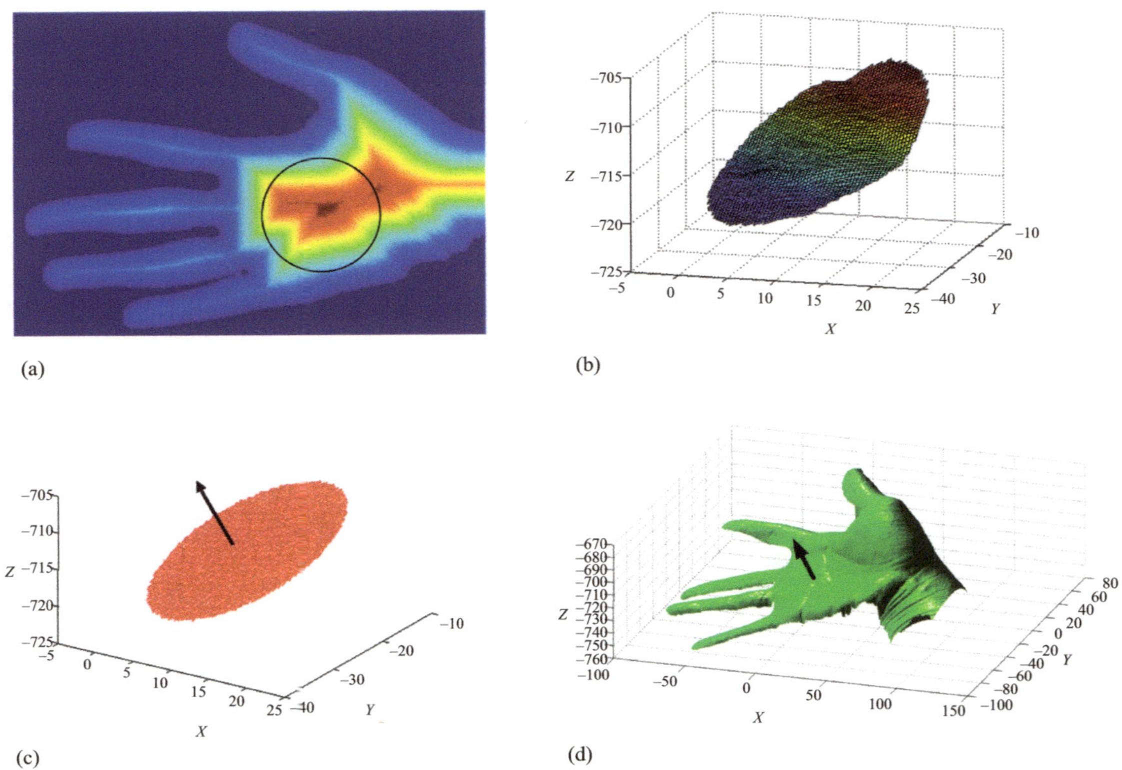

Figure 13.6 *(a) Distance transform of the hand range scan; (b) extracted circular region from the hand palm; (c) oriented plane estimated from the circular palm region; and (d) oriented hand 3D scan [35]*

the matrix R can be expressed as follows

$$R = \begin{pmatrix} \cos\phi_y & 0 & \sin\phi_y \\ \sin\phi_x \sin\phi_y & \cos\phi_x & -\sin\phi_x \cos\phi_y \\ -\cos\phi_x \sin\phi_y & \sin\phi_x & \cos\phi_x \cos\phi_y \end{pmatrix}. \tag{13.6}$$

The same rotation matrix can be used to perform pose correction of the intensity image as well. Having the intensity image with grid coordinates x and y and intensity values I, it can be represented as

$$H_{2D} = \begin{pmatrix} x_1 & x_2 & \cdots & x_n \\ y_1 & y_2 & \cdots & y_n \\ I_1 & I_2 & \cdots & I_n \end{pmatrix}, \tag{13.7}$$

and then the pose corrected data is given by

$$H'_{2D} = RH_{2D}. \tag{13.8}$$

As both the 3D and 2D pose corrected data are 3D point clouds, they have to be converted back into the depth map and intensity image, respectively. This is achieved by resampling the sets of pose corrected data on a uniform grid in the x–y plane as described in [35].

13.3 3D feature extraction

Given annotated three-dimensional hand data, we can proceed toward the feature extraction process. The majority of the proposed methods describe the underlying hand surface by means of lengths, distances, normal vectors and surface curvatures [2,7,23,35]. Lately, an alternative approach based on Delaunay triangulation of extracted feature points was proposed [6]. More detailed description of the particular methods follows.

13.3.1 Surface normal estimation

In order to perform the curvature analysis of a 3D surface, it is often needed to compute the surface normal vectors first. Several methods are presented below depending on the surface representation we are given.

Assuming we work on the range data, the neighborhood information is represented by the range map grid structure implicitly, and we can therefore obtain an estimate of the surface normal $n = (n_x, n_y, n_z)$ at point $p_{u,v} = (p_x^{u,v}, p_y^{u,v}, p_z^{u,v})$ efficiently using finite difference method as

$$n' = (p_z^{u+1,v} - p_z^{u-1,v},\ p_z^{u,v+1} - p_z^{u,v-1}, 2.0), \tag{13.9}$$

$$n = \frac{n'}{\|n'\|_2}, \tag{13.10}$$

where u, v are the range map grid coordinates.

Given a triangular mesh instead, a surface normal for a triangle can be computed as cross product of two edges of the triangle. In this case, the order of the vertices in the computation affects the orientation of the normal. For a triangle p_1, p_2, p_3, we can define vectors $u = p_2 - p_1$ and $v = p_3 - p_2$ and calculate the surface normal n in the following way:

$$n = u \times v = (u_y v_z - u_z v_y, u_z v_x - u_x v_z, u_x v_y - u_y v_x). \tag{13.11}$$

This gives us the normal of a triangle. Normal of a vertex can be computed by averaging normals of all its adjacent triangles.

In the most difficult setting, we are given a point cloud without having any information about the relationship between the points. One of the simplest solutions to this problem is its approximation by estimating the normal n_i of the tangent plane of the surface S at a point p_i, which essentially boils down to a least-squares plane fitting problem. The estimation of the surface normal is therefore reduced to principal component analysis of covariance matrix created from the nearest neighbors of the point p_i. There is unfortunately no mathematical way to solve the sign of the normals, and their orientation is therefore ambiguous. That implies necessity for postprocessing of the computed normals in order to ensure their consistent orientation. For more details, we refer the reader to [36].

13.3.2 Analytical curvature analysis

Curvature analysis can be certainly considered the traditional, and still widely used, approach to 3D hand geometry analysis. Many of the presented approaches have chosen the way of computing curvature of surface S at point p by fitting a polynomial to a set of neighboring points S_p.

Woodard and Flynn [2] first estimate the surface normal at point p together with two orthogonal vectors that span the tangent plane centered at p. Next, they fit a bicubic Monge surface of the form

$$z = f(x,y) = ax^3 + bx^2y + cxy^2 + dy^3 + ex^2 + qxy + gy^2 + hx + iy + j \tag{13.12}$$

to the set of points S_p using linear regression. By using the result, one can calculate analytically the partial derivatives $f_x, f_y, f_{xy}, f_{xx}, f_{yy}$ in order to obtain the principal curvatures $\kappa_{\min}$ and $\kappa_{\max}$ expressed as

$$\beta = \frac{f_{xx} + f_{yy} + f_{xx}f_y^2 + f_{yy}f_x^2 - 2f_xf_yf_{xy}}{2\left(1 + f_x^2 + f_y^2\right)^{3/2}}, \tag{13.13}$$

$$\kappa_{\min,\max} = \beta \pm \sqrt{\beta^2 - \frac{f_{xx}f_{yy} - f_{xy}^2}{\left(1 + f_x^2 + f_y^2\right)^2}}. \tag{13.14}$$

These analytical estimates of curvature are generally noisy. One could smooth the range data prior to the curvature estimation, which is however prone to destroying

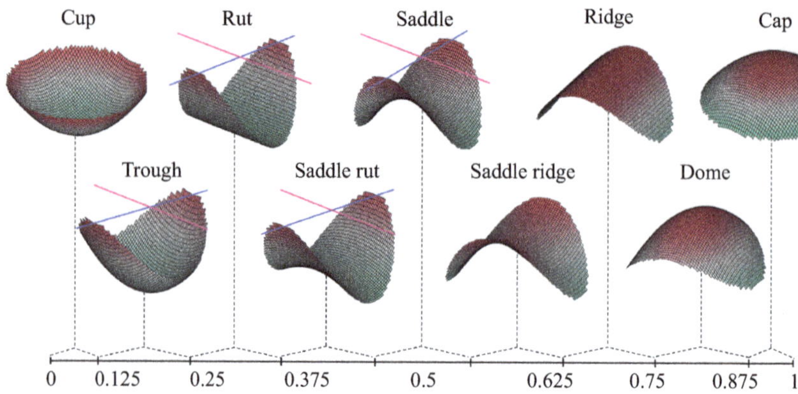

Figure 13.7 Geometrical meaning of different values of shape index (SI) [38]

some fine details on the surface. As a better alternative, the problem can be addressed by choosing relatively large number of points for the Monge surface fit as described in [2]. The obtained $\kappa_{\min}$ and $\kappa_{\max}$ are subsequently used to compute the value of so-called Shape Index (SI), originally introduced by Dorai and Jain [37], at each point p, which is expressed by the following formula:

$$SI = \frac{1}{2} - \frac{1}{\pi} \arctan \frac{\kappa_{\max} + \kappa_{\min}}{\kappa_{\max} - \kappa_{\min}}, \tag{13.15}$$

where $\kappa_{\max} \geq \kappa_{\min}$ and $SI \in [0, 1]$. The values of SI allow for shape classification. As mentioned in [2], in the rare case of $\kappa_{\max} = \kappa_{\min}$, where the SI formula is undefined, it is assigned to 0. That basically indicates that the surface at this point is planar, which is consistent with the case of equal principal curvatures. The meaning of particular values of SI is depicted in Figure 13.7.

Another approach, described by Kanhangad *et al.* [4], is to extract several cross-sectional segments (lines) on the 3D finger surface S and fit a 2D polynomial to each point $p \in S$ in order to compute the mean curvatures and normal vectors. These features can be computed by estimating numerical partial derivatives of the fitted polynomial at each data point p. For this purpose, extension of widely used Savitzky–Golay filters [39] into two dimensions can be used. They employ a 2D polynomial $f(x, y)$ of the form

$$f(x,y) = c_{00} + c_{10}x + c_{01}y + c_{11}xy + c_{20}x^2 + c_{02}y^2, \tag{13.16}$$

where x and y are the two-dimensional coordinates of the point p. The neighborhood of p of size N is represented as a vector $d = (d_0, d_1, \ldots, d_{N-1})^T$. The polynomial fitting can be then formulated in matrix form as

$$Mc = d, \tag{13.17}$$

$$\begin{pmatrix} 1 & x_0 & y_0 & x_0y_0 & x_0^2 & y_0^2 \\ 1 & x_1 & y_1 & x_1y_1 & x_1^2 & y_1^2 \\ \vdots & \vdots & \vdots & \vdots & \vdots & \vdots \\ 1 & x_{N-1} & y_{N-1} & x_{N-1}y_{N-1} & x_{N-1}^2 & y_{N-1}^2 \end{pmatrix} \begin{pmatrix} c_{00} \\ c_{10} \\ c_{01} \\ c_{11} \\ c_{20} \\ c_{02} \end{pmatrix} = \begin{pmatrix} d_0 \\ d_1 \\ \vdots \\ d_{N-1} \end{pmatrix}. \tag{13.18}$$

The above equation is easily solvable for the polynomial coefficients c by least squares approach. Subsequently, the mean curvature of the fitted 2D polynomial at each data point p on a cross-sectional segment can be computed using its coefficients as

$$\kappa_{2D} = \frac{\left(1 + c_{10}^2\right)c_{02} + \left(1 + c_{01}^2\right)c_{20} - c_{10}c_{01}c_{11}}{\left(1 + c_{10}^2 + c_{01}^2\right)^{3/2}}. \tag{13.19}$$

Additionally to the mean curvature, it is possible to compute the surface normal vector at every data point p as well. Assuming parametric surface and the fact that the surface normal is parallel to the cross product of partial derivatives of this surface, the expression for unit normal vector n at each data point in terms of the fitted polynomial coefficients c is expressed as

$$n = \frac{(-c_{10}, -c_{01}, 1)}{\sqrt{1 + c_{10}^2 + c_{01}^2}}. \tag{13.20}$$

13.3.3 Discrete curvature analysis

In 2014, Wang *et al.* [23] proposed an approach to perform discrete curvature analysis of the three-dimensional hand contour.

Let us suppose, there are N points on the 3D contour of the hand $C_h = \{p_1, p_2, \ldots, p_n\}$, where p_i is a contour point.

Given a point $p_i \in C_h$, a geodesic curve around p_i is defined as a point set

$$\Gamma_i^{s_j} = \chi(p_i, s_j) = \{p_k | i - s_j \leq k \leq i + s_j\}, \tag{13.21}$$

where s_j is a step that determines the length span of local geodesic curve $\Gamma_i^{s_j}$. For each contour point p_i, its local geodesic curve set S_i consisting of M steps can be formulated as

$$S_i = \left\{\Gamma_i^{s_j} = \chi(p_i, s_j) | 1 \leq j \leq M\right\}. \tag{13.22}$$

The curvature $\kappa_i^{s_j}$ of a local geodesic curve $\Gamma_i^{s_j}$ on the hand contour is then expressed as

$$\kappa_i^{s_j} = \delta \frac{l_{p_i}^{s_j}}{|p_{i-s_j}, p_{i+s_j}|}, \tag{13.23}$$

where $|p_i, p_j|$ is the Euclidean distance between any points p_i and p_j, and $l_{p_i}^{s_j}$ denotes the geodesic length of the curve $\Gamma_i^{s_j}$

$$l_{p_i}^{s_j} = \sum_{k=i-s_j}^{i+s_j-1} |p_k, p_{k+1}|. \tag{13.24}$$

The δ in (13.23) indicates whether the local curve $\Gamma_i^{s_j}$ is convex or concave

$$\delta = \begin{cases} 1, & [p_{i-s_j}, p_{i+s_j}] \in R_h \\ -1, & \text{otherwise} \end{cases}, \tag{13.25}$$

where $[p_{i-s_j}, p_{i+s_j}] \in R_h$ indicates that the line segment between points p_i and p_j is inside the hand region R_h.

By using the above equations, given a point $p_i \in C_h$, we can represent the corresponding geodesic curve denoted as $\Gamma_i^{s_j}$ and curvature value $\kappa_i^{s_j}$ at a step s_j. For a geodesic curve set S_i, we denote K_i as its corresponding curvature set

$$K_i = \{\kappa_i^{s_j} | 1 \leq s_j \leq M\}. \tag{13.26}$$

Given N contour points $\{p_1, p_2, \ldots, p_n\}$, their curvature set $\{K_1, K_2, \ldots, K_n\}$ forms a 2D curvature matrix with M rows and N columns, denoted as M_c. The curvature matrix can be computed for various steps and then averaged. This new curvature matrix M_h contains the mean value of the curvatures of contour points over different steps. For more details, the reader is referred to [23].

The matrix M_h contains the details of the hand's biometric contour features. Therefore, geometric features can be extracted by analyzing the hand's curvature matrix M_h. First, singular value decomposition is performed, formulated in the following way:

$$M_h = U \begin{pmatrix} \Delta & 0 \\ 0 & 0 \end{pmatrix} V^H, \tag{13.27}$$

where U and V^H are unitary matrices, and Δ is a diagonal matrix with singular values of M_h, denoted as

$$\Delta = diag(\delta_1, \delta_2, \ldots, \delta_r), \qquad \delta_i = \sqrt{\lambda_i}, \quad i = 1, 2, \ldots, r, \tag{13.28}$$

r being the rank of M_h and λ_i a singular value. So-called implicit geometric feature vector v_k can be formulated as normalized vector:

$$v_g = \left(\hat{\delta}_1, \hat{\delta}_2, \ldots, \hat{\delta}_r\right), \quad \hat{\delta}_i = \frac{\delta_i}{\sum_{k=1}^{r} \delta_k}. \tag{13.29}$$

In addition to the implicit geometric feature vector v_g, an explicit geometric feature vector v_f, which encapsulates the finger lengths, is used as well. It is defined as

$$v_f = \left(\hat{l}_f^{(1)}, \hat{l}_f^{(2)}, \ldots, \hat{l}_f^{(n)}\right), \quad \hat{l}_f^{(i)} = \frac{l_f^{(i)}}{\sum_{k=1}^{r} l_f^{(k)}}, \tag{13.30}$$

where $l_f^{(k)}$ is a finger length.

The final geometric feature vector is then obtained as a linear combination of implicit v_g and explicit v_f feature vectors

$$v_{gf} = \omega_g v_g + \omega_f v_f, \quad \omega_g + \omega_f = 1. \tag{13.31}$$

13.3.4 Geodesic distance computation

Taking advantage of the range map, Svoboda *et al.* [7] in 2015 proposed an approach based on analysis of the range image in order to annotate the hand shape followed by 3D geodesic distance computation.

Assuming that the 3D camera provides one-to-one correspondence between range map points and point cloud vertices, they employ 2D hand annotation algorithm proposed in [34] to recover the positions of the fingertips and finger valleys. As depicted in Figure 13.8, the range image is first thresholded in order to obtain binary hand mask. A sequence of morphological dilation and erosion with adequate circular kernel is performed in order to erase the palm area of the hand from the binary mask. This operation should leave major separated blobs representing the fingers and possibly some noise. The five biggest blobs are therefore considered as fingers and kept for further analysis.

Subsequently, the area of each blob is extracted as a binary image and its moments [40] are computed as

$$m_{ij} = \sum_{x,y} I_{x,y} x^j y^i, \tag{13.32}$$

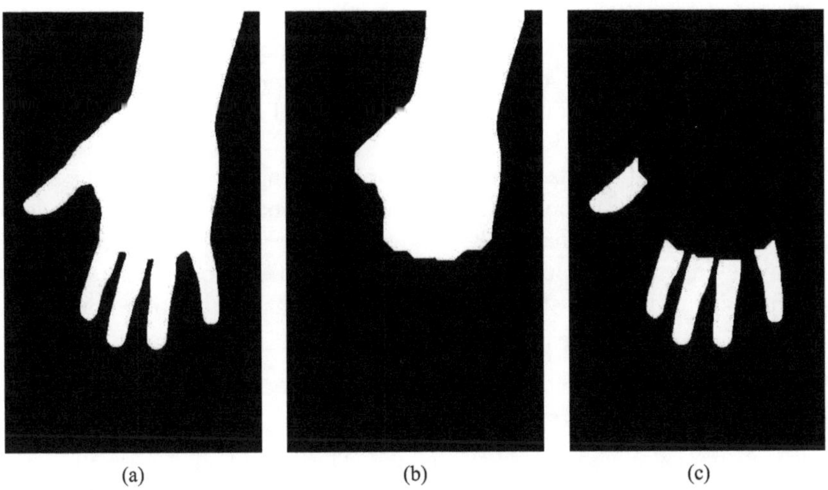

Figure 13.8 (a) Full hand mask before processing; separated parts of the hand by using morphological operations, (b) palm and (c) fingers

where $I_{x,y}$ is the pixel value at a position (x, y). The center of mass of the finger blob $(\bar{x}, \bar{y})$ is then expressed as

$$\bar{x} = \frac{m_{10}}{m_{00}}, \quad \bar{y} = \frac{m_{01}}{m_{00}}. \tag{13.33}$$

Knowing the center of mass, one can easily compute central moments

$$\mu_{ij} = \sum_{x,y} I_{x,y}(x - \bar{x})^j (y - \bar{y})^i, \tag{13.34}$$

which will serve to recover the finger orientation.

The blob orientation can be derived from the 2 × 2 covariance matrix that is constructed using second order central moments

$$\mathrm{cov}(I_{x,y}) = \begin{pmatrix} \mu'_{20} & \mu'_{11} \\ \mu'_{11} & \mu'_{02} \end{pmatrix}, \tag{13.35}$$

where the coefficients are expressed as follows

$$\mu'_{20} = \frac{\mu_{20}}{\mu_{00}}, \quad \mu'_{02} = \frac{\mu_{02}}{\mu_{00}}, \quad \mu'_{11} = \frac{\mu_{11}}{\mu_{00}}. \tag{13.36}$$

By performing eigendecomposition of this matrix, one can get the orientation of the principal axis as the angle between the eigenvector associated with the largest eigenvalue and the axis closest to this eigenvector, which is given as

$$\Theta = \frac{1}{2}\arctan\left(\frac{2\mu'_{11}}{\mu'_{20} - \mu'_{02}}\right), \tag{13.37}$$

considering that the special case of $\mu'_{20} - \mu'_{02} = 0$ is handled.

Performing the above processing for each finger's blob, the orientations of the fingers are obtained. Fingertips are identified as the points of intersection of the finger principal axis and finger contour lying further from the hand center of mass. The second intersection point for each finger is stored for further processing and also used to find the finger valleys.

Having the hand annotated, the fingertips and finger valleys can be simply used to index the array of point cloud vertices as the one-to-one correspondence mentioned earlier is assumed to be known. A finger geodesic length is computed by taking the two points of intersection and employing fast marching algorithm [41] to find the shortest path from the first intersection point to the second one.

13.3.5 Palm area triangulation

An unconventional method of describing the 3D geometry of the hand has been proposed by Genovese *et al.* [6]. The complete work is focused on palmprint recognition; however, as they obtain only a coarse 3D model of the hand surface, the details of the palmprint are not captured in 3D. Nevertheless, the approximate

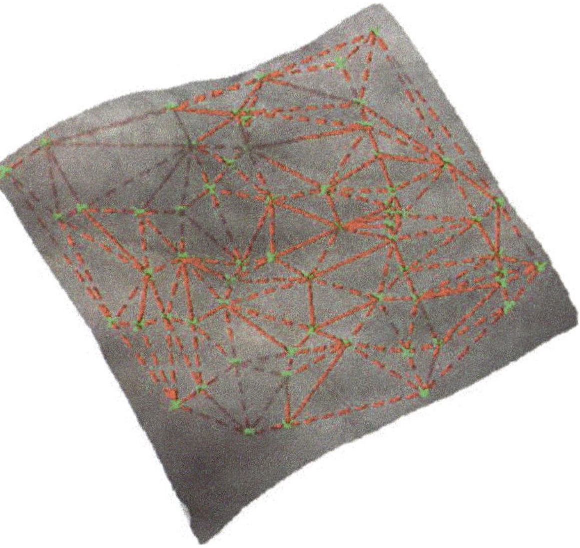

Figure 13.9 Example of Delaunay triangulation of a hand palm geometry [6]

3D structure of the hand palm is captured and provides valuable information for recognition.

At the first stage, matching points of two hand images are computed from the 2D texture images of the hand. Knowing the one-to-one correspondence between 2D texture image and 3D point cloud, those matching points are converted into 3D coordinates.

Next, the corresponding point clouds are registered using the Iterative Closest Point algorithm.

The Delaunay triangulation [42] of the registered point clouds is computed considering only x- and y-coordinates (see Figure 13.9). The z-coordinate is added afterwards to each vertex of the computed triangles.

Finally, features are expressed in terms of the lengths of the sides of the triangles in the 3D space as

$$\begin{aligned} L_{1i} &= \sqrt{(X_{1i}-X_{2i})^2+(Y_{1i}-Y_{2i})^2+(Z_{1i}-Z_{2i})^2}, \\ L_{2i} &= \sqrt{(X_{2i}-X_{3i})^2+(Y_{2i}-Y_{3i})^2+(Z_{2i}-Z_{3i})^2}, \\ L_{3i} &= \sqrt{(X_{1i}-X_{3i})^2+(Y_{1i}-Y_{3i})^2+(Z_{1i}-Z_{3i})^2}, \end{aligned} \tag{13.38}$$

where $(X_{1i}, Y_{1i}, Z_{1i}), (X_{2i}, Y_{2i}, Z_{2i}), (X_{3i}, Y_{3i}, Z_{3i})$ are the 3D vertices of the ith triangle and L_{1i}, L_{2i}, L_{3i} are the lengths of the sides of the ith triangle.

For more details, we refer the reader to [6].

13.4 Feature matching

Different feature extraction approaches have been described in the previous chapter. Depending on the representation of the extracted features, appropriate matching algorithms are applied in order to obtain the desired matching score.

13.4.1 Shape index matching

Woodard and Flynn [2] use SI as the feature representing the 3D surface, obtained as described earlier in Section 13.4.1. The matching score can be computed from SI as the sample normalized correlation coefficient, which is given by the following formula:

$$CC(SI_Q, SI_T) = \frac{\sum_{(i,j)\in valid}(SI_Q(i,j) - \overline{SI}_Q)(SI_T(i,j) - \overline{SI}_T)}{\sqrt{\sum_{(i,j)\in valid}(SI_Q(i,j) - \overline{SI}_Q)^2 \cdot \sum_{(i,j)\in valid}(SI_T(i,j) - \overline{SI}_T)^2}}, \tag{13.39}$$

where $SI_Q(i,j)$, $SI_T(i,j)$ are valid shape index values, and $\overline{SI}_Q$, $\overline{SI}_T$ are the sample mean shape index values of the query and the template images, respectively. More detailed explanation of the computation can be found in [2]. The final matching score is in the interval $[-1, 1]$, and the higher it is, the better the match is.

The above computation is happening for each finger separately. Combination into the resulting matching score is performed by score-level fusion. Authors experimented with different fusion rules, in particular average, median and maximum fusion rule. The maximum fusion rule has shown the best performance among the mentioned. For more elaborate description, the reader is referred to [2].

13.4.2 Matching curvatures and normals

As presented in the previous section, one of the most common techniques for feature extraction is based on curvature analysis and normal vector computation. Metrics to efficiently match those features were proposed by Kanhangad *et al.* [4] in 2009.

Those features are typically extracted from fingers of the hand. The authors propose to match the features for each finger separately and perform the fusion on the matching score level afterward. That should compensate for small of-the-plane rotations of the hand.

First, assume the features are extracted for N_s cross-sectional segments from both template T_i and query Q_i fingers, where i is an identifier of particular finger (little, ring, middle or index). Matching of the curvature features is done using the cosine similarity metric. Authors define the matching score as

$$s_i^c = \frac{1}{N_s}\sum_{j=1}^{N_s} \frac{(\Phi(T_i^j, l_T^j, l_Q^j)\Phi(Q_i^j, l_Q^j, l_T^j))}{|(\Phi(T_i^j, l_T^j, l_Q^j)||\Phi(Q_i^j, l_Q^j, l_T^j))|}, \tag{13.40}$$

where the function Φ ensures the template and the query feature vectors are the same lengths

$$\Phi(F, l_1, l_2) = \begin{cases} F(1{:}\, l_1) & \text{if } l_1 > l_2 \\ F & \text{otherwise} \end{cases}, \tag{13.41}$$

and the l_T^j and l_Q^j are the number of feature points on the jth cross-sectional segment of the template T_i^j and the query Q_i^j fingers, respectively. As the number of feature points l_T^j and l_Q^j cannot be guaranteed to be the same, the authors in addition perform multiple matches by sliding the shorter feature vector over the longer one and taking the best match among all the overlaps as the final matching score.

Matching of the normal vector features is performed by comparing the angle between the vectors at two corresponding features points. The smaller the angle is, the more similar the corresponding normal vectors are. Having template T_i and query Q_i fingers, the score can be computed as

$$s_i^n = \frac{1}{N_s} \sum_{j=1}^{N_s} \cos^{-1}(\Phi(T_i^j, l_T^j, l_Q^j)\Phi(Q_i^j, l_Q^j, l_T^j)). \tag{13.42}$$

The final matching score is then obtained by combining both curvature and normal matching scores for all fingers as follows:

$$S = w \frac{1}{4} \sum_{i=1}^{4} s_i^c + (1 - w) \frac{1}{4} \sum_{i=1}^{4} s_i^n, \tag{13.43}$$

where w is the empirically determined weighting parameter between the curvature and the normal features.

13.4.3 Metric learning approach

Works from Wang *et al.* [23] and Svoboda *et al.* [7] use feature vectors that contain lengths or distances measured on the surface of the hand (e.g., lengths of the fingers, finger widths, etc.). Those features are typically stored as k-dimensional feature vectors, where k is the number of computed distances.

Feature vectors of this type could be easily compared using the traditional squared distance metric

$$d(x, y) = \sqrt{((x - y)^T (x - y))}, \tag{13.44}$$

where x and y are the two compared feature vectors. Such simple metric considers all the features (elements of the feature vector) to be equally important for the final decision. This is however not true in practice, and some sort of weighting of the features is desired in order to give higher importance to the more prominent features. Ideally, one would like to express the feature vectors in a metric that makes the genuine feature vectors very similar and impostor feature vectors distant. And that is where metric learning comes into play.

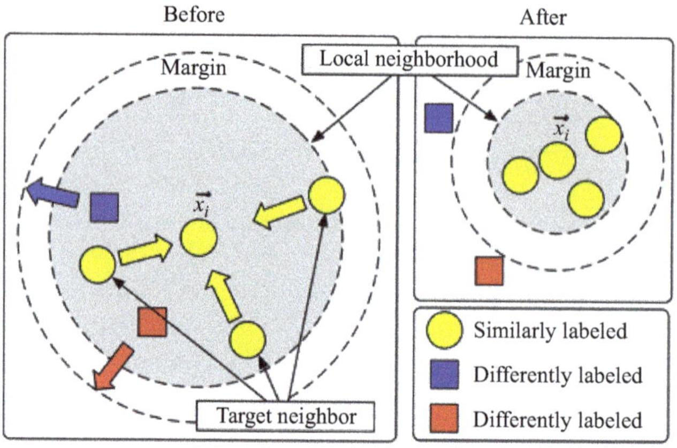

Figure 13.10 Large-Margin Nearest Neighbors (LMNN) metric learning illustration [43]

Our desire is to learn a transformation matrix M, which allows one to compute distances between feature vectors in a different metric defined by M. As shown in Svoboda *et al.* [7], the optimal M can be learnt using Large-Margin Nearest Neighbors metric learning approach described by [43]. Supplying feature vectors representing instances of hands of the same individual (positives) and hand instances of different individuals (negatives), M is found by solving the following optimization problem

$$\min_{M \succeq 0} \sum_{(x,x^+,x^-) \in \mathcal{T}} (1-\mu) d_M^2(x,x^+) + \mu \max\left\{1 + d_M^2(x,x^-) - d_M^2(x,x^+), 0\right\}, \quad (13.45)$$

where (x,x^+,x^-) is a triplet that contains a positive (x,x^+) and negative (x,x^-) pair taken from some training set $\mathcal{T}$ and $\mu \in [0,1]$ is trade-off parameter. The optimization strives to find such M that minimizes the distance between positive pairs (x,x^+) and simultaneously tries to put a unit margin between the positive and negative distances, $d_M^2(x,x^-) \leq d_M^2(x,x^+) + 1$ (see Figure 13.10). In [43], the optimization is solved as a semidefinite programing problem.

Assuming we have the transformation matrix M, which transforms the feature representation into a metric that is better for comparison of two features vectors. The distance in metric M can be expressed in terms of Mahalanobis distance as

$$d_M(x,y) = \sqrt{(x-y)^T M (x-y)}, \quad (13.46)$$

where x and y are the two compared feature vectors.

13.4.4 Triangle similarity matching

Section 13.3.5 describes an unconventional way of extracting features from 3D hand geometry by performing Delaunay triangulation and computing lengths of sides for each triangle. Let us assume that ith triangle has the lengths of its sides L_{1i}, L_{2i}, L_{3i}. Having query and template triangulations, each triangle of the query is compared to every triangle of the template in order to determine whether their sides have similar lengths

$$\begin{aligned} L_{1i} - L'_{1j} &< (t_D L_{1i}) \\ L_{2i} - L'_{2j} &< (t_D L_{2i}) \\ L_{3i} - L'_{3j} &< (t_D L_{3i}), \end{aligned} \tag{13.47}$$

where L_{1i}, L_{2i}, L_{3i} are lengths of the sides of the ith triangle of the query triangulation, and $L'_{1j}, L'_{2j}, L'_{3j}$ are lengths of the sides of the jth triangle of the template triangulation. As mentioned in [6], t_D denotes a fixed threshold that is determined experimentally.

The final matching scores is the number of similar triangles.

13.5 Summary

With the recent developments in low-cost 3D sensing, the interest in employing 3D biometrics has increased. We have wrapped up the current state of the art in 3D hand biometric recognition. Depending on the representation of the acquired geometry, we are offered different feature extraction methods. Typically, the hand scan is stored as range map. The neighborhood structure is well defined on such surfaces, and it is therefore easy to compute approximate geodesic distances as well as perform curvature analysis. Those are, in fact, the paths that most of the researchers have followed in the past. The features are typically stored as *n–D* vectors and are well suited to be compared using a specific distance metric. As we have shown, in order to find the best separation of the subjects, dimensionality reduction and metric learning can be employed to exploit the information better. To encourage future development, we have shown some unconventional representations of the hand geometry as well.

Independently, the field of hand tracking advanced very fast, currently offering fast and precise 3D hand-tracking models. Employing such models in 3D hand biometric recognition is very promising, as it could potentially boost the precision of the hand annotation that would naturally result in better stability and performance of the 3D hand recognition. As the hand-tracking models usually work in real time, new possibilities (e.g., continuous user verification, etc.) come up as well.

References

[1] D. P. Sidlauskas. 3D hand profile identification apparatus, April 5 1988. URL https://patents.google.com/patent/US4736203. US Patent 4,736,203.

[2] D. L. Woodard and P. J. Flynn. Finger surface as a biometric identifier. *Computer Vision and Image Understanding*, 100:357–384, 2005.

[3] S. Malassiotis, N. Aifanti and M. G. Strintzis. Personal authentication using 3-D finger geometry. *IEEE Transactions on Information Forensics and Security*, 1:12–21, 2006.

[4] V. Kanhangad, A. Kumar and D. Zhang. Combining 2D and 3D hand geometry features for biometric verification. In *Proceedings of the IEEE Computer Society Conference on Computer Vision and Pattern Recognition Workshops*, pages 39–44, 2009.

[5] B. Mathivanan, V. Palanisamy and S. Selvarajan. Multi-dimensional hand geometry based biometric verification and recognition system. *International Journal of Emerging Technology and Advanced Engineering*, 2:348–354, 2012.

[6] A. Genovese, V. Piuri and F. Scotti. *Touchless Palmprint Recognition Systems*, volume 60 of *Advances in Information Security*. Cham: Springer International Publishing, September 2014. 978-3-319-10365-5.

[7] J. Svoboda, M. M. Bronstein and M. Drahanský. Contactless biometric hand geometry recognition using a low-cost 3D camera. In *2015 International Conference on Biometrics (ICB)*, pages 452–457, May 2015.

[8] R. Klette, K. Schluns and A. Koschan. *Computer Vision: Three-Dimensional Data from Images*. Singapore: Springer, 1998. ISBN 9813083719.

[9] M. Gelautz, M. Bleyer, L. He and N. Brosch. *Evaluation and design of energy functions for global stereo matching*. 2012. https://www.ims.tuwien.ac.at/projects/stereo-matching/.

[10] S. Zhang and S.-T. Yau. High-speed three-dimensional shape measurement system using a modified two-plus-one phase-shifting algorithm. *Optical Engineering*, 46, 2007. https://doi.org/10.1117/1.2802546.

[11] Basler A. G. Basler's time-of-flight camera. 2016. URL https://www.baslerweb.com/en/products/cameras/3d-cameras/time-of-flight-camera/.

[12] Konica Minolta Inc. Minolta Vivid 910. 2016. URL https://www.konicaminolta.com.

[13] Artec 3D. Artec Eva. 2016. URL https://www.artec3d.com.

[14] Microsoft. Microsoft Kinect. 2016. URL https://www.microsoft.com.

[15] SoftKinetic. Softkinetic ds 325. 2016. URL https://www.softkinetic.com.

[16] Occipital Inc. Structure sensor. 2016. URL https://structure.io.

[17] Orbbec 3D. *Orbbec Astra*. 2016. URL https://orbbec3d.com.

[18] Intel. RealSense. 2016. URL http://www.intel.com/realsense.

[19] M. Bohme, M. Haker, T. Martinetz and E. Barth. Shading constraint improves accuracy of time-of-flight measurements. In *2008 IEEE Computer Society Conference on Computer Vision and Pattern Recognition Workshops*, pages 1–6, June 2008.

[20] Y. Han, J. Y. Lee and I. S. Kweon. High quality shape from a single RGB-D image under uncalibrated natural illumination. In *2013 IEEE International Conference on Computer Vision*, pages 1617–1624, December 2013.

[21] R. Or-El, G. Rosman, A. Wetzler, R. Kimmel and A. M. Bruckstein. RGBD-fusion: Real-time high precision depth recovery. In *2015 IEEE Conference on Computer Vision and Pattern Recognition (CVPR)*, pages 5407–5416, June 2015.

[22] A. K. Jain, A. Ross and S. Pankanti. A prototype hand geometry-based verification system. *Proc. AVBPA*, Washington, DC, pages 166–171, 1999.

[23] C. Wang, H. Liu and X. Liu. Contact-free and pose-invariant hand-biometric-based personal identification system using RGB and depth data. *Journal of Zhejiang University Science C*, 15(7):525–536, 2014.

[24] I. Oikonomidis, N. Kyriazis and A. A. Argyros. Efficient model-based 3D tracking of hand articulations using Kinect. *British Machine Vision Conference*, 1:3, 2011.

[25] A. Tagliasacchi, M. Schröder, A. Tkach, S. Bouaziz, M. Botsch and M. Pauly. Robust articulated-ICP for real-time hand tracking. *Computer Graphics Forum*, 34(5):101–114, 2015.

[26] J. Taylor, L. Bordeaux, T. Cashman, *et al.* Efficient and precise interactive hand tracking through joint, continuous optimization of pose and correspondences. *ACM Transactions on Graphics (TOG)*, 35(4):143.1–143.12.

[27] D. J. Tan, T. Cashman, J. Taylor, *et al.* Fits like a glove: Rapid and reliable hand shape personalization. In *2016 IEEE Conference on Computer Vision and Pattern Recognition (CVPR)*, pp. 5610–5619, June 2016.

[28] R. Szeliski. *Computer Vision: Algorithms and Applications.* Springer-Verlag New York, Inc., New York, NY, USA, 1st edition, 2010.

[29] R. Martin, X. Sun, P. Rosin and F. Langbein. Fast and effective feature-preserving mesh denoising. *IEEE Transactions on Visualization and Computer Graphics*, 13:925–938, 2007. ISSN 1077-2626.

[30] Z. Deng, X. Lu and W. Chen. A robust scheme for feature-preserving mesh denoising. *IEEE Transactions on Visualization and Computer Graphics*, 22(3):1181–1194, March 2016.

[31] Š. Mráček, R. Dvořák, J. Váňa, T. Novotný and M. Drahanský. 3D face recognition utilizing a low-cost depth sensor. In *11th IEEE International Conference on Automatic Face and Gesture Recognition (FG 2015)*, pages 1–1, 2015.

[32] N. Otsu. A threshold selection method from gray-level histograms. *IEEE Transactions on Systems, Man, and Cybernetics*, 9(1):62–66, January 1979. ISSN 0018-9472.

[33] J. Kovac, P. Peer and F. Solina. Human skin color clustering for face detection. In *The IEEE Region 8 EUROCON* 2003. *Computer as a Tool,* volume 2, pages 144–148, September 2003.

[34] A. de Santos-Sierra, C. Sánchez-Ávila, G. B. del Pozo and J. Guerra-Casanova. Unconstrained and contactless hand geometry biometrics. *Sensors*, 11(11): 10143–10164, 2011. ISSN 1424-8220.

[35] V. Kanhangad, A. Kumar and D. Zhang. Human hand identification with 3D hand pose variations. In *2010 IEEE Computer Society Conference on Computer Vision and Pattern Recognition—Workshops*, pages 17–21, June 2010.

[36] R. B. Rusu. Semantic 3D object maps for everyday manipulation in human living environments. *Künstliche Intelligenz*, 24(4):345–348, 2010.

[37] C. Dorai and A. K. Jain. COSMOS-A representation scheme for 3D free-form objects. *IEEE Transactions on Pattern Analysis and Machine Intelligence*, 19(10):1115–1130, 1997.

[38] S. G. Thakku, Y.-C. Tham, M. Baskaran, *et al.* A global shape index to characterize anterior lamina cribrosa morphology and its determinants in healthy Indian eyes LC-GSI to characterize LC morphology. *Investigative Ophthalmology & Visual Science*, 56(6):3604, 2015.
[39] A. Savitzky and M. J. E. Golay. Smoothing and differentiation of data by simplified least squares procedures. *Analytical Chemistry*, 36(8):1627–1639, 1964.
[40] M.-K. Hu. Visual pattern recognition by moment invariants. *IRE Transactions on Information Theory*, 8(2):179–187, 1962.
[41] J. A. Sethian. A fast marching level set method for monotonically advancing fronts. *Proceedings of the National Academy of Sciences of the United States of America*, 93(4):1591–1595, 1996.
[42] B. Delaunay. Sur la sphère vide. A la mémoire de Georges Voronoï. *Bulletin de l'Académie des Sciences de l'URSS*, (6):793–800, 1934.
[43] K. Q. Weinberger and L. K. Saul. Distance metric learning for large margin nearest neighbor classification. *The Journal of Machine Learning Research*, 10:207–244, 2009.

Chapter 14

Antispoofing and multispectral (optical) methods in hand-based biometrics

Mona Heidari[1], Tomáš Goldmann[1], Michal Dvořák[1], and Martin Drahanský[1]

14.1 Introduction

Vulnerability of any given security system is an important aspect that needs to be carefully analyzed during design phase. When dealing with biometric systems, one important source of such vulnerability that needs to be especially considered is its tendency to deception by spoofs. Since a high enough quality spoof can become indistinguishable from the original biometric characteristic by human eye, we may need additional methods of spoof recognition. In this chapter, we will investigate several approaches used as an antispoofing method, with emphasis on approaches utilizing the liveness detection (antispoofing).

14.2 Fingerprint recognition

There was no need for a special hardware for acquiring fingerprint images in the past, because fingerprints were kept on dactyloscopic cards using black ink or other dye. These images then could be digitized via a scanner, this method is called off-line sensing [1].

For live-scan sensing, there is no need for the ink; a person just needs to place his/her finger near the electronic fingerprint scanner. The first generation of on-line sensors was based on optical technology [2], but others such as solid-state or ultrasound were invented.

Fingerprints are physiological characteristic, which means that they can be damaged. A user may (un)intentionally harm his finger and make the recognition impossible [3]. There are also security issues. Most users are not observant to the fact that they leave their latent fingerprints on many places such as glasses, doors,

[1]Faculty of Information Technology, Centre of Excellence IT4Innovations, Brno University of Technology, Czech Republic

etc. Such latent fingerprints can be used to create artificial fingertip to fool the sensor and to obtain unauthorized access to a system.

Therefore, there is no wonder that many researchers put a lot of effort in spoofing fingerprint systems and at the same time, developing antispoofing mechanisms.

14.3 Liveness detection on fingers

The capability of a biometric system to detect whether the provided biometric sample is alive or not is denoted as liveness detection [2], and its main purpose is to detect spoofing.

Liveness detection methods can be divided into three categories [1]:

1. Using only data collected for biometric purposes.
2. Using further-processed information collected in order to generate discriminating features or by sensing the biometric sample in time.
3. Using additional hardware.

Liveness detection uses acquired data to determine whether the input biometric characteristic is alive. Properly performed liveness detection should reject all nonbiologic spoofs and majority of the remaining ones, increasing the security of the system during an acquisition stage.

Basic principle of liveness detection and its role in biometric system can be seen in Figure 14.1. The acquisition of liveness data as well as biometric data should ideally be simultaneous and performed on the same part of the biometry in question; otherwise, the attacker can perform separate attack on the individual parts of the system making the task less complex, as the data would only need to meet the liveness requirement at some time/place and be a match to the template at other, creating a possible vector of attack [4]. There are two ways how to

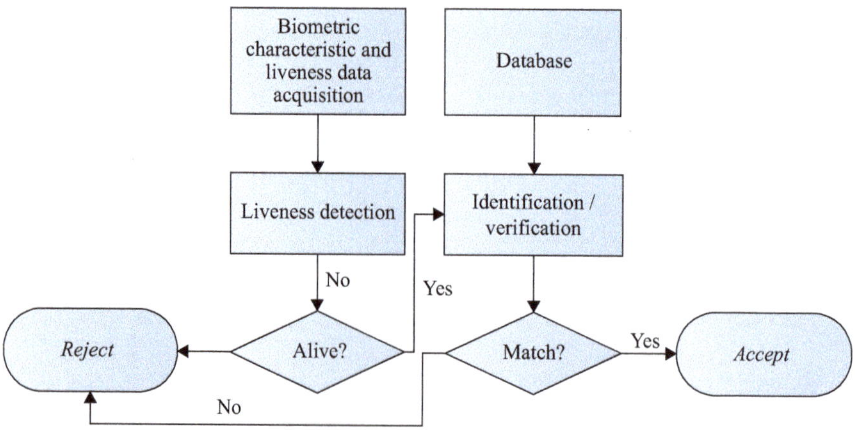

Figure 14.1 Biometric system with liveness detection

determine whether a presented biometric sample is alive or not—liveness detection focuses on unique properties of human body parts and nonliveness detection focuses on typical properties of materials used to fool the system. To discourage potential attackers from presenting a fake finger (i.e., an imitation of the fingertip and the papillary lines) or, even worse, to hurt a person to gain access, the system must be augmented by a liveness detection component [5,6].

Various methods that are used for liveness detection will now be introduced.

14.3.1 Perspiration and sweat pores based detection

One of the several liveness detection methods without any additional hardware is using natural perspiration behavior of skin. Human skin in general, but fingertips especially are covered by sweat glands that excrete sweat fluid. The presence of the pores and of sweat fluid can be used as a mark of a liveness. For the sweat fluid detection, both the optical, chemical and capacitive approach had been explored [7–9]. In Figure 14.2, principle of determining liveness based on perspiration is shown. The spoof incapable of producing sweat fluid would fail to have any change in perspiration measured.

For the pore detection, the systems require, the resolution be high enough to acquire the necessary detail, its viability has been demonstrated as well [8].

14.3.2 Temperature and temperature stimuli based detection

Unlike artificial spoof, which will retain a room temperature without external heat source, human extremities will regulate its own body temperature. The temperature of extremities depends on outside conditions, biological state and stimuli; this temperature will vary in between approximately 25 and 37 °C and can be measured using thermal camera, for example, as can be seen in Figure 14.3.

By measuring this temperature, we can, to an extent, claim that the object in detector is part of human body. However, this method may be compromised, if the hand is undercooled or the room temperature is similar to body temperature. It is also viable to add an active heating element to the spoof, thus bypassing this security feature.

To increase the reliability of this method, we may use additional stimuli. The human skin will react to a stimulus by changing the blood flow to regulate the

Figure 14.2 Temporal perspiration pattern extracted from capacitive sensor [9]

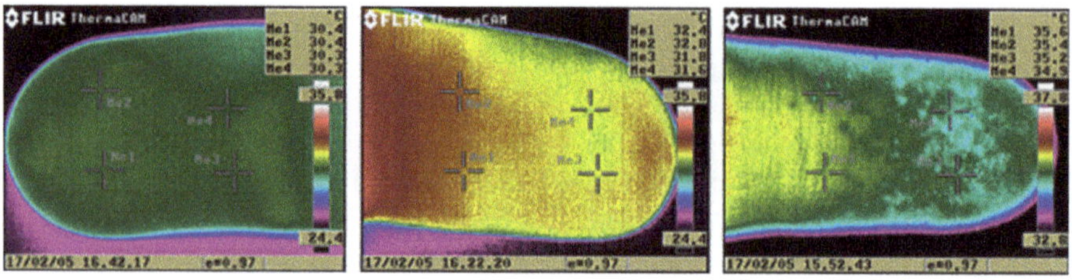

Figure 14.3 Thermographs of human hand showing temperature variation in human finger [10]

temperature. By detecting this change and determining the appropriateness of this change, we may determine the likeliness that observed reaction is that of a human tissue. The downside of this approach is a reaction time. As the stimuli cannot be dangerous, or cause of discomfort, which might compromise the measurement by voluntary or involuntary movement of the subject, the time for change to be observable will be in order of seconds.

14.3.3 Pulse and blood oxygenation based detection

Longitudinal blood pressure wave generated during systolic phase of cardiac cycle, also known as pulse, changes the volume of the blood vessels that can be observed even on the surface of the skin. Observation can be performed by measuring the periodic change in position of human skin [7], or by the minute change in color that occurs due to periodic increases and decreases in the oxygen level of blood.

14.3.4 Multimodal approach

Using additional biometric modalities increases the difficulty to attack the biometric system at the sensor-level [11]. It is more demanding for the attacker, because it needs to elicit additional biometric data from the valid user and create an additional artificial sample, which must be accepted by the attacked biometric system.

14.3.5 Material identification based detection

Human skin, like any material, may undergo various measurements of physical properties due to its nature; these properties may be different from materials used to create a spoof. Electrical properties can be measured, such as resistivity or conductivity, which varies from other materials due to perspiration and salt content in sweat fluid. Reflection of ultrasound presents other source of material analysis, as the ultrasound will penetrate the topmost parts of skin and is then scattered at different frequency, this change can then be measured.

Another approach is to measure the optical characteristics of human skin. This will be investigated in upcoming parts of the chapter.

14.4 Finger vein recognition

There are several ways [11] how to acquire an image of human blood vessels, such as using X-ray, magnetic resonance imaging, ultrasound, far-infrared light, near-infrared (NIR) light or thermal approach [12]. The NIR light approach is used most often in biometrics due its trade-off between cost, power supplement requirements, size of the sensor and quality of resulting image.

The two types of hemoglobin (Hb, HbO_2) have different absorption spectra. Experiments have proved that the permeability of human tissue is high for electromagnetic radiation in the range from 600 to 1,300 nm [11,13]; this range is often called optical window. Especially in the range between 750 and 950 nm [11], the radiation can penetrate deep enough into the skin to reach the superficial arteries and veins, and as a consequence it is absorbed in blood vessels, due to higher absorption coefficient blood. This effect can be captured by a camera on the image that shows where blood vessels are located, are darker. Even though the described method can not only capture veins, but also arteries, this method for capturing blood vessels will be further denoted as a vein characteristic.

There are two main methods of acquiring an image of blood vessels of a finger with NIR approach-reflection and transmission.

14.4.1 Reflection method

The finger is illuminated by NIR light, and the partially reflected radiation is captured by a CCD camera, which is located on the same side of the finger as the light emitter. Because veins are closer to the skin than arteries, veins are usually captured on a resulting image. Simplified schema of a finger vein sensor using reflection method can be seen in Figure 14.4(a).

The main advantages of this approach are the sensor size, compactness and possibility to easily extend existing (fingerprint) systems with finger vein recognition system.

14.4.2 Transmission method

The transmission method is usually used only in finger vein recognition, because fingers are generally thin enough to allow the light to pass through the tissue. The NIR light emitter is in such case located on the opposite side of the finger than the camera. Although the transmission method is leading in finger vein sensing, it must be mentioned that the device itself is usually larger than the one using reflection method. On the other hand, the quality of resulting image is more high-contrasted in comparison to reflection approach [14]. Simplified schema of a finger vein sensor using reflection method can be seen in Figure 14.4(b).

14.4.3 Wavelength of light emitters

The range, where the permeability of human tissue is high, is very wide from 600 to 1,300 nm [11]. The absorbency of hemoglobin also differs. Until today, it has not been proven, which wavelength is generally the best suitable for transmission and

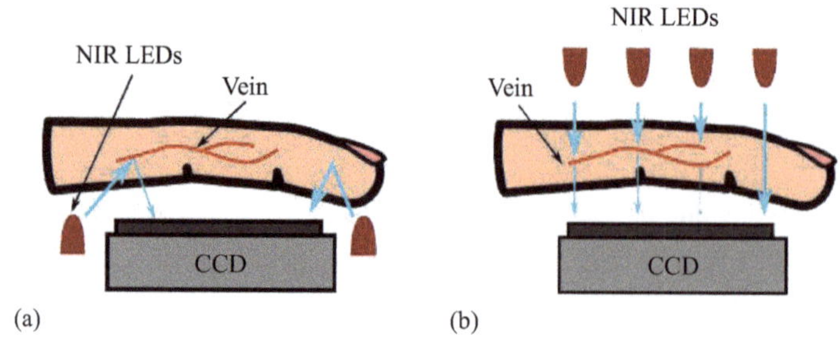

Figure 14.4 Two main approaches for acquiring finger vein pattern using NIR light: (a) refection method and (b) transmission method

reflection method. Yang and Shi [15] describe a finger vein sensor using 760 nm LEDs with declaration that this is the optimal light source wavelength to capture veins.

14.5 Spectral analysis approach to spoof detection

Hand-based biometric systems may have antispoofing methods based on several approaches. One of the approaches is liveness detection, which can be realized with parts of hand. Although it is possible to perform lifetime detection based on skin temperature, electrical resistance, electrical conductivity or bio-impedance, these methods are not reliable [16]. However, there are more reliable methods to help determine if spoofing has occurred.

Like other parts of the human body, heartbeat can be detected in the hand. When heart pumps oxygenated blood to veins, there are volume changes in the veins and vessels. Therefore, the pulse can be detected on the skin of finger. With a very accurate laser rangefinder, these changes can be recorded to determine whether spoofing has taken place or not.

Another aspect that affects the skin's properties is the actual blood flow, which can vary depending on many factors, e.g., body temperature. By analyzing the image obtained by illuminating the skin of different wavelengths, it is possible to get a picture of the different parts beneath the surface of the skin and verify that it is really human skin. It is possible to use several light sources of different wavelengths for a single exposure [17]. This is the basis for multispectral analysis of the human skin. The main advantage is the ability to use multispectral analysis based on image data to assess the health condition epidermal layer of the skin or detect spoofing.

With biometric system using a multispectral analysis, it is very difficult to perform a spoofing attack because homogeneous materials cannot be used.

Human skin is a multilayered organ. Every layer, fulfilling different functions, has different properties; these differences can become evident during optical analysis. There are two major methods to this approach. Either the light directly

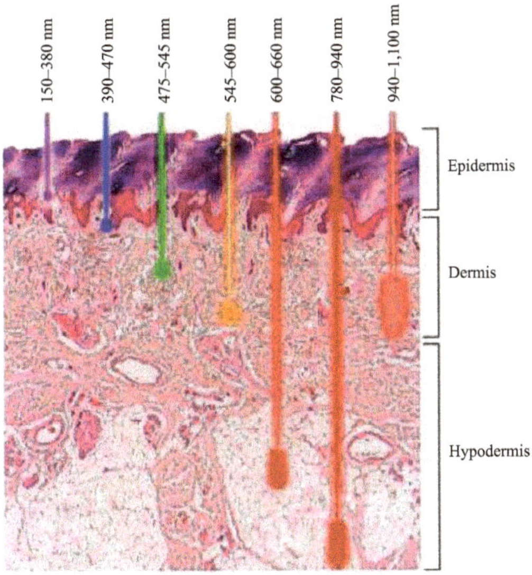

Wavelength (nm)	Color	Penetration (mm)
150–380	Ultraviolet	<0.1
390–470	Violet - cobalt blue	~ 0.3
475–545	Blue - green	~ 0.3–0.5
545–600	Yellow - orange	~ 0.5–1.0
600–650	Red	~ 1.0–2.0
650–950	Carmine red - infrared	2–3
950–1,200	Infrared	1

Figure 14.5 Depth of skin penetration based on wavelength [18]

reflected from the skin is analyzed, or the system analyzes the dispersed light generated after its propagation through the skin.

14.5.1 Using the dispersed light

Due to different composition of different layers of the skin, various wavelengths will experience different absorption rate at different layers. In Figure 14.5, we can see an example of such differences. The image outlines the average penetration distance into the skin based on the wavelength of a chosen light.

Based on these properties, an innovative approach to liveness detection can be proposed. This system will use the variation in absorption as a function of wavelength. For the system to work, a source of light is required and photosensitive parts that will record the rate and characteristics of absorption.

In Figure 14.6, the prototype developed for verifying the hypothesis is presented. It supports placement of light-emitting diodes of various wavelengths, wide spectrum photodiodes for overall absorption rate information and RGBW (Red Green Blue White) detectors, to analyze the absorption at individual color channels [17].

The range of wavelengths has been investigated, such as infrared LEDs of wavelengths 740 and 940 nm, ultraviolet LED with wavelength of 400 nm and RGB LED with dominant wavelengths of 650, 520 and 460 nm.

Via experimentation, it has been determined that measuring the characteristics of dispersion from RGB LED and 740 nm IR LED is a viable method of liveness detection. At these wavelength ranges, the characteristics were distinguishable from the most common materials used for spoof construction.

To attack this antispoof method, a nonhomogeneous, multilayer spoof would need to be developed, and to the knowledge of the authors, no research has been

Figure 14.6 Liveness detector—prototype

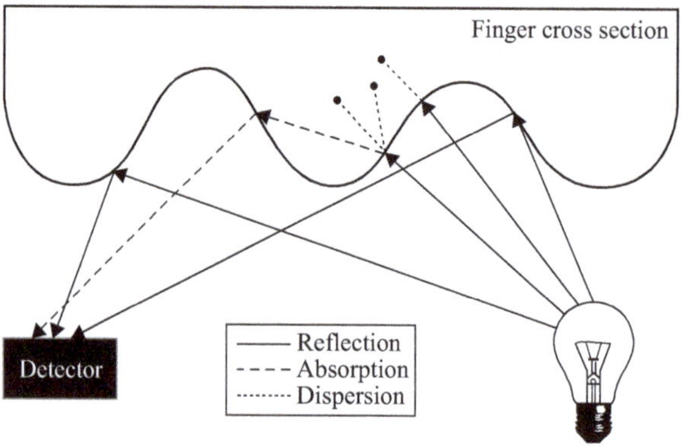

Figure 14.7 Liveness detection based on spectral characteristics [20]

performed in this area of hand spoof development. The UV light presented mixed results, the spectral characteristics of dispersed light was differentiable from most spoofs, but the interclass variability was too large, influenced by a skin pigmentation. Further research needs to be performed in this area.

This method however presents a disadvantage, as the analyzed hand needs to be in contact with the device, to eliminate the effect of reflected light. By forming a seal around the chosen LED, the only source of light that reaches the sensors is the light that propagates through the skin and tissue. This prevents the construction of the touchless biometric system using this technology. To overcome this limitation, another approach to spectral analysis needs to be used. Viable alternative comes in the form of analysis of reflected light.

14.5.2 Using reflected light

In Figure 14.7, the difference between reflection and dispersion is outlined. Whereas the dispersion utilizes only the light that has penetrated the skin, the sensors based on reflection use the directly reflected light and dispersed. The observed characteristics of various wavelengths reaching the detector are then used for the liveness detection [19]. Using this principle, the biometric sensor may be touchless.

To determine liveness based on reflected light properties, an appropriate wavelength must be chosen. To enable this system more for integration into current technologies, the wavelengths in ranges used by current cameras used in a biometric sensor are preferred; for that reason, wavelengths of 400, 410, 470, 525, 550, 570, 590, 635, 700 and 800 nm have been investigated. In Figures 14.8 and 14.9, we can see a finger illuminated using some of these wavelengths.

Liveness detection can be then performed in the following steps:

- During acquisition, a skin sample is sequentially illuminated using chosen wavelengths and the images are being acquired
- The image is blurred to eliminate glares and unwanted artifacts
- Pixels/pixel ranges are extracted in a repeatable manner as features
- Feature vector is constructed
- Classification is performed

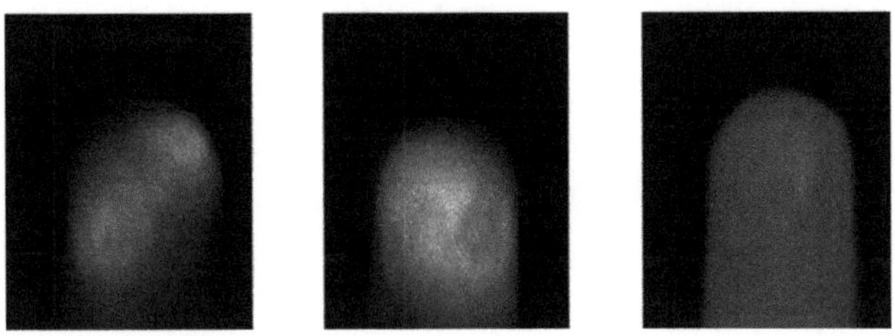

Figure 14.8 Fingertip illuminated using LED of 400 (left), 470 (middle) and 525 nm (right)

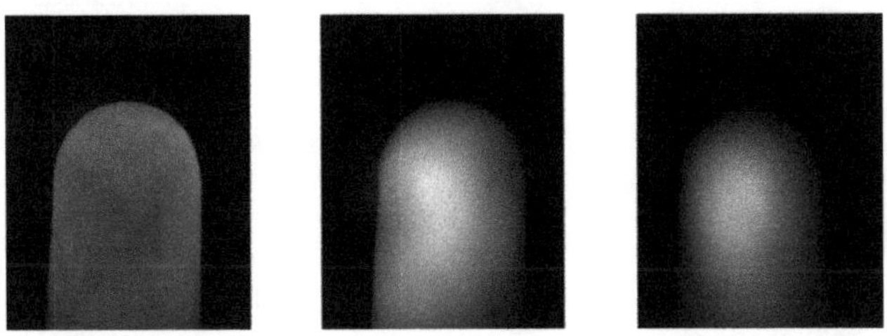

Figure 14.9 Fingertip illuminated using LED of 550 (left), 635 (middle) and 700 nm (right)

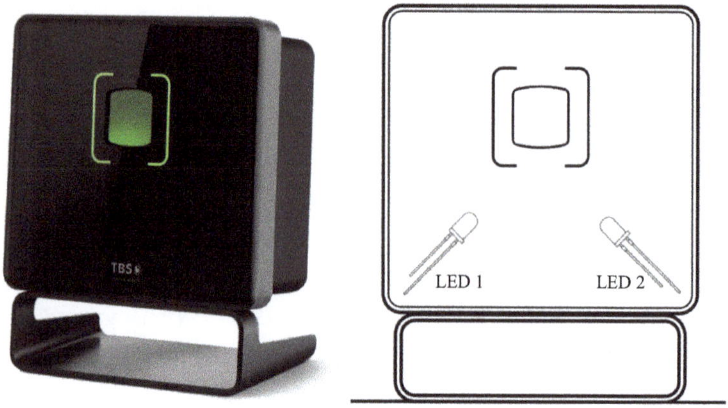

Figure 14.10 Fingerprint acquisition device [21] (left) and hardware modification for liveness detection (right)

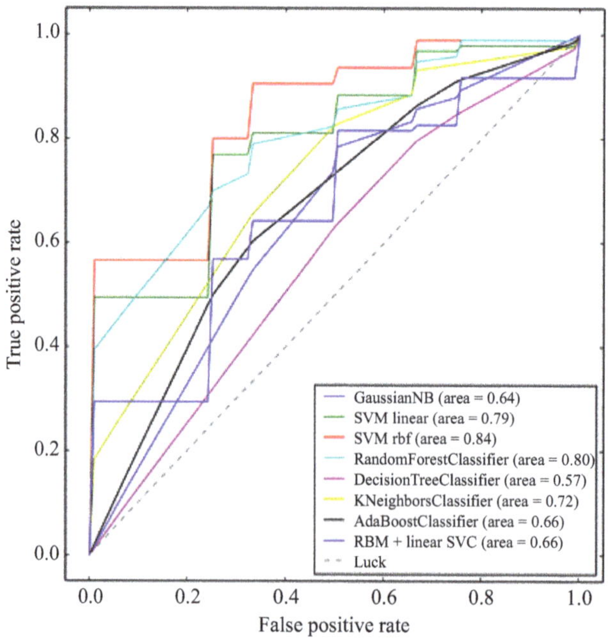

Figure 14.11 ROC of liveness detection based on optical properties using various machine-learning algorithms

As was indicated, the advantage of this approach lies in ease of integration into existing systems. Any optical based biometric system only needs to integrate the light source of chosen wavelengths into itself. As we can see in Figure 14.10, existing biometrics solution can be simply modified to support this method of liveness detection.

Due to feature vector being constructed from potentially large amount of data, an appropriate machine-learning method needs to be utilized to prepare a strong classifier.

Figure 14.11 shows performance of classifiers, constructed using various machine-learning approaches, with strongest being created using the support vector machine with radial basis function kernel.

From the figure we can also see that the underlying principle is valid, but further research needs to be performed to perfect this approach.

14.6 Hematoma (improvement of antispoofing methods in the presence of hematoma)

Hematoma, popularly called bluish or bruise, is caused by bleeding from damaged blood vessels under the skin. The word comes from the Greek words *hema* and *soma*, which means blood and body. There are many cases of hematoma; the hematoma is primarily due to trauma. It is caused by a strong pressure and/or a stroke that causes damage to the bloodstream. After it, the blood leaks from vessels and veins to interstitial space. The human organism has mechanism to stop blood leakage and is based on vessels contraction and hemostatic substances. As a result of these circulatory system mechanisms, the human organism can protect itself against a large loss of blood. In this case, hematoma almost disappears from organism after some time. In some cases, the large hematomas need to be removed surgically.

The other cause of the formation hematoma is fragile meningococcal meningitis, autoimmune diseases such as congenital and acquired disorders blood clotting, absence of vitamin K, liver disease, bone marrow disorders and low blood platelet counts, e.g., thrombocytosis [22]. The hematoma is a common problem encountered by every human. In the field of medicine, hematoma is named by localization on the human body. Accordingly, examples of some hematomas:

- *Subdermal hematoma* is hematoma located below skin (commonly called bruise)
- *Subgaleal hematoma* is boundary accumulation of blood between Epicranial aponeurosis and periosteum [23]
- *Cephalohematoma* is boundary accumulation of blood under the periosteum that covers most of the cranial bone in hairy head [23]
- *Epidural hematoma* is local accumulation of blood between cranial bones and dura mater [24]
- *Subdural hematoma* is hematoma beneath the dura of skull or in the spinal cord [24]
- *Subarachnoid hemorrhage* is extravasation of blood into the subarachnoid space [24]
- *Auricular hematoma* is boundary accumulation of blood located between skin and cartilage of ear [25]

- *Monocular hematoma* occurs when soft tissue is injured or the bones of face are fractured [26], while the binocular hematoma occurs around the fractured bones
- *Perianal hematoma* is acute thrombosis of the anal blood vessels
- *Intermuscular hematoma* occurs between muscles
- *Intramuscular hematoma* occurs into muscle
- *Subungual hematoma* occurs under nail [27]

Of these types of hematomas, the subdermal hematoma can be used as part of antispoofing solution. The subungual hematoma is visible too, but multispectral properties are affected by nail, which overlaps the hematoma. For capturing, multispectral properties are using the devices described in the previous section.

14.7 Hematoma stages

The human body contains a lot of chemical substances; in the case of blood, it is the same too. After a hematoma formation, chemical reactions between the blood components and other chemicals of the human organism begin. Therefore, the properties of hematoma change over time. Time evolution of the hematoma can be divided into five stages according to the hemoglobin conversion. From a multispectral point of view, hemoglobin is an important chemical substance that has a significant influence on light absorption. The color of hemoglobin depends on iron atoms and oxygen [28]. The following stages are divided by hemoglobin conversion to other chemical substances [29]. For us, however, it is important that bruise color is different in each stage [30].

In the biometrics field, multispectral scanning can be used for extraction distribution on veins and vessels. Since the different types of hemoglobin have different absorption coefficient, veins and vessels image can be extracted separately. Since the distribution of veins exhibits a high degree of variability, this can be used as biometrics. The vein authentication device convert image with vessel lines into binary image and then compare with template from the database.

14.7.1 Stage—oxyhemoglobin

After the trauma, the red blood cells (RBCs) are extravasated from the bloodstream into the interstitial space. If trauma causes subdermal hematoma (bruise), a red or blue spot appears on the skin. In the stage, the extravasated hemoglobin (oxyHb) in RBCs can still carry oxygen; therefore, the spot has color depending on the ratio of oxygenated and deoxygenated hemoglobin (deoxyHb). The edges and area of bruise look according to location and how much the blood vessels have been damaged [31]. The oxyhemoglobin has at the beginning of the NIR spectrum (around 700 nm) a lower rate of absorption coefficient than deoxyhemoglobin, shown in Figure 14.12. In the stage, it can be determined whether this is a venous or vascular bleeding by NIR light and photo sensor.

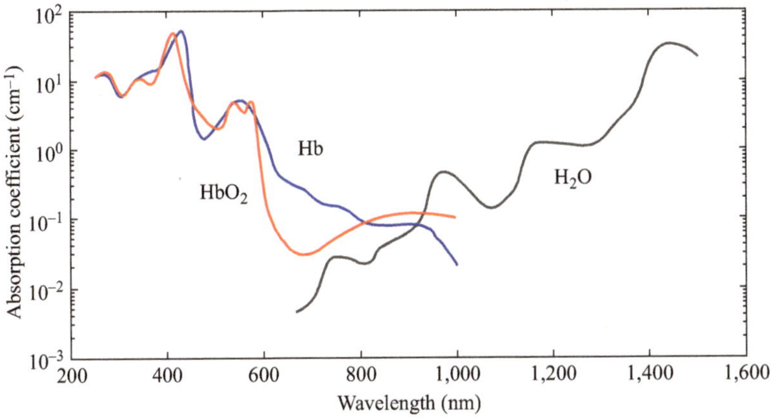

Figure 14.12 Absorption spectra of hemoglobin—Hb, deoxyhemoglobin; HbO_2, oxyhemoglobin; H_2O, water [2]

14.7.2 Stage—deoxyhemoglobin

After 1–5 days, almost all molecules of hemoglobin do not contain oxygen. According to one plausible theory, when the oxygen is not bound in hemoglobin, the color is less red than then color of oxyhemoglobin, and therefore the hemoglobin absorbs redder spectrum [32]. If the red wavelengths are subtracted, the blue spectrum is more predominate; therefore, skin has bluish color [33]. This phenomenon is called cyanosis. Therefore, the color of bruise gradually turns shades of blue. According to previous information, deoxyhemoglobin has different spectral properties against the oxyhemoglobin.

For hematoma detection, the stage provides best conditions, because color difference between area without and with hematoma is the biggest of all stages, as shown in Figure 14.13.

14.7.3 Stage—biliverdin

The hemoglobin breakdown process begins 5 days later. First, the molecule of deoxyhemoglobin is divided into molecules of globin and molecules of heme. Then the molecules of heme are split by enzyme on hemoxygenase and molecule of oxygen into tetrapyrrole, which is called biliverdin. During the process, the equimolar amount of ferro cation (Fe^{3+}) is relaxed too [34]. Because the biliverdin contains conjugated double bond, the biliverdin has a green color.

14.7.4 Stage—bilirubin

After 7 days, the bruise changes color from shades of green to shades of yellow. This is caused by conversion of biliverdin to bilirubin. Conversion from biliverdin

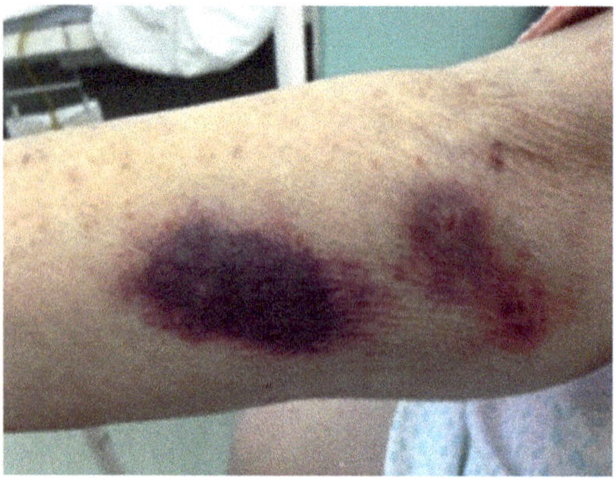

Figure 14.13 Hematoma in the second stage

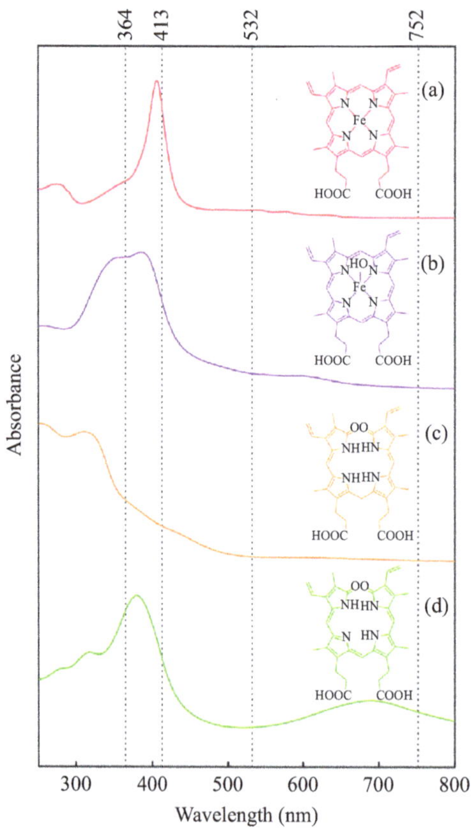

Figure 14.14 Schematic chemical structures and absorption spectra in aqueous solution of the heme chromophores: a—hemoglobin (red), b—hematin (purple), c—bilirubin (orange), d—biliverdin (green)

is carried out using a coenzyme nicotinamide adenine dinucleotide phosphate that occurs as part of the metabolism of organisms. Because bilirubin is the main product of heme degradation, it participates most in the resulting color of the hematoma at this stage [34,35]. According to Figure 14.14, the bilirubin has different spectral characteristics than biliverdin, the greatest difference being in the range from 300 to 364 nm wavelength.

14.7.5 Stage—hemosiderin

After that, all the RBCs may be resorbed, and the skin resumes its normal color. In some cases, ferro cation (Fe^{3+}), which is product of heme catabolism, is taken up by macrophages and degraded into hemosiderin [36]. This causes the coloring of bruise from yellow to brown. The main disadvantage of hemosiderin is that it can cause pigment stains which can last a long time.

However, the stages of hemoglobin conversion may seem unimportant in terms of antispoofing, the opposite is true. The stages of hematoma can be useful information for liveness detection, since spectral properties of hemoglobin and hemoglobin products are quite unique. When hematoma is detected on the hand, we can use multispectral analysis for precision liveness detection. For this reason, hematoma localization is important for the possibility of further evaluation.

14.8 Hematoma detection

The success of hematoma detection depends on the shape, shade of hematoma color and skin color. In the stages 1–4, hematoma has more intense color than normal skin of europoid (light skin). However, when the human has dark skin, it can be assumed that the detection based on skin color will be inaccurate. The process of hematoma detection on light skin can be divided into five stages, shown in Figure 14.15.

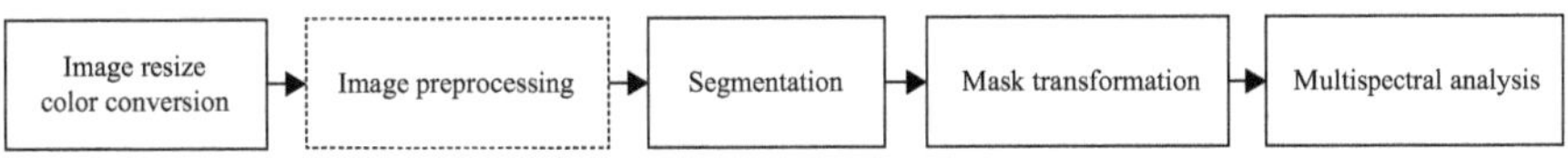

Figure 14.15 Hematoma detection process

14.9 Image resize and color conversion

For faster hematoma segmentation, it is advisable to downscale the size of the input image. If the image has low resolution, this step can be skipped. Most algorithms for hematoma detection are faster with the downscale image than with original image. In the last step, all coordinates of edges of hematoma must recalculate with scale coefficient. The coefficient is determined from the aspect ratio of the original and the reduced image. Almost all images are saved in the RGB color representation; but for some algorithms, it is better to use Hue Saturation Value (HSV) representation [37]. Therefore, color conversion is the useful step for next image processing. Saturation between areas is useful information for hematoma segmentation.

14.9.1 Image preprocessing

During image processing, it is often needed to remove high-frequency areas from the image. In the other words, the noise needs to be reduced. One way to remove noise is to use the convolution mask that represents selected function, e.g., Gaussian function.

14.9.2 Segmentation

After editing the image, the objective is to get a binary image consisting of two areas. One area is made up of an object of interest (hematoma), and the other consists of a background. This part of algorithm is the most important, but also it is the most difficult to implement. The simplest method is thresholding based on the difference in brightness. However, the method has bad performance when the skin color is dark. Advanced methods can be based on texture analysis and thresholding multiple color channels.

In some cases, the binary area, which represents area without hematoma, contains small enclosed areas. Areas that do not belong to the hematoma are scattered across the area, so it is advisable to remove them using a morphological operation [38]. These disadvantages can be removed by morphological operation *Open*. After the operation is used, the binary mask has smooth and straight contours, and background mask does not contain noise.

14.9.3 Scale and border extraction

If edges of hematoma for advanced algorithms are needed, they should be extracted from binary mask. The morphological operation erode is using for creation edges of hematoma. The resulting contour is obtained by subtracting the mask after applying the erosion filter from the input binary mask, as shown in Figure 14.16. If the input image was downscaled, the mask must be upscaled by the ratio from the first step. The output mask defined localization of hematoma and can be used for accuracy liveness detection.

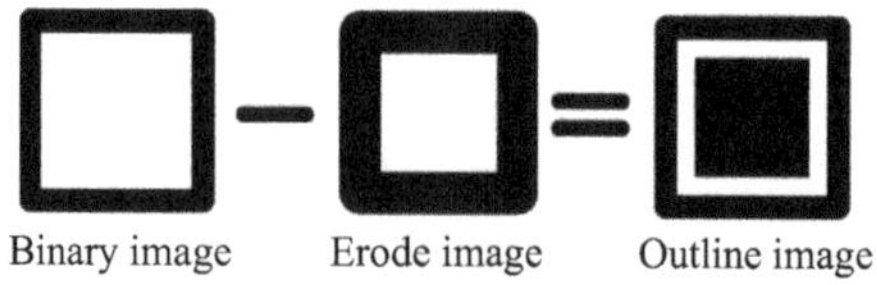

Figure 14.16 Outline extraction

14.10 Multispectral analysis

After localization, the hematoma is analyzed. The easiest solution is creating histogram for all hematoma stages and comparing them. This approach is strongly affected by various hematomas; therefore, it is not as good as antispoofing solution. Enhancements should be achieved using histogram-based feature vectors from different color representations, such as RGB + HSV. A better method can be based on spectral analysis. According to previous information, the hematoma (bruise) has different absorbance across an electromagnetic spectrum in individual phases. This feature can be used to determine whether it is a fake hand or none. Histogram-based algorithms may not detect fake based on artificially created color spot; therefore, the multispectral approaches are better. Among the problems of this detection are the effects of other chemical substances on the body and probably the possibility to use a hand several hours after death. After death, however, hemoglobin degrades to other substances, but conversion from oxyhemoglobin to deoxyhemoglobin is the same from living to dead tissue. Unfortunately, the bruises are affected by other chemical substances and physical phenomena. For example, about 20 hours after death, the pH of blood decreases to 5.5, which affects other chemical processes in the decomposition of the body [39]. Real use of the postmortem hand for spoofing will have to be verified by other experiments.

Despite everything, the multispectral analysis of hematoma to improve the accuracy of antispoofing methods is an unconventional but interesting approach. However, multispectral skin analysis has a great benefit for liveness detection.

14.11 Liveness detection system based on finger vein pattern

The liveness detection can be perceived from a different point of view. The main task is to distinguish presented biometric samples between living and nonliving classes. The requirements on the proposed system are as follows:

- The system is divided into two separated subsystems acquisition of images and image processing.
- The first subsystem is built using existing fingerprint sensor.
- The second subsystem must be hardware independent, and fully automated input image is automatically classified as live or nonlive without any user input.
- The second subsystem supports a "recognition" mode, which, instead of classification into live or nonlive category, verifies registered users.

14.11.1 Image acquisition

First step in liveness detection on fingers is the acquisition of a presented finger. For our experiments in the laboratory, we selected a TBS fingerprint sensor and extended it with liveness detection based on finger vein pattern (Figure 14.17), however this solution is suitable for any touchless fingerprint reader.

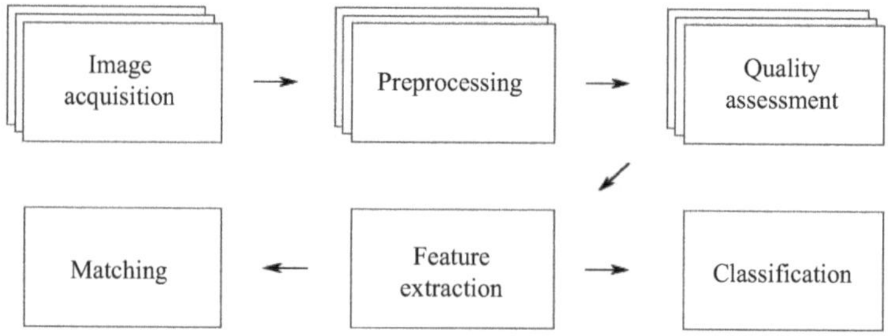

Figure 14.17 Designed architecture of a proposed system

14.11.2 Touchless biometric systems fingerprint sensor

Commercial device TBS 3D-Enroll is a three-dimensional touchless sensor for enrollment and generation of fingerprint templates. There are three cameras inside the sensor as can be seen in Figure 14.10 and several LEDs of various colors. It should be mentioned that the sensor is designed to acquire only a fingertip and therefore a user inserts only this part of the finger into the box. The sizes of both veins and arteries decrease with the distance from heart. For vein recognition, usually the most interesting are parts of finger areas near the middle and proximal phalanges. The blood vessels near distal phalanx are hardly visible, but only image of this finger part is acquirable using this device.

14.11.3 Extension of an existing sensor

Two different methods of sensing are reflection and transmission.

The main disadvantages of transmission method are the size of the device and that a user must put his finger into a "black box." On the other hand, it seems that the transmission method gives images, where the finger veins are more visible and easily distinguishable in comparison to reflection method. Since all these disadvantages are already present in a TBS sensor, the transmission method was chosen to obtain a more high-contrast images. A decision to utilize only the middle camera for this prototype was made; but in future, all three cameras could be used to reconstruct the 3D model of blood vessels of the finger. A board was constructed that five NIR LEDs could be mounted to the top of the fingerprint sensor and the construction allowed effective changing of LEDs and the voltage. The board is powered from the sensor's main power supply, so no additional power supply is needed.

LEDs are connected in parallel to allow one to set different levels of intensity for each LED, because the fingertip thickness and tissue characteristic varies for everyone.

Many finger vein solutions [14,40,41] use automatic brightness control to eliminate such individual variations; thus, the system can be extended with this automated approach. This solution was not incorporated in this work, because it increases both hardware and software demands. After several experiments, we decided to give the resistor R_1 higher resistance in comparison to others ($R_2 = R_3 = R_4 = R_5$) (Figure 14.18).

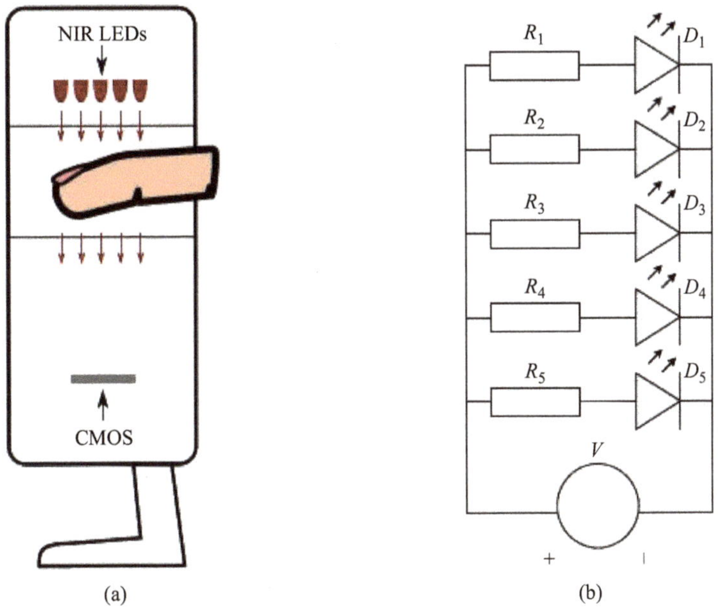

Figure 14.18 Proposed solution of finger vein sensing: (a) overall view and (b) wiring diagram of the extension circuit

The various thicknesses of a finger make it difficult for choosing a general camera setting. As can be seen in Figure 14.19, when a thin (top row) finger is captured, a shorter exposure time is needed in comparison to thicker ones (bottom row). A possible software solution combines images of the finger with different exposure times using high dynamic range imaging (HDR) technique, which produces a single-HDR image with greater dynamic range of luminosity. A different solution is to capture several images with various exposure times and then to use quality evaluation algorithms to choose the best from image series. Since the camera had to be reprogramed each time, the exposure time is changed to capture a sequence of images taking several seconds, and therefore HDR approach could not be utilized due to possible finger posture changes.

14.12 Preprocessing

For the preprocessing of the originally captured images, first, the image is segmented to obtain the finger mask, and after that the elimination of finger posture changes will be described.

14.12.1 Region of interest localization

Conventional algorithms for finger vein region of interest (ROI) localization assume touch-based sensor, and therefore, the ROI of the finger is obtained by applying a fixed window mask. This straightforward approach cannot be used in

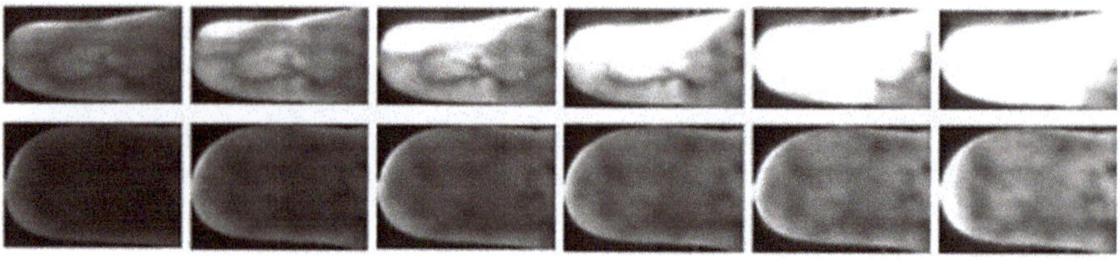

Figure 14.19 *Finger vein images captured with different exposure times. In each row, there are images of the same finger, and appropriate columns belong to the same exposure time. The exposure time grows in the right. All images were enhanced using CLAHE*

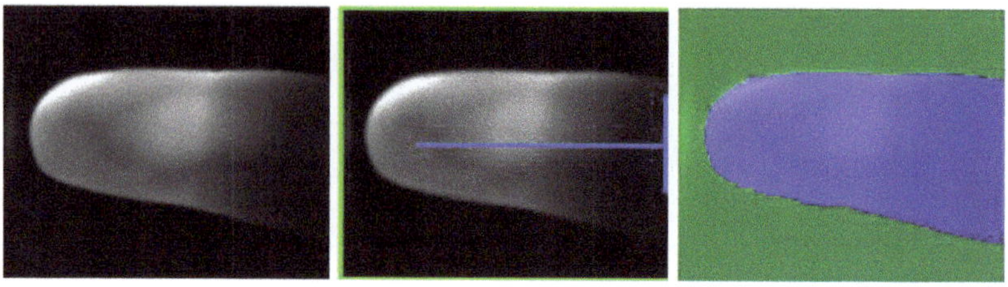

(a) (b) (c)

Figure 14.20 *Example of Watershed procedure. needle blue of part (b) W_f, background green of part (c) W_b. (a) Original image, (b) defined marker sets and (c) segmented image*

most touchless systems, and therefore a method using Watershed for ROI localization will be introduced.

The Watershed algorithm was utilized to distinguish between the finger and the background.

To use the Watershed algorithm, for all classes, there must be preset a set of pixels, which belong to the appropriate class. Let us denote the image size as a tuple (x_m, y_m). The background marker set W_b is defined using (14.1), and the finger marker set W_f using (14.2). Examples of predefined sets can be seen in Figure 14.20(b), and the result of the Watershed algorithm is displayed in Figure 14.20(c).

$$W_b = \{(1,y)|y \in \{1,\ldots,y_m\}\} \cup \{(x,y)|x \in \{1,\ldots,x_m\} \wedge y \in \{1,y_m\}\} \tag{14.1}$$

$$W_f = \left\{(x,y)\middle|x \in \left\{\frac{1}{4}x_m,\ldots,x_m\right\}, y = \frac{1}{2}y_m\right\} \cup \left\{(x_m,y)\middle|y \in \left\{\frac{1}{3}y_m, \frac{2}{3}y_m\right\}\right\} \tag{14.2}$$

According to the segmented image, mask $M(x, y)$ is defined, it returns 1 if (x, y) belongs to the finger segment, 0 otherwise. To reduce possible inaccuracies in segmentation, the morphological opening is applied on finger mask M, followed by morphological closing.

Table 14.1 Defined coordinate system and six finger posture changes based on [42]

Type	**Name**	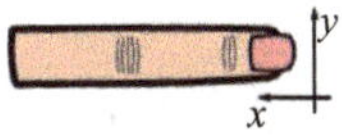	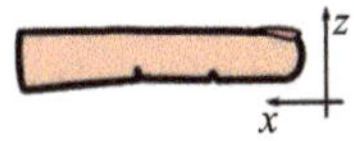	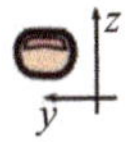
1	Shift along x-axis			
2	Rotation around x-axis			
3	Shift along y-axis			
4	Rotation around y-axis			
5	Shift along z-axis			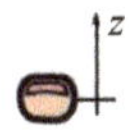
6	Rotation around z-axis	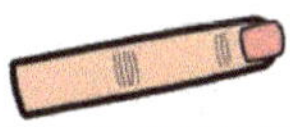		

14.12.2 Correction of finger posture changes

There are six basic finger posture changes described in Table 14.1.

Type 6 correction: Type 6 finger posture change can be eliminated by rotation around the z-axis using rotation matrix $R(\theta)$. To obtain the unknown rotation θ, sets C_{top}, C_{bottom} and $C_{middleline}$ in equation sets below are defined:

$$C_{top} = \{(x,y)|M(x,y) = 1 \wedge M(x,y-1) = 0\}$$
$$C_{bottom} = \{(x,y)|M(x,y) = 1 \wedge M(x,y-1) = 0\}$$
$$C_{middleline} = \left\{(x,y)\middle|\exists(x,y_T) \in C_{top} \wedge \exists(x,y_B) \in C_{bottom} \wedge yB + yT = 2y\right\}$$

where $M(x, y)$ is an image segmentation mask. Examples of these sets are visualized in Figure 14.21(b).

Type 3 correction: When the Type 6 posture change is eliminated, it is possible to crop the finger image according to the finger mask to minimize the image size while preserving the ROI. Consequently, the Type 3 change is eliminated as well. This step is visualized in Figure 14.21(d).

Type 1 correction: The Type 1 posture change is trivially solved, since the position of the fingertip is known, and therefore alignment according to left border solves the Type 1 posture change.

Type 5 correction: To eliminate Type 5 posture change, the finger vein image is normalized to a rectangle using a finger image mask. The image with size (x_{max}, y_{max}) is separated by columns and each column segment, where $M(x, y) = 1$ is linearly stretched to the size $(1, y_{max})$. The result of this operation is displayed in Figure 14.21(e).

Both Types 2 and 4 posture changes are ignored in this work.

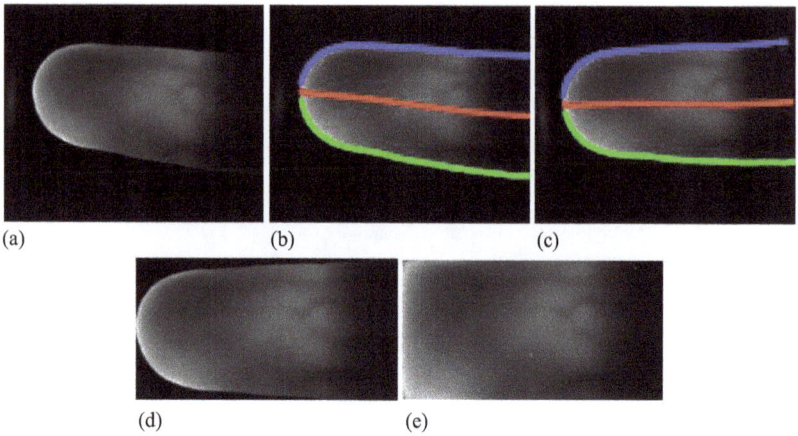

Figure 14.21 Procedure of elimination of finger posture changes—C_{top}, C_{bottom}, $C_{middleline}$ sets: (a) original image, (b) visualized sets C_*, (c) rotated by θ, (d) cropped and (e) stretched

14.12.3 Interphalangeal joint localization

The phalangeal joints of a finger are constituted by several components including synovial fluid [43], the density of which is much lower than the density of bones. Therefore, when using transmission method to acquire a finger vein image, a brighter region can be seen in the acquired image, which corresponds to the interphalangeal joint. This region can be substituted by a single line with a pixel width [43]. The Type 6 change is ignored, and sum of all gray level values for each column (perpendicular to the length of the finger) is computed in (14.3), and the maximum row-sum in (14.4) approximately denotes the position of the interphalangeal joint. After Type 2 correction, it is possible to detect the interphalangeal joint. Better results were obtained using weighted ϕ in (14.5), where smaller weight is given with the length of the finger. This weighting is necessary, due to frequent wrong positioning of the user's finger. The fingertip is then overilluminated and might cause wrongly localized interphalangeal joint.

$$\phi(x) = \sum_{y=1}^{y_{\max}} f(x, y) \tag{14.3}$$

$$r_k = \underset{x \in \{1, \ldots, x_{\max}\}}{\arg\max} \left(\phi(x)\right) \tag{14.4}$$

$$r_k = \underset{x \in \{1, \ldots, x_{\max}\}}{\arg\max} \left(\sqrt{\frac{x}{x_{\max}}} \cdot \phi(x)\right) \tag{14.5}$$

14.12.4 Size normalization

Having a feature vector of a constant size can be achieved by cropping the stretched image followed by scaling. For cropping, the r_k (interphalangeal joint approximation) is used as right border and $r_k/8$ is used as a left border. The height of the image is preserved. After that, the cropped image is scaled to size 300 × 200 pixel.

Due to various thicknesses of fingers of different persons, it is necessary to acquire several images of the same finger with different exposure time. Each finger vein record is contained of a totally ordered set V with six pictures; this set will be denoted as a sequence. There exists a function e: $V \rightarrow N$, which assigns each image an exposure time expressed as a natural number and e is totally ordered on V. The quality estimation function q: $V \rightarrow R$ assigns a score to each image. After that, the image with best score I_{best} is chosen: $I_{best} = \arg\max q(v)$, $(v \in V)$ from the sequence. Several concrete quality estimation approaches can be utilized to choose the best quality image and its score from V and will be presented below.

No heuristic A quality estimation *NoHeuristici*, which statically assigns 1 to the image with the Ith highest exposure time and 0 to others.

Haralick $HaralickF(d, \theta)$ is a quality estimation function, which assigns score to image according to selected Haralick feature F, where $F \in$ {*Energy*, *Contrast*, *Homogenity*, *Entropy*} and displacement vector (d, θ).

An expert can choose a reference image, which histogram will be computed and another image's quality will be estimated by histogram comparison, for example using chi-square method.

14.13 Feature extraction

Various features, which could be extracted from finger vein images, will be described as follows:

- *Local binary patterns* (LBP) is a method for extracting textural features of a grayscale image. The original LBP [44] uses 3*3 nonparametric operator for labeling each pixel (x_c, y_c) by thresholding the eight-neighborhood with the gray level of the center pixel (x_c, y_c) and summing threshold values weighted by powers of two [45].
- *Repeated line tracking:* This method extracts finger vein pattern even from a noised and irregularly shaded image. The method is based on tracking dark lines starting at various pixel positions. The dark line is followed until it is not detectable anymore, and then a new line from different pixel is started.
- *Maximum curvature:* It is claimed to be resistant to various vein widths and its brightness. The algorithm has become a benchmark for the newly developed finger vein extraction and comparison methods [46–48].

The Haralick and Local Binary Patterns Histogram (LBPH) features described were utilized as a texture describing features, which will be mainly used as a discriminative feature for liveness detection.

In the case of Haralick features, all four mentioned (contrast, energy, homogenity and entropy) are concatenated to a single-feature vector. On the other hand, maximum curvature and repeated line tracking are suited for detailed vein evaluation and therefore finger vein recognition. The outcome of both maximum curvature and repeated line tracking is a locus space with probability of being a vein. This locus space is thresholded using threshold t, computed as t-percentile of nonzero locus space values.

The last step in processing pipeline is dependent on the chosen mode: liveness detection or vein recognition. In the classification mode, binary (with classes live and nonlive) nonlinear probabilistic support vector machine (SVM) with radial basis function (RBF) as a kernel is used for final decision. The recognition mode details are dependent on used feature extraction method. To determine score of two LBPH vectors, a chi-square metric is used. In the case of repeated line tracking and maximum curvature, the veins are aligned using iterative closest point (ICP) algorithm, after that the detected veins are emphasized with morphological operations and the final comparison score is based on Hamming distance of intersecting regions.

14.14 Quality annotator

An application for manual quality annotation of finger vein was implemented for objective evaluation of quality-assessment algorithms. It scans for all sets of six images from the same session and then it displays them on screen using QT framework7. An expert can then assign the images a label (good or bad) as can be seen in Figure 14.22. The output of quality annotator can be used in evaluation subsystem to evaluate different approaches of quality assessment.

14.15 Dataset description

To evaluate the proposed system, a dataset containing both live fingers and fingers spoofs was gathered. Six fingers (both index, middle and ring fingers) of each volunteer were captured, and each finger can be perceived as a biometric sample of a unique individual. Each finger was captured six times in two different days to consider possible changes of vein visibility. The nonlive dataset consists of 370 sequences of 59 different materials or material combinations.

14.16 Conclusion

In this chapter, we have introduced multiple general approaches that can and are being used for purposes of liveness detection in hand and finger-based biometrics. Methods based on spectral analysis have been further elaborated on. The principle using the light absorption as well as light reflection has been discussed, and a method of acquiring data described. The advantages, disadvantages and results of mentioned approaches to liveness detection were discussed.

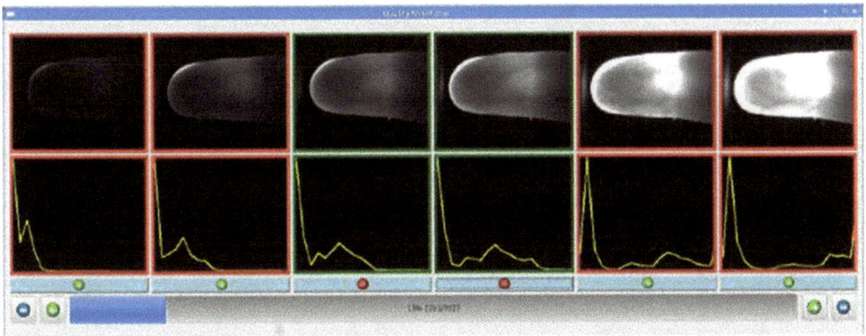

Figure 14.22 Screenshot of quality annotator application. The images highlighted with green color was marked as good; on the other hand, images highlighted with red are marked as poor quality images

Lately, the spoofing attacks have shown that it is possible to fool most commercial fingerprint sensors with artificial fingerprints. On the contrary, finger vein recognition is a relatively new but promising method suitable for biometric authentication achieving high recognition performance. One advantage of vein recognition is that blood vessels are an internal biometric characteristic, and it is harder for an attacker to obtain it from a user. Image of veins is usually acquired using infrared radiation for biometric purposes. When we create a biometric system, which examines both mentioned characteristics at once, it is more demanding to compromise it. An intruder must obtain both fingerprint and image of finger veins of a legitimate user and merge those into a single spoof, which has desired optical characteristics. All implemented methods were thoroughly evaluated and compared. In liveness detection mode, EER equal to 6.5% was achieved using LBPH features. In finger vein verification mode, the best result EER equal to 10.8% was obtained using maximum curvature features. A future development, which may improve the error rates, was outlined including reconstruction of 3D model of blood vessels. As we can see, the error rates are quite high for unimodal system, but in fusion with high precision fingerprint recognition, it might be possible to preclude most of impostor attempts without rejecting any legitimate user.

Acknowledgment

This work was supported by The Ministry of Education, Youth and Sports of the Czech Republic from the National Programme of Sustainability (NPU II); project IT4Innovations excellence in science—LQ1602.

References

[1] Jain, A.K., Flynn, P. and Ross, A.A. *Handbook of Biometrics*. Springer-Verlag New York, Inc., Secaucus, NJ, USA, 2007. ISBN: 978-0387710402.

[2] Sandstrfom, M. *Liveness Detection in Fingerprint Recognition Systems*. Ph.D. thesis, Linköping University, 2004. Available at: http://www.diva-portal.org/smash/get/diva2:19729/FULLTEXT01.pdf.

[3] Feng, J., Jain, A.K. and Ross, A. *Fingerprint Alteration*. Tech. Rep. MSU-CSE-09-30, Department of Computer Science, Michigan State University, East Lansing, Michigan, Dec. 2009.

[4] Drahanský, M. *Liveness Detection in Biometrics*. InTech—Open Access Publisher, Rijeka, 2011, pp. 179–198. ISBN: 978-953-307-487-0.

[5] Drahanský, M., Funk, W. and Nötzel, R. *Liveness Detection based on Fine Movements of the Fingertip Surface*. In: IEEE – The West Point Workshop, West Point, New York, 2006, pp. 42–47. ISBN: 1-4244-0130-5.

[6] Drahanský, M. *Methods for Quality Determination of Papillary Lines in Fingerprints*. NIST, Gaithersburg, 2007, p. 25.

[7] Drahanský, M. and Lodrová, D. *Liveness Detection for Biometric Systems Based on Papillary Lines*. In: 2008 International Conference on Information Security and Assurance (ISA 2008) [online]. IEEE, 2008, pp. 439–444 [cit. 2017-10-13]. DOI: 10.1109/ISA.2008.58. ISBN: 978-0-7695-3126-7.

[8] Shahzad, A. M. *Novel Active Sweat Pores Based Liveness Detection Techniques for Fingerprint Biometrics*. Brunel University, 2012. Available at: http://bura.brunel.ac.uk/bitstream/2438/7060/1/FulltextThesis.pdf.

[9] Shuckers, S., Hornak, L., Norman, T., Derakhshani, R. and Parthnasardi, S. *Issues for Liveness Detection in Biometrics*. West Virginia University, 2006. Available at: http://www.biometrics.org/bc2002/2bc0130 DerakhshabiBrief.pdf.

[10] Drahanský, M. *Liveness Detection in Biometrics*. Chetty, Girija, ed. Advanced Biometric Technologies [online]. InTech, 2011 [cit. 2017-11-12]. DOI: 10.5772/17205. ISBN: 978-953-307-487-0.

[11] Hartung, D. *Vascular Pattern Recognition and its Application in Privacy-Preserving Biometric Online-Banking Systems*. Ph.D. thesis, Gjøvik University College, 2012. ISBN: 978-82-93269-01-4.

[12] Fuksis, R., Greitans, M., Nikisins, O. and Pudzs, M. *Infrared imaging system for analysis of blood vessel structure*. Elektronika ir Elektrotechnika 97, 1 (2010), pp. 45–48. ISSN: 2029-5731.

[13] Anderson, R.R. and Parrish, J.A. *The optics of human skin*. Journal of Investigative Dermatology 77, 1 (1981), pp. 13–19. ISSN: 0022-202X.

[14] Hashimoto, J. *Finger vein authentication technology and its future*. In: VLSI Circuits, 2006. Digest of Technical Papers. 2006 Symposium on (2006), pp. 5–8. ISBN: 1-4244-0006-6.

[15] Yang, J. and Shi, Y. *Towards finger-vein image restoration and enhancement for finger-vein recognition*. Information Sciences 268 (2014), pp. 33–52. ISSN: 0020-0255.

[16] Drahanský, M., Orság, F. and Doležel, M. *Biometrie*, 2011, p. 294. ISBN: 978-80-254-89796.

[17] Drahanský, M., Dvořák, R., Váňa, J., Goldmann, T., Dvořák, M. and Kanich, O. *A multispectral lifecycle detector especially suited for the fingerprint recognition technology*. Utility model no. 31405, Touchless Biometric Systems, 2018.

[18] Leoš, N. *Nové pohledy na neinvazivní laser (New views of non-invasive laser)*. Grada Publishing, 2015. ISBN: 978-80-247-1651-0.

[19] Drahanský, M., Doležel, M., Váňa, J., Březinová, E., Yim, J. and Shim, K. *New optical methods for liveness detection on fingers*. BioMed Research International [online]. 2013 (2013), pp. 1–11 [cit. 2017-10-13]. DOI: 10.1155/2013/197925. ISSN 2314-6133.

[20] Drahanský, M. *Liveness Detection on Fingers*. Advanced Biometric Technologies. InTech—Open Access Publisher, 2011. ISBN: 978-953-307-487-0. Available at: https://www.intechopen.com/books/advanced-biometric-technologies/liveness-detection-in-biometrics.

[21] Touchless Biometric Systems. 3D Enroll—touchless fingerprinting—datasheet [online]. Available at: https://www.tbs-biometrics.com/files/documents/en/TBS_Productsheet_E_3D-ENROLL-07-2017.pdf [Accessed 13 Oct. 2017].

[22] Brown, R. and Davenport J. *Forensic Science: Advanced Investigations*. South-Western Educational Pub, 2011. ISBN: 978-0538450898.

[23] Gleason, C.A. and Juul, S.E. *Avery's Diseases of the Newborn*. E-Book. Elsevier Health Sciences, 2011. ISBN: 9780323401395.

[24] Pryse-Phillips, W. *Companion to Clinical Neurology*. Oxford University Press, 2009. ISBN: 978-0195367720.

[25] Leybell, I. *Auricular Hematoma Drainage* [online]. Available at: https://emedicine.medscape.com/article/82793-overview [Accessed 9 Nov. 2017].

[26] Hardt, N. and Kuttenberger, J. *Craniofacial Trauma: Diagnosis and Management*. Springer Science & Business Media, 2010, ISBN: 978-3-540-33041-7.

[27] Reich, D., Psomadakis, C.E. and Buka, B. *Subungual Hematoma*. In: Top 50 Dermatology Case Studies for Primary Care. Springer, 2017, pp. 125–130, ISBN: 978-3-319-18626-9.

[28] Davis, C.P. *Hemoglobin (Low and High Range Causes)* [online]. Available at: https://www.medicinenet.com/hemoglobin/article.htm [Accessed 9 Nov. 2017].

[29] Thunderpath Media Inc. (2013). *The Stages of Contusion* [online]. Available at: http://www.baronerocks.com/index.php/mnemonics/467-the-stages-of-a-contusion [Accessed 8 Nov. 2017].

[30] Vanezis, P. *Interpreting bruises at necropsy*. Journal of Clinical Pathology 54, 5 (2001), pp. 348–355.

[31] Lecomte, M.J., Holmes, T., Kay, D.P., Simons, J.L. and Vintiner, S.K. *The use of photographs to record variation in bruising response in humans*. Forensic Science International 231, 1 (2013), pp. 213–218.

[32] Clark, V.L. and Kruse, J.A. *Clinical methods: the history, physical, and laboratory examinations*. The Journal of the American Medical Association 264, 21 (1990), pp. 2808–2809.

[33] McGee, S. *Evidence-Based Physical Diagnosis*. E-Book. Elsevier Health Sciences, 2016. ISBN: 978-0323392761.

[34] Kaneko, J.J., Harvey, J.W. and Bruss, M.L. (ed.). *Clinical Biochemistry of Domestic Animals*. Academic Press, 2008. ISBN: 978-0123704917.

[35] Stave, U. (ed.). *Perinatal Physiology*. Springer Science & Business Media, 2013. ISBN: 9781468423167.

[36] Damjanov, I. *Pathology Secrets*. Elsevier Health Sciences, 2009. ISBN: 978-0323055949.

[37] Huang, Z.K. and Liu, D.H. *Segmentation of Color Image Using EM Algorithm in HSV Color Space*. In: 2007 International Conference on Information Acquisition, Seogwipo-si, 2007, pp. 316–319. DOI: 10.1109/ICIA.2007.4295749.

[38] Solomon, C. and Breckon, T. *Fundamentals of Digital Image Processing: A Practical Approach with Examples in Matlab*. John Wiley & Sons, 2011. ISBN: 978-0470844731.

[39] Donaldson, A.E. and Lamont, I.L. *Biochemistry changes that occur after death: potential markers for determining post-mortem interval*. PLoS ONE 8, 11 (2013), p. e82011.

[40] Huang, B., Dai, Y., Li, R., Tang, D. and Li, W. *Finger-vein authentication based on wide line detector and pattern normalization*. In Pattern Recognition (ICPR), 2010 20th International Conference on (2010), IEEE, pp. 1269–1272. ISBN: 978-1-4244-7542-1.

[41] Miura, N., Nagasaka, A. and Miyatake, T. *Feature extraction of finger-vein patterns based on repeated line tracking and its application to personal identification*. Machine Vision and Applications 15, 4 (Oct. 2004), pp. 194–203. ISSN: 0932-8092.

[42] Huang, B., Liu, S. and Li, W. *A finger posture change correction method for finger-vein recognition*. In: Computational Intelligence for Security and Defence Applications (CISDA), 2012 IEEE Symposium on (Jul. 2012), pp. 1–7. ISBN: 978-1-4673-1416-9.

[43] Yang, J. and Shi, Y. *Finger-vein ROI localization and vein ridge enhancement*. Pattern Recognition Letters 33, 12 (2012), pp. 1569–1579. ISSN: 0167-8655.

[44] Park, K.R. *Finger vein recognition by combining global and local features based on SVM*. Computing and Informatics 30, 2 (2011), pp. 295–309. ISSN: 1335-9150.

[45] Ahonen, T., Matas, J., He, C. and Pietikäinen, M. *Rotation invariant image description with local binary pattern histogram Fourier features*. In: Proceedings of the 16th Scandinavian Conference on Image Analysis (Berlin, Heidelberg, 2009), SCIA '09, Springer-Verlag, pp. 61–70. ISBN: 978-3-642-02229-6.

[46] Zhang, H., Liu, Z., Zhao, Q., Zhang, C. and Fan, D. *Finger vein recognition based on Gabor filter*. In: Intelligence Science and Big Data Engineering, C. Sun, F. Fang, Z.-H. Zhou, W. Yang, and Z.-Y. Liu, Eds., Vol. 8261 of Lecture Notes in Computer Science. Springer Berlin Heidelberg, 2013, pp. 827–834. ISBN: 978-3-642-42056-6.

[47] Raghavendra, R., Raja, K., Surbiryala, J. and Busch, C. *A low-cost multi-modal biometric sensor to capture finger vein and fingerprint*. In: Biometrics (IJCB), 2014 IEEE International Joint Conference on (Sep. 2014), pp. 1–7. ISBN: 978-1-4799-3585-7.

[48] Ton, B.T. and Veldhuis, R.N. *A high quality finger vascular pattern dataset collected using a custom designed capturing device.* In: Biometrics (ICB), 2013 International Conference on (2013), IEEE, pp. 1–5. ISBN: 978-1-4799-0310-8.

Chapter 15
Spoofing methods in hand-based biometrics

Martin Drahanský[1], *Ondřej Kanich*[1], *and Michal Dvořák*[1]

15.1 Introduction

The biometric systems, oriented in this chapter especially on fingerprints, have been introduced in the previous chapters. The functionality of such systems is influenced not only by the used technology but also by the surrounding environment (including skin or other diseases). Biased or damaged biometric samples could be rejected after revealing their poor quality, or may be enhanced, what leads to the situation that samples, which would be normally rejected, are accepted after the enhancement process. But this process could present also a risk, because the poor quality of a sample could be caused not only by the sensor technology or the environment, but also by using an artificial biometric attribute [imitation of a finger(print)]. Such risk is not limited just to the deceptional technique, but if biometric system is not able to recognize whether an acquired biometric sample originates from a genuine living user or an impostor, we would then scan an artificial fake and try to enhance its quality using an enhancement algorithm. After a successful completion of such enhancement, such fake fingerprint would be compared with a template and if a match is found, the user is accepted, notwithstanding the fact that he can be an impostor! Therefore the need of careful liveness detection, i.e., the recognition whether an acquired biometric sample comes from a genuine living user or not, is crucial.

Each component of a biometric system presents a potentially vulnerable part of such system. The typical ways of deceiving a biometric system are as follows (Figure 15.1) [1–4]:

1. *Placing fake biometrics on the sensor.* A real biometric representation is placed on the device with the aim to achieve the authentication, but if such representation has been obtained in an unauthorized manner, such as making a fake gummy finger, an iris printout or a face mask, then it is considered as a deceiving activity.
2. *Resubmitting previously stored digitized biometric signals (replay attack).* A digitized biometric signal, which has been previously enrolled and stored in

[1]Faculty of Information Technology, Centre of Excellence IT4Innovations, Brno University of Technology, Czech Republic

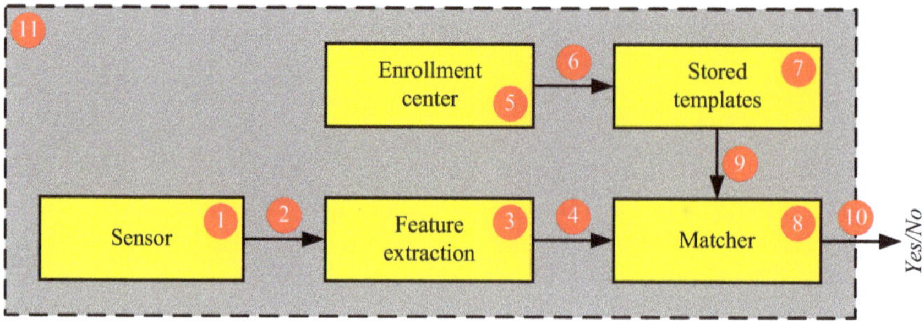

Figure 15.1 Basic components of a biometric system

the database, is replayed to the system, thus circumventing the acquisition device.

3. *Overriding the feature extraction process.* A preselected template is produced in the feature extraction module using a Trojan horse.
4. *Tampering with the biometric feature representation.* During the transmission between the feature extraction and matching modules, a fraudulent feature set replaces the template acquired and processed by the device.
5. *Attacking the enrollment center.* The enrollment module is also vulnerable to spoof attacks such as those described in the previous points 1–4.
6. *Attacking the channel between the enrollment center and the database.* During the transmission, a fraudulent template replaces the template produced during the enrollment.
7. *Tampering with stored templates.* A template, previously stored in the database (distributed or not), can be modified and used afterward as corrupted template.
8. *Corrupting the matcher.* A preselected score is produced in the matching extraction module using a Trojan horse.
9. *Attacking the channel between the stored templates and the matcher.* During the transmission between the database and the matching module, a fraudulent template replaces the template previously stored.
10. *Overriding the final decision.* The result of the decision module can be modified and then used for the replacement of the output obtained previously.
11. *Attacking the application.* The software application can also be a point of attack and all possible security systems should be used to reduce the vulnerability at this level.

From the above list of possible attacks, we can deduce that most security risks or threats are quite common and could be therefore resolved by traditional cryptographic tools (i.e., encryption, digital signatures, public key infrastructure authentication of communicating devices, access control, hash functions, etc.) or by having vulnerable parts at a secure location, in tamper-resistant enclosure or under constant human supervision [1].

When a legitimate user has already registered his finger in a fingerprint system, there are still several ways how to deceive the system. In order to deceive the fingerprint system, an attacker may put the following objects on the fingerprint scanner [3,5,6]:

- *Registered (enrolled) finger.* The highest risk is that a legitimate user is forced, e.g., by an armed criminal, to put his/her live finger on the scanner under duress. Another risk is that a legitimate user is compelled to fall asleep with a sleeping drug in order to make free use of his/her live finger. There are some deterrent techniques against similar crimes, e.g., to combine the standard fingerprint authentication with another method such as a synchronized use of PINs or identification cards; this can be helpful to deter such crimes.
- *Unregistered finger (an impostor's finger).* An attack against authentication systems by an impostor with his/her own biometrics is referred to as a noneffort forgery. Commonly, the accuracy of authentication of fingerprint systems is evaluated by the false rejection rate and false acceptance rate (FAR) as mentioned in the previous chapters. FAR is an important indicator for the security against such method (because a not enrolled finger is used for authentication). Moreover, fingerprints are usually categorized into specific classes [7]. If an attacker knows what class the enrolled finger is, then a not enrolled finger with the same class (i.e., similar pattern) can be used for the authentication at the scanner. In this case, however, the probability of acceptance may be different when compared with the ordinary FAR.
- *Severed fingertip of enrolled finger.* A horrible attack may be performed with the finger severed from the hand of a legitimate user. Even if it is the finger severed from the user's half-decomposed corpse, the attacker may use, for criminal purposes, a scientific crime detection technique to clarify (and/or enhance) its fingerprint.
- *Genetic clone of enrolled finger.* In general, it can be stated that identical twins do not have the same fingerprint, and the same would be true for clones [5]. The reason is that fingerprints are not entirely determined genetically but rather by the pattern of nerve growth in the skin. As a result, such pattern is not exactly the same even for identical twins. However, it can be also stated that fingerprints are different in identical twins, but only slightly different. If the genetic clone's fingerprint is similar to the enrolled finger, an attacker may try to deceive fingerprint systems by using it.
- *Artificial clone of enrolled finger.* More likely attacks against fingerprint systems may use an artificial finger. An artificial finger can be produced from a printed fingerprint made by a copy machine or a DTP technique in the same way as forged documents. If an attacker can make then a mold of the enrolled finger by directly modeling it, he can finally also make an artificial finger from a suitable material. He may also make a mold of the enrolled finger by making a 3D model based on its residual fingerprint. However, if an attacker can make an artificial finger which can deceive a fingerprint system, one of the countermeasures against such attack is obviously based on the detection of liveness.

- *Others.* In some fingerprint systems, an error in authentication may be caused by making noise or flashing a light against the fingerprint scanner, or by heating up, cooling down, humidifying, impacting on or vibrating the scanner outside its environmental tolerances. Some attackers may use such error to deceive the system. This method is well known as a "fault-based attack" (e.g., denial of service) and may be carried out by using one of the above-mentioned techniques. Furthermore, a fingerprint image may be made protruding as an embossment on the scanner surface, if we spray some special material on such surface.

Many similar attacks are documented in the literature, including all the above-mentioned types. In this chapter, however, we will focus only on finger(print) fakes. One example of the attack on fingerprint technology has been presented in [8]. Hackers in the club-magazine "Die Datenschleuder" (4,000 copies in one edition) have printed a fingerprint of the thumb from the right hand of the German minister of the interior—Dr. Wolfgang Schäuble, and invited readers to make a fake finger(print) of the minister and to try to pretend that their identity is those of the minister. This could be considered as a bad joke, as a fingerprint also serves as a conclusive proof of a person's identity. A hacker has acquired this fingerprint from a glass after some podium discussion. Nevertheless, biometric travel documents (issued in Germany starting from 2007, issued in the Czech Republic from 2009), enforced not only by Dr. Schäuble, should be protected just against this situation. The implementation of fingerprints into the travel documents was prescribed by a direction of the European Union in 2004.

It is clear from [5] that the production of a fake finger(print) is very simple [9]. First step is acquiring fingerprint of desired person. Fingerprints are left on everything the person touched. By using dactyloscopic dust (there are many types of them) or special vapor (dusting) techniques [10] fingerprints can be made visible from many, even unexpected materials. And from almost all of them, it can be acquired and used for fake fingerprint production.

Our own experiments have shown that to acquire some images (e.g., from glass, CD, film or even paper) is not very difficult and, in addition, such image could be enhanced and postprocessed, what leads to a high-quality fingerprint. The following production process of a fake finger(print) is simple and can be accomplished in several hours. After that, it is possible to claim the identity as an impostor user and common (nearly all) fingerprint recognition systems confirm this false identity supported by such fake finger.

Therefore, the application of liveness detection methods is a very important task, and should be implemented (not only) in all systems with higher security requirements, such as border passport control systems, bank systems, etc. The biometric systems without the liveness detection could be fooled very easily and the consequences might be fatal.

The security of a biometric system should never be based on the fact that biometric measurements are secret, because biometric data can be easily disclosed. Unlike typical cryptographic measures, where a standard challenge–response protocol can be used, the security of a biometric system relies on the difficulty of replicating biometric samples [11]. This quality known as the liveness ensures that the measured characteristics come from a live human being and are captured at the time of verification. We should realize that any testing of liveness is worthless unless the capture device and communication links are secure. Due to the fact that a biometric system uses physiological or behavioral biometric information, it is impossible to prove formally that a capture device provides only genuine measurements. Consequently, it cannot be proven that a biometric system as a whole is foolproof [11]. Each solution of this problem has its own advantages and disadvantages; it is more suitable for a certain particular type of the biometric system and environment than for other. Some solutions are software-based; others require a hardware support. Methods that combine both approaches can also be used.

15.2 Liveness detection

Securing automated and unsupervised fingerprint recognition systems used for the access control is one of the most critical and most challenging tasks in real world scenarios. Basic threats for a fingerprint recognition system are repudiation, coercion, contamination and circumvention [12,13]. A variety of methods can be used to get an unauthorized access to a system based on the automated fingerprint recognition. If we neglect attacks on the algorithm, data transport and hardware (all these attacks demand good IT knowledge), one of the simplest possibilities is to produce an artificial fingerprint using soft silicon, gummy and plastic material or similar substances [5,14]. A fingerprint of a person enrolled in a database is easy to acquire, even without the user's cooperation. Latent fingerprints on daily-use products or on sensors of the access control system itself may be used as templates.

To discourage potential attackers from presenting a fake finger (i.e., an imitation of the fingertip and the papillary lines) or, even worse, to hurt a person to gain access, the system must be augmented by a liveness detection component [12,13]. To prevent false acceptance, we have to recognize if the finger on the plate of the fingerprint sensor (also referred to as fingerprint scanner) is alive or not. There are decent amount of liveness detection methods in fingerprint recognition. That is because producers are trying to push their technology to the top. Hence, recent topics for research are usually aimed at liveness detection. Some methods are simple and cheap, others are more demanding but also more reliable. Each producer chooses liveness detection that suits their products. In the following subsections, various liveness detection in fingerprint recognition principles will be described [9].

15.2.1 Perspiration

A noninvasive biomedical measurement for determination of the liveness for use in fingerprint scanners was developed by the Biomedical Signal Analysis Laboratory at Clarkson University/West Virginia University [15]. This software-based method processes the information already acquired by a capture device and the principle of this technique is the detection of perspiration as an indication of liveness (see Figure 15.2).

It is worth noting that the outmost layer of the human skin houses around 600 sweat glands per square inch [15]. These sweat glands diffuse the sweat (a dilute sodium chloride solution) on to the surface of the skin through pores. The position

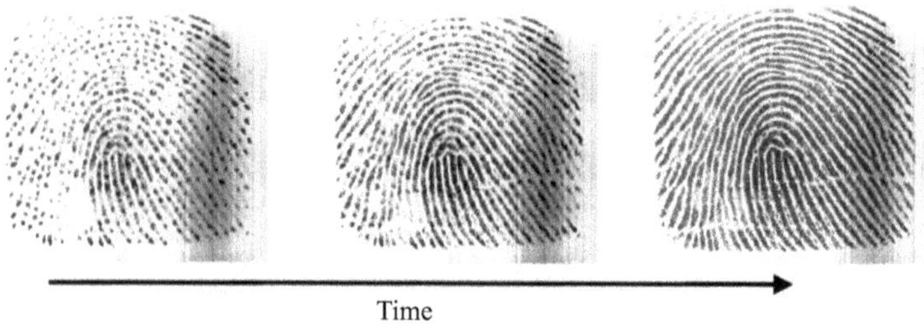

Figure 15.2 Example of live fingerprint images acquired some time apart [15]

Figure 15.3 Ascent of sweat from sweat pores on a fingertip (4× zoomed)

of skin pores does not change over time and their pore-to-pore distance is approximately 0.5 mm over fingertips.

The perspiration method is based on a high difference in the dielectric constant and electrical conductivity between the drier lipids that constitute the outer layer of the skin and the moister sweaty areas near the perspiring pores. The dielectric constant of sweat is around 30 times higher than the lipid, so the electrical model of the skin thanks to perspiration can be created. If we can detect activity of sweat pores then we can also detect if presented finger is alive. First, finger is pushed to sensing part of fingerprint scanner. After that, it is left on the scanner for approximately 5 s. Meanwhile, droplets of sweat start to leak from the sweat pores in ridges. Because of that ridges became darker in the fingerprint image (or final image). Example of this behavior is shown in Figure 15.3 where 4× zoom factor was used and 10 s time delay between each fingerprint scan.

15.2.2 Spectroscopic characteristics

The technology discussed in this section was developed by the Lumidigm company [11,16] from Albuquerque and is based on the optical properties of human skin. This hardware method may be regarded not only as a liveness detection mechanism but also as an individual biometric system with an inherent liveness capability.

Living human skin has certain unique optical characteristics due to its chemical composition, which predominately affects optical absorbance properties, as well as its multilayered structure, which has a significant effect on the resulting scattering properties [16,17]. By collecting images generated from different illumination wavelengths passed into the skin, different subsurface skin features may be measured and used to

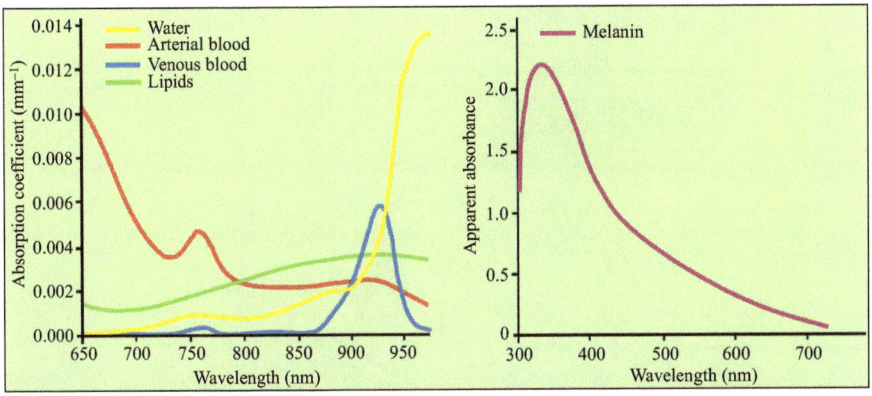

Figure 15.4 Spectrographic properties of different components of living tissue (suitable for detection of spoofing attacks on iris recognition) [18]

ensure that the material is living human skin. When such a multispectral sensor is combined with a conventional fingerprint reader, the resulting sensing system can provide a high level of certainty that the fingerprint originates from a living finger.

The principle of this technique lies in passing light of various wavelengths through a sample and measuring the light returned, which is affected by the structural and chemical properties of the sample. Different wavelengths have to be used to measure the sample satisfactorily, because diverse wavelengths penetrate to different depths into the sample and are differently absorbed and scattered [11]. For example, when we put a flashlight against the tip of a finger, only the red wavelengths can be seen on the opposite side of the finger. This is because shorter (mostly blue) wavelengths are absorbed and scattered quickly in the tissue, unlike longer (red and very near infrared) ones, which penetrate deep into the tissue. In addition, there is an interesting fact about this—each user can have its own profile because skin of each person is reacting on different wavelengths a little bit differently. Differences are so subtle that using spectral properties as biometrics itself is not possible; however for liveness detection, it is more than enough.

The measurements can be transformed into a graph (Figure 15.4) that shows the change in all measured wavelengths after interacting with a sample and is known as a spectrum. Next, the proper analysis of tissue spectra, based on multivariate mathematical methods, has to be done to provide correct results.

Figure 15.5 shows the principle of multispectral fingerprint sensor. The light sources are LEDs of various wavelengths spanning the visible and short-wave infrared region. Crossed linear polarizers may be included in the system to reduce the contribution of light that undergoes a simple specular reflection to the image, such as light that is reflected from the surface of the skin. The crossed polarizers ensure that the majority of light seen by the imaging array has passed through a portion of skin and undergone a sufficient number of scattering events to have randomized the polarization. The imaging array is a common silicon complementary metal–oxide–semiconductor (CMOS) or CCD detector.

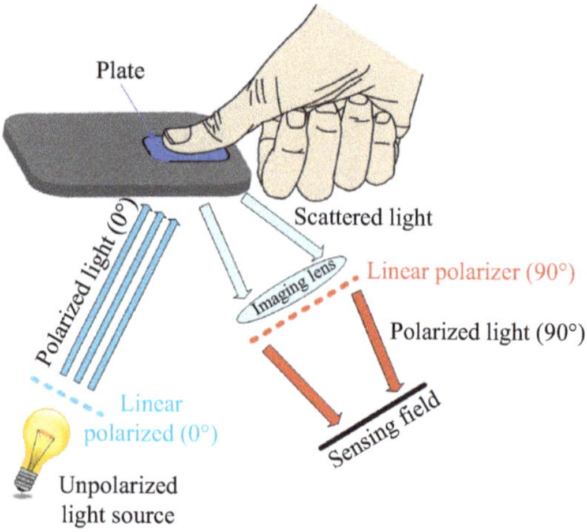

Figure 15.5 Multispectral fingerprint sensor (reworked from [17])

A highly realistic artificial finger made by Alatheia Prosthetics [16] was one of a number of different spoof samples used to test a multispectral imager's ability to discriminate between real fingers and spoofs. Figure 15.6 shows the results of a multivariate spectral discrimination performed to compare the consistency of the spectral content of a multispectral image of a real finger with both a second image of a real finger and a prosthetic replica of the same finger. The imager's ability to distinguish between the two sample types is clear.

Another approach of the liveness detection using the wavelet analysis in images is presented in [19].

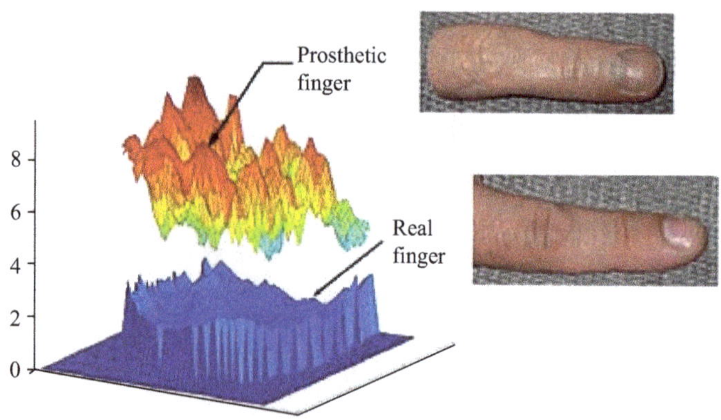

Figure 15.6 Multispectral image data can clearly discriminate between a living finger and an ultrarealistic spoof. The graphs on the left side show how similar the spectral content of each image is to that expected for a genuine finger [16,18]

15.2.3 Ultrasonic technology

In this paragraph, a biometric system using an ultrasonic technology with inherent liveness testing capability will be described. This technique was first developed by the company Optel from Poland and is based on the phenomenon called contact scattering. Another ultrasonic biometric device is offered by the company Qualcomm [20] (former Ultra-Scan) from the United States. Companies Sonavation [21] and InvenSense [22] have also some products in this sector [23].

Standard ultrasonic methods [11] use a transmitter, which emits acoustic signals toward the fingerprint, and a receiver, which detects the echo signals affected by the interaction with the fingerprint (Figure 15.7). A receiver utilizes the fact that the skin (ridges) and the air (valleys) have difference in acoustic impedance; therefore the echo signals are reflected and diffracted differently in the contact area. This approach with inherent liveness testing capability among its foremost principles uses the fact that sound waves are not only reflected and diffracted but are also subject to some additional scattering and transformation. This phenomenon is called contact scattering [11] and it was discovered that this scattering is, to a significant extent, affected by the subsurface structure of the acquired object. Hence, the class corresponding to the live tissue could be modeled and whenever the received acoustic waves are inconsistent with this class, they are rejected. This can easily detect fingerprint spoof because reflected signals from one (spoof material) layer have different characteristics than signals reflected from several layers of living tissue.

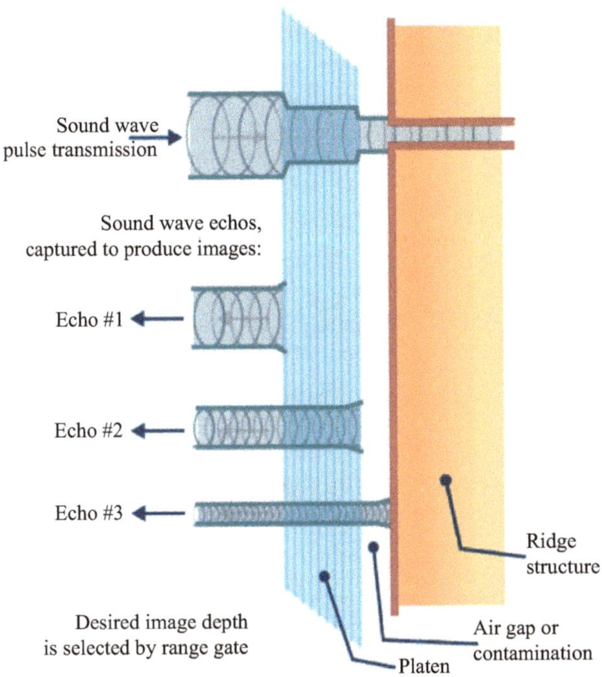

Figure 15.7 Schematic of ultrasonic pulse/echo principle [24]

The main problem here is not to obtain clear signals but to analyze and to make a reconstruction of internal structures from signals which are very difficult to interpret.

The ultrasonic device reached the following conclusions [11,25]:

- As the inner structure of the live skin compared with spoof samples differs, the character and the amplitude of acoustic signals also differ significantly. Hence, it is possible to distinguish between live and artificial fingers.
- There is no need to deal with the problem known as latent print reactivation because the signal level from the latent print is at least 30 dB lower than the signal given by the real finger. Even when the soot or metal powder is used in order to enhance the quality of signal, the previous is true.
- This method is much less sensitive to dirt, grease and water compared with other methods. In addition, fingers with damaged surface give a relatively clear image, because their inner structure seems to be visible.
- Since this approach scans the inner structure of the object, it has the ability to check for pulse by measuring volumetric changes in the blood vessels [25].

15.2.4 Physical characteristics

Physical characteristics belong to the easiest methods for liveness detection. They are usually based on various measurable phenomena connected to the skin tissue.

15.2.4.1 Temperature

This simple method measures the temperature of the epidermis during a fingerprint acquisition. According to our measurement, in different environment conditions including different seasons, mental and physical state of the user, it was determined that temperature of human finger is in the range of 25–37°C. Unfortunately, temperature also change while scanning so intraclass variability is huge and interclass variability is small.

In addition, there are many people who have problems with blood circulation, a fact which leads to deviations in the body's temperature and hence to wrong liveness module decision. The only way how to improve such a situation is to make the working range broader again or simply warm the user's finger. The former will increase the likelihood that the system will be deceived while the latter can also be applied to fake samples. In the case where an attacker uses a wafer-thin artificial fingerprint glued on to his finger, this will result in a decrease by a maximum of 2°C [26] compared with an ordinary finger. Since the difference in temperature is small, the wafer-thin sample will comfortably fall within the normal working margin. In consequence, this method is not a serious security measure at all (Figure 15.8).

15.2.4.2 Hot and cold stimulus

This technique is based on the fact that the human finger reacts differently to thermal stimuli compared with other artificial, nonliving material.

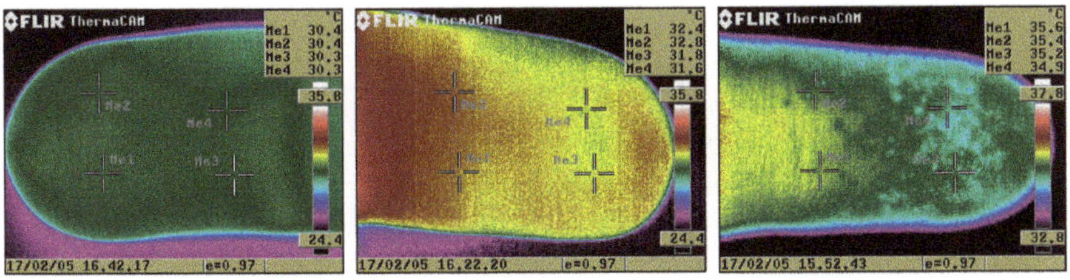

Figure 15.8 Thermo-scans of the fingertips acquired using a thermo-camera FLIR

The designed liveness testing module [11,27] is working as follows. A stimulus-giving section gives a stimulus (it may cover a cool and a hot stimulus) to the finger by a contact plate with which the finger makes contact. Next, typical information could be measured by an organism information-measuring section, which is produced by the live finger in response to the stimulus. Concretely, the amount of the fluctuation for the flow rate of the blood flowing in the peripheral vascular tracts varies according to the stimuli. Hence, as peripheral vascular tracts of the tip of the finger are extended or contracted, the amplitude value of the blood flow is measured and processed by an organism information-measuring section. Under hot stimulus the amplitude of the blood flow increases, while it decreases under cool stimulus. Moreover, according to the autonomic nervous system, the amplitude is delayed a little with respect to the application of the stimulus. Since these facts are typically observed when the live fingers are measured, they could be employed to distinguish live among artificial and dead samples. After the processing phase, such information is transferred to a determining section, where together with the other information related to stimulus (i.e., the time intervals, the strength of stimuli, etc.) is evaluated. Finally, a determining section analyses how the amplitude of the blood flow fluctuates in response to the stimulus to make the right decision.

Since the human peripheral nervous system is very sensitive, it is able to react to weak cool and hot stimuli without being noticed by the person whose fingerprint is checked. This fact should also reduce success spoofing ratio. More information about the method discussed here can be found in [27].

15.2.4.3 Pressure stimulus

The principle of this method lies in some changes in characteristics of the live skin, which are realized due to pressure applied to the finger [11,28]. Since the structure and the characteristics of artificial and dead samples are different, when compared with a live finger, this phenomenon could not be seen if such samples were used.

The color of the live skin of the unpressured finger is usually reddish but becomes whitish when pressure is applied to the skin of the finger. This change of coloration is because skin is reacting to increasing pressure by pushing out blood from pressured spot.

It has been shown that the spectral reflectance of the light in the red spectral range (i.e., the light wavelength of approximately 640–770 nm) [28] does not show a substantial difference between the pressed state and the nonpressed state. On the other hand, the spectral reflectance of the light in the blue and green spectral range (i.e., the light wavelength of approximately 400–600 nm) [28] in the not pressed state is much smaller than in the pressed state. Hence, for the purposes of the device discussed in this section, it is suitable to measure the spectral reflectance in the blue and green spectral range (see Figure 15.9).

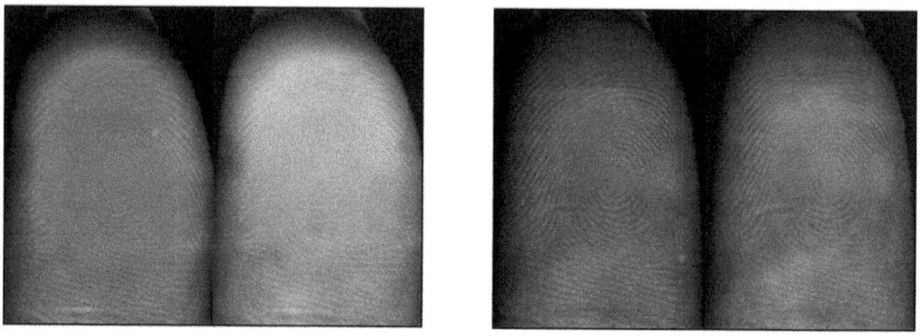

Figure 15.9 Images of the fingertips pressed tightly (left) and slightly (right) to the sensor [29]

A liveness testing module is proposed in [28] and consists of a transparent plate, a light source, a light detection unit and a determining section. Since the light source and the light detection unit are placed under the plate, this plate has to be transparent to enable light to be sent toward the finger and receiving the reflected light. The light source projects a light beam toward the surface of the placed finger. Next, depending on the pressure or nonpressure state, the reflected light is measured by the light detection unit.

Based on such measurements, the determining section returns the right decision, i.e., as the finger changes its state from nonpressure to pressure, the color of the skin changes from reddish to whitish, what leads to a change in the spectral reflectance. As a result, the light detection unit can detect that the spectral wavelength of the spectral ranges is increased.

Another method using pressure-based characteristics is discussed in [30], but unlike the method described in the previous paragraph, this technique employs the change in fingerprint ridges width. That is because ridges are elastic. When applying pressure, they tend to widen themselves.

When the fingerprint changes its state from nonpressure to pressure, the fingerprint ridges change, i.e., as the pressure becomes stronger, the fingerprint ridges flatten out, and therefore their change of width could be measured. Only objects which demonstrate the typical change in fingerprint ridge width due to pressure could be determined as live ones.

A new approach to the fake finger detection based on skin elasticity analysis has been introduced in [31]. When a user puts a finger on the scanner surface, the scanner captures a sequence of fingerprint images at a certain frame rate. The acquired image sequence is used for the fake finger detection. One or more of them (see Figure 15.10) can be used for fingerprint authentication.

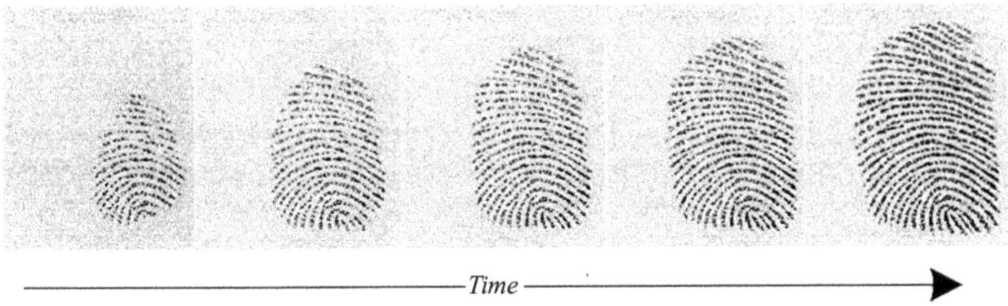

Figure 15.10 A sequence of fingerprint images describing the deformation of a real finger [11]

15.2.4.4 Electrical properties

Some methods of liveness testing are based on the fact that the live human skin has different electrical properties compared with other materials [11]. The suitable fingerprint recognition system could be extended by an electrode system and an electrical evaluation unit. These sections are the main parts of the liveness testing module where the electrical evaluation unit can evaluate the change in the state in the electrode system. The sensing of the electrical change should take place simultaneously with the recognition of the fingerprint. Therefore, these parts of biometric systems should be designed in such a way that two simultaneous measurements cannot disturb each other. Furthermore, such a system may be able to measure more than one of the fingerprint liveness characteristics related to electrical properties (e.g., conductivity, dielectric constant).

The conductivity [11] of the human skin is based on humidity, which is dependent on people's biological characteristics and environmental conditions: some people have dry fingers and others have sweaty ones; also during different seasons, climatic and environmental conditions, humidity differs significantly. As a result, the span of permissible resistance levels has to be big enough to make the system usable. In such a situation, it is quite easy for an intruder to fool the system. Moreover, the intruder can use a salt solution of a suitable concentration or put some saliva on the fake finger to imitate the electric properties of the real finger.

The relative dielectric constant (RDC) [11] of a specific material reflects the extent to which it concentrates the electrostatic lines of flux. Many advocates claim that the RDC has the ability to distinguish between real and artificial samples. However, the RDC is highly dependent on the humidity of the sample, so the same situation as in the case of conductivity arises. To fool this method, an attacker can simply use an artificial sample and dip it into a compound of 90% alcohol and 10% water. In [32], we can read that the RDC values of alcohol and water are 24 and 80, respectively, while the RDC of the normal finger is somewhere between these two values. Since the alcohol will evaporate faster than the water, the compound will slowly turn into the water. During evaporation, the RDC of spoof samples will soon be within the acceptance range of the sensor.

We have run a small test series with ten people, each finger, horizontal and vertical measurement strips, and five measurements per finger—conductivity (resistance) measurements. The range of values we found was from 20 kΩ to 3 MΩ [26]. A paper copy or an artificial finger made of nonskin-like material has higher electrical resistance, but for example, soft silicon (moisturized) shows resistance values close to the range found in our experiments.

15.2.4.5 Bio-impedance

Bio-impedance [33–35] describes the passive electrical properties of biological materials and serves as an indirect transducing mechanism for physiological events, often in cases where no specific transducer for that event exists. It is an elegantly simple technique that requires only the application of two or more electrodes. The impedance between the electrodes may reflect "seasonal variations in blood flow, cardiac activity, respired volume, bladder, blood and kidney volumes, uterine contractions, nervous activity, the galvanic skin reflex, the volume of blood cells, clotting, blood pressure and salivation."

Impedance Z [34] is a general term related to the ability to oppose AC (alternating current) flow, expressed as the ratio between an AC sinusoidal voltage and an AC sinusoidal current in an electric circuit. Impedance is a complex quantity because a biomaterial, in addition to opposing current flow, phase-shifts the voltage with respect to the current in the time-domain.

The conductivity of the body is ionic (electrolytic) [34], because of the presence of, e.g., Na^+ and Cl^- in the body liquids. The ionic current flow is quite different from the electronic conduction found in metals: the ionic current is accompanied by a substance flow. This transport of substance leads to concentrational changes in the liquid: locally near the electrodes (electrode polarization), and in a closed-tissue volume, during prolonged direct current flow.

The body tissue is composed of cells with poorly conducting, thin-cell membranes. Therefore, the tissue has capacitive properties [34]: the higher the frequency, the lower the impedance. The bio-impedance is frequency-dependent, and impedance spectroscopy, hence, gives important information about tissue and membrane structures as well as intra- and extracellular liquid distributions.

Figure 15.11 shows three most common electrode systems. With two electrodes, the current carrying electrodes and signal pick-up electrodes are the same. If the electrodes are equal, it is called a bipolar lead, in contrast to a monopolar lead. With 3-(tripolar) or 4-(tetrapolar) electrode systems, separate current carrying and signal pick-up electrodes are used. The impedance is then transfer impedance [34]: the signal is not picked up from the sites of current application.

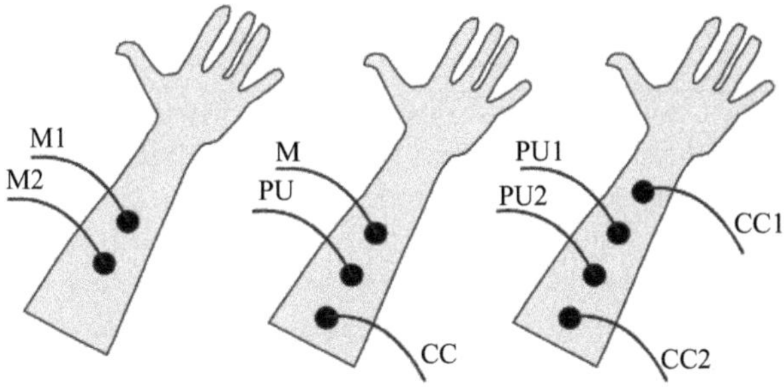

Figure 15.11 Three skin surface electrode systems on an underarm [34]. Functions: M—measuring and current carrying, CC—current carrying, PU—signal pick-up

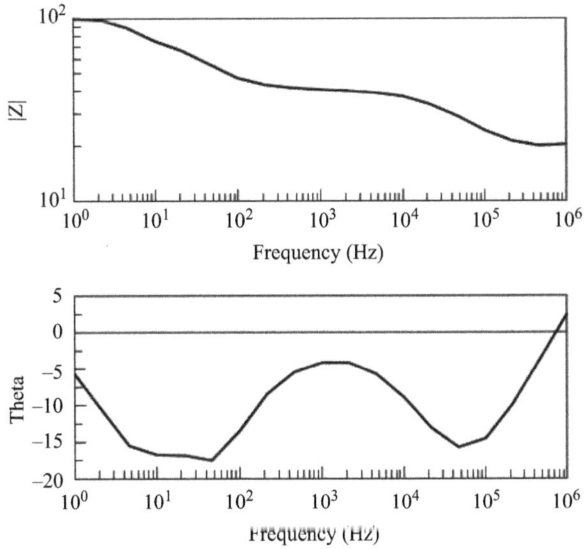

Figure 15.12 Typical impedance spectrum obtained with four equal electrodes attached to the skin of the underarm [34]

Figure 15.12 shows a typical transfer impedance spectrum obtained with the 4-electrode system from Figure 15.11. It shows two dispersions [34]. The transfer impedance is related to, but not solely determined by the arm segment between the polyurethane (PU) electrodes. The spectrum is determined by the sensitivity field of the 4-electrode system as a whole. The larger the spacing between the electrodes, the more the results are determined by deeper tissue volumes. Even if all the electrodes are skin surface electrodes, the spectrum is, in principle, not influenced by skin impedance or electrode polarization impedance.

The necessity of this method is to apply electrodes on user forearm which is something that majority of users would not be willing to do.

15.2.4.6 Pulse

Scanners based on this technique try to detect whether the scanned object exhibits characteristics of the pulse and blood flow consistent with a live human being [11]. It is not very difficult to determine whether the object indicates some kind of pulse and blood flow, but it is very difficult to decide if the acquired characteristics are coincident with a live sample. As a result, it is difficult to create an acceptance range of the sensor, which would lead to small error rates. The main problem is that the pulse of a human user varies from person to person—it depends on the emotional state of the person and also on the physical activities performed before the scanning procedure. In addition, the pulse and blood flow of the attacker's finger may be detected and accepted when a wafer-thin artificial sample is used.

One of the sensors usually detects variation in the levels of the reflected light energy from the scanned object as evidence of the pulse and blood flow [11]. First, the light source illuminates the object and then a photo-detector measures the light energy reflected from the object. Finally, there is the processing instrument (which also controls the light source) which processes the output from the photo-detector. Since there are some ways how to simulate pulse and blood flow characteristics (e.g., by flashing the light or by motion of the scanned object), scanners should have a deception detection unit [11].

Our skin is semipermeable for light, so that movements below the skin (e.g., blood flow) can be visualized. One example of an optical skin property is the skin reflection [12,36]. The light illuminating the finger surface is partly reflected and partly absorbed (Figure 15.13). The light detector acquires the reflected light which has been changed in phase due to dispersion and reflection and thus has a slightly different wavelength compared to the original light source. One can try to link the change in wavelength to the specific characteristics of the skin with respect to light dispersion and reflection to detect whether the light has been scattered and reflected only from the fingerprint skin, or if there is some intermediate layer between the finger skin and the light source or detector.

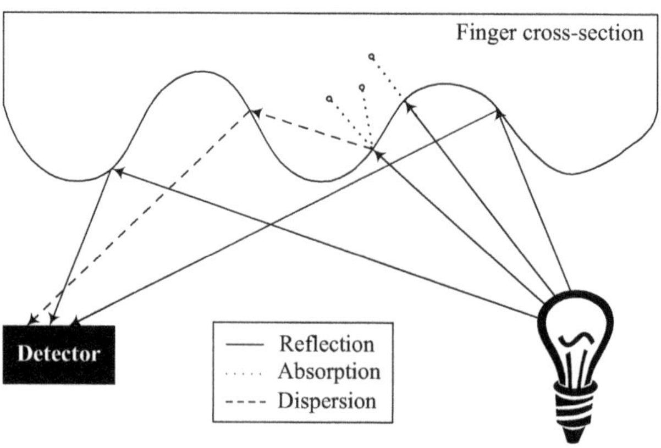

Figure 15.13 Light absorption, dispersion and reflection by a fingerprint [12]

Another example for optical skin feature is the saturation of hemoglobin [12,36], which binds oxygen molecules. When blood comes from the heart, oxygen molecules are bound to the hemoglobin, and vice versa, when blood is flowing back to the heart, it is less saturated by oxygen. The color of oxygenated blood is different from that of nonoxygenated blood. If we use a light source to illuminate the finger skin, we can follow the blood flow based on the detection of oxygenated and nonoxygenated blood, respectively [12,36]. The blood flow exhibits a typical pattern for a live finger, i.e., the analysis of blood flow is well suited for finger liveness detection.

In both above-mentioned examples, it is shown that the human skin has special characteristics which can be used for the liveness testing. It can be argued that it is possible to confuse such system, e.g., by using a substance with similar optical characteristics as a human skin, or, in the second example, to simulate the blood flow. Even though the argument is correct, obviously the effort to be exerted for these attacks is much higher than for the other physical characteristics presented so far.

Another solution is proposed in [12,36] based on the analysis of movements of papillary lines of the fingertips and measurements of the distance of the fingertip surface to a laser sensor, respectively. The system is compact enough to be integrated with optical fingerprint sensors.

One advantage of this implementation is that the finger is not required to be in contact with a specific measuring device, and so it can be integrated with standard fingerprint sensors. Moreover, the implementation could be acceptably low. This is of particular importance, as in most cases, the liveness detection will be an add-on that augments already existing robust and field-tested fingerprint scanners.

The method presented in [12,36] requires the analysis of at least one heart activity cycle, thus both the camera and the laser measurement method sketched in this section would add an extra time of at least 1 or 2 s to the overall authorization process interval.

In [12,36], two approaches for measuring of fine movements of papillary lines, based on optical principles, are suggested (Figure 15.14). The first solution is based on a close-up view of the fingertip acquired with a charge-coupled device (CCD) camera; the second one is distance measurement with a laser sensor. It should be noted that adding the proposed liveness detection solution (either camera or laser-based) to a fingerprint recognition system, as proposed in Figures 15.15 and 15.16, may significantly influence the hardware requirements imposed on the complete system.

Camera solution

The camera solution scheme is outlined in Figure 15.15. The main idea is that a small aperture (approximately 6 mm) is created in the middle of a glass plate with an alternately functioning mirror below the plate.

First, during the fingerprint acquirement phase, the whole fingerprint is stored and the system operates as a classical fingerprint acquisition scanner (mirror permeable) by projecting the fingerprint on the CCD/CMOS camera. Next, in the liveness detection phase, the mirror is made impermeable for light and a part of the fingertip placed on the aperture is mirrored to the right and projected on the CCD/CMOS camera by a macro lens. This macro lens has to magnify minimally 10–12× of original. The latter part of the system is used to acquire a video sequence for the liveness detection analysis.

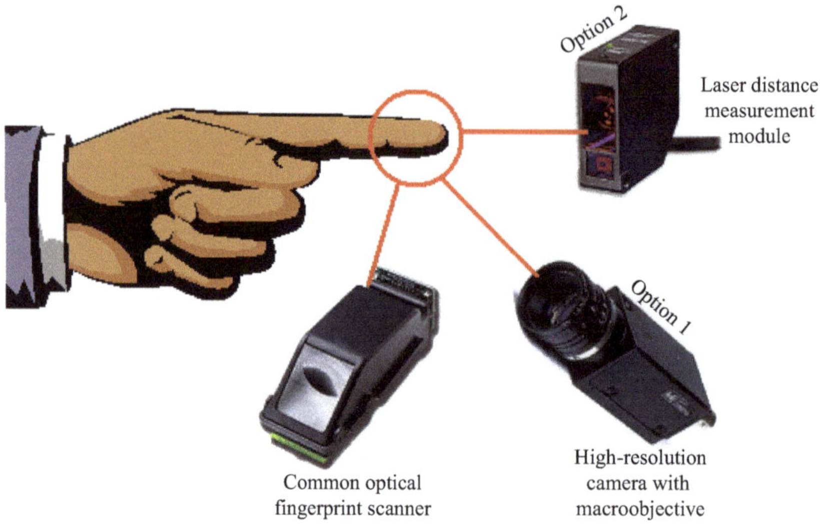

Figure 15.14 Integrated liveness detection—scanner + optical and laser solution [37]

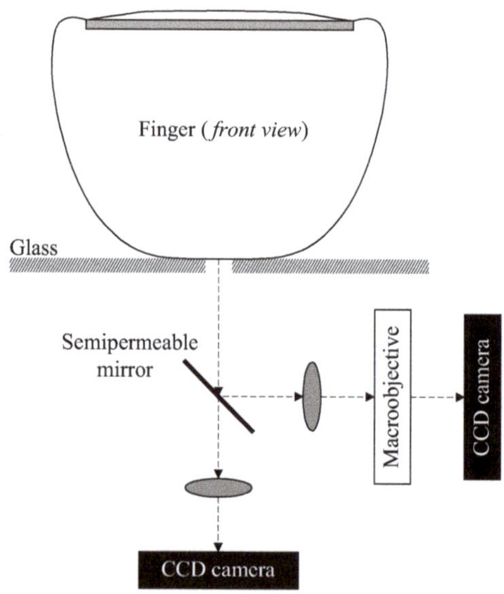

Figure 15.15 Possible integration of a camera-based measurement system for liveness detection with optical fingerprint sensor (CCD/CMOS camera) [12]

Laser solution

The second optical method for the liveness testing is based on laser distance measurements [12,36]. Figure 15.16 outlines the laser distance measurement module, which could be integrated with a standard optical fingerprint sensor. The optical lens system and CCD camera for acquisition of the fingerprint are the same as in Figure 15.15. However, unlike the solution shown in Figure 15.15, the laser distance measurement module is placed to the right side of the glass plate, which is L-shaped here. The user places his finger in such a way that it is in contact with the horizontal and the vertical side of the glass plate.

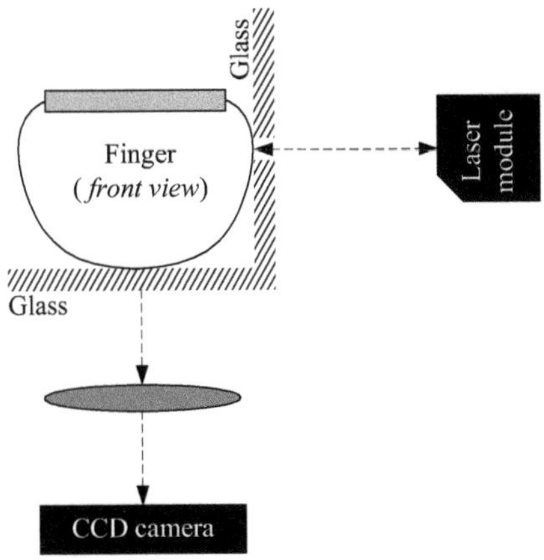

Figure 15.16 Possible integration of laser distance measurement for liveness detection with optical fingerprint sensor (CCD/CMOS camera; aperture approx. 6 mm) [12]

The underlying physical measurement principle is the same as in the video camera solution. We assume volume changes (expansion and contraction) due to the heart activity, which causes fine movements of the skin. The laser sensor is able, based on the triangulation principle, to measure very small changes in distance down to several μm.

The comparison of the computed curve and a normalized standard curve (the template) will reveal whether the measurement corresponds to a standard live fingerprint or indicates a fake finger or another attempt of fraud. For example, the comparison between both curves can be realized by the normalization followed by the cross correlation.

There are other liveness detection methods based on optical principles, see [28,30]. They coincide in principles (both are optical) but differ in monitored physical characteristics.

15.2.4.7 Blood oxygenation

Sensors that measure blood oxygenation [11] are mainly used in medicine and have also been proposed for use in liveness testing modules.

The technology involves two physical principles. First, the absorption of light having two different wavelengths by hemoglobin differs depending on the degree of hemoglobin oxygenation. The sensor for the measurement of this physical characteristic contains two LEDs: one emits visible red light (660 nm) and the other infrared light (940 nm). When passing through the tissue, the emitted light is partially absorbed by blood depending on the concentration of oxygen bound on hemoglobin. Important thing is that in case of live finger, there has to be reaction for both type of light. This liveness detection can also very easily distinguish between amputated and live finger. Former contain only deoxygenated blood. Second, as the volume of arterial blood changes with each pulse, the light signal obtained by a photo-detector has a pulsatile component which can be exploited for the measurement of pulse rate.

The sensors mentioned above are able to distinguish between artificial (dead) and living samples but, on the other hand, many problems remain. The measured characteristics vary from person to person and the measurement is strongly influenced by dyes and pigments (e.g., nail varnish).

15.2.4.8 Blood pressure

There are some other methods based on the medical science characteristics which have been suggested for liveness testing purposes [11]. Nonetheless, they are mostly inconvenient and bulky. One example can be the measurement of blood pressure [12] but this technology requires to perform measurement at two different places on the body, e.g., on both hands.

We distinguish between the systolic and diastolic blood pressure [12,38]; these two levels characterize upper and lower blood pressure values, respectively, which depend on heart activity. For a healthy person, the diastolic blood pressure should not be lower than 80 mmHg (lower values mean hypotension) and the value of the systolic blood pressure should not be below 120 mmHg (again, lower values mean hypotension). People with hypertension have higher blood pressure values, with critical thresholds 140 mmHg for the diastolic blood pressure and 300 mmHg for the systolic blood pressure. In fact, diastolic and systolic blood pressure values are bound up with the ranges from 80 to 140 mmHg and from 120 to 300 mmHg, respectively [38]. On the one hand, blood pressure values outside these normal ranges can indicate a fake fingerprint [12]. On the other hand, we can think of configurations, where the blood pressure measurement of a fake fingerprint glued to the finger which significantly lowers the measured blood pressure value can still give us a measurement value within the accepted range. An attacker with hypertension would be accepted as a registered person in such configuration [12].

15.2.5 Other methods

There are specific solutions that use or combine previously mentioned methods. Also secret commercial solutions exist where even if the principle was known to authors for obvious reason they cannot be explained. It is literally impossible to list all combinations and variations of liveness detection methods.

15.3 Fingerprint spoofs

Efforts to break into fingerprint access systems are probably as old as these systems. With the boom of personal devices secured by fingerprint technology, there is more attention focused on the ways of bypassing them. The motivation for doing so may be lawful or unlawful. This section is an introduction to production and usage of fingerprint spoofs. First, there is description of what kind of preparation is needed, after that, materials and production of a spoof itself is described. Finally, the last subsection is about usage of fingerprint spoofs, experience with different fingerprint sensors, materials of spoofs and liveness detection [39,40].

15.3.1 Preparation phase for spoof production

Generally, there are several methods of spoofing fingerprint access system. They can be seen in Figure 15.17. There is cooperative branch which presumes that a mold can be done directly with the cooperation of the person which is impersonated. That is usually not the case and there are four other possibilities left. Either the latent fingerprint can be reactivated on the device which is possible for some fingerprint sensor technologies or a cadaver finger or cut finger can be used. All these options presume some cooperation or very specific actions of the impersonated user [39,40].

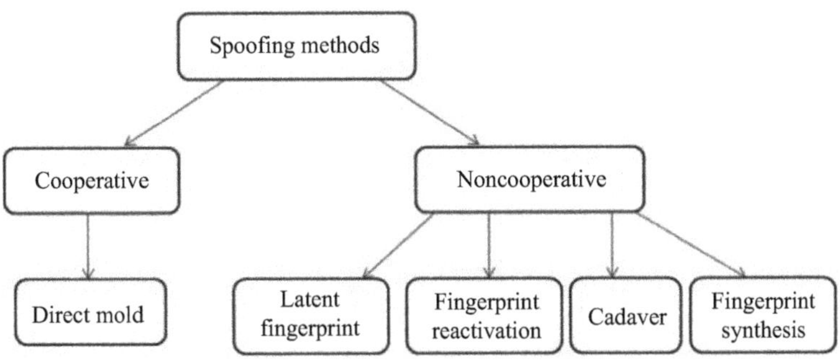

Figure 15.17 Overview of various spoofing methods [39]

15.3.1.1 Desired fingerprint acquirement

Without this cooperation or specific actions from the original user, the only remaining action presumes that the latent fingerprint can be obtained. Last part, the fingerprint synthesis is also somewhat relevant to this part because latent fingerprint often needs quality enhancement. This can be done by automatic or manual tools for fingerprint reconstruction. Fingerprint synthesis can be classified as an automatic method, but rather it should be seen as another type of reconstruction—the aim is from whatever information is present to create a perfect synthetic fingerprint. One method or another, the focus is on getting the best possible quality digitized latent fingerprint [39,40].

As it was stated in section 15.1, process usually starts with securing desired objects with latent fingerprints on them. After that, vapor (dusting) techniques are used to make these fingerprints visible. Next step is scanning the fingerprints to get digitized images. Final step is to get the best quality out of input data.

Process of fingerprint acquirement is the most difficult task in the spoof production. From this fingerprint a mold will be produced. Final spoof will never have better quality than quality of the mold. Acquiring several latent fingerprints and combining them together, patient reconstruction of ridges and minutia all these tasks are important for the final image quality. Note the difference between perfect synthetic fingerprint and real fingerprint in Figure 15.18.

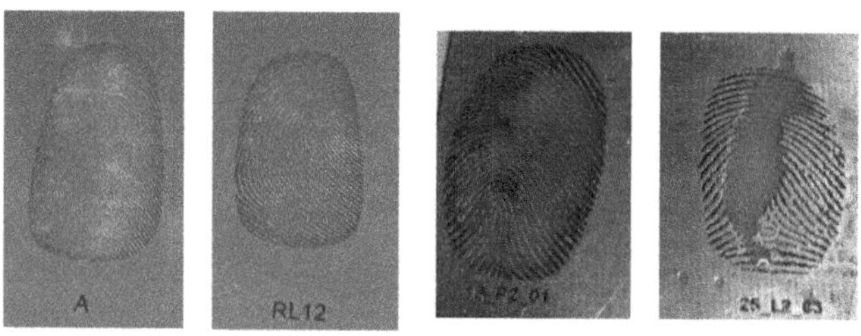

Figure 15.18 Synthetic (left) and real (right) fingerprint molds

15.3.1.2 Production of the mold

Mold can be made from various materials. Basically, it can be anything which will (a) form the spoof material, (b) allow easy removal of spoof without damage and (c) optionally be used multiple times. Ideal mold material also depends on material which is used for spoofs. Soft materials as latex can be used on relatively fragile material. Some play-doh which need steady pressure to form needs more durable one.

One of the most durable and also most commonly used mold is made as printed circuit board (PCB). Digitized fingerprint image is converted to the PCB drawing file. File is transferred to the company dealing with PCBs which create the cuprextit fingerprint mold. Metal is durable—it can withstand physical damage, temperature changes both hot and cold without damaging the fingerprint image. It can be easily cleaned and use of separators or others special materials is usually not a problem. Some spoof materials need special treatment which includes previously stated dangers for mold. Presumably, this type of mold was chosen, next step is production of the spoof itself [39–41].

15.3.2 Materials for spoof production

There are dozens of materials which can be used for fingerprint spoof. Ideal spoof material would be something with same properties (i.e., physical, electrical, etc.) as human skin and it should be easy to work with. From properties perspective, the most important is elasticity, greasiness and durability. That is because these are almost necessary to scan spoof with any sensor. Other properties (like color, conductivity, resistance, pulse, sweat, etc.) are usually important when facing a specific sensor and/or liveness detection technology. From production perspective, the most important factors are (a) how easy it is to copy mold image, (b) how easy it is to remove material from mold, (c) how long does this process take and (d) whether some special treatment is needed and difficulty of using this treatment.

It is almost impossible to list all materials which were ever used. Also, it is very difficult to sort the best material because all of them lack some important parts which were defined in previous paragraph. Often the material is either not durable and can be used only few times before image on the material disappear or is durable and elastic but special treatment is needed when producing. Nevertheless, description and figures of some of the most used or often mentioned materials are given below.

Play-doh can be categorized as modeling clay designed mainly for children. It is soft and malleable and it cannot be hardened (this is not true for some clays). Spoofs are not durable but they are very easy to use. In Figure 15.19, spoof and its image can be seen [40].

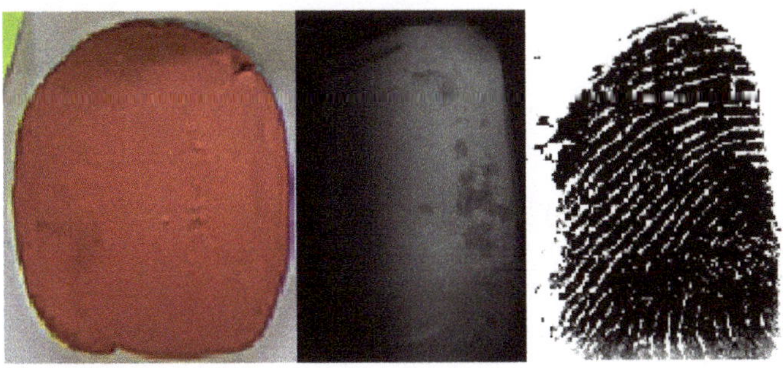

Figure 15.19 Play-doh spoof [40]

Latex is a white fluid which can be found inside of some plants. It is a base for natural rubber, tires, etc. Latex can be seen as category of rubbers or similar materials. When making spoofs, the biggest challenge is to make an equally thick layer. This layer should be at the same time as thin as possible, however thick enough to be easily removed. It is fairly durable, elastic but not so easy to make. Example of latex spoof is shown in Figure 15.20 [40].

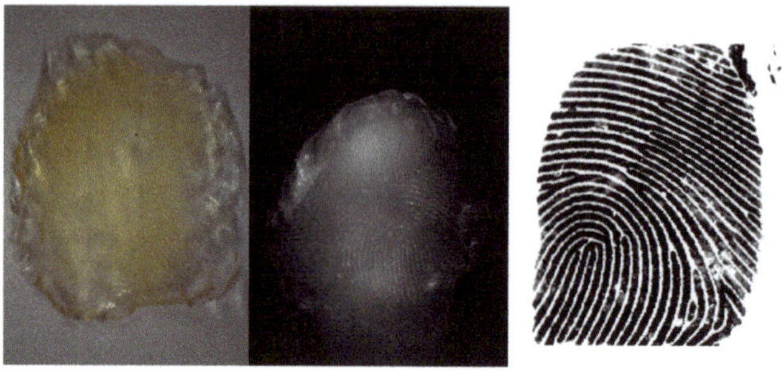

Figure 15.20 Later spoof [40]

Gelatin belongs to food industry spoofs category. Spoofs made from gelatin have to stick to a transparent glue tape during removing them from the mold. Gelatin is too soft and fragile material to be removed directly without the glue tape, otherwise during the removing the fingerprint spoofs could be damaged. They are usually made soft by heating and harden when they cool off. Spoofs are elastic, easy to make, not so easy to remove and also not durable. In Figure 15.21, gelatin spoof can be seen [39].

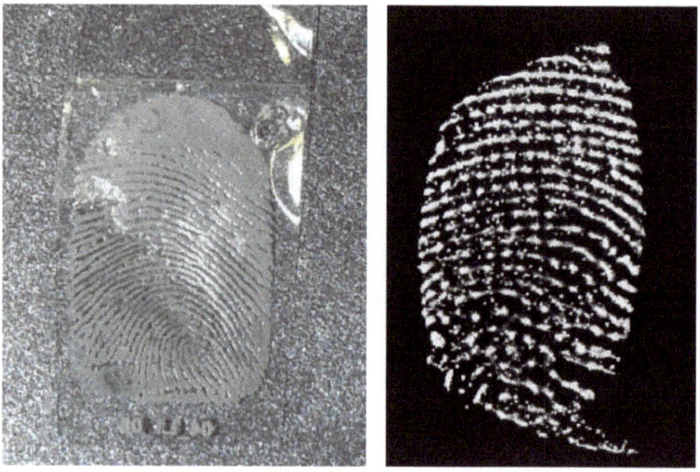

Figure 15.21 Gelatin spoof [39]

Aquarium silicone can be categorized as glues and other industrial materials. Spoofs are nicely soft after drying, flexible and durable with clearly visible ridges, where the minutiae points could be precisely located. On the other hand, it is very hard to remove them from mold, not so easy to be made without air bubbles. Examples can be seen in Figure 15.22 [39].

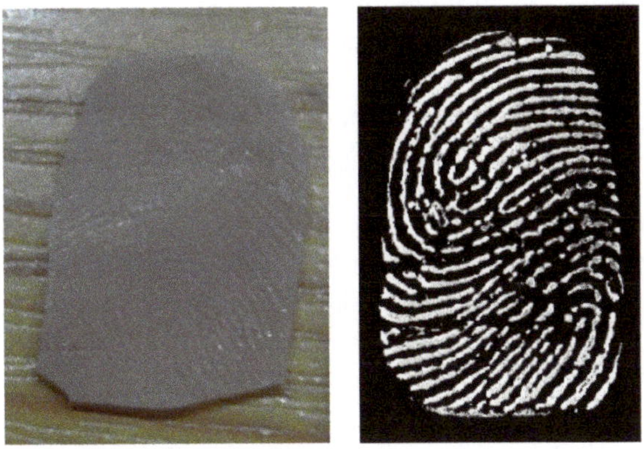

Figure 15.22 Aquarium silicone spoof [39]

Glass colors could form a special "others" category. These are special colors which can be used on glass, windows, mirrors, bathroom tiles, etc. The colors are applied to a plastic foil after that they can be peeled off and stuck to another similar surface. The mold can be filled with a really thin layer by pouring a small drip and spreading the color with a knife. If it is too thin, then it stretches out and can even tear apart while removing it from the mold. Spoofs can be done also by just pouring the color until it fits the mold. Spoofs are relatively durable, relatively elastic, very easy to make and the only problematic part is the removal. Spoofs from this material can be seen in Figure 15.23 [40].

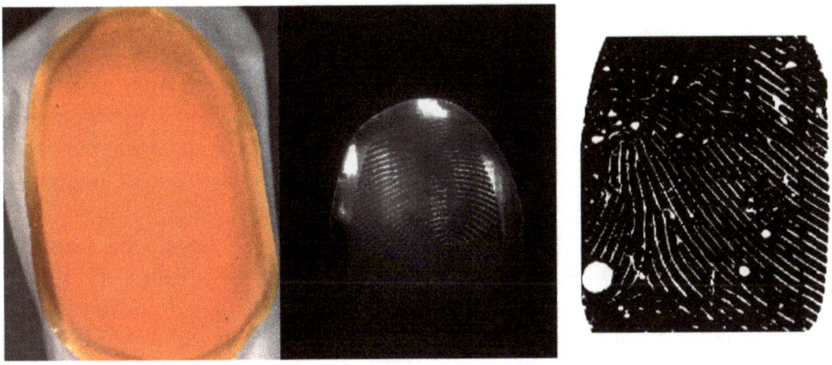

Figure 15.23 Glass colors spoof [40]

15.3.3 Spoof usage

Spoof usage is generally simple—spoof is used instead of a finger on the fingerprint scanner. There are two main choices how to present the spoof, either use it as a new layer on finger or to present it with other material (e.g., fake rubber finger). Now the specific attributes of the spoof material, sensor technology and (if present) liveness detection are determining how successfully spoof will perform.

If the spoof material is not durable, then it will (a) be used with real finger to use appropriate strength, (b) be almost unable to be used with swipe sensor technology, (c) be even with careful usage used only few times and (d) probably leave some residue on the sensor.

If the spoof material has not skin like properties, then it will possibly need special technique to present to sensor like using strong pressure, cropping spoof to exactly match sensing area, etc. There will probably exist sensor technology or liveness detection based on missing properties. If that is the case then usually combination of real finger and spoof have to be used. That means trying to use real finger for liveness detection and spoof for sensing the image. That generally results on very thin spoofs glued to real finger or using a real finger on a small part of sensing area and spoof in the remaining regions of sensing area.

Some technologies are acquiring fingerprint images from deeper layer of the skin. If that is the case, then it can also ignore the spoof and scan the real fingerprint below the spoof. In that case, it is better not to use real finger with the spoof.

Each combination of material and sensor is unique and needs different approach to use effectively. Some spoof materials can be enhanced with other materials to make them better when using specific sensor (e.g., adding coloring to spoof which has not naturally a skin color to overcome optical technology). To our best knowledge, there is not a sensor which cannot be overcome (although using special enhanced spoofs and excellent present technique).

15.4 2D/3D hand spoofs

The large advantage of hand geometry-based biometry technologies is the affordability. The current commercially available devices such as HandKey and HandPunch series from Schlage use single downward facing camera as the sensory input and store the features in 9 kB template. While this in comparison with retina scanner for example allows for lower cost of device, the system can be fooled even via trivial means.

In [42] a spoof was created using a simple paper silhouette. This spoof was then successfully accepted by HandKey system. From this paper, we can infer that the system does not perform any form of liveness detection. For even a texture analysis would prevent the spoofing of the system using this way (Figure 15.24).

The current generation of hand-based biometry sensors could be spoofed using this technique, however, in the upcoming generation, this approach should prove to

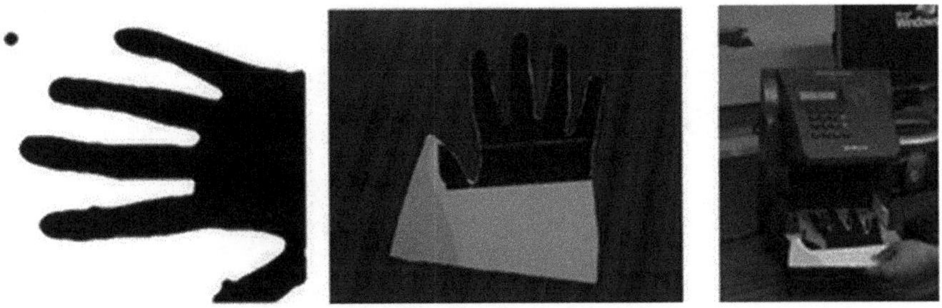

Figure 15.24 Paper silhouette spoof used to bypass commercial device [26]

be obsolete. As the liveness detection is covered in another part of this chapter, we will concentrate on spoofs based on increase in detail.

If the resolution of acquisition is increased substantially, the level of detail acquired increases as well. When the level of detail is low, such as in low resolution hand geometry verification, the matter of gathering the biometric information form uncooperative even unaware subject is a trivial matter. However, if the detail is increased enough during data gathering, the retrieval of source data from uncooperative subject becomes nontrivial, since the position, orientation, even movement can make the data gathering process fail. The spoof development then needs to integrate data from multiple acquisitions for spoof to contain all the data that are retrieved during calibrated acquisition.

Other direction that can increase the complexity of spoof development is an increase in dimensionality of used biometric. With increased computing power, it is now possible to perform a 3D reconstruction in real time, this allows for expansion of previously 2D biometric features into 3D feature, or using both of them. The increase in computing power, however, makes the creation of high detail spoof possible as well.

In Figure 15.25, we can see an example of mold casted spoof, this kind of spoof would not only overcome the 2D technologies, but also low detail 3D technologies. The shortcoming of this spoof lies in the acquisition process, which requires cooperation from the target, especially if the scanning devices, relies on pin guided approach, this would make it further complex, as the target's hand would need to be in a precise position during mold creation.

The more universal approach however, utilizes 3D printing technologies. In this case, no mold creation is required, as the data can be acquired, and the model reconstructed without the knowledge of the target, either by using the 3D scanning or by reconstructing the 3D model from an array of 2D images.

In Figure 15.26, we can see a model of hand printed using a commercially available low-cost 3D printer, with nozzle size of 0.3 mm and layer size of 0.125 mm, we can achieve a precision of up to 200 DPI in one axis. And on industrial scale, with infrared-laser-based 3D printer, the resolution can be as low as 15 μm [43]. If the texture analysis is not performed, a surface finish can be applied.

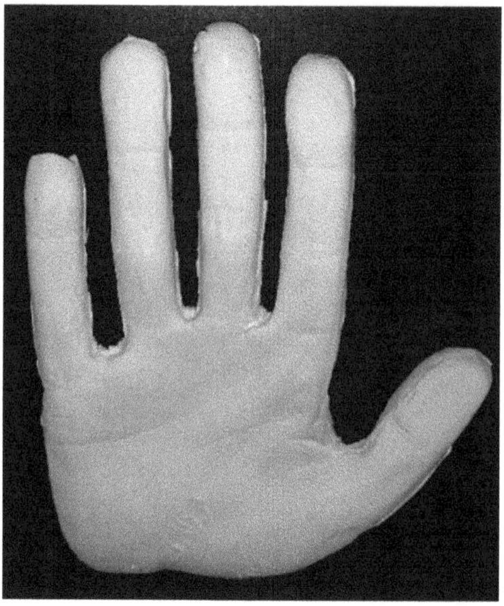

Figure 15.25 Mold casted hand spoof

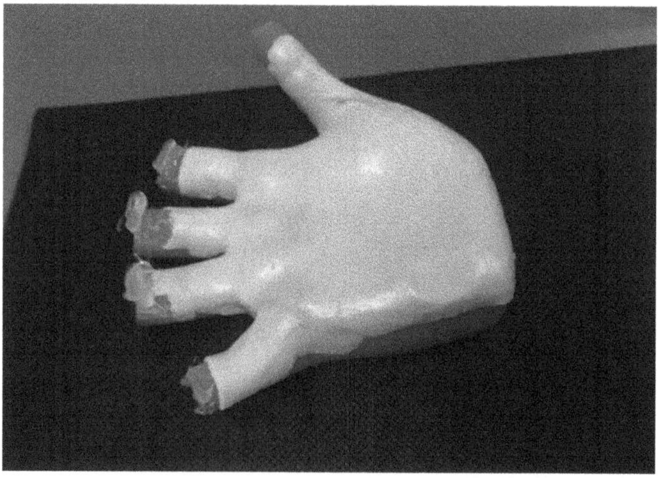

Figure 15.26 Hand spoof from 3D printer

The capabilities of these technologies need to be considered when designing the security system. Otherwise, there is a danger that newly created system is obsolete from the beginning, as the technology to create spoofs offers higher level of detail than the security technology can successfully verify or even observe.

To the knowledge of the authors, there is not at the moment a commercially available micrometric 3D printing technology, capable of simultaneous support of multiple printing materials. This would suggest that for creation of 3D spoof that integrates layers of material with different spectral absorption characteristics, necessary to overcome a vein detection-based liveness verification system on a 3D hand geometry-based biometrics system, new technology would have to be developed.

15.5 Conclusion

In this chapter, we discussed antispoofing and spoofing methods for hand-based biometrics. Generally, the methods which are based on skin properties could be used not only on fingers, but on the whole hand or even body.

It is very easy to produce a finger(print) fake, just based on fingerprint acquired from any object the person of interest touched. There are three ways how to produce such finger(print) fakes—2D/3D print, use a form (e.g., PCB) or to use a laser burner. The quality and use possibilities very strongly depend on the technology, which is used for antispoofing. Some of the fakes degrade after some time, i.e., are not usable after some days, weeks or months.

Regarding the antispoofing methods, a lot of them were proposed; however, on nearly all of them, at least one attack have been found. The most promising technology is very probably the multispectral illumination and sensing. This technology acquires real skin reaction to the illumination with a concrete wavelength, which is hard to simulate for a combination of more wavelengths. Another very promising technology is to use the electrical properties of the human finger, i.e., combination of charging and turbulent flow (induction). However, some fakes have been found for this technology as well. Also, some fingerprint acquisition technologies include automatic antispoofing detection – these are for example, ultrasound scanning and optical tomography. These methods penetrate beneath the skin and acquire the underlying structures of the skin. Therefore if there is any fake glued on the fingertip surface, these technologies show artifacts on the image, which means that a finger(print) fake is in use.

It is fully clear that antispoofing methods need to be used in a real world. The possibilities of how to attack a biometric system using fakes are wide. Therefore these attack attempts could be avoided just using such antispoofing methods.

Acknowledgment

This work was supported by the Ministry of Education, Youth and Sports of the Czech Republic from the National Programme of Sustainability (NPU II); project IT4Innovations excellence in science—LQ1602.

References

[1] Dessimoz, D., Richiardi, J., Champod, C., Drygajlo, A.: Multimodal Biometrics for Identity Documents, Research Report, PFS 341-08.05,

Version 2.0, Université de Lausanne & École Polytechnique Fédérale de Lausanne, 2006, p. 161.

[2] Jain, A.K.: Biometric System Security, Presentation, Michigan State University, East Lansing, MI, 2005, p. 57.

[3] Ambalakat, P.: Security of Biometric Authentication Systems, In: 21st Computer Science Seminar, SA1-T1-1, 2005, p. 7.

[4] Galbally, J., Fierrez, J., Ortega-Garcia, J.: Vulnerabilities in Biometric Systems: Attacks and Recent Advances in Liveness Detection, Biometrics Recognition Group, Madrid, Spain, 2007, p. 8.

[5] Matsumoto, T., Matsumoto, H., Yamada, K., Hoshino, S.: Impact of Artificial "Gummy" Fingers on Fingerprint Systems, In: Proceedings of SPIE Vol. 4677, Optical Security and Counterfeit Deterrence Techniques IV, 2005, p. 11.

[6] Roberts, C.: Biometric Attack – Vectors and Defences, 2006, p. 25.

[7] Collins, C.G.: Fingerprint Science, Copperhouse/Atomic Dog Publishing, p. 192, 2001, ISBN 978-0-942-72818-7.

[8] LN: Němečtí hackeři šíří otisk prstu ministra (German Hackers Distribute the Minister's Fingerprint), Lidové noviny (newspaper), March 31, 2008.

[9] Drahanský, M.: Fingerprint Recognition Technology: Liveness Detection, Image Quality and Skin Diseases, Habilitation Thesis, Brno, CZ, 2010, p. 153.

[10] Straus, J., Porada, V., *et al.*: Kriminalistická daktyloskopie (*Criminalistics dactyloscopy*), Police academy of Czech Republic and Criminalistic Institute Prague of the Police of Czech Republic, Praha, 2005, p. 286, ISBN 80-7251-192-0.

[11] Kluz, M.: Liveness Testing in Biometric Systems, Master Thesis, Faculty of Informatics, Masaryk University Brno, CZ, 2005, p. 57.

[12] Drahanský, M., Funk, W., Nötzel, R.: Liveness, Detection Based on Fine Movements of the Fingertip Surface, In: IEEE – The West Point Workshop, West Point, New York, USA, 2006, pp. 42–47, ISBN 1-4244-0130-5.

[13] Drahanský, M.: Methods for Quality Determination of Papillary Lines in Fingerprints, NIST, Gaithersburg, USA, 2007, p. 25.

[14] Tan, B., Lewicke, A., Schuckers, S.: Novel Methods for Fingerprint Image Analysis Detect Fake Fingers, SPIE, 10.1117, 2.1200805.1171, 2008, p. 3.

[15] Schuckers, S., Hornak, L., Norman, T., Derakhshani, R., Parthasaradhi, S.: Issues for Liveness Detection in Biometrics, CITeR, Presentation, West Virginia University, Morgantown, WV, 2003, p. 25.

[16] Rowe, R.K.: A Multispectral Sensor for Fingerprint Spoof Detection, www.sensormag.com, January 2005.

[17] Rowe, R.K.: Spoof Detection, In: Summer School for Advanced Studies on Biometrics for Secure Authentication, Alghero, Italy, 2008, p. 43.

[18] Toth, B.: Biometric Liveness Detection, In: Information Security Bulletin, Vol. 10, 2005, pp. 291–297, retrieved from on 2017-08-13: www.chi-publishing.com.

[19] Abhyankar, A., Schuckers, S.A.C.: A Wavelet Based Approach to Detecting Liveness in Fingerprint Scanners. In: Proceedings of the SPIE Vol. 5404, Biometric Technology for Human Identification, 2004, p. 9.

[20] Qualcomm, Qualcomm Fingerprint Sensors, 2017-10-04, retrieved from on 2017-10-08: https://www.qualcomm.com/products/features/security/fingerprint-sensors.

[21] Sonavation, Ultrasound Biometric Sensor, 2017-10-04, retrieved from on 2017-10-08: http://www.sonavation.com/ultrasound-biometric-sensor/.

[22] InvenSense, InvenSense and GLOBALFOUNDRIES Collaborate on Industry-Leading Ultrasonic Fingerprint Imaging Technology, 2017-10-04, retrieved from on 2017-10-08: https://www.invensense.com/news-media/invensense-and-globalfoundries-collaborate-on-industry-leading-ultrasonic-fingerprint-imaging-technology/.

[23] Mainguet, J.F.: Ultrasonic Fingerprint Sensing, 2017-10-04, http://biometrics.mainguet.org/types/fingerprint/fingerprint_sensors_physics_ultrasound.htm.

[24] UltraScan: The Theory of Live-Scan Fingerprint Imaging (Breaking the Optical Barriers with Ultrasound), UltraScan, USA, 2004, p. 8.

[25] Bicz, W.: The Impossibility of Faking Optel's Ultrasonic Fingerprint Scanners, Optel, Poland, http://www.optel.pl/article/english/livetest.htm, 2008.

[26] Drahanský, M.: Experiments with Skin Resistance and Temperature for Liveness Detection, In: Proceedings of the Fourth International Conference on Intelligent Information Hiding and Multimedia Signal Processing, Los Alamitos, US, IEEE CS, 2008, pp. 1075–1079, ISBN 978-0-7695-3278-3.

[27] U.S. Patent 6,314,195 – Organism Identifying Method and Device, November 2001.

[28] U.S. Patent 5,088,817 – Biological Object Detection Apparatus, February 1992.

[29] Drahanský, M., Lodrová, D.: Liveness Detection for Biometric Systems Based on Papillary Lines, In: Proceedings of Information Security and Assurance, 2008, Busan, KR, IEEE CS, 2008, pp. 439–444, ISBN 978-0-7695-3126-7.

[30] U.S. Patent 6,292,576 – Method and Apparatus for Distinguishing a Human Finger From a Reproduction of a Finger, September 2001.

[31] Jia, J., Cai, L., Zhang, K., Chen, D.: A New Approach to Fake Finger Detection Based on Skin Elasticity Analysis, In: S.-W. Lee and S.Z. Li (Eds.): ICB 2007, LNCS 4642, 2007, pp. 309–318, Springer-Verlag Berlin Heidelberg, 2007, ISSN 0302-9743.

[32] Putte, T., Keuning, J.: Biometrical Fingerprint Recognition: Don't Get Your Fingers Burned, In: IFIP TC8/WG8.8 4th Working Conference on Smart Card Research and Advanced Applications, Kluwer Academic Publishers, 2000, pp. 289–303.

[33] Martinsen, O.G., Grimnes, S., Haug, E.: Measuring Depth Depends on Frequency in Electrical Skin Impedance Measurements, In: Skin Research and Technology No. 5, 1999, pp. 179–181, ISSN 0909-752X.

[34] Grimnes, S., Martinsen, O.G.: Bioimpedance, University of Oslo, Norway, Wiley Encyclopedia of Biomedical Engineering, John Wiley & Sons., Inc., 2006, p. 9.

[35] Das BIA-Kompendium – Data Input GmbH, Body Composition, 3rd Edition, 2007, p. 70, retrieved from on 2017-09-17: www.data-input.de.

[36] Drahanský, M., Funk, W., Nötzel, R.: Method and Apparatus for Detecting Biometric Features, International PCT Patent, Pub. No. WO/2007/036370, Pub. Date 05.04.2007, Int. Application No. PCT/EP2006/009533, Int. Filing Date 28.09.2006, retrieved from on 2017-10-10: http://www.wipo.int/pctdb/en/wo.jsp?wo=2007036370&IA=WO2007036370&DISPLAY=STATUS.

[37] Lodrová, D., Drahanský, M.: Methods of Liveness Testing By Fingers, In: Analysis of Biomedical Signals and Images, Brno, CZ, VUTIUM, 2008, p. 7, ISBN 978-80-214-3612-1, ISSN 1211-412X.

[38] Online Canadian Pharmacy Drugstore, retrieved from on 2017-11-10: www.healthandage.com.

[39] Spurný, J., Doležel, M., Kanich, O., Drahanský, M., Shinoda, K.: New Materials for Spoofing Touch-Based Fingerprint Scanners, In: Proceedings of International Conference on Computer Application Technologies 2015, IEEE, Matsue, 2015, p. 15, ISBN 978-1-4673-8211-3.

[40] Kanich, O., Mézl, M., Drahanský, M.: Use of Creative Materials for Fingerprint Spoofs, In: The 11th IAPR International Conference on Biometrics, 2018, Queensland, p. 8.

[41] Schuckers, S.A.C.: Spoofing and Anti-Spoofing Measures, Information Security Technical Report, Vol. 7, No. 4, 2002, pp. 56–62. DOI doi:10.1016/S1363-4127(02)00407-7.

[42] Chen, H., Valizadegan, H., Jackson, C., Soltysiak, S., Jain, A.K.: Fake Hands: Spoofing Hand Geometry Systems, Biometric Consortium 2005, Washington DC, 2005.

[43] DMP60 Series. 3D MicroPrint, retrieved from on 2017-11-10: http://www.3dmicroprint.com/products/machines/dmp60-series/.

Chapter 16

Antispoofing methods in hand-based biometrics

Stephanie Schuckers[1] and Robert Rowe[2]

16.1 Introduction

When any technology starts to become pervasive, malicious parties will always search for methods of circumvention. One of methods of circumvention in biometric recognition is a presentation attack (or spoofing), where an individual attempts to masquerade as someone else to either avoid being identified or to fraudulently gain access where they have no right to. A presentation attack is defined as "presentation to the biometric data capture subsystem with the goal of interfering with the operation of the biometric system" [1]. This definition includes attempts at avoiding identification, where the individual attempts to portray a nonexistent identity. This may include presentation of synthetic biometric characteristics or the alteration of one's own biometric characteristics in such a way as to portray a new unique identity. To mitigate the risk to biometric recognition technology, automated methods, known as presentation attack detection (PAD), are used to detect such presentation attacks. This chapter will provide an overview of presentation attacks, PAD, and associated international standards.

16.2 Presentation attacks

16.2.1 Framework of spoof attacks

This section provides a framework for creation of spoof attacks using artificial biometric samples, or as termed in ISO standards, a presentation attack instrument (PAI). Presentation attacks require a source of the biometric characteristic, either from the target individual and another source if seeking to hide their own identity. Table 16.1 lists several sources of the biometric characteristics, including cooperative sources, latent, recordings, template regeneration, synthetic, and impersonation. Table 16.2 describes common approaches for production of the PAI. Artifacts, i.e., objects made by a human being, can use the biometric characteristics

[1]Department of Electrical and Computer Engineering, Clarkson University, USA
[2]HID Global, USA

Table 16.1 Sources of biometric characteristics

	Description	Examples
Cooperative	Biometric characteristics captured directly from another individual with assistance	Finger mold, hand mold, face mask
Latent	Biometric characteristics captured indirectly through a latent sample	Latent fingerprint, latent palm print, hair, skin, body fluid
Recording	Biometric characteristics captured directly from an individual onto media	Photograph, video recording, audio recording
Template regeneration	Use of information from a template to synthetically regenerate a biometric characteristic	Fingerprint regeneration, face
Synthetic	Creation of synthetic biometric characteristics (not mapped to a real person)	Synthetic fingerprint, iris, face, voice synthesized sample
Impersonation	Conversion of natural biometric to another individual's biometric with artificial assistance	Computer-assisted voice conversion

generated from the sources in Table 16.1 or may not have biometric characteristics present (e.g., ski mask). Note that the tables contain information applicable to many biometric modalities and are included in ISO part 3 [2]. Many or most could be applied for a presentation attack for hand-based biometrics recognition system. The next sections give more specific examples for modalities within hand-based biometrics.

16.2.2 Presentation attacks: fingerprint

Presentation attacks for fingerprint recognition are well documented in academic literature. One of the most cited of these studies is that of Matsumoto *et al.* [3]; however, other studies have presented comparable methods [4,5]. These methods mainly focus on attacks on automated biometric recognition systems using live scan technology. The basic concept is that an impostor desires to present or deposit biometric characteristics either belonging to another person or synthetically generated. This process requires multiple steps, which are described below.

First, the impostor must have a source of biometric characteristics. For the case where one desires to masquerade as a specific person, that person's biometric characteristics must be stolen or acquired somehow. For fingerprints, this is not too difficult of a task, as people leave their fingerprints on most surfaces that they touch or hold. The surface of a fingertip contains friction ridge skin (FRS), a pattern of ridges and furrows, which when imaged becomes a fingerprint. In FRS, the ridges are lined with pores called eccrine glands, which secrete a solution composed primarily of water and sodium chloride. When the FRS of a fingertip comes in contact with an object or surface, residue of these natural secretions is deposited on the surface, leaving a mark behind. Given the nature of FRS, the ridges will come in

Table 16.2 Production of artifacts

Physical reproduction		**Description**	**Examples**
	Cast (two step mold/cast process)	**1. Molding**—Three-dimensional representation of the biometric characteristic	Face mold captured by body double; finger mold captured by of dental material, molding plastic, modeling clay, printed circuit board, printed transparency
		2. Casting—Reproduction created from mold	Theatrical mask of face, finger spoof made of modeling clay, gelatin, silicone, latex, wood-glue, glycerin
	Direct rendering	**Printing 2D**	Photo of iris or face, fingerprint printed on a transparency/paper
		Printing 3D	Contact lens printed with pattern, prosthetic hand printed with vein pattern
		Etching	Fingerprint etched on metal
		Painting—patterns and colors painted on prosthesis	Ocular prosthetic with painted iris pattern, prosthetic hand painted with vein pattern
	Mask	Modify or conceal biometric characteristics (either partially or completely) with an artifact	Glue on finger, false facial hair, cosmetics, removable implants, opaque lenses, ski mask, Halloween mask, makeup
Digital media			
	Computer screen	Laptop or tablet to present image or video	Face or iris image, face or iris video
	Audio	Recording of voice	Recording of voice

contact with the surface, leaving the furrows raised off of the surface and so the deposited mark will take the form of the FRS pattern from which it was created. These marks (or latent prints) can be enhanced using powder or cyanoacrylate and then either be "lifted" using adhesive and scanned or photographed, resulting in an electronic record of the print. Both of these methods are standard practices for forensic latent print examiners.

Another method of acquiring fingerprint images that has been reported by some researchers [6,7] is by analyzing ordinary high-resolution photographs. In cases where a photograph has been acquired in which the subject is gesturing or otherwise holding their fingers in an appropriate orientation (e.g., making a

Figure 16.1 *Fingerprint mold made of dental material (left), finger fakes made of silicon (middle), and fingerprint mold made of etched PCB and resultant fake (right)*

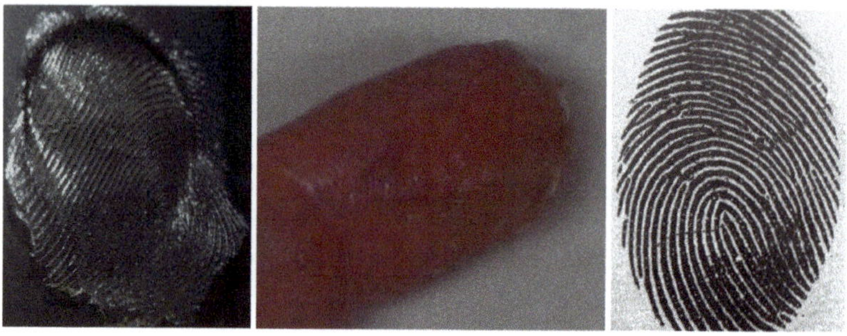

Figure 16.2 *Fingerprint fake (left), fake overlaid on real finger (middle), and resulting imaged fake fingerprint (right)*

"peace" sign), some of the subjects' fingerprint patterns may be able to be extracted. Although this method is not able to reliably extract usable fingerprint images from every photograph, it docs suggest the ability to surreptitiously acquire usable fingerprints in certain cases.

A more trivial method of acquiring someone's fingerprint would be from a direct impression in some mold material (the impression material used by dentists for creating molds of teeth works well for this task). While this method is certainly feasible with cooperation of the victim, it is not particularly practical for covertly acquiring a fingerprint. Examples are shown in Figures 16.1 and 16.2.

Once a fingerprint has been acquired, an impostor is tasked with figuring out a way to present the print to a live scan device. On some devices, this has been as easy as placing a lifted print (on adhesive) directly on the sensor. However, for a more robust method, more work is required. The next step is to create a mold (negative representation) of the print. One basic method to create a fingerprint mold is as follows. First, the fingerprint image is enhanced (e.g., using image editing software such as Adobe Photoshop) to create a two-toned image, distinguishing ridges from valleys. The image is then inverted to make the valleys dark and ridges light and printed out on transparency paper. This printout is placed over photosensitive printed circuit board (PCB) and exposed to UV light. The UV light erodes away the copper not covered by the ink, creating troughs corresponding to the locations of ridge segments. After some cleanup, the PCB becomes a useable fingerprint mold. An alternate, slightly less-reliable method is to just use the printout on transparency paper as the mold, the ink itself providing the relief between ridges and valleys. This method was used by the Chaos Computer Club for their hacking of the Apple iPhone 5s TouchID system when it was released [8].

The next step is to create a fake finger by casting some material in the mold. A wide variety of casting materials can be used, including gelatin, latex, Silicone, Play-Doh, and glue, to name a few. Once the cast solidifies, it can be presented to a live scan device as a finger would be. If it is made thin enough and of the right color, it can even be placed on the tip of the impostor's finger and covertly presented to the device.

16.2.3 Presentation attacks: 2D and 3D hand systems

There are a variety of hand-based biometrics that rely on surface characteristics and geometry of the hand to perform a biometric match. One of the earliest commercial biometric systems of this sort is still being sold today as Schlage Handkey II, which is based on the geometry of the user's hand [9]. This hand geometry system uses a very simple optical system to collect a silhouette of the user's hand when placed on the platen. Although the system is able to perform biometric matching using the silhouette data, there is limited ability to use this same information to distinguish between genuine samples and spoofs. As a consequence, researchers have been able to demonstrate that such systems can be readily spoofed using either molded hands or cardboard silhouettes of enrolled user's hands [10].

2D palm print systems base their biometric determination on the similarity of the creases of the ventral palm. Researchers have shown [11,12] that such systems may be susceptible to being spoofed using simple printed or digitally displayed images of a palm. A 3D print system was developed by [9] using optical imaging based on structured illumination, which was processed to extract the 3D topography of the palm. The researchers showed that such a system was not susceptible to simple 2D images.

16.2.4 Presentation attacks: hand and finger-vein systems

Biometric sensors based on vein patterns present in the hand are generally available in two configurations. The first is designed to collect the vein pattern of a single finger to biometrically match a user. The second common configuration uses the vein pattern of the palm to identify a user. In both cases, infrared light is used to illuminate the subsurface veins, and an appropriate imaging system is then used to acquire the data for subsequent processing and matching. Commercial versions of finger-vein systems are developed and sold by Hitachi [13], and commercial palm vein systems are developed and sold by Fujitsu [14].

Benchtop palm vein systems have been shown [15] to be susceptible to spoofs made using simple printed images of a genuine, enrolled vein image. However, the literature on the topic of palm vein spoofing is sparse, and it is unclear if commercial palm vein systems have the same vulnerability. The manufacturer claims that the commercial palm vein system has built-in liveness detection [16], but this has not been independently verified to our knowledge.

Finger-vein spoofing has also been demonstrated to be susceptible to spoofing using readily available materials and methods. As in the case of palm vein, a benchtop finger-vein system was shown [17] to be vulnerable to spoof attacks that used a printed image of the enrolled finger-vein pattern. Tests performed in one of the authors' labs confirmed that the commercial finger-vein system is susceptible to acquiring a printed image. However, unlike fingerprints, vein patterns are an internal biometric that are not readily observed without specialized equipment and conditions. Therefore, the degree of effort needed to acquire and replicate a genuine vein pattern requires techniques different than commonly used for fingerprints and other external biometrics.

However, the inability of a finger-vein sensor (or palm vein sensor) to effectively discriminate against simple spoof material like printed patterns makes them vulnerable to spoofing during enrollment, even if it is difficult to directly access the acquired vein patterns of enrolled users. In this scenario, a user can enroll an image of an arbitrary vein pattern, and then that same image can be used by others to match the enrolled spoof. Such a vulnerability has been demonstrated by researchers [18]. This form of spoof attack may be of concern for biometrically controlled programs such as medical insurance coverage or social welfare services of various kinds where users may have an incentive to "share" their biometric with friends and family.

16.2.5 Mitigation of presentation attacks

There are numerous ways to reduce the risk of spoofing in an application. These include

- Supervision of the collection of the biometric
- Combinations of factors (card, pin, and biometric)
- Multiple biometric traits
- Process-based methods (e.g., random checks for fake fingerprints)
- Limitations on the number of attempts
- PAD or "liveness" detection

Supervision The risk of spoofing is reduced when the collection of the biometric is supervised. However, it is important to note that some fake biometrics can be quite difficult to spot, e.g., Figure 16.2.

Combinations of multiple factors For verification scenarios, requiring combinations of authentication methods (e.g., biometric + pin + smart phone), the risk associated with the vulnerability of each method on its own is reduced.

Multiple biometric traits Combinations of biometric traits can put a greater burden on the attacker to fake more than one modality. Many multibiometric systems are focused on compensating for variations in quality. If reducing vulnerability is the goal, biometric match scores in a multibiometric system must be combined with this objective as a priority. For example, in positive verification mode, all biometric modalities should each achieve a match score above a threshold (or other suitable fusion method) to prevent someone just spoofing one modality to achieve positive verification. In negative identification mode, a further exploration of an individual may be indicated if there is a match in only in one modality (e.g., iris) but not another (e.g., fingerprint). For example, an individual that does not want to be matched to a watchlist might seek to hide their identity by spoofing one modality but may still be matched if another modality is obtained.

Process-based methods Thorough analysis of the end-to-end process for an application must be performed to assess the risk associated with biometric spoofing. Analysis should be performed as to the individuals most likely to commit this type of fraud and mitigation techniques which can be implemented. For example, supervision for the groups of individuals, which may have the highest risk, should be considered. For individuals of lower risk where unsupervised collection of a biometric is chosen, random checks or using liveness detection may be desired in order to flag persons to go to a secondary inspection.

Limitation of number of attempts Creating a quality PAI can be challenging. It often takes more than one attempt to mount a successful attack. Limitation on the number of attempts is recommended such that when this limitation is reached, the system may lock out the user, require another form of verification, or in supervised scenario, the monitor can check more carefully for presentation attacks.

PAD The goal of liveness testing, or more generally, PAD, is to determine if the biometric being captured is an actual measurement from the authorized, live person who is present at the time of capture [17]. In ISO, it is defined as "automated determination of a presentation attack" [1]. PAD is the subject of the following section.

16.3 PAD

Early overviews of PAD approaches are described in [19,20]. Methods to measure liveness fall into several categories. A taxonomy is presented whereby methods are divided into hardware and software-based methods. In the first category, liveness is captured through additional hardware integrated with the fingerprint sensor. The second category is software-based, which involves further processing the biometric signature to obtain liveness information. For example, this may mean extracting perspiration information from fingerprint image. Additionally, liveness can be an inherent part of the biometric; in other words, the biometric features and liveness are intimately linked, e.g., the electrocardiogram which has been suggested as a biometric [21]. Most systems that have PAD do so through additional hardware and/or software processing.

16.3.1 PAD-fingerprint (hardware based)

The first method uses extra hardware to acquire life signs. Previously developed approaches measure fingertip temperature, pulse, pulse oximetry, blood pressure, electric resistance, odor, multispectral information, or electrocardiogram [22–24]. These methods require dedicated hardware integrated with the fingerprint system. Electrocardiogram is the electrical measurement of the heart collected through electrodes on two skin contact points on the body (e.g., two hands, hand, and foot). Pulse oximetry is the measurement of the oxygen content of the blood through the comparison of absorption of two wavelengths of light by the blood. This measurement requires an LED and photodetector that are both in close contact with the skin, shielded from ambient light, and optically separated. With such an arrangement, the optical absorption varies due to the pulsatile motion of the blood vessels and therefore becomes an indicator for a real, living finger. Because of noise in the measurement as well as typical heart rates, pulse-based measurements may require a few seconds of measurement time to properly determine the presence of a heartbeat.

A critical aspect of any hardware-based PAD method is that the sensor used for PAD is tightly coupled with the sensing of the biometric feature. Ideally, the PAD measurement and the biometric measurement should occur on the same portion of the skin in order to avoid scenarios where a spoof sample is used to acquire the biometric, while genuine skin is used to satisfy the PAD sensor. In cases where the two measurements cannot be made on the same region of skin, the system should be configured in a way that makes it impractical to use a combination of real and spoof samples for one measurement.

In addition to dedicated hardware-based PAD methods, there are some biometric sensing methods that naturally capture liveness information such as fingerprint sensors based on multispectral imaging as well as ultrasound-based fingerprint sensors. They are placed here in the hardware category, because these approaches, while commercially viable, require purchase of a specific scanner and are not applicable to standard fingerprint readers. One commercially available fingerprint sensor (Lumidigm, USA) is based on multispectral imaging using multiple wavelengths of light as well as different optical polarization conditions. The optical configuration of the sensor enables images that contain information from both the surface and subsurface of the skin. As shown in Figure 16.3, the capillaries and other anatomical features of the finger skin have the same general pattern as the surface fingerprint [25,26]. Analysis of the multispectral images extracts the fingerprint pattern from both the surface and subsurface features. As well, further analysis of the multispectral images extracts rich information on the optical characteristics of skin or other material from which the fingerprint pattern is acquired. This information is used in a classifier that has been trained to distinguish between genuine, living skin and other materials that may be used in an attempt to spoof the sensor [23,25].

Similarly, ultrasound measurements have been suggested as a way to measure fingerprint images (e.g., Qualcomm, USA). While ultrasound can image a spoof or cadaver fingerprint itself, analysis of the characteristics of the acoustic energy reflected from the sample can be further used to determine if the sample is genuine or fake [28].

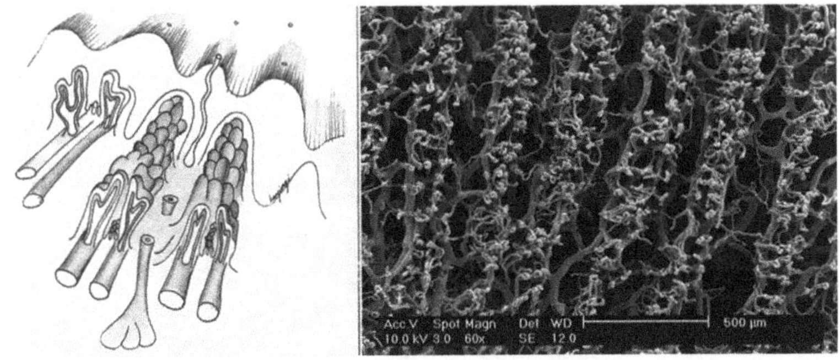

Figure 16.3 *Histology of the skin on the palmar surface of the fingertip. The sketch on the left shows the pattern of the capillary tufts and dermal papillae that lie below the fingerprint ridges. The scanning electron microscope photo on the right side shows the rows of capillary tufts imaged from a portion of an excised thumb after the surrounding skin has been removed [26,27]*

16.3.2 PAD-fingerprint (software based)

The second method uses the information already present in a conventional fingerprint image to detect life signs, for example, skin deformation, pores, power spectrum, or perspiration pattern.

Skin deformation and elasticity Skin deformation techniques use the information regarding how the fingertip's skin deforms when pressed against the scanner surface [29–31]. The studies show that when a real finger moves on a scanner surface, it produces a significant amount of nonlinear distortion. However, fake fingers are more rigid than skin, and the deformation is lower even if they are made of highly elastic materials. One approach quantifies this considering multiple frames of clockwise motion of the finger [30].

Perspiration pattern Previously, one of the author's labs has demonstrated that perspiration can be used as a measure of liveness detection for fingerprint biometric systems. Unlike spoof and cadaver fingers, live fingers demonstrate a distinctive spatial moisture pattern when in physical contact with the capturing surface of the fingerprint scanner. The pattern in fingerprint images begins as "patchy" areas of moisture around the pores spreading across the ridges over time. Image/signal processing and pattern-recognition algorithms have been developed to quantify this phenomenon using wavelet and statistical approaches [32–34]. These approaches require two time-series images, which might be inconvenient for the users. Other methods to quantify this phenomenon have been developed for a single image [35].

Characteristics of spoof and live images A natural extension to the specific categories above is to begin to assess the characteristics that define live and spoof fingers which cover a broad range [36–39]. These include image power spectrum, which reveals stamp fabrication processes, noise residue in the valleys due to the spoof material, and combinations of multiple factors, for example, fusion of perspiration and deformation features. The difference between live and spoof images is mainly due to the mold/cast fabrication process which causes an alteration of details of the fingerprint. The Fourier transform feature can quantify the difference in terms of high frequency information loss for fake fingers.

16.3.3 PAD—palm print systems

The vulnerability of a 2D palm print system to printed and displayed images was evaluated in [10]. The underlying premise of detection was that palm print displayed on a laptop, a tablet, or in printed form has reflectance and textural properties different than a genuine hand. As such, the researchers tested two methods of acquiring and classifying the images based on these characteristics. They developed an initial set of 189 features based on pixel values, wavelet coefficients, and gray-level co-occurrence matrices. They then down selected a subset of features using correlation-based feature selection [40] and then used these features with a support vector machine to classify samples as genuine or spoof. They compared this method to a method that is based on local binary patterns that was successfully employed to detect facial spoofs [41]. The results indicate that the 189 feature-set was able to detect palm print spoofs significantly better than the approach based on local binary patterns (LBP).

As noted earlier, [9] demonstrated the ability to use a 3D palm system to effectively discriminate against simple 2D spoofs. However, they did not test the ability to discriminate against 3D spoof samples (e.g., hand castings).

16.3.4 PAD—finger-vein systems

Software-based PAD for finger-vein systems has been studied by several researchers [42–44]. In [36], researchers used Fourier transforms, Haar wavelets, and Daubechies wavelets to generate features that were then operated on by a support vector machine (SVM) classifier to distinguish between genuine fingers and printed finger-vein images. In [37], steerable pyramids are used to generate texture features that were used in combination with an SVM classifier to distinguish between genuine fingers and two kinds of printed images as well as one displayed on a smart phone. In [38], a number of extensions of the LBP were tested against the standard LBP for a spoof-detection task using printed vein images. The researchers determined that there were no systematic benefits of the LBP extensions over the base LBP feature set.

16.4 Standards and performance evaluation of PAD

Metrics for PAD have been defined as part of the recently published ISO standards ISO/IEC SC 37, WD 30107 Part 3 [2]. This is necessary as there has been quite a bit of confusion around metrics related to spoofing, particularly when simply using the term "false accept rate" which is routinely used for biometric matching.

Associated metrics for assessing the PAD system alone are as follows:

Attack Presentation Classification Error Rate (APCER): "proportion of attack presentations using the same PAI species incorrectly classified as bona fide presentations in a specific scenario" [2]

The PAD module alone is assessed by performing a presentation attack repeatedly and measuring the response of the system. An example would be testing a fingerprint PAD module with a set of spoof fingerprints.

Bona fide Presentation Classification Error Rate (BPCER): "proportion of bona fide presentations incorrectly classified as presentation attacks in a specific scenario" [2]

The PAD module alone is assessed by repeatedly using the module in normal operation. An example would be testing a fingerprint PAD module with a group of live subjects with normal fingerprints (i.e., not a presentation attack). *No PAD test should be performed without also considering BPCER as it is a trade-off with APCER.*

Other metrics are introduced in the ISO standard which consider cases of full system performance.

Impostor attack presentation match rate (IAPMR): "<full-system evaluation of a verification system> proportion of impostor attack presentations using the same PAI species in which the target reference is matched" [2] IAPMR is a trade-off with false accept rate and false reject rate.

Publication of the ISO standards is particularly timely [1,2], as commercial implementations are becoming common place. The assessment of the effectiveness of these systems is critical and is aided by the standards.

In addition, benchmark competitions for liveness detection, e.g., LivDet, which has been held biannually since 2009 (e.g., [45,46]) can also be useful for determining state of the art in this field, as well as usefulness of datasets available for development and further benchmarking after the competition is completed. An example of results from this type of evaluation is shown from the LivDet competitions in Figure 16.4. Competitions such as this are less prevalent for vein or hand-based biometrics; however, this may be changing. For example, the first finger-vein competition was hosted in 2015 [47].

16.5 Summary

In summary, spoofing is shown to be a legitimate threat for hand-based recognition. A fake biometric, capable of presenting identifying biometric characteristics, can, in some cases, be produced with minimal effort using any number of easily accessible materials. PAD systems are being developed and deployed to reduce the vulnerability due to spoofing. However, there is a fundamental trade-off between the security provided by PAD systems and the usability and convenience experienced by the authorized users of the biometric system. The goal of continuing PAD

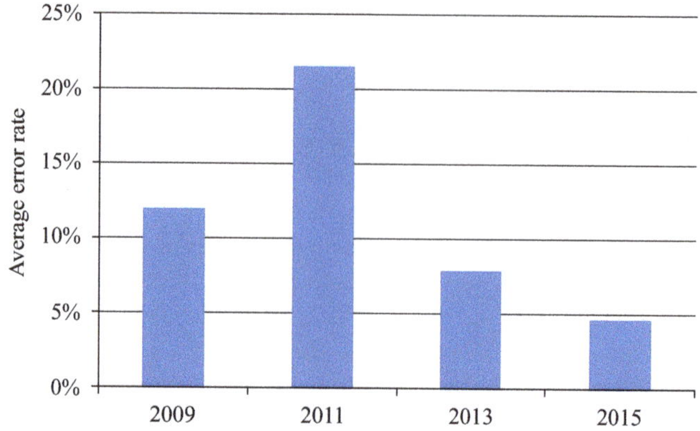

Figure 16.4 Average of APCER and NPCER for top two performers across all databases tested in LivDet competitions hosted 2009, 2011, 2013, 2015

development is to improve security while not degrading the usability to unacceptable levels. Even in cases where a satisfactory trade-off is achieved for a particular biometric technology and application at a particular time, the nature of security systems is that malevolent parties will continue to develop new ways to defeat existing PAD methods. As such, PAD may be boiled down to an attempt to stay one step ahead of those intending to defeat the system through spoofing attacks. For this reason, there is likely to be the need for continuing developments in this area for the foreseeable future.

References

[1] International Standards Organization, ISO/IEC TC JTC1/SC 37, ISO/IEC 30107-1, Information technology – Biometric presentation attack detection – Part 1, 2016.

[2] International Standards Organization, ISO/IEC TC JTC1/SC 37, ISO/IEC 30107-3, Information technology – Biometric presentation attack detection – Part 3: Testing and reporting, 2017.

[3] T. Matsumoto, H. Matsumoto, K. Yamada, and S. Hoshino, "Impact of artificial 'gummy' fingers on fingerprint systems," in Electronic Imaging, 2002, vol. 4677, pp. 275–289.

[4] T. Van der Putte and J. Keuning, "Biometrical fingerprint recognition: don't get your fingers burned," Smart Card Research and Advanced Applications, vol. 31, no. 0, pp. 289–303, 2000.

[5] L. Thalheim, J. Krissler, and P. Ziegler, "Body check: biometric access protection devices and their programs put to the test," c't, vol. 11, p. 114ff, 2002.

[6] https://www.engadget.com/2017/01/13/peace-sign-pics-fingerprints/, retrieved Feb 10, 2018.

[7] http://www.bbc.com/news/technology-30623611, retrieved Feb 10, 2018.
[8] "Chaos Computer Club breaks Apple TouchID," 2013. [Online]. Available: http://www.ccc.de/en/updates/2013/ccc-breaks-apple-touchid, retrieved Feb 10, 2018.
[9] https://us.allegion.com/en/home/products/categories/biometrics/schlage-handkeyII.html, retrieved Feb 10, 2018.
[10] H. Chen, H. Valizadegan, C. Jackson, S. Soltysiak, and A. K. Jain, "Fake hands: Spoofing hand geometry systems," Biometric Consortium, Washington DC, Sep 2005.
[11] D. Zhang, V. Kanhangad, N. Luo, and A. Kumar, "Robust palmprint verification using 2D and 3D features," Pattern Recognition, vol. 43, no. 1, pp. 358–368, 2010.
[12] S. Bhilare, V. Kanhangad, and N. Chaudhari, Pattern Anal Appl, 2017. doi:10.1007/s10044-017-0606-y.
[13] http://www.hitachi.co.jp/products/it/veinid/global/products/index.html, retrieved Feb 10, 2018.
[14] http://www.fujitsu.com/us/solutions/business-technology/security/palmsecure/, retrieved Feb 10, 2018.
[15] P. Tome and S. Marcel, "On the vulnerability of palm vein recognition to spoofing attacks," in 2015 International Conference on Biometrics (ICB), Phuket, 2015, pp. 319–325.
[16] http://www.fujitsu.com/us/about/resources/news/press-releases/2014/ffna-20140401-01.html, retrieved Feb 10, 2018.
[17] P. Tome, M. Vanoni, and S. Marcel, "On the vulnerability of finger vein recognition to spoofing," in 2014 International Conference of the Biometrics Special Interest Group (BIOSIG), Darmstadt, 2014, pp. 1–10.
[18] https://www.youtube.com/watch?v=8HuL6cHJTK0, retrieved Feb 13, 2018.
[19] S. A. C. Schuckers, "Spoofing and anti-spoofing measures," Information Security Technical Report, vol. 7, no. 4, pp. 56–62, 2002.
[20] P. Coli, G. L. Marcialis, and F. Roli, "Vitality detection from fingerprint images: a critical survey," in Advances in Biometrics, 2007, pp. 722–731.
[21] L. Biel, O. Pettersson, L. Philipson, and P. Wide, "ECG analysis: a new approach in human identification," IEEE Transactions on Instrumentation and Measurement, vol. 50, no. 3, pp. 808–812, 2001.
[22] P. Kallo, I. Kiss, A. Podmaniczky, and J. Talosi, "Detector for recognizing the living character of a finger in a fingerprint recognizing apparatus," 2001.
[23] D. Baldisserra, A. Franco, D. Maio, and D. Maltoni, "Fake fingerprint detection by odor analysis," in Advances in Biometrics, 2006, pp. 265–272.
[24] K. A. Nixon and R. K. Rowe, "Multispectral fingerprint imaging for spoof detection," in Defense and Security, 2005, pp. 214–225.
[25] K. A. Nixon, V. Aimale, R. K. Rowe, "Spoof Detection Schemes," Handbook of Biometrics, Editors A.K. Jain, P. Flynn and A.A. Ross, Springer, 2007, pp. 403–423.

[26] S. Sangiorgi, A. Manelli, T. Congiu, *et al.*, "Microvascularization of the human digit as studied by corrosion casting," Journal of Anatomy, vol. 204, no. 2, pp. 123–131, 2004.

[27] S. Sangiorgi. Personal Communication with Matthew Ennis, PhD, Lumidigm. 2005.

[28] M. Agassy, B. Castro, A. Lerner, G. Rotem, L. Galili, N. Altman, US Patent Application 20170231534, Aug 17, 2017.

[29] Y. Chen, S. Dass, A. Ross, and A. Jain, "Fingerprint deformation models using minutiae locations and orientations," in Proceedings – Seventh IEEE Workshop on Applications of Computer Vision, WACV 2005, 2005, pp. 150–155.

[30] A. Antonelli, R. Cappelli, D. Maio, and D. Maltoni, "Fake finger detection by skin distortion analysis," IEEE Transactions on Information Forensics and Security, vol. 1, no. 3, pp. 360–373, Sep 2006.

[31] J. Jia, L. Cai, K. Zhang, and D. Chen, "A new approach to fake finger detection based on skin elasticity analysis," in Advances in Biometrics, 2007, pp. 309–318.

[32] R. Derakhshani, S. A. C. Schuckers, L. A. Hornak, and L. O'Gorman, "Determination of vitality from a non-invasive biomedical measurement for use in fingerprint scanners," Pattern Recognition, vol. 36, no. 2, pp. 383–396, 2003.

[33] A. S. Abhyankar and S. A. C. Schuckers, "A wavelet-based approach to detecting liveness in fingerprint scanners," in *Defense and Security*, 2004, vol. 5404, pp. 278–286.

[34] S. T. V. Parthasaradhi, R. Derakhshani, L. A. Hornak, and S. A. C. Schuckers, "Time-series detection of perspiration as a liveness test in fingerprint devices," IEEE Transactions on Systems, Man, and Cybernetics, Part C (Applications and Reviews), vol. 35, no. 3, pp. 335–343, Aug 2005.

[35] B. Tan and S. A. C. Schuckers, "Liveness detection for fingerprint scanners based on the statistics of wavelet signal processing," in 2006 Conf. Comput. Vis. Pattern Recognit. Work., pp. 26–26, 2006.

[36] Y. S. Moon, J. S. Chen, K. C. Chan, K. So, and K. C. Woo, "Wavelet based fingerprint liveness detection," Electronics Letters, vol. 41, no. 20, p. 1112, 2005.

[37] J. Jia and L. Cai, "Fake finger detection based on time-series fingerprint image analysis," in Advanced Intelligent Computing Theories and Applications. With Aspects of Theoretical and Methodological Issues, 2007, pp. 1140–1150.

[38] P. Coli, G. L. Marcialis, and F. Roli, "Power spectrum-based fingerprint vitality detection," in IEEE Workshop on Automatic Identification Advanced Technologies, 2007, pp. 169–173.

[39] B. Tan and S. A. C. Schuckers, "New approach for liveness detection in fingerprint scanners based on valley noise analysis," Journal of Electronic Imaging, vol. 17, no. 1, p. 011009, Jan 2008.

[40] M. A. Hall, Correlation-based feature selection for machine learning. Dissertation, University of Waikato, 1999.

[41] I. Chingovska, A. Anjos, and S. Marcel, "On the effectiveness of local binary patterns in face anti-spoofing," in 11th International Conference Biometrics Special Interest Group, 2012, pp. 1–7.

[42] D. T. Nguyen, Y. H. Park, K. Y. Shin, S. Y. Kwon, H. C. Lee, and K. R. Park, Fake finger-vein image detection based on Fourier and wavelet transforms, Digital Signal Processing, vol. 23, no. 5, pp. 1401–1413, 2013.

[43] R. Raghavendra and C. Busch, "Presentation attack detection algorithms for finger vein biometrics: a comprehensive study," in 2015 11th International Conference on Signal-Image Technology & Internet-Based Systems (SITIS), Bangkok, 2015, pp. 628–632.

[44] D. Kocher, S. Schwarz, and A. Uhl. "Empirical evaluation of LBP-extension features for finger vein spoofing detection," in Biometrics Special Interest Group (BIOSIG), 2016 International Conference of the. IEEE, 2016.

[45] G. L. Marcialis, A. Lewicke, B. Tan, *et al.*, "First international fingerprint liveness detection competition-LivDet 2009," in Image Analysis and Processing – ICIAP 2009, 2009, vol. 5716, pp. 12–23.

[46] L. Ghiani, D. A. Yambay, V. Mura, G. L. Marcialis, F. Roli, and S. A. Schuckers, "Review of the fingerprint liveness detection (LivDet) competition series: 2009 to 2015," Image and Vision Computing, vol. 58, pp. 110–128, 2017.

[47] P. Tome, R. Raghavendra, C. Busch, *et al.*, "The 1st competition on counter measures to finger vein spoofing attacks," in Biometrics (ICB), 2015 International Conference on. IEEE, May 2015, pp. 513–518.

Index

www.ingramcontent.com/pod-product-compliance
Lightning Source LLC
LaVergne TN
LVHW051048210926
848486LV00001B/46